OXFORD MEDICAL PUBLICATIONS

Understanding Other Minds

UNDERSTANDING OTHER MINDS

Perspectives from Autism

Edited by

SIMON BARON-COHEN

Lecturer in Psychopathology,
University of Cambridge

HELEN TAGER-FLUSBERG

Professor of Psychology,
University of Massachusetts

and

DONALD J. COHEN

Professor of Child Psychiatry and Psychology,
Director, Child Study Center,
Yale University

Oxford   New York   Tokyo
OXFORD UNIVERSITY PRESS

Oxford University Press, Walton Street, Oxford OX2 6DP

Oxford   New York
Athens   Auckland   Bangkok   Bombay
Calcutta   Cape Town   Dar es Salaam   Delhi
Florence   Hong Kong   Istanbul   Karachi
Kuala Lumpur   Madras   Madrid   Melbourne
Maxico City   Nairobi   Paris   Singapore
Taipei   Tokyo   Toronto
and associated companies in
Berlin   Ibadan

Oxford is a trade mark of Oxford University Press

Published in the United States
by Oxford University Press Inc., New York

© S. Baron-Cohen, H. Tager-Flusberg, D. Cohen, and the contributors on pp. xi–xiii.,
1993

First published 1993

First published in paperback 1994

A catalogue record for this book is available from the British Library

Library of Congress Cataloging in Publication Data
Understanding other minds: perspectives from autism/edited by Simon
Baron-Cohen, Helen Tager-Flusberg, and Donald J. Cohen.
p.   cm. – (Oxford medical publications)
Includes bibliographical references and indexes.
1. Autism in children.   2. Philosophy of mind in children.
I. Baron–Cohen, Simon.   II. Tager–Flusberg, Helen.   III. Cohen
Donald J.   IV. Series.
[DNLM: 1. Autism, Infantile.   WM 203.5 U55]
RJ506.A9U5   1993   618.92′8982 – dc20   92–48765
DNLM/DLC
for Library of Congress
ISBN 0 19 262056 8 (Pbk)

Printed in Great Britain by
Bookcraft (Bath) Ltd, Midsomer Norton, Avon

Preface

The child's acquisition of a 'theory of mind' has become a central and exciting domain of enquiry in developmental psychology in the 1990s, as well as being one of the most rapidly growing in this field. Part of this excitement and productivity is due to the important connections that 'theory of mind' has with psychopathology, primatology, and philosophy.

This book is the first to bring together the work from developmental psychopathology, by focusing specifically on the question of whether children with autism are impaired in the development of a theory of mind. Indeed, the book places autism at the centre of the enquiry. It presents a critical examination not only of this important hypothesis, but of its implications for our understanding of the ontogenesis and evolution of this capacity.

Many chapters in this book follow the current convention in using the term 'theory of mind'. However, whilst for some authors this represents a commitment to the idea that in the normal case the child is indeed constructing a *theory* of the relationship between mental states, the world, and action, for other investigators the term is used as a synonym for the development of a 'concept of mind', or our 'folk psychology'. To encompass this range of meanings, we have used the phrase *Understanding other minds* as our title. This title reflects one key function of a theory of mind, namely, to make sense of *others*. The work in this book also has, we hope, considerable relevance for the other side of the coin, that is, developing a concept of one's *own* mind.

This book will be of interest to those concerned with how children come to understand mind, and why children with autism may not. Whilst aimed at students and researchers of child development and cognitive science, discussion of these issues should also be of relevance to clinicians, educators, and others concerned with understanding individuals with autism and related disorders.

London	S.B-C.
Boston	H.T-F.
New Haven	D.J.C.

February 1993

Acknowledgements

We would like to thank the Derek Ricks Memorial Fund for their financial support of the Workshop in Seattle, April 1991, during which first drafts of most of the chapters in this book were presented and debated. We would also like to thank the Mental Health Foundation, the Medical Research Council, the National Institutes of Health, and the National Institute of Mental Health, who supported the editors during the preparation of this book. Finally, we would like to thank the staff of Oxford University Press for their guidance throughout.

The cover illustration shows a sculpture from the Pende, Zaïre. The photograph is reprinted from *L'Art Africain* (Stoullig-Marin, Françoise, 1988, Editions Citadels, Paris). Reproduced with kind permission of Mr Jacques Kerchache. We are grateful to David Holzer for suggesting its use in this book.

Contents

x *Contents*

Contributors

Anthony Bailey
MRC Child Psychiatry Unit, Institute of Psychiatry, Denmark Hill, London SE5 8AF, UK.

Simon Baron-Cohen
Department of Experimental Psychology and Psychiatry, University of Cambridge, Downing St, Cambridge, CB2 3EB, UK.

Jerome Bruner
New York University, Department of Psychology, 6 Washington Place, New York, New York 10003, USA.

Donald Cohen
Child Study Center, Yale University, 333 Cedar Street, New Haven, CT 06510, USA.

Carol Feldman
New York University, Department of Psychology, 6 Washington Place, New York, New York 10003, USA.

Uta Frith
Medical Research Council, Cognitive Development Unit, 17 Gordon St, London WC1H 0AH, UK.

Juan-Carlos Gómez
Dto Psicología Evolutiva, Facultad de Psicología, Universidad Autónoma de Madrid, 28049, Madrid, Spain.

Alison Gopnik
Department of Psychology, University of California, Berkeley, California 94720, USA.

Paul Harris
Department of Experimental Psychology, University of Oxford, South Parks Road, Oxford, OX1 3UD, UK.

Peter Hobson
Adult Department, The Tavistock Clinic, Tavistock Centre, 120 Belsize Lane, London, NW3 5BA, UK.

Patricia Howlin
Department of Psychology, The Medical School, St George's Hospital, Tooting, London, SW17, UK.

Connie Kasari
Neuropsychiatric Institute and Hospital, UCLA Center for Health Sciences, 760 Westwood Plaza, Los Angeles, California 90024–1759, USA.

Ami Klin
Yale Child Study Center, PO Box 3333, 333 Cedar Street, New Haven, CT 06510, USA.

Alan Leslie
Rutgers University Center for Cognitive Science, Busch Campus, Rutgers University, Piscataway, NJ 08854, USA.

Cathy Lord
Greensboro TEACCH Center, Suite #104, 2415 Penny Road, Highpoint, N. Carolina 27265, USA.

Katherine Loveland
University of Texas Health Sciences Center at Houston, Mental Sciences Institute, Department of Psychiatry and Behavioral Sciences, 1300 Moursund, Texas 77030, USA.

Linda Mayes
Child Study Center, Yale University, 333 Cedar Street, New Haven, CT 06510, USA.

Andrew Meltzoff
Department of Psychology, WJ-10, University of Washington, Seattle, Washington 98195, USA.

Peter Mundy
Department of Psychology, Psychological Services Center, University of Miami, PO Box 248185, Coral Gables, Florida 33124, USA.

Josef Perner
Laboratory of Experimental Psychology, University of Sussex, Brighton, BN1 9 QG, UK.

Daniel Roth
Department of Psychology, University of Tel Aviv, Israel.

Michael Rutter
Department of Child and Adolescent Psychiatry, Institute of Psychiatry, Denmark Hill, London SE5 8AF, UK.

Jerry Samet
Department of Philosophy, Brandeis University, Waltham, Massachusetts, 02254 USA.

Encarnación Sarriá
Facultad de Psicología, UNED, Ciudad Universitaria S/N, Madrid 28040, Spain.

Marian Sigman
Neuropsychiatric Institute and Hospital, UCLA Center for Health Sciences, 760 Westwood Plaza, Los Angeles, California 90024–1759, USA.

Beate Sodian
Institut für Empirische Pädagogik und Pädagogische Psychologie, Universität München, Leopoldstrasse 13, 8000 München 40, Germany.

Helen Tager-Flusberg
Department of Psychology, University of Massachusetts–Boston, Harbor Campus, Boston, MA, 02125–3393, USA.

Javier Tamarit
Centro CEPRI, Avda. de la Victoria 63, El Plantio 28023, Madrid, Spain.

Belgin Tunali
University of Texas Mental Sciences Institute, Department of Psychiatry and Behavioral Sciences, University of Texas Medical School at Houston, 1300 Moursund, Texas 77030, USA.

Fred R. Volkmar
Child Study Center, Yale University, 333 Cedar Street, New Haven, CT 06510, USA.

Henry M. Wellman
University of Michigan, Center for Human Growth and Development, 300 N Ingalls Building, 10th Level, Ann Arbor, Michigan 48109–0406, USA.

Andrew Whiten
Psychological Laboratory, University of St Andrews, Fife, KY16 9JU, Scotland, UK.

Part I Introduction

1

An introduction to the debate

HELEN TAGER-FLUSBERG, SIMON BARON-COHEN,
AND DONALD COHEN

This book focuses on a psychological theory of autism that has generated considerable interest in the past decade. It is known as the *theory of mind hypothesis of autism*. By 'theory of mind' is meant the ability of normal children to attribute mental states (such as beliefs, desires, intentions, etc.) to themselves and other people, as a way of making sense of and predicting behaviour. The theory of mind hypothesis of autism holds that in children with autism, this ability fails to develop in the normal way, resulting in the observed social and communication abnormalities in behaviour. We felt that this hypothesis deserved to be subjected to close and critical scrutiny, by leading authorities in the fields of psychology, psychiatry, and related disciplines, for several reasons.

First, if autism is indeed caused by a failure to develop a theory of mind, then studying autism might hold the clues to how this important ability is normally acquired so effortlessly. Secondly, studying autism from this perspective might show us what happens to a child when this ability is not available in the normal way. Thirdly, this hypothesis has been surrounded by fascinating debate about the role of affect and the nature of the cognitive mechanisms involved in supporting a theory of mind. These debates are important because resolving them will teach us about the relevant *processes* in development and pathology. Finally, we felt that subjecting the theory to scrutiny might help to reveal both its strengths and shortcomings, so as to guide future research in the field of autism.

For all of these reasons, we put together a book to debate the theory of mind hypothesis of autism. Most of our contributors presented their chapters in draft form at a two-day workshop in Seattle, in April 1991, with a set of key questions to guide our discussions: How do children acquire a theory of mind? What are the developmental origins of this ability, and what is its evolutionary history? Is autism a syndrome that is best understood in terms of a primary impairment in this capacity?

WHAT IS AUTISM? A NOTE ON DIAGNOSIS

Kanner's (1943) description of the syndrome of autism is a classic example
of the contribution of clinical observation to psychiatric taxonomy. In
his clear prose, he described a group of children who had impoverished
or absent social relations from the very first years of life, and with language
(when it was present) which was distinctively deviant. Although there have
been modifications, Kanner's diagnostic criteria have proved remarkably
robust and have been echoed in all subsequent psychiatric classification
systems. Thus, thirty years after Kanner, Rutter (1978) reviewed the major
studies and highlighted four essential features of autism: impaired social
development; delayed and deviant language; insistence on sameness; and
onset before 30 months. A similar set of features formed the basis of the
diagnostic criteria in the American Psychiatric Association taxonomy in the
1980 edition of the Diagnostic and Statistical Manual (DSM-III), and in the
World Health Organization taxonomy in the (1987) International Classification of Diseases, 9th Edition (ICD-9 1987).

   To provide a more developmental approach, and in order to encompass
the broad range of individuals with autism, the definition of autism was
elaborated in the next edition of the Diagnostic and Statistical Manual
(DSM-IIIR 1987). Thus, although the same three features are retained,
DSM-IIIR provides a range of diagnostic items for each of these. For example, for the social impairment, it ranges from marked lack of awareness of
others (for those with the most severe social impairments) to gross abnormalities in peer relations (for those who are least impaired). The next
planned revisions of diagnostic criteria will appear in DSM-IV and ICD-10,
both due to be published in the mid-1990s.

   Changes in diagnostic criteria are not merely of academic interest: they
have implications for which individuals receive the diagnosis of autism. For
example, DSM-III and ICD-9 criteria encompass most of the children whom
most clinicians would categorize as having autism. In contrast, DSM-IIIR
criteria are broader and include some individuals whom some clinicians
might feel fall outside the usual domain of the concept.* The shifting boundaries of the diagnostic criteria for autism reflect a fundamental problem
for virtually all psychiatric disorders: the absence of an independent and
fully accepted diagnostic 'gold standard' (Volkmar and Cohen 1988a).
Given this limitation, the diagnosis of autism is remarkable for the general
agreement among clinicians, over decades and across nations. There *are*
paradigmatic cases of autism about which all experienced clinicians would
agree. This is reassuring in relation to the use of categorical diagnosis as

---

* Using clinical diagnosis as the standard of comparison, DSM-IIIR appears to have
increased sensitivity and decreased specificity (Volkmar *et al.* 1987).

an anchor in research studies on particular mechanisms, such as those described in the present volume.

There is, however, one major diagnostic issue which deserves special note in relation to autism research. If one collected all the individuals who are diagnosed as having autism in one room, probably the most striking fact would not be their similarity, but how vastly different they are among themselves (Volkmar and Cohen 1988*b*). Included among individuals with autism are three-year-old children and senior citizens, people with profound mental handicap and university graduates, adults who barely have a word of expressive vocabulary (and almost undetectable receptive language) and adults who read encyclopaedias for recreation and speak with pedantic exactitude. Some individuals with autism are self-destructive, while others are over-conscientious about their physical well-being. There are individuals with autism who memorize road maps and train schedules, and others who couldn't make sense of either.

To counter the potential research problems associated with such variability, the majority of studies discussed in this book focus on high-functioning individuals with autism. By 'high-functioning' we mean individuals with only moderate or mild intellectual impairment. In this respect, they constitute the upper 25 per cent of the population with autism (Rutter 1978). The selection of this subject group reflects a research strategy which enables us to identify autism-specific impairments independently of the effects of mental handicap in general. Ultimately, it will be important that research focusing on this group of individuals with autism should be extended to the full range of people with the condition.

A BRIEF HISTORY OF THEORY OF MIND

The literature on the development of a theory of mind has grown exponentially over the last ten years. Studies on children's developing understanding of the mental world arguably began with Piaget (1929). He claimed that children younger than seven years of age were unable to make the ontological distinction between the mental and physical realms. The discussion about the ontogeny of an understanding of minds was reopened in 1978 with the publication of Premack and Woodruff's seminal paper: 'Does the chimpanzee have a theory of mind?' Premack and Woodruff described a series of experiments that suggested to them that their famous chimp Sarah, who had knowledge of a symbol system, was able to predict and interpret a human's actions in terms of mental states such as intentions. They argued that Sarah's success indicated that she had a theory of mind. Commentaries on this paper, especially by Dennett, Pylyshyn, and Bennett, pointed out that it is not until one demonstrates an understanding of false belief (where the mental state conflicts with reality) that one can unequivocally attribute a theory of mind to an individual, human or otherwise.

Within a few years, developmental psychologists began devising ingenious experiments to tap children's understanding of false belief, using the ideas suggested by Dennett and others. Wimmer and Perner (1983) published their important study of three- and four-year-olds' understanding of false belief, involving the now famous Maxi and the chocolate scenario. In this task, an object (a bar of chocolate) is unexpectedly moved whilst the main protagonist, Maxi, is out of the room. The child is then asked to predict where Maxi thinks the chocolate is, or where he will look for it. The main findings, which have been replicated many times, are that only older three-year-olds and over can pass this task. This study set in motion a flurry of research investigating young children's knowledge of false belief, other mental states, and related cognitive and linguistic achievements (see Astington *et al.* 1988; Butterworth *et al.* 1991; Frye and Moore 1991, and Whiten 1991, for recent collections of papers).

The first extension of this line of work to the study of autism which utilized Wimmer and Perner's (1983) false-belief test was carried out by Baron–Cohen *et al.* (1985). They used this test in order to ask the question 'Does the autistic child have a theory of mind?' This study, and subsequent replications, provided strong evidence that children with autism have a specific impairment in their understanding of false belief.* Given that it had been argued in the philosophy of mind and language (Dennett 1978; Grice 1975) that a theory of mind was *necessary* for social understanding and communication, it seemed plausible that a deficit in this area might account for at least two of the core symptoms in autism.

Earlier, Hobson (1981) had proposed the theory that children with autism have a primary problem in the development of a concept of other persons, specifically in coming to understand that people have minds. His approach was to investigate the understanding of expressions of emotion by children with autism, which grew out of his view that affective impairments (especially a relative lack of empathic responsiveness to others) could lead to impairments in conceptual development. The 1985 paper of Baron–Cohen *et al.* placed the emphasis on a primary *cognitive* deficit, and the debate about the primacy of affect or cognition in this domain continues to fuel new research ideas.

---

* Inevitably, many contributors in this book make reference to this early study. Whilst this may create some redundancy, each chapter uses this simply as a starting point for their own empirical and theoretical directions.

THE DEBATES

In this volume there are several fascinating theoretical debates that resurface with vigour between contributors, each time from a different angle. Here we mention the key issues, indicating in which chapters they are taken up:

The first set of debates focuses on underlying processes and developmental origins of a theory of mind: Does a theory of mind require a capacity for metarepresentation? If so, what is meant by metarepresentation (Leslie and Roth; Perner)? Does a theory of mind, or metarepresentation, arise *de novo* in the second to third year of life, or are there infancy precursors to either or both of these (Wellman)? If there are infancy precursors, what are these? Imitation (Meltzoff and Gopnik)? Joint-attention (Baron–Cohen; Mundy, Sigman, and Kasari)? Narrativity (Bruner and Feldman)? Affective sensitivity (Hobson)?

The second set of debates focuses on what consequences one would expect if a theory of mind was impaired: What would the effects be on language and communication (Tager-Flusberg; Loveland and Tunali) and on social development (Lord)? Is an inability to deceive a cardinal example of theory of mind failure (Sodian and Frith)? And what are the clinical implications of such deficits (Baron–Cohen and Howlin)?

The third set of debates focuses on alternative theories of the data from autism: Is task performance better understood in terms of executive control systems? If so, do deficits in the latter make better sense of the lack of imaginative play in autism, than the metarepresentation theory of autism (Harris)? Which symptoms of autism are successfully explained by the theory of mind hypothesis, and which cannot be (Klin and Volkmar)? Is the notion of a theory of mind a mistaken notion? If so, is it more appropriate to emphasize the development of self? (Samet); or social desire? (Mayes, Cohen, and Klin).

A final set of debates centre on what can be learnt about autism and the development of a theory of mind from the study of non-human primates (Gómez, Sarriá, and Tamarit; Whiten), or from philosophy of mind (Samet), or from a psychoanalytic perspective (Mayes *et al.*).

THE THEORY OF MIND HYPOTHESIS OF AUTISM: A PARADIGM CASE IN THE APPLICATION OF DEVELOPMENTAL PSYCHOPATHOLOGY

This volume exemplifies the significance of the field of developmental psychopathology, a field that uses theories and research on normal populations to advance our understanding of atypical children, and in turn acknowledges the influence that the study of atypical populations can have on our understanding of normal development (Cicchetti 1984; Sroufe and Rutter 1984).

As our brief historical review shows, it is straightforward to recognize the influence of developmental psychology, and more broadly cognitive science, on the evolving theoretical and empirical work in the field of autism during the past ten years, and these are widely represented and acknowledged in numerous chapters in this volume. The other side of the equation — the contribution of the study of psychopathology to our understanding of normal development — can, with hindsight, also be discerned. Indeed, in having influenced theories of normal development it stands out as exceptional. Without the study of autism it is debatable whether the field would have been focusing on the significance of joint-attention in the development of a theory of mind, for example, or would have considered the modularity of a theory of mind in neuropsychological terms. Indeed, even the link between theory of mind and pretend play owes much to the associated deficits uncovered in autism. Finally, the burning questions of the primacy of affect and cognition in the development of a theory of mind can be seen as strongly influenced by the parallel debates about the primary impairment in autism.

ORGANIZATION OF THIS BOOK

In Part I, we begin with introductory chapters that review the development of a normal theory of mind, and the core social abnormalities characteristic of autism, since it is these that the theory of mind hypothesis set out to explain. Parts II and III then take up the central debates. Part II includes chapters that advance the theory of mind hypothesis of autism, all taking a cognitive approach, though there is by no means unanimous agreement among the contributors in this section on the nature of the deficit in autism. In Part III, a range of critical perspectives on the cognitive approach to the theory of mind hypothesis are presented. In Part IV, the debates are broadened still further to include philosophical, evolutionary, psychoanalytic, and developmental theories of autism and the theory of mind. In the final chapters, the implications of the theory of mind hypothesis of autism for clinical issues and for future research are considered.

We hope that this volume will provide an impetus for future work that will bring us closer to a more complete psychological understanding both of autism and of the normal development of a theory of mind.

REFERENCES

Astington, J., Harris, P., and Olson, D. (1988). *Developing theories of mind.* Cambridge University Press, New York.

Baron–Cohen, S., Leslie, A.M., and Frith, U. (1985). Does the autistic child have a 'theory of mind'? *Cognition*, **21**, 37–46.

Butterworth, G., Harris, P., Leslie, A., and Wellman, H. (1991). *Perspectives on the child's theory of mind.* Oxford University Press/British Psychological Society.

Cicchetti, D. (1984). The emergence of developmental psychopathology. *Child Development*, **55**, 1–7.

Dennett, D. (1978). *Brainstorms: philosophical essays on mind and psychology.* Harvester Press.

DSM-III-R (1987). *Diagnostic and statistical manual of mental disorders.* (3rd rev. edn). American Psychiatric Association, Washington DC.

Frye, D., and Moore, C. (1991). *Children's theories of mind.* Lawrence Erlbaum Associates, Hillsdale, NJ.

Grice, H.P. (1975). Logic and conversation. In *Syntax and semantics: speech acts* (ed. R. Cole and J. Morgan). Academic Press, New York. (Original work published in 1967).

Hobson, R.P. (1981). The autistic child's concept of persons. In *Proceedings of the National Society for Children and Adults with Autism, International Annual Conference* (ed. D. Park). Washington.

ICD-9 (1987). *International Classification of Diseases* (9th edn). World Health Organization, Geneva.

Kanner, L. (1943). Autistic disturbance of affective contact. *Nervous Child*, **2**, 217–50. Reprinted in Kanner, L. (1973). *Childhood psychosis: initial studies and new insights.* Wiley, New York.

Piaget, J. (1929). *The child's conception of the world.* Harcourt Brace, New York.

Premack, D., and Woodruff, G. (1978). Does the chimpanzee have a 'theory of mind'? *Behavioral and Brain Sciences*, **4**, 515–26.

Rutter, M. (1978). Diagnosis and definition. In *Autism: a reappraisal of concepts and treatment* (ed. M. Rutter and E. Schopler). Plenum, New York.

Sroufe, A., and Rutter, M. (1984). The domain of developmental psychopathology. *Child Development*, **55**, 17–29.

Volkmar, F., and Cohen, D. (1988*a*). Diagnosis of pervasive developmental disorders. In *Advances in clinical child psychology* (ed. B. Lahey and A. Kazdin). Plenum, New York.

Volkmar, F., and Cohen, D. (1988*b*). Classification and diagnosis of childhood autism. In *Diagnosis and assessment in autism* (ed. E. Schopler and G. Mesibov). Plenum, New York.

Volkmar, F., Sparrow, S., Goudreau, D., Cicchetti, D., Paul, R., and Cohen, D. (1987). Social deficits in autism: an operational approach using the Vineland Adaptive Behavior Scales. *Journal of the American Academy of Child and Adolescent Psychiatry*, **26**, 156–61.

Whiten, A. (ed.) (1991). *Natural theories of mind.* Basil Blackwell, Oxford.

Wimmer, H., and Perner, J. (1983). Beliefs about beliefs: representation and constraining function of wrong beliefs in young children's understanding of deception. *Cognition*, **13**, 103–28.

2

Early understanding of mind: the normal case

HENRY M. WELLMAN

'Theory of mind' describes one approach to a larger topic: everyday naïve psychology. The essential endeavour is one of characterizing and charting our everyday understanding of people; how do we ordinarily understand each others' actions, thoughts, and lives? For example, on analogy to traditions or theories within scientific psychology, what sort of theoretical tradition characterizes our everyday psychological thought? According to 'theory of mind', our ordinary psychology is mentalistic. We construe people in terms of internal *mental* states such as their beliefs, desires, intentions, and emotions. Moreover, our everyday understanding of people in these terms has a notable coherence. Because an actor has certain beliefs and desires, he or she engages in certain intentional actions, the success and failure of which result in certain emotional reactions. So, according to this approach, theory of mind is a proper shorthand for referring to our everyday psychology and for capturing two of its essential features: coherence and mentalism.

This ordinary psychology is not 'theoretical' in the sense of being removed from practical application in our lives. On the contrary, the assumption is that this understanding guides all social action and interaction. Here are a few applications of our everyday psychology: you want to make someone happy, so you give him something he wants; you want to have something that someone else also wants, so you deceive her into mistakenly believing that it is unavailable or broken; you know something that someone else does not, so you tell him or show him. As in these examples, our everyday interactions are founded on considerations of people's wants, beliefs, knowledge, and emotions. Consider, as a contrast, how differently we would interact with one another if we were behaviourists instead of mentalists — attempting to classically or operantly condition one another's behaviour, manipulating and arranging stimuli and responses rather than manipulating and assessing beliefs and desires. 'Theory of mind' endeavours to characterize the everyday understanding of people that frames and determines our social world and our social acts.

A critical question in this endeavour concerns how and when our everyday theory of mind arises. This has recently become a central problem for developmental psychology, inspiring a flurry of research on children's developing understanding that they and others have minds, that mental states such as beliefs and desires guide human action, and even the understanding of such notions as brains, intelligence, epistemology, and the like. Developmental psychologists pursuing such research initially concentrated on testing normal children's understandings. More recently, researchers have considered the question of whether some children may never achieve an understanding of mind. To continue the thought experiment of the last paragraph, just what would social cognition and action be like for an organism that did not share in our everyday mentalism? An especially intriguing attempt to explore this question focuses on the study of autistic children (see for example Baron-Cohen *et al.* 1985) — children whose reasoning about and interaction with people is impaired and for whom an initial case can be made that they lack a theory of mind, or something central to it. Just how good a case, and just what autism shows about developing theories of mind, is the special topic of this volume. I believe that these two approaches — the normal and abnormal — go hand in hand; but in this chapter I concentrate on the normal case. What do unimpaired children ordinarily come to understand about the mind?

I begin with a few words about what our ordinary theory of mind might be like, to give a framework for examining children's understanding. Then I briefly review current knowledge about three- and four-year-olds' theory. Classically it was claimed that children of this age are ignorant of the mind (Piaget 1929); but now it seems that this is the age when children's understanding can first be shown to be recognizably like our own. Finally, I concentrate on still younger children. Theorizing about very early development is increasingly important for understanding acquisition of everyday psychology. Furthermore, characterizing the early foundations for a theory of mind seems an especially fitting endeavour for this volume. Knowing the precursors of ordinary understanding would help in the task of characterizing those who fall short of ordinary acquisitions.

A SKETCH OF OUR EVERYDAY THEORY OF MIND

There are two intuitive aspects of our understanding of the mental states of self and other; I will call these the 'ontological' and 'causal' aspects of mind (as in Wellman 1990).* The ontological aspect picks out mental contents,

---

* I have argued (Wellman 1990) that our theory of mind is a special kind of theory, namely a broad framework theory (rather than a specific detailed theory). And I argue that framework theories characteristically specify an ontology for a domain — that is, define the sorts of entities and processes to be considered — and provide causal-explanatory frameworks for thinking about how entities and processes in a domain interact. Thus it is no accident that our everyday mentalism, as a framework theory of action and thought, encompasses ontological and causal aspects.

states, and processes as a domain of things to consider, and distinguishes that mental world from the real world of physical objects, material states, and mechanical or behavioural processes. The essence of the ontological aspect, therefore, is our ordinary understanding of the difference between thoughts or ideas on the one hand and objects or overt behaviour the other. For example, a thought about a house is not the same sort of thing as a house. Indeed, a thought about a house is not even the same sort of thing as a shadow of a house, or a picture of a house. The contents and states of the mind are internal, mental, and subjective, whereas the contents and states of the world are external, substantial, and objective.

Moreover, according to our everyday theory, our thoughts, beliefs, and ideas are not only distinct from the physical world of objects and behaviour, but they are also causally related to that physical–behavioural world. Causal influence goes from mind to world and from world to mind: mental states cause actions in the world and the world causes mental states. A useful and typical shorthand is to divide our causal mental states into two generic sorts: beliefs and desires. A phrase like *belief–desire psychology* as a label for our everyday psychology highlights this causal aspect.

What, in a bit more detail, does this everyday theory look like? It includes reasoning such as the following:

Why did Jill go to that restaurant? She *wanted* to eat quickly and *thought* that it was a fast-food place.

Thus one essential, causal idea is that people engage in actions because they *believe* those actions will achieve certain *desires*. However, as noted, the world also influences one's mental state, leading to a more complicated and extensive set of causal connections. For example, belief-desire psychology also includes reasoning such as:

Why did Jill go to that restaurant? She was *hungry* and wanted vegetarian food, and thought she'd *seen* vegetarian selections on that restaurant's menu. Boy, is she going to be *disappointed*.

Figure 2.1 depicts something of the organization of these related constructs in our everyday theory. Briefly and simplistically, according to this everyday theory, physiological states and basic emotions ground one's desires. Beliefs, on the other hand, are often derived from perceptual experiences. Moreover, one's actions lead to outcomes in the world, and these outcomes lead to emotional reactions of predictable sorts. At least two classes of emotional reactions are encompassed by the theory: reactions dependent on desires and reactions dependent on beliefs. Thus (1) the outcome of an action can satisfy or fail to satisfy the actor's desires, leading, generically, to happiness reactions. You want something and get it so you are happy; you want it and fail to get it so you are sad or angry. Also, (2) the outcome of

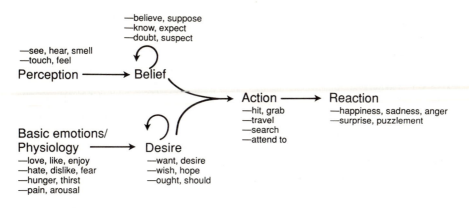

Fig. 2.1. Belief–desire psychology: a schema of the organization of the constructs of our everyday theory of mind.

an action can match or fail to match an actor's beliefs, leading to what may be generically termed surprise reactions. You think something will happen, but it does not; so you are surprised or puzzled.

This sketch does not provide anything like a comprehensive account of adult belief–desire reasoning; such an account is elaborated more fully in D'Andrade (1987) or Wellman (1990, Chapter 4). But even this outline provides a useful perspective for asking when children understand what about the mind. In the next sections I describe various understandings as characterizing certain ages, such as three- or four-year-olds. This is intended as a convenience only; a device to capture a series of accomplishments rather than an attempt to specify precise ages of attainment.

THEORY OF MIND IN THREE- AND FOUR-YEAR-OLDS

Researchers of children's understanding of the mind differ on a great many questions. Adopting a broad level of analysis, however, a sizeable battle has been won amid the many remaining skirmishes. Specifically, most researchers in this area now agree that a rather sophisticated level of reasoning about the mental states of self and others, resting on a basic awareness of the very existence of minds and thoughts, is evident in three- to five-year-olds. This contemporary consensus overthrows an earlier wisdom that asserted, in contrast, that young children were ignorant about the mind, misconstruing internal mental states and contents as external physical ones until six or seven years of age (see, for example, Piaget 1929). I cannot describe here the methodological and theoretical advances that I believe are responsible for these new conclusions, but can outline what now seems evident.

With regard to ontology, children as young as three years firmly divide the mental and physical worlds. For example, if told about one boy who has a dog and another one who is thinking about a dog, they correctly judge which 'dog' can be seen, touched, and petted, and which cannot. More precisely yet, if told about someone who has a dog that has run away, versus someone who is thinking of a dog, three-year-olds know that neither 'dog' can be seen or petted, but that none the less one is mental ('just in his mind', 'only imagination'), whereas the other is physically real, and merely unavailable. In these ways, young children appropriately distinguish between real and mental entities (Wellman and Estes 1986; Estes *et al.* 1989; Harris *et al.*, 1991). Such young children also understand the subjectivity of thoughts. In appropriately simple tasks they are able to state, for example, that they can 'see' their own mental images but that others cannot (Estes *et al.* 1989), or that while they think a particular cookie tastes yummy, someone else could think it's yucky (Flavell *et al.* 1990). Young children can, of course, make mistakes about specific instances. For example, waking from a vivid dream they may not know whether that particular experience was dream or reality. None the less they recognize the importance of the basic distinction to be made, that between mental contents and states versus real-world ones.

What about the causal aspect of mind? A great deal of research has investigated young children's understanding that persons have beliefs and how such beliefs guide behaviour. Beliefs have received extensive attention because the mental state of belief is central to our everyday understanding of how the mind influences action (Bill takes his umbrella because he believes it is raining) and of how the mind reflects the world (Bill believes it is raining because he saw rain outside). To understand mental causation it is essential that children realize that people live their lives with regard to beliefs about the world, not with regard to the world itself. Bill's belief that it is raining induces him to take his umbrella, even if it is not really raining. This example illustrates that understanding false beliefs provides especially intriguing and useful evidence for an understanding of belief, and of causal mental states more generally. Because of this, children's understanding of false beliefs is well researched, and a plethora of studies now show that by four years children are quite proficient at reasoning about false beliefs (see for example Perner *et al.* 1987). Other studies show an even earlier understanding of belief in three-year-olds via their understanding of true and discrepant beliefs (Wellman and Bartsch 1988) and via their natural-language use of common mental terms such as *think* and *know* (Shatz *et al.* 1983). Indeed several recent studies show, I believe, an initial awareness of even false beliefs in three-year-olds (Bartsch and Wellman 1989; Siegel and Beattie 1991; Lewis and Osbourne 1990; Moses 1990; Hala *et al.* 1991; Wellman and Banerjee 1991).

In this same age-range young children also demonstrate understanding that perception is instrumental in the acquisition of beliefs (Wimmer *et al.* 1988; O'Neill and Gopnik 1991), that beliefs and desires together cause action (Wellman and Bartsch 1988; Bartsch and Wellman 1989), and that emotions such as happiness and surprise are organized and caused, in part, by actors' underlying beliefs and desires (Yuill 1984; Stein and Levine 1989, Harris 1989). Again, recent studies suggest that some significant initial understanding of the links between perception and belief (Pillow 1989; Pratt and Bryant 1990), between belief and desire (Wellman and Woolley 1990), and between desires and emotions such as happiness, as well as between beliefs and emotions such as surprise (Wellman and Banerjee 1991) occurs in three-year-olds as well in four- and five-year-olds.

The status of three-year-olds' understanding of false belief and of such related phenomena as surprise, lies, and deception remains one of the heated controversies in the area. My current opinion is that while three-year-olds genuinely understand that persons can have beliefs and at times false beliefs, they do not yet realize that persons' actions are always, necessarily, framed by their beliefs. Four-year-olds more clearly appreciate the centrality of beliefs, and therefore their understanding is importantly different from that of three-year-olds. For four-year-olds, as for adults, actors' actions are filtered through their beliefs, false or true; to reiterate, people live their lives with regard to beliefs about the world, not with regard to the world itself directly. Three-year-olds, I believe, do not yet see beliefs as inextricably central to human action, while none the less realizing that people sometimes have such mental states as belief and even false belief.

This characterization of the similarities and differences between three- and four-year-olds is by no means proven; it is undoubtedly controversial. But such controversies should not overshadow the substantial consensus wherein almost all researchers in the field now characterize four-year-olds, if not three-year-olds, in belief–desire reasoning terms. Indeed, by four years, normal children's understanding of the mental mediation of experience and behaviour is so robust that they understand that even such a seemingly veridical experience as perception itself (for example, my perception that that is a flower) is a mental representational state akin to belief rather than to fact. That is, by four years children understand that persons are prey not only to false beliefs but also to false perceptions (Flavell *et al.* 1986).

In the discussion thus far I have said little about children's understanding of desire, although desire complements belief in our belief–desire psychology. An emphasis on belief reflects the state of the empirical literature. Only recently has children's understanding of desire become more focal. Emerging research on this issue shows that three- and four-year-olds also know quite a bit about desire, as is to be expected if they are engaging in belief–desire reasoning. What is most interesting in this research, however, is that it shows that for the youngest children, three-year-olds, understandings of desire are more advanced than similar understandings of belief. Three-year-olds who struggle with false beliefs have little trouble with false or unfulfilled desires (Gopnik and Slaughter 1991); three-year-olds who inconsistently evidence understanding of how beliefs are involved in emotional reactions such as surprise consistently understand how desires are involved in emotions such as happiness (Wellman and Banerjee 1991); three-year-olds who know that two people may hold different desires may reject the idea that mental states of belief can similarly differ subjectively across people (Flavell *et al.* 1990; Wellman and Woolley 1990); and, in explaining actions, three-year-olds who consistently mention the character's desires may only rarely, if ever, mention beliefs (Bartsch and Wellman 1989; Moses and Flavell 1990). Indeed, in cases where three-year-olds fail to show the sort of reasoning about beliefs, surprises, and deceptions that only slightly older children do, it may well be that this is because they have focused instead more directly on characters' desires, ignoring beliefs.

These findings suggesting three-year-olds' concentration on desires, then, may explain in part their evanescent grasp of belief. Moreover, such findings sponsor an intriguing hypothesis about the psychological reasoning of still younger children. Quite simply, the proposal is that younger children, two-year-olds for example, may construe persons in terms of mental states revolving around desire, with no complementary understanding of belief. Along these lines, I have proposed that before becoming belief–desire psychologists young children are simple *desire psychologists* (Wellman 1990; Wellman and Woolley 1990). According to this proposal, two-year-olds understand quite well that people have desires for things, and that because of their desires they act to try to get those things, and also react emotionally if their desires are achieved or not (by being happy or sad). For them, desires characterize people and explain their actions; beliefs play no role. Thus, two-year-olds fail to conceive of people as even having beliefs, let alone that beliefs inevitably guide action. It is worth discussing this proposal a bit more for two reasons. First, contrasting an understanding of belief to a hypothetical simplified understanding of desire helps to emphasize what a momentous and distinctive acquisition it is for three-year-olds to have achieved an understanding of belief at all, irrespective of whether this is integrated into a mature belief–desire psychology or not. Second, describing a simplified understanding of desire makes a useful transition to a discussion of infants and toddlers, by beginning to consider what the precursors to ordinary belief–desire mentalism might be like.

Figure 2.2 attempts to depict a contrast between an understanding of simple desires versus ordinary beliefs. In this figure I try to portray something like the right-hand person's construal of the left-hand person's mental state. Our ordinary conception of beliefs is that beliefs are representational

Desire (wants an apple)

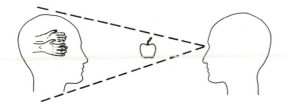

Belief (thinks that that is an apple)

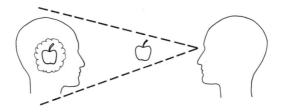

Fig. 2.2. The constrast between the understanding of another person as desiring something and the understanding of another person as believing something, in terms of the mental representation of the objects of desire and of belief.

in an informal but important sense. To attribute to someone the thought, 'that is an apple', involves construing that person as representing an apple in his mind — something like what is illustrated at the bottom of Fig. 2.2. Beliefs, thoughts, and the like, of course, are not literally pictures in the head, but refer to representational mental content. But note how that construal of beliefs contrasts with a conceivable, albeit quite crude or simple, concept of desire. *Simple* desires in the sense I am depicting and discussing them here, require no attribution to the target person of a representation. In this simple understanding to say that 'he wants an apple' attributes to the other a subjective longing for, an urge toward attaining, a very real apple. Of course, the apple need not literally be present in arm's reach — it can be unavailable (hidden inside that locked cupboard); but it is none the less a real apple. And the object need not be a physical object; it could be a situation (to be outside) or an action (to climb the mountain). I show here only an easily depicted object.

The essential contrast I want to capture, therefore, is this: construing someone as having simple desires need involve no conception of mental contents or representations, but just of an internal orientation toward external contents or objects. To understand beliefs as representations, on the other hand, requires conceiving of two sorts of contents—those in the world and those in the mind. Four-year-olds clearly understand the existence of representational mental states such as beliefs, and so, I believe, do three-year-olds. But a simpler non-representational understanding of mental states is possible (see also Flavell 1988; Perner 1988), and that understanding may characterize essentially children younger than three (Wellman 1990).

INFANTS AND TODDLERS

Suppose we accept, as a rough portrait, the picture I have drawn in the preceding section as to the normal three- or four-year-olds' belief–desire understanding. Now I want to back up, begin at infancy, and address the developmental precursors to those achievements. In doing so I am returning to questions of the sort asked ten years ago by Bretherton *et al.* (1981): when or in what sense might infants acquire an understanding of persons' internal subjective lives? There is no definitive consensus on these issues; they are only beginning to be considered in the literature. Still, various findings exist; indeed, a great many researchers have been interested in one aspect or another of infants' social understandings. It is possible to tell an orderly developmental story from these various findings, and that is what I attempt to do here.

As a preface, I assume that infants come prepared to learn about people in some way. It seems clear that human infants are born into a social species, into a world of social objects and experiences that evolution has equipped them to survive in and learn about. Moreover, by acquiring anything like a belief–desire psychology by the age of three or four, children have acquired a complex system of understanding extremely quickly, *on a timetable as rapid as initial language-learning*. It would be impossible to acquire so much so quickly (or even at all) if infants began as a blank slate in this domain. The raw data of social experience underdetermine a theory of mind unless the cognizer comes prepared in some manner to parse, attend to, and think about these data in some fashions rather than others. This sort of thinking about infant cognition does not, I believe, require the conclusion that humans must be born with an understanding of mind in terms of beliefs and desires (Fodor 1987), but it does require some rich precursors. This sort of thinking also does not require that the infant's task should be one of solitary theory-building; far from it. Initial social understandings are necessary, in part, to enable the infant to enter vigorously into the complex and instructive social life of its family and culture. Based on some initial hypotheses, the infant embarks on a certain sort of collaborative research programme, the results of which are, proximally, the three-year-old's theory of mind and, distally, the adult's folk psychology of his or her culture.

Young infants

Even young infants (six months and younger) are complex social creatures. Not necessarily immediately at birth, but quickly in the months thereafter, they cry, smile, attend to faces, imitate others, become attached, and interact in dyadic face-to-face routines. Moreover, at a rough level of description, it is clearly appropriate to describe infants as organisms that *have* beliefs and desires. They reach for and look at things they want, form representations about objects and people, remember occurrences and relations, show frustration at thwarted desires and surprise at unexpected occurrences. But having beliefs and desires (or something like them) does not count as understanding beliefs and desires. The question here is whether and when infants attribute anything like such mental states to others or recognize them in themselves? Below I will hypothesize that this is an acquisition of older infancy; but it is useful to mention briefly how certain capacities of younger infants might prepare the way for this later achievement.

For three-year-olds (I claim) as for adults (surely), a theory of mind is closely tied to and manifest in two aspects — an ontological aspect that picks out certain entities for consideration (say, intentional acts and mental states), and a causal aspect that frames explanations about how such entities interact (say, belief–desire causation). Early infant social understanding can be construed as encompassing two parallel, precursory aspects — mechanisms for picking out social objects for special processing, and early understandings of animate, human causation.

The young infant seems to be an organism specially attuned to attend to and interact with a social set of objects. Infants preferentially attend to faces (or, at first, arrays with face-like configurations and features) (Sherrod 1981); they discriminate facial expressions of emotion at an early age (Nelson 1987); they imitate some human actions (Meltzoff and Moore 1983). Infants find bodily contact with and being held by others desirable, and seek out such contacts with the onset of voluntary movement. Infants attend preferentially to human speech over other sounds — especially to female speech, and even specifically their own mother's voice (DeCasper and Fifer 1980). Theories of attachment (for example, Bowlby 1969) suggest good reasons for, and research on parent–infant interaction (for example Stern 1985) amply documents, certain proclivities that help the infant pick out, attend to, and represent people as special, significant, and information-laden entities. Several infant investigators have noted that from about one to three or four months of age infants are intensely interested in people, spending much time in dyadic social interaction. This is followed by a period in the middle of the first year where infants turn away from people, in part, and toward a world of objects (Collis 1981; Kaye and Fogel 1980; Bakeman and

Adamson 1984). But this relative change of focus serves to emphasize the early social orientation.

What about infant understanding of causation? By six months of age (Leslie and Keeble 1987) infants demonstrate some rudimentary knowledge of mechanical causation among physical objects. It seems easy to imagine that they also and equally rapidly come to understand certain special aspects of personal or animate causation. Young infants discriminate animate-biological motions versus random or artificial ones (Bertenthal *et al.* 1985), and older infants, by thirteen to sixteen months for example, distinguish between the sorts of internally generated and self-propelled movements possible to animals and people versus the transmission of external forces necessary for the movement of physical objects such as balls and chairs (Poulin-Dubois and Shultz 1988, Golinkoff 1983). Indeed, Premack (1990) has recently argued (on the basis of thought experiments rather than empirical ones) that young infants will probably evidence a rich biologically-prepared tendency to discriminate animate self-propelled movements in contrast to inanimate externally-caused physical ones.

To understand that certain objects move on their own whereas others do not seems an important distinction, useful to the infant for separating animate from inanimate objects and occurrences (Gelman and Spelke 1981). And paying special attention to animate objects and movements could aid infants, who live in groups of human animals, in acquiring considerable information about people. But conceiving of certain entities and behaviours as animate requires no special conception of the *mental*. After all, earth-worms are animate entities and sneezes are animate behaviours, but neither require understanding via a theory of mind. A conception of mental states and mentally caused actions requires something more, beyond self-propulsion. It requires something like an understanding of intentionality, as philosophers use that term.

The hallmark of intentionality in this sense, is '*aboutness*', or object-directedness. Intentional states such as desires and beliefs are of or toward or about some 'object'. Consider the depictions of simple desire and ordinary belief in Fig. 2.2. Both a desire for an apple and a belief that that is an apple are object-specific; they are about apples. In this sort of analysis an ordinary intentional act — deliberately reaching for an apple — manifests intentionality in two related senses. In the everyday sense it is intentional because it is 'on purpose'. In the wider sense it manifests an internal intentional state such as a goal (to get the apple), a desire (for the apple), or a belief (that's an apple). An intentional act in this sense is very different from a merely self-propelled motion; it is at the very least a self-propelled motion toward a specific object. Attributing intentionality therefore requires more than attributing a capacity for self-propelled movement; it requires attributing internal states (or attitudes) directed toward (or about) specific objects (or contents). Self-propulsion may divide the animate from the inanimate, but intention marks the mental. Even simple desires, such as that depicted in Fig. 2.2, are intentional in this rudimentary sense.

Premack (1990) suggests that when young babies discriminate self-propelled motion, they also automatically ('are hardwired to') interpret such motions as intentional. But no research with young infants that I know of shows them construing humans and human action intentionally. In fact, in contrast, many infant researchers describe a transition in infant social interaction first evident in the period from nine to twelve months wherein the infant seems to see self and others in notably different terms. This change has been termed the advent of a sense of subjectivity (Stern 1985), of secondary intersubjectivity (Trevarthen and Hubley 1978), of intentional communication (Bates *et al.* 1979), of triadic awareness (Adamson and Bakeman 1985), and even of an implicit theory of mind (Bretherton *et al.* 1981). I think that it can be cogently argued that the older infant manifests a transition to now understanding people (self and others) as intentional in the rudimentary but significant sense of having certain experiences *about* or *of* external objects or events.

Older infants

The young infant attends to and interacts with people and a bit later attends to and manipulates objects. At about nine to twelve months the older infant begins to attend to both simultaneously. This is evident in studies of gaze, where older infants attend back and forth between an object and another person (see, for example, Bakeman and Adamson 1984). Beyond evidencing a more sophisticated ability to juggle attention between two foci, these interactions begin to reveal a conception of people and objects as connected together — the person directed toward the object. Conceiving of people as connected to objects in this directional fashion is, at the least, a step toward an understanding of intentionality. Beyond that, in the second year, I believe that infants come to recognize persons in intentional terms: the person as experiencing something *about* an object, that object as an object *of* attention or intention. Of course infants are unlikely to understand in even the most rudimentary way all the sorts of intentional states that we as adults distinguish. But there is evidence that in the second year they may well understand three foundational intentional aspects of persons: a simple but intentional understanding of desire, of perception, and of emotion. That is, they come to see themselves and others as attending to the world, seeking to attain things or experiences from it, and reacting to it affectively.*

Consider, first, early understanding of perception or attention. Gestures such as pointing and showing emerge between nine and thirteen months or so (Bates *et al.* 1979; Lempers *et al.* 1977; Butterworth 1991; Masur 1983; Zinober and Martlew 1985). Some early sorts of pointing could be simple actions of the infant toward the object itself (something like attenuated poking or reaching). But in the second year research consistently documents pointing and showing directed at getting the other to attend to something. For example, if the other's eyes are covered by his or her hands, eighteen-month and two-year-olds will move the hands or try to place a to-be-shown object between the hands and eyes (Lempers *et al.* 1977). Other behaviours show that early in the second year infants understand that a person sees not only in some vague eyes-open sense but more specifically in the sense of attending to *an object*. Most simply, in the second year the infant points at an object but also in increasingly sophisticated ways looks at the other to check that person's gaze as well (Lempers et al. 1977, Masur 1983). In compelling cases the infant will begin to point only when the other is attending to him or her, execute the point, check the other's visage, keep pointing or augment the behaviour if the other is not correctly directed, and only stop when the other orients towards or comments on the object (Bates *et al.* 1979; Butterworth 1991).

Research on older infants' manual gestures also demonstrates that babies at about one year understand that they and others not only direct their gaze at objects, but also desire them, at least in the rudimentary sense of seeking to attain them, hold them, and manipulate them. Infants at this age express their own wants for objects to other people, for example, by extending a hand toward an object and opening and closing the hand (Masur 1983). Or the infant at this age can also use a palm-away gesture to refuse something he or she doesn't want (Zinober and Martlew 1985). Of course such gestures could be merely learned routines by infants, routines produced because they 'worked' in the past by attaining an attractive object for the infant. But if so, the infant need never look to the other. It would be sufficient for the child

---

* This description of early intentional understanding extends my earlier proposal that before a belief–desire psychology, toddlers utilize a simple desire psychology. The current description is meant to improve on and eventually replace that one. I still believe that an early understanding of desire is central to early simple intentional understanding, in something like the sense that belief–desire reasoning is central to later mentalism. Understanding persons as wanting things is especially important, for example, to understand how intentional states cause and explain actions. But this does not mean (in this or in the earlier proposal) that the older infant is limited to an understanding of desire alone. Indeed, as will be noted in a later footnote, desires encompass only one of two important directions of fit between intentional states and the world, so it is important to emphasize that the infant's simple intentional understanding encompasses both of those directions, which are evident in simple intentional understanding of perception and emotions in addition to that of desire.

to produce the behaviour and stop when the toy is attained. But for one-year-olds such gestures are often directed at another 'triadically' or with 'dual-direction': the gesture toward the object, the child's attention toward the other (Clark 1978; Masur 1983).

Extended observations like the following make it difficult to regard the infant as simply producing learned routines that have been successful in the past, as opposed to communicating a desire toward having or seeing another object:

Marta is unable to open a small purse, and places it in front of her father's hand (which is resting on the floor). F does nothing, so M puts the purse in his hand and utters a series of small sounds, looking at F. F still does not react, and M insists, pointing to the purse and whining. F asks, 'What do I have to do?' M again points to the purse, looks at F, and makes a series of small sounds. Finally, F touches the purse clasp and simultaneously says, 'Should I open it?' Marta nods sharply (Bates *et al.* 1975, p. 219).

The infant not only expresses such simple wants but is able to read them in others. For example, Masur (1983) found in her study of nine to eighteen-month-olds that spontaneous giving and extending objects to the mother (beginning at about twelve or thirteen months) 'frequently occurs as responses to maternal behaviours, such as mother's palm-up requesting gestures'. If someone points or gestures to an object and the infant takes it to her, this suggests an incipient understanding, not just of attention to the object, but that the person seeks to have it. Similarly, Reddy (1991) documents instances where at about one year infants seem to play and tease with someone else's desires. An infant may hold out an object to another, who reaches to take it, but at the last second will pull it away, watching the other and smiling. In such teasing, the other not only attends to an object but seeks to attain it; and attainment of the object is begun, and then jokingly thwarted, begun again, and then thwarted again. Construing the other as seeking to attain objects (and distinguishing between the seeking and the attaining) constitutes a very simple construal of desire for objects.

Beyond these demonstrations of infants construing persons as seeing some objects and not others, and wanting to attain some objects and not others, older infants evidence understanding that persons have several emotional reactions directed toward objects seen and attained. Adamson and Bakeman (1985), for example, document one-year-olds' expressions of, and attention to others' expressions of, affect toward objects in triadic situations where the infant ensures self and other are attending to the same object. However, infants' intentional understanding of affect has been most clearly shown in the literature on social referencing. Early social referencing studies showed that infants from about ten or twelve months on not only attended to others' emotional expressions but also read them as meaningful in some sense. For example, when a baby is placed across a visual cliff from its mother, if the mother smiles, the baby is likely to cross, but if she appears fearful the baby is likely to stay where she is (Sorce *et al.* 1985). Of course, such attention and reaction to others' emotional displays may or may not imply an intentional understanding, may or may not involve construing the other's emotion as *about* something, for example about the cliff. Instead, seeing a mother frown or smile could simply and directly alter the baby's mood (Feinman 1985) or action; perhaps the infant sees the mother's fear expression and simply freezes. Potentially, however, social referencing could work by the infant reading the others' emotional expression as about or directed toward a particular situation or object. Indeed, Hornick *et al.* (1987) seem to have demonstrated just that sort of intentional understanding by one-year-olds. In that research, mothers posed expressions of delight or disgust toward a particular toy, in a situation containing several alternative toys. The twelve-month-olds in that study were selective in reaction, interpreting mother's expression as about a particular toy. For example, they avoided a toy toward which the mother expressed disgust, but showed no change in overall mood, and approached and played with other non-target toys.

At this point I want to be purposefully vague, to be open as yet between two different developmental stories. One story would be that at nine to twelve months infants start out with a 'behavioural' understanding of intentionality. When they evidence an understanding that others can see things, they are evidencing only an understanding that a person's eyes are externally open and directed toward a certain object. When they understand that the other 'wants' something they understand only, simply and literally, that the person is reaching for, grasping, taking, or chasing after an object. When they understand that the other is emoting about something, they understand nothing more than that the person is overtly displaying distinctive facial expressions towards certain objects. If infants' initial understandings are of this sort, then it represents a needed and genuine developmental advance when they later come to see people as not just behaviourally oriented toward objects, but as having inner *experiences* of those objects: not just orienting their sense organs toward an object, but in some sense perceptually experiencing it; not just reaching for or grasping after some object, but wanting it; not just displaying certain expressions toward objects, but liking or fearing them, or at the least experiencing some sort of affective valence. One possible story, therefore, would require these two distinct chapters.

Another story could be told, however, whereby even at nine to twelve months infants' interpretation of intentional orientation toward objects is not just one of reference, but also of experience. In this story infants' first intentional understandings are that persons experience objects, not just orient toward them. I read the arguments for secondary inter-subjectivity, intentional communication, and social referencing in nine- to twelve-month-olds as more in line with a story of this second sort, a story that in my terms attributes to these infants an initial intentional understanding that encompasses reference *and* experience (rather than an initial more 'behavioural' construal followed only later by an understanding of inner experience). In fact, I lean toward an account that sees infants' first understanding of intentionality as encompassing reference *and* experience, but the data are far from decisive. What I want to focus on instead, therefore, is the conclusion that if either of these stories is true, then by late in the second year older infants' understanding of intentionality is of a psychological–experiential sort, and not just behavioural. By eighteen to twenty-four months, at least, older infants construe others as intentional in the sense of having internal mental *experiences* of external objects (situations, actions). The data for such a conclusion are also less than decisive, but I believe more compelling.

I believe that infants achieve such a construal in the second year because I think that in this year infants increasingly evidence an understanding of someone else's gaze toward, *plus* desire for attainment of, *and* reactions such as happiness or pleasure with the same object in extended sequence. Such extended sequences — encompassing a person's not only looking at but being pleased to see, not only getting but being happy to have — increasingly suggest that they understand persons as having personal experiences regarding the objects in their visual space. Similarly, the data on older infants' attempts to comfort or hurt others suggest to me that they understand something important about persons' internal experiences with, and not just their behavioural orientations toward, certain objects, persons, and situations (see for example Harris 1989, Chapter 2). Finally, the attainment of additional competencies on the infant's part helps underwrite attributions of this sort of intentionality to infants before the second birthday. Specifically, in the months after their first birthday, children quickly reveal in their emerging verbal language at least a rudimentary understanding of a handful of intentional–experiential states, notably seeing, wanting and reacting emotionally. For example (from Bretherton *et al.* 1981): 'I see a car' (20 months), 'let me see' (20 months), 'want juice' (20 months), 'I wanna take a nap' (20 months), 'I no like celery' (22 months), 'those ladies scare me' (25 months), 'I sad I popped it' (25 months). The early language of infants in their second year thus often concerns itself with, and reveals an understanding of certain simple intentional states — subjective *experiences* about objects or situations. Indeed, Bloom (see for example Bloom *et al.* 1988) recently has proposed what she terms an intentional theory of language-acquisition based on the premise that first-language-acquisition is motivated by a need to, and designed primarily to, communicate about personal, intentional experiences — for example, to communicate inner emotional experiences of the world. In short, by a year and a half or so, the infant's understanding of others and objects together seems of this intentional form, including reference toward and experience of objects.*

The intentional understandings described above rest on examples of communicative interchange, first in an early protolinguistic form — the child pointing, giving, requesting objects, as well as reading affect about objects — and then in early linguistic form. Early protolinguistic communications of these kinds have been studied by researchers interested in language development, many of whom agree that beginning at about nine to twelve months, infants begin to communicate intentionally in the sense of trying to do so (for example Golinkoff 1983; Stern 1985; Bates *et al.* 1979). Bretherton *et al.* (1981) argued that the onset of intentional communication showed that infants had an implicit theory of mind. Because infants at this age intentionally point and gesture, but also gesture with alternating eye contact between goal-objects and intended listener, and augment or amend their gestures until the goal has been obtained, their early communications do seem purposive. But beyond this Bretherton *et al.* argued that implicit in such early intentional communication, as in mature communication, 'is the fact that the infant recognizes a partner's capacity to *understand* a message. In other words, the infant attributes an internal state of *knowing* and *comprehending* to the mother as he or she communicates' (p. 339). This goes unnecessarily far, I believe. Knowing and comprehending are cognitively complex informational states of mind akin to believing and thinking. Infants need make no such attributions to others in order to communicate simply about gaze toward, and seeking to attain, or reacting fearfully about, *that* object. The infant need not construe people as having 'interfaceable minds', as Bretherton *et al.* put it — only as having a few simple attitudes toward objects, such as seeing, wanting, and fearing them.

The relation between intentional conceptions and intentional communication is an important but tricky one. It is not the case, I believe, that the very existence of early intentional communication licenses us to attribute certain intentional conceptions — such as conceptions of knowledge or comprehension — to the infant. Rather, only if we understand the nature of

---

* That older infants' understanding of simple intentional states includes attending, *and* wanting, *and* emotionally reacting to objects has several implications. These three states together, for example, seem to include both important directions of fit (Searle 1983) between internal states and the world: efferent and afferent. As adults we tend to construe some internal states as directed toward changing the world, making world fit mind. Desires, for example, lead to (and thus underlie) acts such as reaching for, searching for, and asking for objects. Other states, on the other hand, instead accommodate themselves to objects — mind fits world. Most simply, perception is of this sort; but so also are certain emotional reactions such as liking something or not. The older infant's understanding of intentional states thus seems to encompass a significant duality: an initial understanding of both intentional action toward the world and intentional experience of it. This foreshadows an important aspect of our adult naïve psychology, whereby internal states occupy an important middle ground, receptive to the world via perception, but also intruding into the world via intentional action.

infants' first intentional conceptions can we correctly characterize the sense in which they intentionally communicate. Their intentional communications are based on conceiving of people as having only some experiential states, and not others. Thus in early language children use words for perception (see), desire (want), and emotion (happy) before words for belief and imagination (think, know) (Bretherton *et al.* 1981). And if such talk is carefully scrutinized and coded not just for mention of certain terms but for genuine reference to mental states, then reference to genuine states of desire ('he wants an apple') precedes genuine reference to thoughts or beliefs ('he thinks that's an apple') by many months (Bartsch and Wellman 1990; Brown and Dunn 1991). As indicated earlier, references to thoughts and beliefs emerge only in very old two-year-olds, immediately before or at the third birthday (Shatz *et al.* 1983; Bartsch and Wellman 1990).

Recall that in my description of simple desire in Fig. 2.2, I noted that even simple desires are not just desires for currently present visible objects in the world. It is possible, however, that infants' first understanding of intentional states is literally object-specific in this sense. In the demonstrations with nine-to twelve-month olds, the intentional object is a physical object, literally present in the attention of both parties. The infant considers only several states of the person toward *that* object. The intentional object can also be an action (for example, a person's laughter, a nap); but still, at first, only here-and-now actions or occurrences. Adults, of course, in stark contrast, understand the objects of intentional states much more broadly, indeed propositionally. We construe intentional states as propositional attitudes, attitudes toward a proposition such as 'that it snow', including the possibility of mental attitudes toward imaginary or even counterfactual states—'that it snow in July in Florida'. There is suggestive evidence that in the course of the second year the infant's understanding of intentional objects expands. At the least it expands to encompass an understanding that intentional states can be about absent or unavailable objects or behaviours. Butterworth (1991), for example, concludes that pointing is at first understood by the infant as toward a specific object at a location in front of him or her. By about eighteen months, however, infants are able to interpret such attentional gestures as toward an object not currently visible because it is behind them. Language may facilitate, but also demonstrates, this expansion. It is generally conceded that one power of words in comparison to pointing, say, is that words can easily refer to absent, unavailable objects and states of affairs. Gopnik (1984) has shown that infants begin to comment on unavailable objects quite early in language-acquisition, in the second year, in utterances like 'all gone' and 'bye-bye'. Moreover, studies of children's early use of emotion terms document that at least by twenty-eight months children also talk about past and future emotions (Bretherton and Beeghly 1982).

This sort of early expansion in the range of intentional objects would still, I believe, be anchored within the infant's conception of real potentially available objects and states of affairs. The intentional object (the thing, action, or situation) in question, as the older infant construes it, really exists in some sense, although it is not necessarily available to the intentional agent right now. Consider an intentional object such as the apple at the top of Fig. 2.2. The apple may be under lock and key, withheld by a parent, or even thrown away and gone. But the character is seen as wanting a *real* apple. Whether he gets it or not, whether it indeed becomes available or never does, means the character will be happy or sad. Older infants, therefore, construe people as wanting (for example) real objects, but understand that the real world of objects is a world extended beyond the here and now in time and space. Similarly for intentional objects such as actions or states of affairs. Take such objects of desire as desired actions (to hop) and states (to nap). Again at any place only a small sample of such actions and states is available at one time (I can't hop and nap at the same time); but none the less these are still real-world actions and states, albeit currently available or unavailable.

By eighteen to twenty-four months, then, I believe that children understand people as having inner experiences of certain types (principally perceiving, desiring, and certain emotions) about external objects, situations, and actions. These objects, actions, and situations that are the targets of intentional states may not be presently available, experienceable, or do-able, but they are nonetheless real, in the sense of potentially available, as opposed to imaginary, counterfactual, or false.

Pretence

If this sort of proposal proves true, then the older infants' understanding of persons is intriguingly intentional, but their grasp of intentionality is still quite different from our own. It is different from our own precisely because it is limited to a construal of persons as directed toward *actual* (available or unavailable) objects and states of affairs. In contrast, adults understand that the objects (actions, situations) of intentional experience may not only be unavailable but may be fictional, imaginary, or false. At least our construal of mental states such as thoughts, beliefs, ideas, and images is of this sort — the contents of such states are mental contents, representations, and not directly contents of the world itself.

Consider our ordinary conception of belief a bit further. As adults, our conceiving and reasoning about someone's beliefs, for example Bill's belief that it is raining, seems to involve (at least) two interwoven aspects. First, the contents of a person's thoughts ('it is definitely raining') do not necessarily match the contents of the world (it is not raining), so reasoning about beliefs requires some ability to reason about objects or contents or states of affairs that are opaque,* that is, independent of, or bracketed from (or decoupled from), the world itself; not just unavailable at the moment, for example, but potentially not true or not real at all. We see a person as having attitudes about such bracketed contents (Bill thinks 'it is raining' when there is no rain, Bill hopes 'there are unicorns' even though there are no such things as unicorns).

Note, however, that we might construe something like a photograph or a drawing as presenting a special bracketed set of contents, too. And thus when we think of someone as liking the drawing's contents, for example, we construe him or her as having a certain subjective reaction to that set of contents, a set of contents that is out there impersonally in the world, specifically in that drawing. This drawing example shows that our ordinary understanding of beliefs, thoughts, and the like, in contrast to drawings or photographs, includes a second aspect as well. Conceiving of someone's thoughts, for adults, involves seeing that person as related to a bracketed set of contents, but in addition it involves construing the person as internally, subjectively *having* those contents. Mental contents are not, for example, in some external objective photograph; the person not only has attitudes toward bracketed contents, but has the mental contents themselves, in some personalized subjective sense — 'in the mind', as it were. The depiction in Fig. 2.2 of ordinary beliefs captures both of these aspects by showing a bracketed parallel set of contents (in the cloud) as 'in the person's head'. We do not literally, *spatially*, construe mental contents as *inside* of the head (and neither do three-year-olds: Estes *et al.* 1989); but we do construe them as mentally 'held' by the person or agent involved.

By three or four years children also see things in this fashion: people possess mental contents, namely their thoughts, ideas, beliefs 'in the mind'. How do infants evolve from thinking of persons as subjectively reacting to external actual contents to thinking of them as having personalized bracketed mental contents? The available evidence suggests that it takes at least two steps to achieve this later conception of mind. Specifically, there is evidence of an initial ability to reason about bracketed contents evident in toddlers' pretend play, but only later an understanding of subjective *mental* contents themselves, evident in three- and four-year-olds' beginning understanding of beliefs, mental images, and imaginings.

---

* The term 'opaque' here stems from the notion of logical opacity. Our reasoning about belief and other similar sorts of mental states evidences logical opacity. In the real world, if it is true that that is an apple and it is true that it is green, then it follows that it is a green apple; and if the apple is crushed, then a green apple is crushed. These sorts of consequences and logical inferences do not hold in reasoning about thoughts, however. If Bill thinks that is an apple, and it is a green apple, it is not necessarily the case that Bill *thinks* it is a green apple. The apple in Bill's thoughts does not simply, transparently, reflect the apple in the world. Reasoning about Bill's thoughts does not go through to the world. Thoughts as mental states are in this sense opaque.

Alan Leslie (1987) has most cogently pointed to the ability to reason about pretend scenarios as momentous for the psychological understanding of people. Pretend play evidences an important capacity for reasoning about person in terms of their attitudes toward bracketed contents, that is, pretend objects and actions. In pretence, the object is really a banana but I pretend it is a telephone. When you join in this pretence with me I understand your pretence as about a 'telephone', although we are playing with a banana. In shared pretence of this sort toddlers exhibit an important ability to reason about persons' pretend attitudes toward pretend objects, all the while keeping track of the pretence and the reality. In negotiating this sort of pretence, the child must construe persons as seeing, reaching for, manipulating, liking or not liking, bracketed contents: the pretend-phone. As Leslie argued, negotiating this sort of pretence shows that the child by two years has the computational competence necessary to deal with logical opacity (see the last footnote) — the same sort of logical opacity required to reason about beliefs, for example.

But having the capacity to reason about bracketed contents does not mean that the child as yet actually conceives of beliefs or thoughts, etc., and reasons about them. In pretending, the child need not construe people as representing, mentally 'in their minds', an idea or image or mental content: the 'phone'. The child need only understand that he and the other are acting as if, for now, this banana is a pretend-phone, a fake phone, a phone-substitute. Harris (1991) makes a similar distinction (as does Perner 1991). He asks if pretending requires understanding of another's mental state:

Early shared pretence can operate in a much simpler way. An adult acts out a piece of make-believe. To join in, the child needs to recognize what the other person is pretending to do, and also that what they are pretending is not for real. For example, the child needs to recognize that the other person is pretending to pour tea, but that there is no real tea. This may be a lot, but it does not require any insight into the mind of the person pouring the pretend tea. In the same way, a theatre-goer need only respond to the play *as if* it were real while acknowledging that it is not; he or she can ignore the mental processes by which the actors produce their performance (Harris 1991, p. 302).

This analysis, separating reasoning about bracketed pretend contents from a conception of mental contents, can be confusing because as adults we tend to construe pretence mentalistically, as someone imagining in his head a different fictional state of affairs. But the two-year-old need not construe pretence mentalistically. He or she need only be able to stipulate alternative pretend objects or actions and reason out how persons could interact with those.

It is a potentially separable conceptual understanding, then, to be able to reason about bracketed contents as against understanding that thoughts or beliefs are bracketed *mental* contents. And there is evidence to suggest that toddlers make the first step but not the second. Young two-year-olds engage in pretence and reasoning about pretence of the requisite complicated sort, but there is no evidence that they conceive of people as having beliefs, thoughts, ideas, mental images, or the like. There is no evidence for this until just about the age of three, and then the evidence begins to abound. Three-year-olds but not two-year-olds sensibly talk about contentful mental states via words such as *think, know*, and *remember* (see for example, Shatz *et al.* 1983); they evidence an initial understanding of belief in at least some forms (as reviewed earlier in this chapter); and they evidence an understanding of mental images, including an acceptance that such images are contentful yet private, and internal in the sense of 'in the head', 'in the imagination', and 'not in the real world' (Estes *et al.* 1989).

SUMMARY

This outline of early developments is as yet as much story as fact. Still it indicates a variety of steps along the way to a mature theory of mind. It is important to have some sense, however speculative, of these prior steps in order to frame research into early developments and in order to think more precisely about deficits and delays. For these purposes the story outlined here pinpoints, I believe, some important early progressions. To summarize, by the age of three young children see people as possessing beliefs as well as desires; as having ideas, thoughts, and images, as well as emotional reactions. Moreover, by the age of four they see people as living their lives within a world of mental content that determines how they behave in the world of real objects and acts. These achievements are preceded by three earlier phases, in the normal case. There is a first phase of biologically prepared picking out and attention to people as entities and to personal–human causation as different from physical–mechanical causation. There is a second phase of simple intentional understanding of action and experience, evident in construing people as gazing at, seeking to attain, and emotionally reacting to real-world objects and actions. This constitutes first evidence of a rapidly developing understanding of persons as having the intentional subjective experiences of perceiving, wanting, and emotionally experiencing objects. In a third development, toddlers become able to pretend and, more importantly, to understand themselves and others as pretending. This requires, and evidences, an ability to reason about people's relation to alternative pretend objects, actions, and situations. Engaging in pretence is a notable advance over simple intentional understanding because the child reasons about people as relating to bracketed pretend contents, and not just to real contents in the world. This ability shows, as Leslie first noted, that the child has much of the needed logical–cognitive wherewithal to reason about beliefs.

However, this early understanding of pretence does not mean that the young child sees people as having a personal subjective mental world of thoughts, ideas, and beliefs. Understanding mental worlds in this fashion, and in the end construing people's real-world actions as *inevitably* filtered through representations of the world rather than linked to the world directly, begins to emerge at about three years.

This description is on firm ground when it comes to the portrayal of four-year-olds. It is increasingly well-documented when it comes to describing three-year-olds as understanding mental representational states such as beliefs and imaginings, and when it comes to portraying normal children as understanding desire more easily and earlier than beliefs. The description is admittedly speculative, but still on track, I think, in conveying the earlier emergence of a simple intentional understanding of action and experience in infants, and in noting some milestones in that early understanding as infants progress toward the three- and four-year-olds' theory of mind. To reiterate, in this description I have used ages such as two, three, and four years only as landmarks; not the ages but the sequence or progression of conceptions is what matters most.

This outline is admittedly descriptive. I attempt to begin to characterize a progression of conceptual states, and not yet the mechanisms whereby development proceeds from one to the next. The mechanisms of cognitive development, as best we know, are a hybrid lot, including biologically pre-pared understandings and skills, innately given constraints on the path that developmental change can take, learning, and conceptual reorganization. I am not sure exactly how these weave together to produce theory of mind. For example, initial intentional understandings may represent activation of a biologically prepared conceptual stance toward human experience and action, whereas the step from understanding desires to understanding beliefs may represent conceptual change alone. Still, I believe that conceptual reorganization, whereby later ideas stem from earlier ones because we revise our concepts to be increasingly effective, coherent, and responsive to the 'data', constitutes a crucial part of the story. (In Wellman (1990) I begin to lay out how the shift from understanding desires to beliefs may proceed in this fashion.) Theory of mind evidences theory-change. To the extent that this is so, to the extent that earlier conceptions engender later ones, then it will be possible to describe an increasingly rich progression of unfolding conceptions in the child's theorizing. In part, that is what I have tried to do.

CONCLUSIONS

Charting early origins is an important task for understanding normal development of a theory of mind. Furthermore, knowledge of these very early steps should prove fruitful for understanding autism as well. This is not a new idea. Leslie's proposal (1987) of a theory of mind module first evident in pretence, Hobson's (1990) ideas about the early direct perception of affect, and Mundy and Sigman's (1989) research on early joint-attention deficits all make use of knowledge culled from normal infant development to attempt to understand better the onset and course of autism.

A refined understanding of the early developmental course of normal children could bear fruit in several different ways. For example, as I understand it, there are currently two major classes of hypotheses as to how a theory of mind impairment may account for autism. One is the proposal that autism represents a defective theory of mind module—a *deficit* in understanding all or some mental states (for example, pretence and belief). The second is the proposal that the theory of mind itself is best seen as a special stream of human reasoning with a normal course of developments (rather than a single specific module). Autism, under this proposal, would represent a serious *delay* in the normal progression of this sort of thinking. There are, of course, many variations on these themes. It is possible to propose, for instance, that mentalistic reasoning actually encompasses several types of developing understanding, and that autism is a delay in only one of these types and not others (see, for example, Tan and Harris 1991). Or it is possible to propose that social interaction and understanding are delayed (or even completely impaired) in autism, but that 'theory of mind' is an inappropriate, even dangerously mistaken, way to conceive of the essential sort of competence involved (see for example Hobson 1991). Still, the two general accounts I have contrasted represent two important but different classes of proposal. If *either* were the case we would benefit from better understanding of the normal course of early development.

Suppose, on the one hand, that autism does represent a defective theory of mind module, a module that ordinarily comes on line at a certain point in development. If so, then pinpointing the emergence of that special competence within a sequence of related developments would clarify just what abilities should be impaired in autistic children, as well as what earlier and parallel abilities should be unimpaired. Alternatively, suppose instead that autism represents more a serious delay in the normal progression of a special stream of human reasoning. If so, then charting the early course of that stream is even more important. Early milestones within that stream would be good places to test the waters for initial diagnostic signs of autism. Moreover, characterizing normal acquisitions in advance of the sort of reasoning available to four-year-olds could help to characterize capacities that are intact in autistic children of varying ages. This seems an especially important task. Describing autism as a theory of mind deficit helps depict dramatically just how serious and specific such a syndrome might be. But we need to achieve a positive characterization of such persons' understanding as well in order to communicate with them, train them, and simply understand them. From the currently available data it is clearly inaccurate to characterize autistic children as confined to a merely physicalistic understanding of persons, or even a behaviouristic understanding (Baron–Cohen 1989). Their understanding seems richer than that, while still falling short of a four-year-old's theory of mind. If so, then normally acquired earlier understandings give us a richer sense of alternative characterizations. For example, it might prove to be the case that most high-functioning autistic children achieve a simple intentional understanding of persons (and hence an understanding that persons can see available objects, want available objects, and experience simple happiness and sadness about direct situations) while still failing more specifically to understand pretence and beliefs. In general a richer understanding of normal development provides a better platform for achieving a truly developmental perspective on this abnormal condition.

The reverse also holds. A developmentally sensitive understanding of autism will not only aid understanding autism itself, but also understanding of normal development. If it ever was, it is no longer possible to understand development of a theory of mind without weaving together inspiration and evidence from normal and impaired children. In this chapter I have attempted to describe normal development only, with recourse to data on normal unimpaired children alone. This has been extremely difficult to do. For example, the literature on joint-attention behaviour in autism (see the articles in Cicchetti 1989) is part of what leads me to believe that normal infants' acquisition of a simple intentional construal of persons at about one year or so is an important precursor of later developments. The evidence that both pretence and understanding of belief are impaired in autistic children is an important reason for linking abilities to pretend with abilities to understand propositional attitudes. The emerging findings as to autistic children's good understanding of false photographs in conjunction with poor understanding of false beliefs (Leekam and Perner 1991; Leslie and Thaiss 1992) provides important evidence for thinking that our theory of mind springs in part from specially evolved attention to and understanding of people. We are fortunate in *not* being limited solely to studying the normal case.

Along these lines the available literature on autism can be utilized to make an important, but often overlooked point. Studies of autistic children's theory of mind inevitably compare autistic to normal children. But they also often utilize other impaired comparison groups as well. For example, comparable tasks have been used with normal four-year-olds, high-functioning autistic children, Down's syndrome individuals with a mental age of about four (Baron–Cohen *et al.* 1985), mentally handicapped persons (Baron–Cohen 1989), and language-delayed individuals (Perner *et al.* 1989). These studies show essentially normal development of a theory of mind in all subjects save those with autism. Early straightforward acquisition of a theory of mind, evident in three- and four-year-olds, *is* the normal case, I believe.

The collaboration on understanding theory of mind now obvious between researchers of normal and autistic children represents just two strands of what is necessarily a still larger multidisciplinary tapestry. This larger braid-work would weave together several analyses of the nature and development of our mentalistic construal of persons and actions: comparative and evolutionary analyses of the human propensity for complex social collaboration, coalition, and collusion; neuropsychological and computational analyses of our capacity for processing information about mental states and intentional systems; and anthropological and cross-linguistic analyses of universality and variation in a core mentalism, as well as in distinctly different world-views about people. A theory of mind is biologically prepared, originates in early childhood, and emerges from capacities available in infants; but it is also strongly socially nurtured, resulting in universal understanding as well as distinctive folk psychologies. Acquiring a theory of mind is a foundational human development, the impairment of which leads to very serious consequences, as the evidence from normal and autistic children begins to make clear.

Acknowledgements

This chapter was written while the author was a Fellow at the Center for Advanced Study in the Behavioral Sciences. I am grateful for financial support provided by the Center, NICHD, and the MacArthur Foundation.

REFERENCES

Adamson, L. B., and Bakeman, R. (1985). Affect and attention: Infants observed with mothers and peers. *Child Development*, **56**, 582–93.

Baillargeon, R., Spelke, E. S., and Wasserman, S. (1985). Object permanence in five-month-olds. *Cognition*, **20**, 191–208.

Bakeman, R., and Adamson, L. B. (1984). Coordinating attention to people and objects in mother–infant and peer–infant interaction. *Child Development*, **55**, 1278–89.

Bates, E., Camaioni, L., and Volterra, V. (1975). The acquisition of performatives prior to speech. *Merrill-Palmer Quarterly*, **21**, 205–26.

Bates, E., Bonigni, L., Bretherton, I., Camaioni, L., and Volterra, V. (1979). *The Emergence of symbols: cognition and communication in infancy*. Academic Press, New York.

Baron–Cohen, S. (1989). Are autistic children behaviorists? An examination of their mental–physical and appearance–reality distinctions. *Journal of Autism and Developmental Disorders*, **19**, 579–600.

Baron-Cohen, S., Leslie, A.M., and Frith, U. (1985). Does the autistic child have a 'theory of mind?' *Cognition*, **21**, 37–46.

Bartsch, K., and Wellman, H.M. (1989). Young children's attribution of action to beliefs and desires. *Child Development*, **60**, 946–64.

Bartsch, K., and Wellman, H.M. (1990). Everyday talk about beliefs and desires. Portions of this paper were presented by Bartsch at the Piaget Society meetings, Philadelphia, and by Wellman at the National Developmental Conference, Perth, Australia.

Bertenthal, B.I., Proffit, D.R., Spetner, N.B., and Thomas, M.A. (1985). The development of infant sensitivity to biomechanical motions. *Child Development*, **56**, 531–43.

Bloom, L., Beckwith, R., Capatides, J.B., and Hafitz, J. (1988). Expression through affect and words in the transition from infancy to language. In *Lifespan development and behavior* (ed. P. Baltes, D. Featherman, and R. Lerner). Erlbaum, Hillsdale, NJ.

Bowlby, J. (1969). *Attachment and loss*, Vol. 1: Attachment. Basic Books, New York.

Bretherton, I., and Beeghly, M. (1982). Talking about internal states: The acquisition of an explicit theory of mind. *Developmental Psychology*, **18**, 906–21.

Bretherton, I., McNew, S., and Beeghly-Smith, M. (1981). Early person knowledge as expressed in gestural and verbal communication: when do infants acquire a 'theory of mind?' In *Social cognition in infancy* (ed. M. Lamb and L. Sherrod). Erlbaum, Hillsdale, NJ.

Brown, J.R., and Dunn, J. (1991) 'You can cry, mum': the social and developmental implications of talk about internal states. *British Journal of Developmental Psychology*, **9**, 237–56.

Butterworth, G.E. (1991). The ontogeny and phylogeny of joint visual attention. In *Natural theories of mind* (ed. A. Whiten). Basil Blackwell, Oxford.

Cicchetti, D. (ed.) (1989). Special issue: 'Theory of mind and autism'. *Development and Psychopathology*, **1**, 171–217.

Collis, G.M. (1981). Social interaction with objects: a perspective on human infancy. In *Behavioral development* (ed. K. Immelman, G. Barlow, L. Petrinovich, and M. Main). Cambridge University Press.

D'Andrade, R. (1987). A folk model of the mind. In *Cultural models in language and thought* (ed. D. Holland and N. Quinn). Cambridge University Press.

DeCasper, A.S., and Fifer, W.P. (1980). Of human bonding: newborns prefer their mothers' voices. *Science*, **208**, 1174–6.

Estes, D., Wellman, H.M., and Woolley, J.D. (1989). Children's understanding of mental phenomena. In *Advances in child development and behavior* (ed. H. Reese). Academic Press, New York.

Feinman, S. (1985). Emotional expression, social referencing and preparedness for learning in infancy. In *The development of expressive behavior: biology-environment interactions* (ed. G. Zivin). Academic Press, New York.

Flavell, J.H. (1988). The development of children's knowledge about the mind: from cognitive connections to mental representations. In *Developing theories of mind* (ed. J. Astington, P. Harris, and D. Olson). Cambridge University Press, New York.

Flavell, J.H., Green, F.L., and Flavell, E.R. (1986). Development of knowledge about the appearance–reality distinction. *Monographs of the Society for Research in Child Development*, **51**, Serial No. 212.

Flavell, J. H., Flavell, E. R., Green, F. L., and Moses, L. J. (1990). Young children's understanding of fact beliefs versus value beliefs. *Child Development*, **61**, 915–28.

Fodor, J. A. (1987). *Psychosemantics: the problem of meaning in the philosophy of mind*. Bradford Books/MIT Press, Cambridge, Mass.

Gelman, R., and Spelke, E. (1981). The development of thoughts about animates and inanimates: implications for research on social cognition. In *New directions in the study of social-cognitive development* (ed. J. H. Flavell and L. Ross). Cambridge University Press.

Golinkoff, R. M. (1983). Infant social cognition: self, people, and objects. In *Piaget and the foundations of knowledge* (ed. L. Liben) Erlbaum, Hillsdale, NJ.

Gopnik, A. (1984). The acquisition of *gone* and the development of the object concept. *Journal of Child Language*, **II**, 273–92.

Gopnik, A., and Slaughter, V. (1991). Young children's understanding of changes in their mental states. *Child Development*, **62**, 98–110.

Hala, S., Chandler, M., and Fritz, A. S. (1991). Fledgling theories of mind: deception as a marker of 3-year-olds' understanding of false belief. *Child Development*, **62**, 83–97.

Harris, P. L. (1989). *Children and emotion*. Basil Blackwell, Oxford.

Harris, P. L., Brown, E., Marriot, C., Whithall, S., and Harmer, S. (1991). Monsters, ghosts and witches: testing the limits of the fantasy–reality distinction in young children. *British Journal of Developmental Psychology*, **9**, 105–23.

Hobson, R. P. (1990). On acquiring knowledge about people, and the capacity to pretend: response to Leslie. *Psychological Review*, **97**, 114–121.

Hobson, R. P. (1991). Against the theory of 'theory of mind'. *British Journal of Developmental Psychology*, **9**, 33–51.

Hornick, R. Risenhoover, N., and Gunnar, M. (1987). The effects of maternal positive, neutral, and negative affective communications and infant responses to new toys. *Child Development*, **58**, 937–44.

Kaye, K., and Fogel, A. (1980). The temporal structure of face-to-face communication between mothers and infants. *Developmental Psychology*, **16**, 454–64.

Leekam, S. R., and Perner, J. (1991). Do autistic children have a metarepresentational deficit? *Cognition*, **40**, 203–18.

Lempers, J. D., Flavell, E. R., and Flavell, J. H. (1977). The development in very young children of tacit knowledge concerning visual perception. *Genetic Psychology Monographs*, **95**, 3–53.

Leslie, A. M. (1987). Pretense and representation: the origins of 'theory of mind.' *Psychological Review*, **94**, 412–26.

Leslie, A. M., and Thaiss, L. (1992). Domain specificity in conceptual development: neuropsychological evidence from autism. *Cognition*, **43**, 225–51.

Lewis, C. and Osbourne, A. (1990). Three-year-olds' problems with false belief: conceptual deficit or linguistic artifact? *Child Development*, **61**, 1514–19.

Masur, E. F. (1983). Gestural development, dual-directional signaling, and the transition to words. *Journal of Psycholinguistic Research*, **12**, 93–109.

Meltzoff, A. N., and Moore, M. K (1983). Newborn infants imitate adult facial gestures. *Child Development*, **54**, 702–19.

Moses, L. J. (1990). *Young children's understanding of intention and belief*. Unpublished Ph.D. dissertation, Stanford University.

Moses, L. J., and Flavell, J. H. (1990). Inferring false beliefs from actions and reactions. *Child Development*, **61**, 929–45.

Nelson, L. A. (1987). The recognition of facial expressions in the first two years of life: mechanisms of development. *Child Development*, **58**, 889–909.

O'Neill, D. K., and Gopnik, A. (1991). Young children's ability to identify the sources of beliefs. *Developmental Psychology*, **27**, 390–9.

Perner, J. (1988). Developing semantics for theories of mind: from propositional attitudes to mental representations. In *Developing theories of mind* (ed. J. Astington, P. Harris, and D. Olson). Cambridge University Press, New York.

Perner, J. (1991). *Understanding the representational mind*. MIT Press, Cambridge Mass.

Perner, J., Leekam, S. R., and Wimmer, H. (1987). Three-year-olds' difficulty with false belief. *British Journal of Developmental Psychology*, **5**, 125–37.

Perner, J., Frith, U., Leslie, A. M., and Leekam, S. (1989). Exploration of the autistic child's theory of mind: knowledge, belief, and communication. *Child Development*, **60**, 689–700.

Piaget, J. (1929). *The child's conception of the world*. Harcourt, Brace, New York.

Pillow, B. H. (1989). Early understanding of perception as a source of knowledge. *Journal of Experimental Child Psychology*, **47**, 116–29.

Poulin-Dubois, D., and Shultz, T. R. (1988). The development of the understanding of human behavior: from agency to intentionality. In *Developing theories of mind* (ed. J. Astington, P. Harris, and D. Olson). Cambridge Unviersity Press, New York.

Pratt, C., and Bryant, P. E. (1990). Young children understand that looking leads to knowing (so long as they are looking into a single barrel). *Child Development*, **61**, 973–82.

Premack, D. (1990). The infant's theory of self-propelled objects. *Cognition*, **36**, 1–16.

Reddy, V. (1991). Playing with others' expectations: teasing and mucking about in the first year. In *Natural theories of mind* (ed. A. Whiten). Basil Blackwell, Oxford.

Searle, J. R. (1983). *Intentionality*. Cambridge University Press, New York.

Shatz, M., Wellman, H. M., and Silber, S. (1983). The acquisition of mental verbs: a systematic investigation of first references to mental state. *Cognition*, **14**, 301–21.

Sherrod, L. R. (1981). Issues in cognitive–perceptual development: the special case of social stimuli. In *Infant social cognition: empirical and theoretical considerations* (ed. M. Lamb and L. Sherrod). Erlbaum, Hillsdale, NJ.

Siegel, M., and Beattie, K. (1991). Where to look first for children's understanding of false beliefs. *Cognition*, **38**, 1–12.

Sorce, J. F., Emde, R. N., Campos, J. J., and Klinert, N. D. (1985). Maternal emotional signaling: its effect on the visual cliff behavior of 1-year-olds. *Developmental Psychology*, **20**, 195–200.

Stein, N. L., and Levine, L. J. (1989). The causal organization of emotional knowledge: a developmental study. *Cognition and Emotion*, **3**, 343–78.

Stern, D. N. (1985). *The interpersonal world of the infant*. Basic Books, New York.

Tan, J., and Harris, P. L. (1991). Autistic children understand seeing and wanting. *Development and Psychopathology*, **3**, 163–74.

Trevarthen, C., and Hubley, P. (1978). Secondary intersubjectivity: confidence, confiders, and acts of meaning in the first year of life. In *Action, gesture and symbol* (ed. A. Lock). Academic Press, New York.

Wellman, H.M. (1990). *The child's theory of mind*. MIT Press, Cambridge, Mass.

Wellman, H.M., and Banerjee, M. (1991). Mind and emotion: children's understanding of the emotional consequences of beliefs and desires. *British Journal of Developmental Psychology*, **9**, 191–214.

Wellman, H.M., and Bartsch, K. (1988). Young children's reasoning about beliefs. *Cognition*, **30**, 239–77.

Wellman, H.M., and Estes, D. (1986). Early understanding of mental entities: a reexamination of childhood realism. *Child Development*, **57**, 910–23.

Wellman, H.M., and Woolley, J.D. (1990). From simple desires to ordinary beliefs: The early development of everyday psychology. *Cognition*, **35**, 245–75.

Yuill, N. (1984). Young children's coordination of motive and outcome in judgments of satisfaction and morality. *British Journal of Developmental Psychology*, **2**, 73–81.

Zinober, B., and Martlew, M.C. (1985). Developmental changes in four types of gesture in relation to acts and vocalizations from 10 to 21 months. *British Journal of Developmental Psychology*, **3**, 293–306.

3

Social development in autism: historical and clinical perspectives

FRED R. VOLKMAR AND AMI KLIN

Autism has captivated the imagination and research endeavour of investigators in disciplines as diverse as ethology and neurophysiology. In the search for the 'Rosetta stone' of social development, many researchers have studied autism with the intent of unravelling the very essence of human social relatedness and culture. Many decades ago, a similar endeavour brought about the once fashionable anthropological quest for 'the savage in a state of nature' (Zingg 1940). Such a savage, it was thought, would show us which aspects of social and cultural behaviours were innate and which were acquired. The search for isolated humans living outside society led to the description of so-called 'feral children', who allegedly grew up in the wild, reared by mammals other than man (Gesell 1949; Maclean 1977). These descriptions were blends of small amounts of fact and large amounts of fancy; it appears that the great majority of feral children were congenitally abnormal children deliberately abandoned in the wild to die (cf. Lévi-Strauss 1949; Bettelheim 1967).

Although there are no reliable cases of children having grown up outside society, there are, due to a tragic accident of nature, children with autism, who live in society, but who for some as yet ill-understood reasons, cannot profit much from the social stimulation provided by loving and caring parents. Efforts to understand the roots of their social impairment have been as difficult as our predecessors' quest for 'the man without culture'. As our predecessors failed to understand that man is biologically a social and cultural animal, we as yet have failed to understand what exactly this biology consists of (Volkmar 1987). The social disabilities of autistic individuals remain the most striking, and probably the least understood, aspect of the autistic syndrome.

Social encounters with autistic individuals illustrate the severity of their social deficits, as well as the complex issues posed by developmental changes and the heterogeneity of the syndrome. Young autistic children may fail to respond differentially to a strange person and may act as if other people, including their parents, are of little or no interest. This is in stark contrast to their often exquisite sensitivity to the inanimate environment, as they may become profoundly distressed in response to minor deviations or changes in seemingly trivial routines. The older autistic child or adult may, on the other hand, approach others in odd or idiosyncratic ways, typically making use of stereotyped and one-sided patterns of social interaction (Howlin 1986; Volkmar 1987).

This chapter provides a general summary of both the history of research on autistic social dysfunction and the available clinical evidence regarding its characterization. The role of a normative developmental framework must be emphasized, since it highlights the distinctiveness of social disabilities in autism, as well as the various points of continuity with more normative developmental processes.

HISTORICAL BACKGROUND

Research in autism has undergone several major shifts over the nearly five decades since Kanner's initial (1943) report of the syndrome. Although the importance of disturbed social relationships (autism) for syndrome definition has continually been emphasized, the various shifts in research emphases and in conceptualizations of the disorder have, somewhat paradoxically, impeded research on just these aspects of the disorder.

Kanner's initial (1943) report emphasized the centrality of social dysfunction as a pathognomonic feature of the disorder; moreover, by contrasting the limited social skills of his first cases with social skills which normally emerge very early on in development, Kanner was careful to place this observation explicitly within a developmental context. Although Kanner's phenomenological description of the condition has proved to be remarkably enduring, other aspects of his report suggested false leads for research. For example, while his initial report emphasized the apparently congenital nature of the disorder, it also noted both the unusual degrees of personal achievement of the parents and their unusual interactional styles with the child. At the time, of course, there was little understanding of the potential contributions of a deviant child to deviant parent–child interaction (see for example, Bell and Harper 1977); subsequent reports emphasized the role of experiential factors in pathogenesis. Such notions were congruent with the then current emphasis on psychodynamic factors in psychopathology, and suggested that children developed the disorder as a result of a confusing, perplexing, and noxious psychosocial environment. Thus, descriptions of 'refrigerator mothers' were common, and the emphasis was on very fundamental disturbances in 'object relations' as central aspects of syndrome pathogenesis and for remediation (see for example Bettelheim 1967, and Mayes *et al.* this volume, Chapter 20). Accordingly, the early emphasis was on removing the child from the deviant environment and providing a comprehensive psychotherapeutic programme to remedy the presumed deficits. Additional sources of confusion arose regarding the independent validity of autism as distinct from other conditions, notably childhood schizophrenia; other aspects of Kanner's original report suggested that the disorder was not associated with 'organic' conditions nor with mental retardation. Essentially, the two decades following Kanner's original report were devoted to clarifying these issues.

Given the early emphasis on psychodynamic factors, it is understandable that early reports based on clinical work with autistic children tended to view all behaviour of the child as imbued with considerable intentionality and intrapsychic meaning. For example, deficits in performance on traditional IQ tests were viewed as reflecting 'negativism' rather than basic cognitive disturbances, and echolalia was viewed as an attempt by the child to avoid social–communicative interaction.

Various lines of evidence, including longitudinal data, were helpful in establishing the centrality of social deficits for the definition of the syndrome and for clarifying the role of experiential factors in pathogenesis. It became apparent, for example, that parents did not exhibit particular deficits in child care nor in parental psychopathology (McAdoo and DeMyer 1978). Similarly, it became clear that even very adverse experiences early in life do not typically lead to autism (Fein *et al.* 1986). Although the emphasis on social deficits had initially suggested some role of psychosocial factors in pathogenesis, it became more reasonable to view the child, rather than the parents, as the source of dyadic deviation. Similarly, longitudinal information such as the frequency of seizure disorders in autistic individuals, their relatively consistent and poor performance on tests of intelligence, and the frequent association of autism with a host of medical conditions was more congruent with a definite, if ill-defined, 'organic' etiology.

The growing consensus on the validity of the syndrome led various investigators to propose categorical definitions (for example Rutter 1978). Such definitions, consistent with Kanner's original report, emphasized the primacy of social factors, at least for purposes of syndrome definition. Typically, however, theoretical models of the condition continued to emphasize other aspects of development as 'primary' for purposes of syndrome pathogenesis. Disturbances in such varied aspects of development as perception (Ornitz and Ritvo 1968), language (Rutter *et al.* 1971), cognition (Prior 1979), and arousal (Richer 1976) were presumed to be central in the development of the condition. Thus, although viewed as 'primary' for purposes of syndrome definition, disturbances in social development were viewed as secondary to other processes. Several lines of evidence, and one major assumption, were consistent with this view.

Firstly, it was clear that some social skills emerged over the course of development (see for example Howlin 1986; Volkmar 1987), so that, for example, patterns of differential social behaviour to familiar adults were observed (see for example Donnellan *et al.* 1984; Sigman and Ungerer 1984; Sigman *et al.* 1987; Rogers *et al.* 1991). This observation seemed consistent with the notion that cognitive, rather than social factors, were 'primary'. Secondly, several carefully conducted observational and experimental studies clearly suggested that social responsiveness in autistic individuals could be increased by various means such as increased adult attention, peer modeling, etc. (Churchill and Bryson 1972; McHale 1983; Charlop *et al.* 1983; Volkmar *et al.* 1985). Finally, an implicit 'cognitive primacy hypothesis' (Cairns 1979) was often assumed; such an hypothesis assumes that children's cognition is the primary determinants of their behaviour, and tends to de-emphasize the importance of social aspects of development. The resurgence of interest in social development in autism over the past decade, and particularly during the past five years, has reflected an increased awareness of the limitations of these arguments.

While some social skills do develop, these are invariably highly deviant, and both quantitatively and qualitatively abnormal, even in the highest-functioning individuals (Howlin 1986; Volkmar 1987; Mundy and Sigman 1989). Clearly the fact that *some* social interest and some social skills develop need not, necessarily, imply that cognitive factors are 'primary'. Similarly, certain cognitive skills may be relatively preserved, although this does not suggest that cognitive factors are less important to syndrome pathogenesis. The observation that some social skills emerge may just as parsimoniously be taken to suggest the importance of attempting to disentangle precisely those aspects of social development that are most uniquely disordered in autism (Rogers and Pennington 1991).

The potential importance of social development is also suggested by the considerable body of work on infant sociability. Social transactions appear to provide the framework for subsequent communicative and cognitive skills, for example symbolization (see for instance Bates *et al.* 1979; Piaget 1962; Wolf and Gardner 1981). The strength of these processes is suggested by the development of selective attachments even in children who are severely neglected and/or abused (Egeland and Sroufe 1981), as well as in children with other very severe developmental disabilities, for example Down's syndrome (Berr *et al.* 1980). As Hobson (1989) notes, it appears at least as reasonable to assume, along with Kanner, that the fundamental problem in autism is indeed a lack of 'affective contact' (see also Hobson, this volume, Chapter 10). Accordingly, the explication of disturbances in specific developmental processes (Ungerer 1989) and the attempt to provide more truly operational definitions of autistic social dysfunction have assumed increased importance.

SOCIAL DEVIANCE AS A CENTRAL DEFINING FEATURE

Both categorical and dimensional definitions of the condition have emphasized the centrality of social dysfunction as a hallmark of autism. Rutter's
(1978) synthesis of Kanner's original report and subsequent research proved
highly influential. Rutter (1978) suggested that the social development of the
autistic individual was deviant even when developmental factors (mental
age) were taken into account. Despite the general agreement on the centrality
of social deficits in syndrome definition it has proved somewhat difficult to
derive simple, readily applied procedures for operationalizing this diagnostic
construct. For example, DSM-III (APA 1980) accorded autism 'official'
diagnostic status in the American psychiatric system for the first time; by
definition individuals with infantile autism had to exhibit 'pervasive' social
deficits, i.e., presumably to exhibit them in most situations and contexts.
This definition proved most applicable to younger children, and was problematic in relation to older and higher-functioning individuals, who exhibited
rudimentary, if highly deviant, social skills.

To address this problem revisions of this definition were undertaken in
DSM-III-R (APA 1987). These revisions were much influenced by a rather
broader view of the condition (see for example Wing and Gould 1979) and,
not surprisingly, resulted in a rather broader diagnostic concept (Volkmar
et al. 1988). Despite the apparent differences between various categorical
approaches to diagnosis of the condition, it does appear (see for example
Siegel *et al.* 1989) that it is the social criteria which, taken individually, most
robustly predict diagnosis.

In contrast to the categorical approach to diagnosis, an alternative
approach has relied on assessment of dimensions of dysfunction. This
approach has been exemplified by various instruments explicitly developed
for assessment of individuals with autism (see Parks 1983 for a review). Such
instruments also emphasize social deficits as defining features of the autistic
syndrome. In theory such approaches have numerous advantages over categorical diagnostic criteria, in that dimensional approaches may more adequately and accurately characterize social dysfunction in autistic individuals.

Unfortunately, several factors have complicated the development and
use of such instruments. In the first place, there is a tremendous range of
syndrome expression, and the instruments developed have understandably
tended to focus on selected subgroups of the autistic population. Secondly,
the developmental problems of many autistic individuals are such that
they cannot be directly interviewed; accordingly, instruments typically rely
either on direct observation of behaviours or on parental or teacher report,
thus raising issues of reliability, instrument development, and instrument
standardization. Instrument development is complicated by the nature of
developmental deviation — for example, since highly deviant behaviours are

sampled, issues of standardization can be problematic. This issue would not, of course, apply to instruments which were truly normative in nature. Finally it is unclear precisely which aspects of social development are most characteristically disordered of the many that are.

Although a considerable body of research exists on the nature of social development of infants and young children, relatively few truly normative instruments for assessing social skills exist. Several recent studies have employed a newly revised assessment instrument, the Vineland Adaptive Behavior Scales (Sparrow *et al.* 1984). These scales use a semi-structured interview administered to parent or caregiver of the individual (child or adult), and provide assessments of communicative (receptive, expressive, and written) and social (interpersonal relationships, play and leisure time, and coping) skills based on a very large, normative sample representative of the United States. A series of studies using this instrument (Volkmar *et al.* 1987; Freeman *et al.* 1988; Loveland and Kelly 1988) have documented that, consistent with Rutter's (1978) definition, social skills in autism are indeed deviant relative to mental age. Similarly, we (Volkmar *et al.* 1990) have extended this approach by developing a series of regression equations, based on the Vineland standardization sample, which predict social and communicative skills on the basis of mental age and other relevant variables. When applied to autistic and non-autistic developmentally disordered samples, social skills in autistic individuals were typically more than two standard deviations below the scores predicted on the basis of mental age alone. Individual item analyses (Klin *et al.* 1992) have also revealed that autistic children typically fail to exhibit a range of social behaviours which are normatively exhibited within the first year of life; the absence of such behaviours was striking even when effects of associated mental retardation were controlled.

CLINICAL ASPECTS OF SOCIAL DEVELOPMENT IN AUTISM

Developmental perspectives

Social development in autism is of a kind both qualitatively and quantitatively deviant from that seen in other childhood disorders (Rutter and Garmezy 1983). Although clinical descriptions of the autistic child have historically emphasized 'autistic aloneness', this view is probably most applicable to younger and more severely impaired individuals. Although failures in language development are typically the presenting concerns of parents at the time of first diagnosis, these usually have been preceded by early, and profound, social deviance. While available data are generally based on parental retrospection and must, accordingly, be viewed with some caution, it is typically the case that parents report very early deviance in the development of quite basic interpersonal skills (Ornitz *et al.* 1977), including failure to make eye-contact and to use gaze to regulate interaction, failures to engage in the social games of early infancy, a preferential interest in the inanimate, as opposed to the social environment, and a relative failure in developing the typically robust patterns of differential attachments to parents (Mundy and Sigman 1989). In stark contrast to what occurs in normally developing infants, the human face appears to hold little interest or have little salience for the autistic child (Volkmar 1987). Typical forms of early non-verbal interchange are deviant, so that, usually, very early emerging forms of 'intersubjectivity' (Trevarthen 1979; Stern 1985) are absent, and young autistic children do not display a differential preference for maternal speech (Klin 1991; 1992). Affected children may not seek physical comfort from parents, and may be difficult to hold (Ornitz *et al.* 1977).

Some social skills develop over time, so that by the ages four or five some evidence of differential social responsiveness to familiar adults is exhibited (see for instance Sigman and Ungerer 1984), although the quality of such behaviours is usually highly deviant (Mundy and Sigman 1989). Similarly, differential patterns of vocalization or facial expression may be observed, although, typically, these tend to be rather idiosyncratic (Ricks 1979). Evidence of visual self-recognition may be observed (e.g. Dawson and McKissick 1984), although usually associated affective responses appear to be absent or deviant (Spiker and Ricks 1984).

Social skills continue to develop as autistic children enter later childhood and adolescence. However, social responsivity remains a source of considerable disability even for higher-functioning autistic children, whose attempts at social interaction fail as a result of their difficulties in pragmatic communication and empathy and their failures to integrate various sources of information relevant to interaction (Langdell 1978; Baron–Cohen *et al.* 1985). Normal peer relationships do not develop, and even when some social relationships develop these tend to be with adults rather than with other children (Volkmar 1987). Even very low-functioning autistic adolescents appear capable of processing at least some forms of socially relevant information (Volkmar *et al.* 1989a), although their poor capacity to use such information remains a source of significant disability (Hobson 1989), and aspects of non-verbal communication such as gaze (Volkmar and Mayes 1990) and facial expression (Yirmiya *et al.* 1989) are highly deviant.

In adulthood a range of social outcome is observed. In a majority of cases individuals continue to exhibit marked deficits in social skills, and never become capable of sustaining an independent existence (DeMyer *et al.* 1981). Even in the very highest-functioning autistic adults, residual social impairments are observed. Such higher-functioning individuals are self-described 'loners', who may exhibit a desire for social contact, although they are typically incapable of it (Kanner *et al.* 1972; Bemporad 1979; Volkmar and Cohen 1985). In many instances, such individuals are aware of their disability, and develop a number of coping strategies, typically revolving around learning concrete rules for mediating social interaction (Kanner *et al.* 1972). This observation is of some interest for the theory of mind hypothesis (Baron–Cohen *et al.* 1985), since it suggests some rudimentary 'metacognitive' skill (Beal and Flavell 1982).

Social subtypes

Given the marked range of syndrome expression observed in autism, it is not surprising that various attempts have been made to identify specific subtypes of the condition. Early distinctions, for example between 'primary' and 'secondary' autism (absence/presence of associated medical conditions), have now been largely abandoned. Subsequent attempts to subtype have been related to clinical features such as age of onset, associated biological findings (for instance, hyperserotonaemia), and IQ. Clearly, individuals with higher IQs are more likely to be verbal and to have better long-term outcomes (Rutter and Garmezy 1983), although the prepotence of IQ as a predictor is not specific to autism. More recent attempts have used various methods to identify specific subtypes based on patterns of cognitive skills (see for example Fein *et al.* 1985), historical and behavioural data (Siegel *et al.* 1986), and social features (for example Wing and Atwood 1987).

Wing's subtypology (Wing and Atwood 1987) is based on clinical identification of three distinctive patterns of social interaction observed in an epidemiological study of the condition (Wing and Gould 1979) and clinical work. Three subtypes are proposed: (1) *aloof* individuals avoid interaction actively; (2) *passive* individuals passively accept social interaction but do not seek it; and (3) *active but odd* individuals accept social interaction but interact in an odd or eccentric fashion. Wing and her co-workers have provided descriptions of the three subtypes, and attempts to elaborate more specific criteria for these subtypes have been made (for example Wing and Atwood 1987; Prizant and Schuler 1987). This system is of considerable interest, since it makes use of an essential diagnostic feature, is applicable to individuals of different ages and with different levels of associated mental retardation, and has potential implications for clinical management as well as research.

Empirical data employing this classification scheme have not, however, been common. In one study (Volkmar *et al.* 1989*b*) clinicians were noted to be able to classify autistic cases into the three subtypes with relatively high reliability, and the three subtypes were observed to differ on a number of relevant measures. The 'aloof' group was younger and more developmentally delayed than the active but odd group; the 'passive' group was intermediate between these two extremes. However, it appeared that the differences between the three types predominantly reflected mental age. The observation that some aspects of social interaction are related to overall developmental level is not, of course, unexpected; but it does suggest a potential line of inquiry focusing on those aspects of social development which remain uniquely impaired in autism throughout development.

Effects of mental age

Kanner's early (1943) impression that autistic individuals had normal intellectual potential has now clearly been shown to be incorrect (Sigman *et al.*, 1987). Although patterns of IQ distribution vary somewhat between centres (Volkmar and Cohen 1988), it is clear that the majority, and perhaps as many as 80 per cent, of autistic individuals function within the range of mental retardation. Accordingly, observed social deficits must be interpreted within the context of any associated mental retardation (Rutter 1978).

While it is clear that social skills (as variously defined and examined) are related to mental age in important respects, it is also the case, consistent with Rutter's 1978 definition, that observed social deficits are not solely a function of mental retardation. Such deficits are observed in autistic adults of normal intelligence (Volkmar and Cohen 1985), and are also observed when explicit metrics of social skills are employed (Volkmar *et al.* 1987). Moreover, while deficits in symbolic thinking and abstract reasoning are clearly established (Sigman *et al.* 1987), differences in sensorimotor aspects of intelligence seem much less striking (Morgan *et al.* 1989). In reality, issues of assessment are complex, particularly in very severely retarded and mute autistic individuals. But even in such instances social skills are less than would be expected given the very early ages at which basic social processes are observed. The developmental context remains important, however, in interpreting various hypothesized mechanisms of social dysfunction. As Hobson (1989) suggests, it appears more helpful to regard the social disturbance as one which has various manifestations over the course of development, but which remains a source of significant disability throughout life. As described elsewhere (see Klin and Volkmar, this volume, Chapter 15), this issue has particular relevance to the theory of mind hypothesis.

NEUROBIOLOGICAL PERSPECTIVES

Considerable evidence in favor of an underlying, if somewhat enigmatic, organic factor or factors responsible for the pathogenesis of autism now exists. This evidence is impressive in terms of its variety — for example, persistence of primitive reflexes; delayed development of hand dominance, increased frequency of seizure disorder, and so forth. Various pathophysiological models have been proposed which locate the 'site' of dysfunction at various points within the central nervous system or in specific neurotransmitter systems (for example, Panskep 1985). Unfortunately no single 'site' of dysfunction is consistently observed, none of the proposed mechanisms have proved readily testable, and none sufficiently account for observed social deficits. Given the early plasticity of the central nervous system and the marked alterations in its structure occurring over the first years of life, it is possible that CNS alterations in autism are reflected in changes in aspects of the fine structure of the brain rather than in specific, readily localized, neuroanatomic sites or in specific neurotransmitter systems.

On the other hand, neuropsychological studies do provide some evidence which suggests the importance of a focus precisely on the social aspects of autism. As Fein *et al.* (1986) note, the assumption of a primary disturbance in social relatedness may provide a more fruitful approach for neuropathological models, since such a procedure might more parsimoniously account for observed neurocognitive deficits. For example, isolated areas of cognitive strength in typical neuropsychological profiles might more parsimoniously be understood as reflecting areas in which social skills have relatively less importance.

The biological bases, and broader evolutionary bases, of social skills have not been sufficiently encompassed by existing research on the social deviance of autism, given that, speaking teleologically, human beings are social creatures for important reasons (Brothers 1989). Various behaviours, for example attachment, imitation, identification, and co-regulatory behaviours are important for both infant survival and ultimate reproduction. As with other primate species, human infant sociability appears to be an evolved characteristic important to infant survival (Freedman 1974; Richards 1978); it is possible that the dearth of apparent cases before Kanner's first description of the syndrome reflects some aspect of differential lack of survival of autistic infants who were, presumably, at greater risk for early death. Ultimately, of course, the capacity to use shared symbol systems is central to the development of culture and the ability for us to function within the context of a long and rich interpersonal and societal history. Unfortunately, the underlying neurobiological basis of this sociability remains unclear (Kling and Stelkis 1976; Fein *et al.* 1986). Clearly, research in this area is impeded by various methodological problems, the lack of good animal models of the condition, and our rather limited understanding of central nervous system aspects of social functioning.

ISSUES FOR RESEARCH

Although research on the social aspects of autism has increased dramatically in recent years, various problems complicate the interpretation of available research studies. These problems include differences in definition of the syndrome, the small samples of subjects typically studied, aspects of developmental change in syndrome expression, etc. While the study of certain subgroups — for example, verbal subjects — is understandable, it is also important to realize that results obtained are not necessarily applicable to other subgroups. Similarly, various matching procedures are typically used to derive required control and contrast groups; unfortunately most matching procedures typically involve matching on some aspect of cognitive, rather than social, skill. In general, the lack of adequate metrics for studying social development has been a severe impediment in research, since the most relevant issue relates to finding just those social skills which are uniquely impaired in autism.

Although young autistic children might be expected, in some sense, to present the 'purest' examples of the disorder, difficulties in diagnosis and ascertainment make such samples difficult to recruit. Various factors act to delay case detection (Siegel, *et al.* 1988), and investigators interested in the earliest aspects of social development have typically been forced to rely on retrospective information, with all the problems inherent in such data. However, the study of such children remains an area of considerable interest, since deficits in older children and adults presumably reflect even more complex interactions of various factors.

Experimental procedures may be highly artificial or may induce unintended confoundings; and the ecological context of relevant social behaviour may be underappreciated (Lytton 1973). The theoretical bias of the investigators — for example belief in the primacy of cognitive or linguistic factors in the pathogenesis of the syndrome — further complicates the interpretation of much available research. An additional problem has been the tendency to equate capacity with actual use. For example, although autistic children listen to sounds as much as mentally retarded children, they fail to exhibit preferential listening to the mother's voice (Klin 1991; 1992); although they may be able to use perceptions of the human face accurately in solving specific problems, the human face may lack general salience to them (Volkmar *et al.* 1989a); and although very-high-functioning autistic adults can be taught to solve specific social problems, their ability to abstract more general rules for social interaction remains significantly impaired.

The theory of mind hypothesis (Baron–Cohen *et al.* 1985) has the considerable advantage of focusing increased attention on the fundamental nature of social deficits in autism. It is also clear that at least some of the observed social deficits (for example in gaze behaviour) might stem from a lack of theory of mind, and that even autistic individuals with lower levels of this capacity lack the ability to make more complex, i.e. 'second order', belief attributions (Baron–Cohen 1989). The failure of verbal autistic individuals to exhibit these basic capacities seems reasonably clear; what is less clear is whether the hypothesis can sufficiently account for social deficits in their entirety and through the tremendous range (in age and developmental level) of syndrome expression (see Klin and Volkmar, this volume, Chapter 15; and Lord, this volume, Chapter 14).

Kanner's original (1943) notion was that the social deficits of autism were congenital in nature. Unless the origins of a theory of mind are to be traced back to birth, the nature and severity of these very early deficits cannot be accounted for. The issue of possible precursors of theory of mind capacities clearly represents an important topic for future research. Many basic processes (for example attention, perception, cognition) are involved, even in what otherwise appear to be very early emergent social activities. Similarly many different processes are subsumed under overarching terms like social development, and there has been a tendency to equate certain aspects of social development (for example, affective development) with social development as a whole. In some sense this latter process would be equivalent to equating performance on one selected kind of cognitive skill, for instance visual–spatial orientation, with the entirety of cognitive development. It is clear that the social world differs in a host of ways from the non-social environment: which is to say that, given that relationships with people, rather than things, are involved, issues of affective expression and understanding, cultural and personal context, and the fact that other people (rather than other things) are being understood, are all factors involved in reciprocal social interaction.

Despite the general consensus on the centrality of social deficits in the definition of autism, it is precisely this aspect of the syndrome that has been until recently the focus of the least systematic research. The recent resurgence of interest in this topic is, accordingly, particularly welcome. Studies in this area have successfully redirected our attention to Kanner's original hypothesis about the nature of social development in autism.

REFERENCES

APA (American Psychiatric Association) (1980) (Revised 1987.) *Diagnostic and Statistical Manual*, 3rd edn. APA Press, Washington, DC.

Baron–Cohen, S. (1989). The autistic child's theory of mind: a case of specific developmental delay. *Journal of Child Psychology Psychology and Psychiatry*, **30**, 285–97.

Baron–Cohen, S., Leslie A. M., and Frith, U. (1985). Does the autistic child have a 'theory of mind?' *Cognition*, **21**, 37–46.

Bates, E., Benigni, L., Camaioni, L., Bretherton, I., and Volterra, V. (1979). *The emergence of symbols: cognition and communication in infancy*. Academic Press, New York.

Beal, C.R., and Flavell, J.H. (1982). Effect of increasing the salience of message ambiguities on kindergartner's evaluation of communicative success and message adequacy. *Developmental Psychology*, **18**, 43–8.

Bell, R.Q., and Harper, L.V. (1977). *Child effects on adults*. Erlbaum, New York.

Bemporad, J.R. (1979). Adult recollections of a formerly autistic child. *Journal of Autism and Developmental Disorders*, **9**, 179–197.

Berry, P., Gunn, P., and Andrews, R. (1980). Behavior of Down's Syndrome infants in a strange situation. *American Journal of Mental Deficiency*, **85**, 213–18.

Bettelheim, B. (1967). *The Empty Fortress*. Free Press, New York.

Brothers, L. (1989). A biological perspective on empathy. *American Journal of Psychiatry*, **146**, 10–19.

Cairns, R.B. (1979). *Social development: the origins and plasticity of interchanges*. Freeman, San Francisco.

Charlop, M.J., Schreibman, L., and Tryon, A.D. (1983). Learning through observation: the effects of peer modeling on acquisition and generalization in autistic children. *Journal of Abnormal Child Psychology*, **11**, 355–66.

Churchill, D., and Bryson, C.Q. (1972). Looking and approach behavior of psychotic and normal children as a function of adult attention or preoccupation. *Comprehensive Psychiatry*, **13**, 171–7.

Dawson, G., and McKissick, F. (1984). Self-recognition in autistic children. *Journal of Autism and Developmental Disorders*, **14**, 383–94.

DeMyer, M.K., Hingtgen, J.N., and Jackson, R.K. (1981). Infantile autism reviewed: a decade of research. *Schizophrenia Bulletin*, **7**, 388–451.

Donnellan, A.M., Anderson, J.L., and Mesaros, R.A. (1984). An observational study of stereotypic behavior and proximity related to the occurrence of autistic child–family member interactions. *Journal of Autism and Developmental Disorders*, **14**, 205–10.

Eglenand, B., and Sroufe, L.A. (1981). Attachment and early maltreatment. *Child Development*, **52**, 44–52.

Fein, D., Waterhouse, L., Lucci, D., and Snyder, D. (1985) Cognitive subtypes in developmentally disabled children. *Journal of Autism and Developmental Disorders*, **15**, 77–95.

Fein, D., Pennington, B., Markowitz, P., Braverman, M., and Waterhouse, L. (1986). Towards a neuropsychological model of infantile autism: are the social deficits primary? *Journal of the American Academy of Child Psychiatry*, **25**, 109–212.

Freedman, D.G. (1974) *Human infancy: an evolutionary perspective*. Erlbaum, New York.

Freeman, B.J., Ritvo, E., Yokata, A., Childs, J., and Poll, J. (1988). WISC-R and Vineland Adaptive Behavior Scale scores in autistic children. *Journal of the American Academy of Child and Adolescent Psychiatry*, **27**, 428–9.

Gesell, A. (1940). *Wolf child and human child*. Harper, New York.

Hobson, R.P. (1989). Beyond cognition: a theory of autism. In *Autism: nature, diagnosis, and treatment* (ed. G. Dawson). Guilford Press, New York.

Howlin, P. (1986). An overview of social behavior in autism. In *Social behavior in autism*) (ed. E. Schopler and G. Mesibov). Plenun, New York.

Kanner, L. (1943). Autistic disturbances of affective contact. *Nervous Child*, **2**, 227–50.

Kanner, L., Rodriguez, A., and Ashden, B. (1972). How far can autistic children go in matters of social adaptation. *Journal of Autism and Childhood Schizophrenia*, **23**, 9–33.

Klin, A. (1991). Young autistic children's listening preferences in regard to speech: a possible characterization of the symptom of social withdrawal. *Journal of Autism and Developmental Disorders*, **12**, 29–42.

Klin, A. (1992). Listening preferences in regard to speech in four children with developmental disabilities. *Journal of Child Psychology and Psychiatry*, **33**, 763–69.

Klin, A., Volkmar, F. R., and Sparrow, S. S. (1992). Some limitations of the theory of mind hypothesis. *Journal of Child Psychology and Psychiatry*, **33**, 861–76.

Kling, A., and Stelkis, H. D. (1976). A neural substrate for affiliative behavior in nonhuman primates. *Brain Behavior and Evolution*, **13**, 216–38.

Langdell, T. (1978). Recognition of faces: an approach to the study of autism. *Journal of Child Psychology and Psychiatry*, **19**, 255–68.

Lévi-Strauss, C. (1949). *Les structures élémentaires de la parenté*. Presses Universitaires de France, Paris.

Loveland, K. A., and Kelly, M. L. (1988). Development of adaptive behavior in adolescents and young adults with autism and Down Syndrome. *American Journal of Mental Retardation*, **93**, 84–92.

Lytton, H. (1973). Three approaches to the study of parent–child interaction: ethological, interview, and experimental. *Journal of Child Psychology and Psychiatry*, **14**, 1–17.

McAdoo, W. G., and DeMyer, M. Y. (1978). Personality characteristics of parents. In *Autism: a reappraisal of concepts and treatment* (ed. M. Rutter and E. Schopler). Plenum, New York.

McHale, S. (1983). Social interactions of autistic and non-handicapped children during free play. *American Journal of Orthopsychiatry*, **52**, 81–91.

Maclean, C. (1977). *The wolf children*. Macmillan, London.

Morgan, S. B., Cutrer, P. S., Coplin, J. W., and Rodriguez, J. R. (1989). Do autistic children differ from retarded and normal children in Piagetian sensorimotor functioning? *Journal of Child Psychology and Psychiatry*, **30**, 857–64.

Mundy, P., and Sigman, M. (1989). Specifying the nature of the social impairment in autism. In *Autism: nature, diagnosis and treatment* (ed. G. Dawson). Guilford Press, New York.

Ornitz, E. M., and Ritvo, E. R. (1968). Perceptual inconstancy in early infantile autism. *Archives of General Psychiatry*, **18**, 76–98.

Ornitz, E. M., Guthrie, D., and Farley, A. H. (1977). Early development of autistic children. *Journal of Autism and Childhood Schizophrenia*, **7**, 207–29.

Panskep, J. (1979). Towards a neurochemical theory of autism. *Trends in Neuroscience*, **2**, 174–7.

Parks, S. L. (1983). The assessment of autistic children: A selective review of available instruments. *Journal of Autism and Developmental Disorders*, **13**, 255–67.

Piaget, J. (1962). *Play, dreams and imitation in childhood*. Norton, New York.

Prior, M. R. (1979). Cognitive abilities and disabilities in infantile autism: a review. *Journal of Abnormal Child Psychology*, **7**, 357–80.

Prizant, B. M., and Schuler, A. L. (1987). Facilitating communication—

pre-language approaches. In *Handbook of autism and pervasive developmental disorders* (ed. D. J. Cohen and A. Donnellan), pp. 301-15. Wiley, New York.

Richards, M. P. M. (1978). The biological and the social. In *Action, gesture and symbol: The emergence of language* (ed. A. Lock). Academic Press, London.

Richer, J. (1976). The social avoidance behavior of autistic children. *Animal Behavior*, **24**, 898-906.

Ricks, D. (1979). Making sense of experience to make sensible sounds. In *Before speech: the beginning of interpersonal communication* (ed. M. Bullowa), pp. 245-68. Cambridge University Press, New York.

Rogers, S. J., and Pennington, B. F. (1991). A theoretical approach to the deficits in infantile autism. *Development and Psychopathology*, **3**, 137-62.

Rogers, S. J., Ozonoff, S., and Maslin-Cole, C. (1991). A comparative study of attachment behavior in young children with autism or other psychiatric disorder. *Journal of the American Academy of Child and Adolescent Psychiatry*, **30**, 483-8.

Rutter, M. (1978). Diagnosis and definition. In *Autism: a reappraisal of concepts and treatment* (ed. M. Rutter and E. Schopler), pp. 1-25. Plenum, New York.

Rutter, M., and Garmezy, N. (1983). Developmental psychopathology. In *Handbook of Child Psychology*, Vol. 4., (ed. E. M. Hetherington). Wiley, New York.

Rutter, M., Bartak, L., and Newman, S. (1971). Autism — a central disorder of cognition and language. In *Autism: concepts, characteristics and treatment* (ed. M. Rutter). Churchill-Livingstone, Edinburgh.

Siegel, B., Anders, T., Ciaranello, R., Beinenstock, R., and Kraemer, C. (1986). Empirically derived subclassification of the autistic syndrome. *Journal of Autism and Developmental Disorders*, **14**, 231-44.

Siegel, B., Pliner C., Eschelr, J., *et al.* (1988). How children with autism are diagnosed: Difficulties in identification of children with multiple developmental delays. *Journal of Developmental and Behavioral Pediatrics*, **9**, 199-204.

Siegel, B., Vucicevic, J., Elliot, G., and Kraemer, H. (1989). The use of signal detection theory to assess DSM-III-R criteria for autistic disorder. *Journal of the American Academy of Child and Adolescent Psychiatry*, **28**, 542-8.

Sigman, M. and Ungerer, J. (1984). Attachment behaviors in autistic children. *Journal of Autism and Developmental Disorders*, **14**, 231-44.

Sigman, M., Ungerer, J. A., Mundy, P., and Sherman, T. (1987). Cognition in autistic children. In *Handbook of autism and pervasive developmental disorders* (ed. D. J. Cohen and A. Donnellan). Wiley, New York.

Sparrow, S., Balla, D., and Cicchetti, D. V. (1984). *Vineland Adaptive Behavior Scales*. American Guidance Service, Circle Pines, Minnesota.

Spiker, D., and Ricks, M. (1984). Visual self-recognition in autistic children: developmental relationships. *Child Development*, **55**, 214-25.

Stern, D. (1987). *The interpersonal world of the human infant*. Basic Books, New York.

Trevarthen, C. (1979). Communication and cooperation in early infancy: a description of primary intersubjectivity. In *Before speech: the beginning of interpersonal communication* (ed. M. Bullowa). Cambridge University Press, New York.

Ungerer, J. A. (1989). The early development of autistic children: implications for defining primary deficits. In *Autism: nature, diagnosis, and treatment* (ed. G. Dawson). Guilford Press, New York.

Volkmar, F. R. (1987). Social development. In *Handbook of autism and pervasive developmental disorders* (ed. D. J. Cohen and A. Donnellan). Wiley, New York.

Volkmar, F. R., and Cohen, D. J. (1985). A first person account of the experience of infantile autism by Tony W. *Journal of Autism and Developmental Disorders*, **15**, 47–54.

Volkmar, F. R., and Cohen, D. J. (1988). Diagnosis of pervasive developmental disorders. In *Advances in clinical child psychology*, Vol. 11 (ed. B. B. Lahey and A. E. Kazdin). Plenum, New York.

Volkmar, F. R., and Mayes, L. C. (1990). Gaze behavior in autism. *Developmental Psychopathology*, **2**, 61–70.

Volkmar, F. R., Hoder, E. L., and Cohen, D. J. (1985). Compliance, 'negativism', and the effect of treatment structure on behavior in autism: a naturalistic study. *Journal of Child Psychology and Psychiatry*, **26**, 865–77.

Volkmar, F. K., Cohen, D. J., and Paul, R. (1986). An evaluation of DSM-III criteria for infantile autism. *Journal of the American Academy of Child Psychiatry*, **25**, 190–7.

Volkmar, F. R., Sparrow, S. A., Goudreau, D., Cicchetti, D. V., Paul, R., and Cohen, D. J. (1987). Social deficits in autism: an operational approach using the Vineland Adaptive Behavior Scales. *Journal of the American Academy of Child Psychiatry*, **26**, 156–61.

Volkmar, F. R., Bregman, J., Cicchetti, D. V., and Cohen, D. J. (1988). Diagnosis of Autism: DSM III and DSM III-R. *American Journal of Psychiatry*, **145**, 1404–8.

Volkmar, F. R., Sparrow, S. A., Rende R., and Cohen, D. J. (1989a). Facial perception in autism. *Journal of Child Psychology and Psychiatry*, **30**, 591–8.

Volkmar, F. R., Bregman, J., Cohen, D. J., Hooks, M., and Stevenson, J. (1989b). An examination of social typologies in autism. *Journal of the American Academy of Child and Adolescent Psychiatry*, **28**, 82–6.

Volkmar, F. R., Carter, A., Sparrow, S., and Cicchetti D. V. (1992). Towards an operational definition of autism. *Journal of the American Academy of Child and Adolescent Psychiatry* (in press).

Wing, L., and Atwood, A. (1987). Syndromes of autism and atypical development. In *Handbook of autism and pervasive developmental disorders* (ed. D. J. Cohen and A. Donnellan). Wiley, New York.

Wing, L., and Gould, J. (1979). Severe impairments of social interaction and associated abnormalities in children: epidemiology and classification. *Journal of Autism and Developmental Disorders*, **9**, 11–30.

Wolf, D., and Gardner, H. (1981). On the structure of early symbolization. In *Early language: aquisition and intervention* (ed. R. Schiefelbusch and D. Bricker). University Park Press, Baltimore.

Yirmiya, N., Kasari, C., Sigman, M., and Mundy, P. (1989). Facial expressions of affect in autistic, mentally retarded, and normal children. *Journal of Child Psychology and Psychiatry*, **30**, 725–35.

Zingg, R. M. (1940). Feral man and extreme cases of isolation. *American Journal of Psychology*, **53**, 487–517.

Part II The theory of mind hypothesis of autism: the cognitive approach

4

From attention–goal psycho to belief–desire psychology development of a theory of mind, and its dysfunction

SIMON BARON-COHEN

In this chapter I begin by reviewing the literature relevant to the *theory of mind hypothesis* of autism: the idea that in autism there is a failure to develop a normal understanding that people have minds and mental states, and that mental states relate to behaviour. As has been discussed in the Introduction to this volume, the interest in testing if children with autism develop a theory of mind stems from the notion that if they do not, this might account for the abnormalities they show in both social interaction and communication, since a theory of mind is held to be necessary for both of these (Dennett 1978; Grice 1975; Baron–Cohen 1988). In reviewing the literature relevant to this hypothesis, I focus on six different classes of mental state, individually. These are: belief, desire, knowledge, pretence, perception, and emotion. I then review what is known about 'non-mental state' social cognition in autism, and some conceptual ramifications of the theory of mind deficit. This review closes with a look at some evidence relevant to assessing if these deficits are both deviant and delayed relative to normal development. Following this review, I then extend a theoretical claim I have made elsewhere (Baron–Cohen 1989*a*, 1991*a*) that these deficits have their origins in the first year of life, in the toddler's understanding of goals and attention.

1. THE THEORY OF MIND HYPOTHESIS OF AUTISM: A REVIEW OF THE EXPERIMENTS

Understanding beliefs

Dennett (1978) argued that understanding *false belief* might constitute a litmus test of a theory of mind, in that in such cases it becomes possible to distinguish unambiguously between the child's (true) belief and the child's

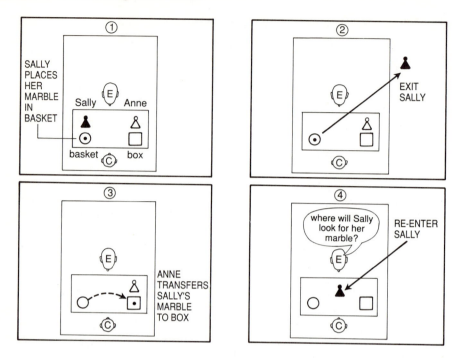

Fig. 4.1. The Sally–Anne test (reprinted from Baron-Cohen *et al.* 1985, with permission).

awareness of someone else's different (false) belief. Wimmer and Perner (1983) acted on this suggestion by designing a 'false belief test', which they used with normal children. They showed that around 3–4 years of age, normal children pass such a test. Baron–Cohen *et al.* (1985) adapted this test for use with children with autism and with Down's Syndrome, as well as clinically normal children. The test is illustrated in Fig. 4.1.

As can be seen, the test involves appreciating that since Sally was absent when her marble was moved from its original location, she won't *know* it was moved, and therefore must still *believe* it is in its original location. On the Belief Question (*Where will Sally look for her marble?**) 23 out of 27 normal children (85 per cent) and 12 out of 14 children with Down's Syndrome (86 per cent) passed on two trials, but only 4 out of 20 children with autism (20 per cent) did so. All subjects passed a memory control

---

* In other replications of this study, the test question has taken the more explicit form of 'Where does Sally *think* her marble is?', although this modification does not seem to make the test any easier for children with autism. Eisenmajer and Prior (1991) found that, as Siegel and Beattie (1991) had demonstrated for normal children, the inclusion of the term 'first' in the test question resulted in a larger proportion of children with autism passing. The significance of this modification deserves further exploration.

question (*Where was the marble in the beginning?*) and a reality control question (*Where is the marble really?*), as well as a Naming Question (*Which doll is Sally?*), thus ruling out the possibilities that failure on the Belief Question was due to either inattention, memory overload, linguistic or motivational factors. Given that the children with autism had a higher CA and MA than either of the two control groups, this initial study gave strong support to the notion that in autism the mental state of belief is poorly understood.

This pattern of results has been replicated by Leekam and Perner (1991), Leslie and Thaiss (1992), Baron–Cohen (1989*b*), and Reed and Peterson (1990) using a similar puppet story, by Leslie and Frith (1988) using real people instead of puppets (thus ruling out that the deficit in understanding beliefs was due to not understanding the nature of puppets), by Swettenham (1992) using computer-graphic presentation of the same story (thus ruling out the possibility that the failure was due to social factors in the laboratory), by Shaw (1989), using the child's mother as experimenter (thus ruling out stranger-anxiety as a cause of the child's failure), and by Mitchell (1990), using a prompting technique at key points in the story. Finally, Perner *et al.* (1989) replicated the finding using a totally different test (the 'Smarties' test), which makes fewer informational demands. The robustness of this finding suggests that in autism there is a genuine inability to understand other people's different beliefs.

In a subsequent study, Baron–Cohen *et al.* (1986) set up a test to contrast understanding of mental and non-mental understanding. We used a largely non-verbal technique, a picture-sequencing test, in which 15 picture stories (each 4 frames long) depicted, if correctly sequenced, either a character's false belief, or a character's interpersonal behaviour, or a character's causal actions on an inanimate object. Again, the children with autism performed very poorly (in fact, at chance level) on the stories involving an understanding of a person's belief, although they were as good or better than Down's Syndrome or normal controls at sequencing the stories involving interpersonal behaviour (which did not necessarily require understanding mental states) and the stories involving physical causality (which again did not require the child to understand mental states). This demonstrated an autism-specific deficit in understanding psychological, but not physical, causality. This study also ruled out a *general* sequencing deficit, contrary to earlier reports (Rutter 1978).

It should be noted that two studies have failed to replicate this pattern of results with the picture-sequencing test (Oswald and Ollendick 1989; Ozonoff *et al.* 1991). The first of these, however, did not report if all 15 of the original stories were used, which casts some doubt on the validity of their replication. The latter did use the full set, and found that the children with autism performed significantly better on the physical causal stories, as

Baron–Cohen *et al.* (1986) found, but did not replicate the finding of an autism-specific deficit in the mental-state condition, the control groups also performing poorly. It is clear that further studies are needed with this paradigm to establish its reliability.

In all the studies reviewed above, it is clear that while most children with autism fail tests of belief-understanding, a minority of them do pass. This subgroup ranges from 20 to 35 per cent in different samples. Moreover, these subjects tend to be the *same* subjects in different tests, which leads to the conclusion that, on the face of it, they have an intact understanding of belief. A later study, however, showed this to be a result of a ceiling effect, in that most false-belief tests are set at an equivalent mental age of about four years old, and the children with autism who pass have a mental age well above this. When given a more taxing test of belief understanding, comprising understanding nested beliefs, or *beliefs about beliefs* (for example, those of the form 'Anne thinks that Sally thinks X') — these being well within the comprehension of normal six- to seven-year-old children (Perner and Wimmer 1985) — teenagers with autism failed outright, despite a mean receptive verbal MA of over seven years old (Baron–Cohen 1989*b*). This result has been replicated by Ozonoff *et al.* (1991). So it appears that, while most children with autism do not even understand beliefs at the equivalent level of normal four-year-old children, some do; but even these show impaired understanding of beliefs at the equivalent level of normal six- to seven-year-old children.

Understanding desire

Desire is often thought to be *the* other key mental state, next to belief, in our folk psychology (Dennett 1978). With beliefs and desires, all kinds of behaviour become interpretable. Consider this example. Why did the burglar look behind the painting? He *wanted* some money and he *thought* the safe might be behind the painting. For normal children, desire is understood earlier than belief — in fact, desire is even understood by normal two-year-olds (Wellman, this volume, Chapter 2). The 'terrible twos' have been interpreted as evidence of this age-group's growing awareness of the frustrating difference between their own and their parents' desires (Wellman 1990). Can children with autism understand desire?

Several studies suggest they can — at least as well as mentally handicapped controls, although not as well as normal children (Baron–Cohen 1991*b*). When children with autism were asked how story characters would feel when given the cereal they like, or how they would feel if they were given one they did not like, 57.4 per cent of them judged correctly that the characters would feel happy in the first case, and sad in the second. While this was significantly less than a normal control group (92.1 per cent of whom made correct judgements), it was remarkably similar to the performance of a mentally handicapped control group, only 59.4 per cent of whom made correct judgements. This pattern of results was also found by Tan and Harris (1991).

Clearly, this experiment only taps one aspect of desire, namely liking something, and predicting emotional responses on the basis of being given something one likes or dislikes. Naturally, there is much more to desire than just this. Astington and Gopnik (1991), for example, distinguish desires as representations from desires as internal states. The former is not clearly understood until four years old (Astington and Lee 1989), while the latter is within two-year-olds' competence (Wellman 1990). Would children with autism understand desires as representations, in this more complex sense? Secondly, Astington and Gopnik (1991) stress the difference between desires and desirability. Again, there are no data yet on whether children with autism would understand the latter – that is, that the desirability of an object rests on subjective, not objective factors. This distinction appears fragile in normal children until four years of age (Gopnik and Seager, in press).

The results from the studies cited earlier suggest that for children with autism, understanding of simple desires as internal states is not *specifically* impaired, relative to a mental-age-matched control group without autism. If their understanding of the more complex aspects of desire is impaired, however, this would be important, as this would further compromise their understanding of the social world.

Understanding knowledge

Understanding knowledge appears to be easier than understanding belief for normal children (Wellman 1990). Quite why this should be is not completely clear, but most authors give the reason that knowledge is true belief, and this should be simpler than false belief, as misrepresentation is not involved. Leslie and Frith (1988) tested understanding knowledge by children with autism. The subject was shown an actor watching the experimenter hiding a counter. When the actor left, the experimenter asked the child to put a second counter in a second hiding-place. The subject was then asked where the actor would look for a counter on her return. They found that 44 per cent of the group with autism passed this test by indicating the place the actor *knew* about, not the place she was ignorant about. They replicated this result in a later study (Perner *et al.* 1989). Although they did not include a control group in either of these knowledge tasks, given that only 27 per cent of them passed a false-belief task in the 1988 study (and only 17.4 per cent of them did so in the 1989 study), these findings suggest that understanding knowledge is slightly easier than belief for children with autism too, but that the majority of them show deficits in comprehension of both mental states. Reed and Peterson (1990) went on to replicate these findings, this time including a mentally handicapped control group. They also found severe deficits in both understanding knowledge (23 per cent of the autistic group passing) and belief (15 per cent passing), again suggesting that knowledge is marginally easier than belief for subjects with autism, but that understanding of both these mental states is impaired relative to subjects with mental handicap.

Understanding pretence

Most studies that have looked at pretence in autism have not directly tested the subject's comprehension of this mental state, but addressed the question indirectly by using the following logic. In order to produce pretence, one must understand how pretending is different from not pretending; therefore, observe if children can *produce* pretend scenarios in their play (Wing and Gould 1979). Ungerer and Sigman (1981), in their study, made important distinctions between three types of play: **sensorimotor** (exploring the physical properties of objects); **functional** (using objects in accordance with their intended or conventional function); and **pretend** (mainly object-substitution). They found that subjects with autism did not differ from controls in the production of the first two types of play, but did in the third, producing significantly less of it.

In a replication of this, but using even tighter controls against the small but real risks of modelling that crept into their procedure, I found a similar set of results (Baron–Cohen 1987). Using Leslie's (1987) criteria, pretend play was scored whenever a child (1) used an object as if it were another object; and/or (2) attributed properties to an object which it did not have; and/or (3) referred to absent objects as if they were present. Only 20 per cent of children with autism in this study produced any pretend play, and even these instances were limited and ambiguous. In contrast, 80 per cent of children with Down's Syndrome produced unambiguous pretend play, these instances being both rich and generative. So did 90 per cent a normal control group. Such results therefore suggest that the mental state of pretence is also largely beyond the understanding of many children with autism.

Some controversy surrounds this conclusion, however. First, some studies (see the review of these by Mundy *et al.*, this volume, Chapter 9; and see also Lewis and Boucher 1988) have found lower levels of functional play. Given that functional play in autism appears to be normal in other studies, however, this result is unlikely to reflect any structural abnormalities. In contrast, the deficits in production of pretend play seem severe, and likely to be structural in origin (Leslie 1987). Secondly, Lewis and Boucher (1988) found that under conditions of *elicitation or instruction*, children with autism did show pretend play, while they did not do this in *spontaneous* play.

This result is open to multiple interpretations (Baron–Cohen 1989c, 1990; Boucher and Lewis 1990; Harris, this volume, Chapter 11).

Understanding perception

Belief, knowledge, desire, and pretence are all *opaque* mental states. That is, they suspend normal truth-conditions governing the propositions they prefix. Thus, 'I believe it is Wednesday' may be true even if in fact it is Tuesday, so long as I believe it is Wednesday. Such truth-suspension is a key feature of what philosophers refer to as the *opacity test*, which picks out whether a mental-state term possesses these logical properties. This was first identified by Frege (1892/1960). In contrast, 'I saw a mouse' is true only if I did indeed see one. Thus perception is transparent, not opaque.

Flavell *et al.* (1981) report that 'Level 1' understanding of perception is well within normal two-year-olds' ability: they can easily judge if someone saw something or not, and what conditions must pertain for someone to see something (for example, that an unobstructed 'line of sight' must be possible). In contrast, not until about three to four years old do children pass Level 2 perceptual role-taking tasks – understanding *how* something will appear to someone from their different visual perspective. Children with autism have been tested at both levels of perceptual role-taking (Hobson 1984; Baron–Cohen 1989d, 1991c; Leslie and Frith 1988; Tan and Harris 1991; Reed and Peterson 1990), and show no deficits. It would appear, then, that transparent mental states such as perceiving may be unproblematic for children with autism – their difficulties may be restricted to opaque mental states.

Although understanding seeing appears intact in autism, it does not follow that understanding the principle that *seeing leads to knowing* is also intact. Indeed, given the evidence reviewed earlier that understanding knowledge seems beyond most children with autism, one might expect this principle to also be beyond them. Perner *et al.* (1989) tested this by showing the subject an object being hidden, but not showing a confederate. They then asked the child who *knew* what was hidden, and who had been allowed *to look*. They found that whilst the vast majority of children with autism passed the looking question (69.6 per cent for self, 74 per cent for other), only about half of them passed the knowledge question (56.5 per cent for self, 43.5 per cent for other).

Goodhart and Baron–Cohen (1994) replicated this, using a strikingly simple method developed by Pratt and Bryant (1990) with normal three-year-olds, which just involves asking subjects which of two actors (or story characters) knows what is in a box, after they have seen one of them *look* into the box, and the other one simply *touch* the box. This paradigm thus controls for the child simply choosing the character who did *something* to the box. Only 33 per cent of children with autism passed this test, despite a verbal MA in excess of 3.5 years of age, and despite passing a control condition of identifying which character had been given which colour token. In contrast, 75 per cent of children with a mental handicap passed.

Additional indirect evidence that this principle poses difficulties for children with autism comes from a naturalistic study of deception in autism (Baron–Cohen, 1992*a*) in which subjects were asked to hide a penny in one of their hands — the Gratch (1964) technique. Across 12 trials, the subjects with autism succeeded in keeping the object 'out of sight' in their closed fist, but failed to hide the (*visible*) tell-tale clues that would enable the guesser to infer (*know*) the whereabouts of the penny (for example, they omitted to close the empty hand, or hid the penny in full view of the guesser, or showed the guesser where the penny was before he had guessed). In contrast, subjects with mental handicap and normal three-year-old children made far fewer errors of this sort. For them, the game was fun if they succeeded in keeping information about the whereabouts of the penny 'out of (the guesser's) mind'. This study adds to the data on deception-deficits in autism (Oswald and Ollendick 1989; Sodian and Frith, this volume, Chapter 8).

The apparent difficulty children with autism have in understanding the seeing-leads-to-knowing principle may point to impairments in at least one aspect of understanding seeing: Dretske (1969) distinguishes between epistemic and non-epistemic seeing, and to date only the latter has been examined in autism, and found to be intact. One might predict that given their difficulty in understanding epistemic states like belief, epistemic seeing may prove hard for children with autism. The case of epistemic seeing is the exception to my claim earlier that perception is transparent, not opaque. Thus, 'I saw a mouse' can be true while 'I saw Bridget's pet' can be false, even if the mouse was in fact Bridget's pet — if I didn't know that the mouse was her pet. Testing comprehension of epistemic seeing in autism is needed.

Emotion

One key set of mental states that has been a major focus of some studies (Hobson, Chapter 10, this volume) is emotion. In his early studies, Hobson (1986*a*, *b*) found that subjects with autism performed significantly worse than control groups on emotion–expression matching tasks. In later studies, these differences were not found when groups were matched on *verbal* mental age (Hobson 1988*a*, *b*, 1989; Tantam *et al.* 1989; Braverman *et al.* 1989; Prior *et al.* 1989; Ozonoff *et al.* 1990). Furthermore, since emotion-recognition deficits are also found in a range of other clinical disorders, such as schizophrenia (Cutting 1981; Novic *et al.* 1984), mental handicap (Gray *et al.* 1983), abused children (Camras *et al.* 1983), deaf children (Odom *et al.* 1973) and prosopagnosia (De Kosky *et al.* 1980; Kurucz *et al.* 1979), the status of this deficit as an explanation is called into question.

Some studies have focused not on emotion-recognition, but emotion-prediction. The aim in these studies is to establish how much children with autism understand about the *causes* of emotion — how a person will feel, given a set of circumstances. Harris *et al.* (1989) showed that normal three- to four-year-old children understand that emotion can be caused by *situations* (for example, nice situations make you feel happy, nasty ones make you feel sad) and *desires* (for example, fulfilled desires make you feel happy, unfulfilled ones make you feel sad). They also showed that by four to six years old, normal children understand that *beliefs* can affect emotion (for example, if you *think* you're getting what you want, you'll feel happy, and if you think you're not, you'll feel sad — irrespective of what you're actually getting).

In a recent study (Baron–Cohen 1991*b*) subjects with autism were easily able to judge a story character's emotion when this was caused by a situation, and were as good as a group with mental handicap at predicting the character's emotion given her desire. However, they were significantly worse at predicting the character's emotion given her belief than either normal five-year-old children or subjects with mental handicap. They were also impaired in their recognition of the facial expression of surprise, compared to the facial expression of happiness or sadness (Baron-Cohen, Spitz, and Cross, 1993). The implication is that 'simple' emotions may be within the understanding of people with autism, whilst 'cognitive' or belief-based emotions (Wellman 1990) may pose considerable difficulty for them. Another implication of this work is that, pursuing an earlier argument, their understanding of desires as internal states may be intact, whilst their understanding of desires as representations may not be. However, these experiments only tap this indirectly. Whether the earlier generalization regarding opaque versus transparent mental states holds for understanding emotion by children with autism deserves consideration.

'Non-mental state' social cognition in autism

So far, we have only considered comprehension of mental states. While this is a key part of social cognition, it is by no means the only part. One plausible objection to the theory of mind hypothesis of autism is that since these children interact with others less than other kinds of children, perhaps *all* their knowledge about the social world is poor (Hobson, 1992), and not just their knowledge about mental states in particular. Several studies allow us to examine this *general* social cognition deficit hypothesis against the more specific theory of mind hypothesis.

First, children with autism have been tested for their comprehension of person permanence (Sigman and Mundy 1989), that is, their understanding that when other people are no longer in view they nevertheless continue to exist. Sigman and Mundy's study (Experiment 2) showed that they did understand this: young children with autism search for their mother when she disappears behind a screen. Whilst this is a very basic aspect of social cognition, it is useful to know it is normal in these children, just as it is in young normal infants.

Secondly, children with autism have been tested for their ability to recognize themselves as distinct from other people — another low-level aspect of their self-concept. This has been examined using self-recognition tests, which looks at how children respond to their mirror or video images. These studies have all found that children with autism are able to recognize themselves in this way (Flannery 1976; Neumann and Hill 1978; Ferrari and Matthews 1983; Spiker and Ricks 1984; Dawson and McKissick 1984; Baron–Cohen 1985), either by touching a red mark on their faces when they first notice it in the mirror image, or by naming themselves when asked who is in the image. Some of these studies have however documented a distinct lack of shyness or embarrassment in front of the mirror, which may indicate abnormalities at 'higher' levels of their self-concept (Baron–Cohen 1985; Neuman and Hill 1978; Spiker and Ricks 1984).

Thirdly, children with autism have been tested for their ability to recognize important features of other people, such as their age, gender, identity, and relationship to others. Gender recognition seems to be in line with mental age in autism (Weeks and Hobson 1987; Abelson 1981), as does identity recognition, as judged from photographs of faces alone (Langdell 1978; Goode 1985; Hobson *et al.* 1988a; Volkmar *et al.* 1989). In the latter, however, it is worth noting that children with autism are less prone to the inversion-effect, suggesting that faces may have a different significance for them than for other people. I will return to discuss this anomaly at the end of the chapter. In relationship perception, children with autism appear able to recognize simple categories of dyad (such as mother–child, father–child, peer, or husband–wife) on the basis of age and gender cues in drawings of people (Baron–Cohen 1991*d*). Examples of this test are shown in Fig. 4.2. The task involves the subject matching each of the four relationships depicted in (b) with each of the four 'targets' in (a). This result fails to support an earlier study showing problems in age-recognition by children with autism (Hobson 1987).

Finally, children with autism seem able to make the fundamental conceptual distinction between animate and inanimate objects, when asked to sort 20 randomly presented pictures of different objects (animals, plants, people, domestic objects, mobile objects, and toy creatures) into two piles, Alive and Not-Alive (Baron–Cohen 1991*d*). The criteria they use for making these judgements also indicate they have a good grasp of this distinction. From a neuropsychological point of view, this suggests that their deficits are dissociable from those shown by brain-damaged patients who do have specific impairments in recognizing the animate-inanimate distinction (Farah *et al.* 1991). This body of evidence points to the intact nature of many aspects of social cognition in autism, and highlights the specificity of the deficits these subjects show in the development of their theory of mind.

(a)

(b)

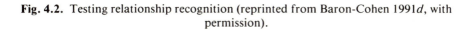

Fig. 4.2. Testing relationship recognition (reprinted from Baron-Cohen 1991*d*, with permission).

Conceptual ramifications of the theory of mind deficit

From the above pattern of results, several serious ramifications might be expected to ensue, at the level of cognitive development. First, one might expect the deficits in theory of mind to interfere radically with the child's growing understanding of the brain as an organ with *mental* functions. Secondly, one might expect this deficit to destabilize the child's developing conceptual distinction between mental and physical entities — disturbing the child's basic ontology. Finally, one might expect it to lead the child into difficulties in precisely those situations in which they have to take into account their *own* mistaken beliefs, such as when confronted by objects with misleading appearances. This might then undermine their ability to distinguish appearance from reality.

In a series of studies, I examined these three possibilities (Baron–Cohen 1989*e*). In one experiment, after establishing that they knew the location of the brain, subjects with autism were asked what they thought the brain was for. In reply, 70.6 per cent of them referred to its role in behaviour ('It makes you move', etc.) while only 23.5 per cent of them referred to its mentalistic role ('It's for thinking', etc.), even after considerable prompting. In contrast, 68.8 per cent of a group of subjects with mental handicap and 84.2 per cent of normal five-year-olds referred to its mental function (dreaming, remembering, keeping things secret, etc.), while only about 31 per cent of both groups referred to a behavioural function of the brain. Such findings confirm Johnson and Wellman's (1982) findings for the precocity of mentalistic understanding of the function of the brain in normal children, and extend these to children with a mental handicap, but reveal almost the opposite pattern in children with autism: a well-developed understanding of the behavioural function of the brain, but a poorly developed understanding of its mental functions. These three groups did not, in contrast, differ in their understanding of the function of another unobservable biological organ, the heart, attributing to it exclusively physical functions ('It pushes your blood', or 'It keeps you alive', etc.).

What of the ontological distinction between mental and physical entities? Wellman and Estes (1986) showed that normal three-year-olds have a stable grasp of this distinction. In the second experiment in this series (Baron–Cohen 1989*e*), Wellman and Estes's method was adapted for use with children with autism: the subject was told a story about two characters, one of whom *has* an object, the other of whom is *thinking* (or dreaming/ pretending/remembering) about an object. Following each story, the subject was asked to make judgements about which character could perform an action on the object. The results showed that 78.9 per cent of the normal subjects, and 68.8 per cent of the subjects with mental handicap scored 7 or more out of a possible 8 points across 4 such stories. In contrast, only 23.5 per cent of the subjects with autism did so, despite passing a Memory Question and a Language Comprehension Test. Their understanding that mental and physical entities differ seemed to be completely undeveloped, their performance on this test being at chance level. This result was replicated by Ozonoff *et al.* (1991)

What of the final problem? Might they have difficulty in distinguishing appearance and reality (A–R)? Flavell *et al.* (1986) showed that, when presented with misleading objects such as a sponge painted to look like a rock, children between four and six years of age can say not only what it looks like (a rock) but also what it really is (a sponge). In doing so, they distinguish between their initial (perceptually-based) belief about the object, and their current *knowledge* about it. How would children with autism perform on A–R tests? Experiment 3 of this series (Baron–Cohen 1989*e*) found that while 78.9 per cent of the group with mental handicap and 81.3 per cent of the normal group were able to answer an Appearance and a Reality Question correctly on 3 out of 4 tests, once again only 35.5 per cent of the group with autism were able to do so. Thus, when shown a stone that looked like an egg, or a magnified penny that looked bigger than a 10p coin, most children without autism were able to say 'It looks like an egg, but really it's a stone', or 'The penny looks bigger, but really it's smaller'. In contrast, most subjects with autism made largely 'phenomenist' errors, saying 'It looks like an egg', and 'It really is an egg', etc. They seemed to be dominated by their perception, and unable to consider their knowledge. This has been replicated by Ozonoff *et al.* (1991), and can be seen to mirror the results from Perner *et al.* (1989) using the misleading appearance (Smarties) task.

In discussing the Appearance–Reality distinction, Flavell *et al.* (1986) wrote:

It is probably a *universal* outcome in our species. This knowledge seems so necessary to everyday intellectual and social life that one can hardly imagine a society in which normal people would not acquire it . . . Knowledge about the distinction seems to presuppose the explicit knowledge that human beings are *sentient, cognizing subjects* (italics added) . . . It is part of the larger development of our conscious knowledge about our own and other minds (pp. 1–2).

The theory of mind deficit appears, then, to have widespread, but highly specific and predictable conceptual ramifications for children with autism. The three discussed here are undoubtedly only a small subset of these. The complete consequences for their cognitive development remain to be explored.

Deviance and delay

Before finishing this literature review, I shall briefly consider some evidence relevant to the question of whether the theory of mind deficit in autism is deviant and delayed compared to normal development. When a range of different mental-state tests are given to the *same* subjects, normal children (Gopnik and Slaughter 1991) and subjects with mental handicap (Baron–Cohen 1991c) find perception, imagination, and pretence easiest, desire slightly more difficult, and belief the hardest mental state of all* (see Fig. 4.3). In contrast, children with autism show a different pattern (see Fig. 4.4). They too find perception the easiest, and desire the next easiest. Like the other groups, too, belief is the most difficult mental state for them. But unlike the other two groups, they do not find imagination and pretence as easy as perception—they find these even more difficult than desire.

If we take degree of difficulty on these tests as reflecting *order of acquisition* of these mental-state concepts (Baron–Cohen 1992b), then these data suggest that children with autism may be both deviant and delayed in their acquisition of a theory of mind. They appear deviant in the sense that they seem to pass through a different *sequence* in understanding mental states to that seen in the other two groups. And they appear delayed in the sense that the youngest child with autism who passed the belief test in this experiment was ten years old, and this is almost six years later than the age at which normal children pass it. They even seem delayed relative to their mental age. (The lowest verbal mental age of a subject with autism who passed the belief test in this study was six years.)

From this and other studies, the developmentally earliest appearance of deviance appears to be at the point of understanding pretence and imagination. From studies of the onset of these skills in normal children (reviewed by Leslie 1987) we can therefore estimate that autism-specific abnormalities must be present from at least the mental-age equivalent of about eighteen months. But could they be present even earlier than this? I consider this possibility in the next section.

II. ORIGINS OF A THEORY OF MIND: UNDERSTANDING ATTENTION AND GOAL

Understanding attention

In earlier work (Baron–Cohen 1989a,d, 1991a) I put forward the claim that understanding the mental state of *attention* may be an early precursor in the development of a theory of mind. Normal infants from about nine months of age[†] reveal this understanding of another attention person's attention

---

* The tests used by Gopnik and Slaughter (1991c) Baron–Cohen (1991c) require the child to recall its own prior mental state, once the experimenter has manipulated events so that the child's mental stage changes.

† In fact, Scaife and Bruner found gaze-monitoring was present even in 30 to 40 per cent of two- to seven-month-olds, but by eight to eleven months was clearly in place. Protodeclarative pointing has a slightly later onset (nine to fourteen months: Butterworth 1991).

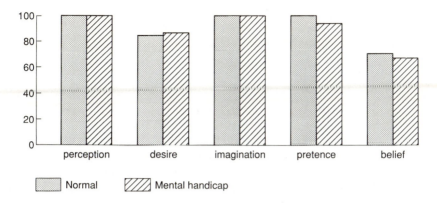

Fig. 4.3. Percentage of normal and mentally handicapped children passing each mental-state test (reprinted from Baron-Cohen 1991c, with permission).

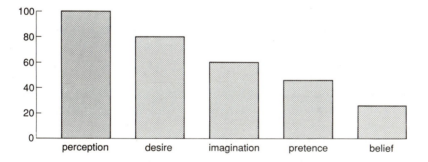

Fig. 4.4. Percentage of children with autism passing each mental-state test (reprinted from Baron-Cohen 1991c, with permission).

during gaze-monitoring (Scaife and Brune 1975; Butterworth 1991), and through gestures such as protodeclarative pointing (Bates *et al.* 1979). In gaze-monitoring, children appear not only to check where someone is looking, but also how the person is *evaluating* what they see (is it interesting? safe? funny? etc.), as is evident in the phenomenon of social referencing (Sorce *et al.* 1985). In protodeclarative pointing, children appear to use the pointing gesture not only to direct another person's attention to an object, but also to *comment* on it as a topic of interest, concern, fun, etc. (Tomasello 1988; Baron–Cohen 1991a).

Strikingly, children with autism show less gaze-monitoring (Sigman *et al.* 1986; Mundy *et al.*, this volume, Chapter 9) and almost no proto-declarative pointing (Baron–Cohen 1989*d*). They do, however, show some protoimperative pointing — using the index-finger gesture to *request* objects. Unlike its protodeclarative counterpart, protoimperative pointing does not necessarily entail the manipulation of someone else's attention *for its own sake*, as it may be used simply as an instrumental strategy for obtaining an object. Given the later deficits in their theory of mind, I argued that it is tempting to consider that in so far as joint-attention deficits reflect an impairment in understanding the mental state of attention, these may be an early precursor of the broader theory of mind deficit in autism (see also Baron–Cohen and Howlin, this volume, Chapter 21; Goméz *et al.*, this volume, Chapter 18).*

Understanding goals

I would like to extend this theory in the following way. First, consider a set of premises, all of which I expect are relatively uncontroversial:

1. That people's attention is normally *directed towards a target*.
2. That one key way in which children compute a person's state of attention is from the *direction of the person's gaze*.
3. That people's actions are normally *goal-directed*.
4. That people normally *look* at (attend to) the object they act on.

It follows from these premises that eye-direction provides information not only about the target of another person's attention, but also about the *goal* of a person's action. It may be then, that one reason toddlers make eye-contact is to facilitate goal-detection.

At what age might such goal-detection via face-processing begin? Phillips *et al.* (1992) investigated this with normal toddlers ranging from nine to eighteen months. The subject was presented with either an ambiguous or an unambiguous action. One ambiguous action consisted of offering an object to the child, but then at the last minute teasingly withdrawing it, just as the child began to reach for it. The unambiguous action consisted of simply giving an object to the child. This study found that all the toddlers responded to the ambiguous actions by instantly looking at the adult's eyes (on at least half of the trials, and within the first five seconds after the action), while only 39 per cent of them did so following the unambiguous action. This was also

---

* The existence of joint-attention deficits may also make some sense of at least part of the language delays seen in autism, given the role of joint-attention in language-acquisition (Tomasello 1988).

true of a group of young mentally handicapped children (mean CA: 60.4m, sd: 11.7) with very little language (mean verbal MA: 38.3m, sd: 14.4): 78 per cent of those looked at the adult's eyes immediately following the tease, but only 12 per cent did so immediately following the 'give'.

This striking phenomenon is consistent with the claim that, under conditions in which the goal of an adult's action is uncertain, the first place toddlers look for information to disambiguate the adult's goal is their eyes. We need further experiments to establish exactly which aspects of the adult's eyes are conveying the relevant information; but, from the assumptions set out earlier, we might expect this to be eye-*direction*, since by checking the direction of the adult's eyes, the toddler can instantly confirm the adult's goal. For example, if the toddler finds that the adult is looking away during the action, the toddler can then interpret the ambiguity as being due to the adult being distracted (where distraction = lack of attention to the target of action). If, however, the toddler finds that the adult is looking at him or her during the act, the toddler cannot then simply explain the ambiguity in terms of the adult's distraction. The toddler presumably then needs to search the adult's face (or the wider context) for other clues as to the nature of the ambiguous act. For example, the adult's facial expression and vocalization may provide the necessary information for the toddler to code the act as a playful tease, a threat, etc.

Phillips *et al.* (1992) also tested very young children with autism (mean CA: 53m, sd: 10.9, and with as little language as the mentally handicapped group) using the ambiguous and unambiguous actions. Less than 11 per cent of the children with autism made eye-contact with the adult in the five second time-window after any of the actions, in either condition. This was not due to gaze-*aversion*, as baseline data revealed they did make eye-contact with the adult at other times. Other studies also rule out the idea that there is gaze aversion in autism (Sigman *et al.* 1986; Hermelin and O'Connor 1970). Rather, the children with autism seemed not to *use* eye-contact for the same function as the normal or mentally handicapped children. Following the analysis given earlier, this may have been because they lack the concept of people's actions being goal-directed. The finding that the children with autism did not use eye-contact in the same way as control groups echoes the marked difference of the significance of the face (and particularly the eyes) to children with autism, compared to normal and mentally handicapped children, mentioned in the earlier literature review (Mirenda *et al.* 1983; Volkmar and Mayes 1991; Baron–Cohen, in press; Tantam, 1992). And given the other theory of mind deficits in autism, one possibility is that deficits in acquiring early concepts such as attention and goal may be developmentally related to the later deficits in the development of a theory of mind, reviewed in Section I of this chapter.

The claim that very young children with autism may not recognize people's actions as goal-directed seems at first glance to be almost incredible. After all, their own actions seem goal-directed just as a normal infant's are. They reach and select things. Surely this implies some awareness of their own goals? One possibility is that much of the time, normal toddlers do not represent actions in terms of goals, but simply focus on the external situation. It may be that it is only under relatively unusual conditions, such as when one's actions lead to unexpected outcomes, or when another person's actions are ambiguous, that the toddler needs to explicitly represent an action with reference to the actor's goal. The teasing test of Phillips *et al.* (1992) described earlier may fall into this category.

The idea that very young children with autism may not recognize people as being goal-directed may make some sense of an early-occurring and widely reported behaviour, namely, literally *treating the adult as an object*: climbing up on to the adult to reach things, taking the adult's hand and putting it on a door-knob, etc. (Kanner 1943; Phillips, Gómez *et al.* 1992). Whether older children with autism go on to develop concepts of attention and goal, while still falling short of developing concepts of false belief, needs further investigation. Certainly, under the specific developmental delay hypothesis (Baron-Cohen 1991*a*), we might expect that, with age, many children with autism might progress to the stage of understanding attention and goal — and get to 'first base', as it were. The relatively good performance by older children with autism on the Behavioural stories of our picture-sequencing task (Baron–Cohen *et al.* 1986) suggests that by late childhood such concepts are indeed firmly in place; but this point reminds us that autism is a *developmental* disorder, and as such we need to keep track of developmental changes within it. Deficits in understanding attention and goal may be restricted to the very youngest of children with autism.

These ideas of course beg many questions. First, what do we mean by a goal, and how is a goal different to an intention? Clearly, given the difficulty even normal three-year-olds have in understanding intentions (Astington and Lee 1991), the concept of goal must be very much simpler, if it is in place in nine- to eighteen-month-old toddlers. At the very least, a goal can be conceptualized as the *target* of an action (or of gaze). An intention, on the other hand, is a mental state that is in principle separable from the action itself (Astington and Lee 1991). Whether the normal toddler's concept of goal is a mentalistic one, right from the start, as Premack (1990) suggests, needs further investigation. Secondly, how does the toddler's understanding of the early states of attention and goal relate to their later understanding of mental states? Interestingly, attention and goal both possess *intentionality*,* in being directed at something else (Brentano 1874). Such early concepts may therefore be the gang-plank into acquisition of more complex

* From the Latin *intendere*, aiming at.

intentional concepts such as desire, knowledge, and belief. Specifying these relationships is an important priority for future research.

In closing, I would like to emphasize, first, that concepts of attention and goal may function for the nine- to twelve-month-old normal child as a simple *theory* of mind, in that, at the very least, from such concepts one can *predict* a person's next action. Secondly, I would like to propose that an **attention-goal psychology** may precede what Wellman (1990) calls a **desire psychology**, which in turn may precede a fully-fledged **belief–desire psychology** (Dennett 1978). Testing for the existence of an attention–goal psychology in nine- to twelve-month-olds, and clarifying the nature of these concepts in toddlers, will be an interesting project for future work.

Acknowledgements

During the writing of this work I was supported by grants from the Bethlem–Maudsley Research Fund, the Mental Health Foundation, and the Medical Research Council. I am grateful to Wendy Phillips, Mike Rutter, Juan-Carlos Gómez and Helen Tager-Flusberg for valuable discussions of the ideas in this work.

REFERENCES

Abelson, A. G. (1981). The development of gender identity in the autistic child. *Child Care, Health, and Development*, 7, 347–56.

Astington, J., and Gopnik, A. (1991). Developing understanding of desire and intention. In *Natural theories of mind* (ed. A. Whiten). Basil Blackwell, Oxford.

Astington, J., and Lee, E. (1991). What do children know about intentional causation? Paper presented at the SRCD Conference, Seattle, Washington.

Baron–Cohen, S. (1985). *Social cognition and pretend play in autism*. Unpublished doctoral thesis, University College, University of London.

Baron–Cohen, S. (1987). Autism and symbolic play. *British Journal of Developmental Psychology*, 5, 139–48.

Baron–Cohen, S. (1988). Social and pragmatic deficits in autism: cognitive or affective? *Journal of Autism and Developmental Disorders*, 18, 379–402.

Baron–Cohen, S. (1989a). Joint attention deficits in autism: towards a cognitive analysis. *Development and Psychopathology*, 1, 185–9.

Baron–Cohen, S. (1989b). The autistic child's theory of mind: a case of specific developmental delay. *Journal of Child Psychology and Psychiatry*, 30, 285–98.

Baron–Cohen, S. (1989c). The theory of mind hypothesis of autism: a reply to Boucher. *British Journal of Disorders of Communication*, 24, 199–200.

Baron–Cohen, S. (1989d). Perceptual role-taking and protodeclarative pointing in autism. *British Journal of Developmental Psychology*, 7, 113–27.

Baron–Cohen, S. (1989e). Are autistic children behaviourists? An examination of their mental–physical and appearance–reality distinctions. *Journal of Autism and Developmental Disorders*, 19, 579–600.

Baron–Cohen, S. (1990). Instructed and elicited play in autism: a reply to Lewis and Boucher. *British Journal of Developmental Psychology*, **8**, 207.

Baron–Cohen, S. (1991*a*). Precursors to a theory of mind: understanding attention in others. In *Natural theories of mind* (ed. A. Whiten). Basil Blackwell, Oxford.

Baron–Cohen, S. (1991*b*) Do people with autism understand what causes emotion? *Child Development*, **62**, 385–95.

Baron–Cohen, S. (1991*c*). The development of a theory of mind in autism: deviance and delay? *Psychiatric Clinics of North America*, **14**, 33–51.

Baron–Cohen, S. (1991*d*). The theory of mind deficit in autism: how specific is it? *British Journal of Developmental Psychology*, **9**, 301–14.

Baron–Cohen, S. (1992*a*). Out of sight or out of mind? A naturalistic study of deception in autism. *Journal of Child Psychology and Psychiatry,* **33**, 1141–55.

Baron–Cohen, S. (1992*b*). On modularity and development in autism: a reply to Burack. *Journal of Child Psychology and Psychiatry*, **33**, 623–9.

Baron–Cohen, S. (in press). Face-processing and theory of mind: how do they interact in development and psychopathology? In *Manual of developmental psychopathology* (ed. D. Cicchetti and D. J. Cohen). Wiley, New York.

Baron–Cohen, S., Leslie, A. M., and Frith, U. (1985). Does the autistic child have a 'theory of mind'? *Cognition*, **21**, 37–46.

Baron–Cohen, S., Leslie, A. M., and Frith, U. (1986). Mechanical, behavioural and Intentional understanding of picture stories in autistic children. *British Journal of Developmental Psychology*, **4**, 113–25.

Baron–Cohen, S., Spitz, A., and Cross, P. (1993). Can children with autism recognize surprise? *Cognition and Emotion*, **7**, 507–16.

Bates, E., Benigni, L., Bretherton, I., Camaioni, L., and Volterra, V. (1979). Cognition and communication from 9–13 months: correlational findings. In *The emergence of symbols: cognition and communication in infancy* (ed. E. Bates). Academic Press, New York.

Boucher, J., and Lewis, V. (1990). Guessing or creating? A reply to Baron–Cohen. *British Journal of Developmental Psychology*, **8**, 205–6.

Braverman, M., Fein, D., Lucci, D., and Waterhouse, L. (1989). Affect comprehension in children with pervasive developmental disorders. *Journal of Autism and Developmental Disorders*, **19**, 301–16.

Brentano, F. von (1870,1970) *Psychology from an empirical standpoint.* (ed. O. Kraus trans. L. L. MacAllister). Routledge and Kegan Paul, London.

Butterworth, G. (1991). The ontogeny and phylogeny of joint visual attention. In *Natural theories of mind*, (ed. A. Whiten). Basil Blackwell, Oxford.

Camras, L. A., Grow, G., and Ribordy, S. C. (1983). Recognition of emotional expression by abused children. *Journal of Child Psychology and Psychiatry*, **12**, 325–8.

Cutting, J. (1981). Judgement of emotional expression in schizophrenics. *British Journal of Psychiatry*, **139**, 1–6.

Dawson, G., and McKissik, F. C. (1984). Self-recognition in autistic children. *Journal of Autism and Developmental Disorders*, **14**, 383–94.

De Kosky, S. Heilman, K., Bowers, M., and Valenstein, F. (1980) Recognition and discrimination of emotional faces and pictures. *Brain and Language*, **9**, 206–14.

Dennett, D. (1978) *Brainstorms: philosophical essays on mind and psychology.* Harvester Press, USA.

Dretske, F. (1969). *Seeing and knowing.* University of Chicago Press.

Eisenmajer, R., and Prior M. (1991) Cognitive linguistic correlates of 'theory of mind' ability in autistic children. *British Journal of Developmental Psychology*, **9**, 351–64.

Farah, M., McMullen, P., and Meyer, M. (1991). Can recognition of living things be selectively impaired? *Neuropsychologia*, **29**, 185–93.

Ferrari, M., and Matthews, W.S. (1983). Self-recognition deficits in autism: syndrome-specific or general developmental delay? *Journal of Autism and Developmental Disorders*, **13**, 317–24.

Flannery, C.N. (1976). *Self-recognition and stimulus complexity preference in autistic children*. Unpublished Master's thesis, University of Orleans.

Flavell, J.H., Everett, B.A., Croft, K., and Flavell, E.R. (1981). Young children's knowledge about visual perception: further evidence for the Level 1 – Level 2 distinction. *Developmental Psychology*, **17**, 99–103.

Flavell, J.H., Green, and Flavell, E.R. (1986). Development of knowledge about the appearance–reality distinction. *Monographs of the Society for Research in Child Development*, **51**.

Frege, G. (1892). On sense and reference. *Zeitschrift für Philosophie und philosophische Kritik*, **100**, 25–50. Reprinted in *Translation from the philosophical writings of Gottlob Frege* (ed. P. Geach and M. Black). Basil Blackwell, Oxford, (1970).

Gómez, J.C. (1991). Visual behaviour as a window for reading the mind of others in primates. In *Natural theories of mind* (ed. A. Whiten). Basil Blackwell, Oxford.

Goode, S. (1985). *Recognition accuracy for faces presented in the photographic negative: a study with autistic adults*. Unpublished M.Phil thesis, Institute of Psychiatry, University of London.

Goodhart, F., and Baron–Cohen (1994). Do children with autism understand how knowledge is acquired? The Pratt and Bryant probe. *British Journal of Developmental Psychology*.

Gopnik, A., and Seager, W. Young children's understanding of desires. *Child Development*. (In press).

Gopnik, A., and Slaughter, V. (1991). Young children's understanding of changes in their mental states. *Child Development*, **62**, 98–110.

Gratch, G. (1964). Response alteration in children: a developmental study of orientations to uncertainty. *Vita Humana*, **7**, 49–60.

Gray, J.M., Frazer, W.L., and Leudar, I. (1983). Recognition of emotion from facial expression in mental handicap. *British Journal of Psychiatry*, **142**, 566–71.

Grice, H.P. (1975). Logic and conversation. In *Syntax and semantics: speech acts* (ed. R. Cole and J. Morgan). Academic Press, New York. (Original work published in 1967.)

Harris, P., Johnson, C.N., Hutton, D., Andrews, G., and Cooke, T. (1989). Young children's theory of mind and emotion. *Cognition and Emotion*, **3**, 379–400.

Hermelin, B., and O'Connor, N. (1970) *Psychological experiments with autistic children*. Pergamon Press, London.

Hobson, R.P. (1984) Early childhood autism and the question of egocentrism. *Journal of Autism and Developmental Disorders*, **14**, 85–104.

Hobson, R.P. (1986a) The autistic child's appraisal of expressions of emotion. *Journal of Child Psychology and Psychiatry*, **27**, 321–42.

Hobson, R.P. (1986b). The autistic child's appraisal of expressions of emotion: a further study. *Journal of Child Psychology and Psychiatry*, **27**, 671–80.

Hobson, R.P. (1987). The autistic child's recognition of age-related features of people, animals and things. *British Journal of Developmental Psychology*, **1**, 343–52.

Hobson, R.P. (1992). Social perception in high-level autism. In *High-functioning individuals with autism* (ed. E. Schopler and G. Mesibov). Plenum, New York.

Hobson, R.P., Ouston, J., and Lee, T. (1988*a*). What's in a face: the case of autism. *British Journal of Psychology*, **79**, 441–53.

Hobson, R.P., Ouston, J., and Lee, A. (1988*b*). Emotion recognition in autism: coordinating faces and voices. *Psychological Medicine*, **18**, 911–23.

Hobson, R.P., Ouston, J., and Lee, T. (1989). Naming emotion in faces and voices: abilities and disabilities in autism and mental retardation. *British Journal of Developmental Psychology*, **7**, 237–50.

Johnson, C.N., and Wellman, H. (1982). Children's developing conceptions of the mind and brain. *Child Development*, **53**, 222–34.

Kanner, L. (1943). Autistic disturbance of affective contact. *Nervous Child*, **23**, 217–50. Reprinted in Kanner, L. (1973) *Childhood psychosis: initial studies and new insights*. Wiley, New York.

Kurucz, J., Feldmar, G., and Werner, W. (1979). Prosopo-affective agnosia associated with chronic organic brain sydrome. *Journal of the American Geriatrics Society*, **27**, 91–5.

Langdell, T. (1978). Recognition of faces: an approach to the study of autism. *Journal of Child Psychology and Psychiatry* **19**, 225–38.

Leekam, S., and Perner, J. (1991). Does the autistic child have a theory of representation? *Cognition*, **40**, 203–18.

Lempers, J.D., Flavell, E.R., and Flavell, J.H. (1977). The development in very young children of tacit knowledge concerning visual perception. *Genetic Psychology Monographs*, **95**, 3–53.

Leslie, A.M. (1987). Pretence and representation: the origins of 'theory of mind'. *Psychological Review*. **94**, 412–26.

Leslie, A.M., and Frith, U. (1988). Austistic children's understanding of seeing, knowing and believing. *British Journal of Developmental Psychology*, **6**, 315–24.

Leslie, A.M., and Thaiss, L. (1992). Domain specificity in conceptual development: neuropsychological evidence from autism. *Cognition*, **43**, 225–51.

Lewis, V., and Boucher, J. (1988). Spontaneous, instructed and elicited play in relatively able autistic children. *British Journal of Developmental Psychology*, **6**, 325–39.

Mirenda, P., Donnellan, A., and Yoder, D. (1983). Gaze behaviour: a new look at an old problem. *Journal of Autism and Developmental Disorders*, **13**, 397–409.

Mitchell, S. (1990). Exploring the autistic child's impaired theory of mind. Unpublished MS, University of Oxford, Department of Experimental Psychology.

Neuman, C.J., and Hill, S.D. (1978). Self-recognition and stimulus preference in autistic children. *Developmental Psychobiology*, **11**, 571–8.

Novic, J., Luchins, D.J., and Perline, R. (1984). Facial affect recognition in schizophrenia: is there a differential deficit? *British Journal of Psychiatry*, **144**, 533–7.

Odom, P.B., Blanton, R.L., and Laukhuf, C. (1973). Facial expressions and interpretations of emotion-arousing situations in deaf and hearing children. *Journal of Abnormal Child Psychology*, **1**, 139–51.

Oswald, D.P., and Ollendick, T. (1989). Role taking and social competence in autism and mental retardation. *Journal of Autism and Developmental Disorders*, **19**, 119–8.

Ozonoff, S., Pennington, B., and Rogers, S. (1990). Are there emotion perception deficits in young autistic children? *Journal of Child Psychology and Psychiatry*, **31**, 343–63.

Ozonoff, S., Pennington, B., and Rogers, S. (1991). Executive function deficits in high-functioning autistic children: relationship to theory of mind. *Journal of Child Psychology and Psychiatry*, **32**, 1081–1106.

Perner, J. (1991). *Understanding the representational mind*. MIT Press, Cambridge, Mass.

Perner, J., and Wimmer, H. 1(985). 'John *thinks* that Mary *thinks* that . . .'. Attribution of second-order beliefs by 5–10 year old children. *Journal of Experimental Child Psychology*, **39**, 437–71.

Perner, J., Frith, U., Leslie, A.M., and Leekam, S. (1989). Exploration of the autistic child's theory of mind: knowledge, belief, and communication. *Child Development*, **60**, 689–700.

Phillips, W., Baron–Cohen, S., and Rutter, M. (1992). The role of eye-contact in goal-detection: evidence from normal toddlers and children with autism or mental handicap. *Development and Psychopathology*, **4**, 375–84.

Phillips, W., Gómez, J.-C., Baron–Cohen, S., Làa, V., and Rivière, A. (1994). Treating people as objects, agents, or subjects. How young children with and without autism make requests. Unpublished MS, Institute of Psychiatry, London.

Pratt, C., and Bryant, P. (1990). Young children understand that looking leads to knowing (so long as they are looking into a single barrel). *Child Development*, **61**, 973–83.

Premack, D. (1990). Do infants have a theory of self-propelled objects? *Cognition*, **36**, 1–16.

Prior, M., Dahlstrom, B., and Squires, T. (1990). Autistic children's knowledge of thinking and feeling states in other people. *Journal of Child Psychology and Psychiatry*, **31**, 587–602.

Reed, T., and Peterson, C. (1990). A comparative study of autistic subjects' performance at two levels of visual and cognitive perspective taking. *Journal of Autism and Developmental Disorders*, **20**, 555–68.

Rutter, M. (1978). Language disorder and infantile autism. In *Autism: a reappraisal of concepts and treatment* (ed. M. Rutter and E. Schopler). Plenum, New York.

Scaife, M., and Bruner, J. (1975). The capacity for joint visual attention in the infant. *Nature*, **253**, 265–6.

Shaw, P. (1989). Is the deficit in autistic children's theory of mind an artefact? Unpublished MS, Dept of Experimental Psychology, Oxford University, South Parks Rd, Oxford, UK.

Siegel, M., and Beattie, K. (1991). Where to look first for children's knowledge of false beliefs. *Cognition*, **38**, 1–12.

Sigman, M., and Mundy, P. (1989). Social attachments in autistic children. *Journal of the American Academy of Child and Adolescent Psychiatry*, **28**, 74–81.

Sigman, M., Mundy, P., Ungerer, J., and Sherman, T. (1986). Social interactions of autistic, mentally retarded, and normal children and their caregivers. *Journal of Child Psychology and Psychiatry*, **27**, 647–56.

Sorce, J., Emde, R., Campos, J., and Klinnert, M. (1985). Maternal emotional signalling: its effect on the visual cliff behavior of 1 year olds. *Developmental Psychology*, **21**, 195–200.

Spiker, D., and Ricks, M. (1984). Visual self-recognition in autistic children: developmental relationships. *Child Development*, **55**, 214–25.

Swettenham, J. (1992). *The autistic child's theory of mind: a computer-based investigation*. Unpublished Ph.D. thesis, University of York.

Tan, J., and Harris, P. (1991). Autistic children understand seeing and wanting. *Development and Psychopathology*, **3**, 163–74.

Tantam, D. (1992). Characterizing the fundamental social handicap in autism. *Acta Psychologica Scandinavica*, **55**, 88–91.

Tantam, D., Monaghan, L., Nicholson, H., and Stirling, J. (1989). Autistic children's ability to interpret faces: a research note. *Journal of Child Psychology and Psychiatry*, **30**, 623–30.

Tomasello, M. (1988). The role of joint-attentional processes in early language acquisition. *Language Sciences*, **10**, 69–88.

Ungerer, J., and Sigman, M. (1981). Symbolic play and language comprehension in autistic children. *Journal of the American Academy of Child Psychiatry*, **20**, 318–37.

Volkmar, F., and Mayes, L. (1991). Gaze behaviour in autism. *Development and Psychopathology*, **2**, 61–9.

Volkmar, F., Sparrow, S., Rende, R. D., and Cohen, D. J. (1989). Facial perception in autism. *Journal of Child Psychology and Psychiatry*, **30**, 591–8.

Weeks, S. J., and Hobson, R. P. (1987). The salience of facial expression for autistic children. *Journal of Child Psychology and Psychiatry*, **28**, 137–52.

Wellman, H. (1990). *The child's theory of mind*. Bradford Books/MIT Press, Cambridge, Mass.

Wellman, H., and Estes, D. (1986). Early understanding of mental entities: a reexamination of childhood realism. *Child Development*, **57**, 910–23.

Wimmer, H., and Perner, J. (1983). Belief about beliefs: representation and constraining function of wrong beliefs in young children's understanding of deception. *Cognition*, **13**, 103–28.

Wing, L., and Gould, J. (1979). Severe impairments of social interaction and associated abnormalities in children: epidemiology and classification. *Journal of Autism and Developmental Disorders*, **9**, 11–29.

5

What autism teaches us about metarepresentation

ALAN LESLIE AND DANIEL ROTH

In the psychological study of autism, there are two interrelated approaches one may pursue. The first tries to encompass the broad *clinical* picture, accounting for the main behavioural characteristics of the syndrome. In this approach, we ask: What, psychologically speaking, is the essential impairment or set of impairments that produces the distinctive syndrome as a clinical entity? The second approach focuses on the cognitive neuropsychological implications of the disorder. In this approach, we ask: What does autism teach us about the *normal* structure of the cognitive system? The 'theory of mind' hypothesis is relevant to both of these approaches. In the first, it can provide a framework for understanding some of the main, and until now, theoretically recalcitrant features of autism (see for example Frith 1989). In the second, this hypothesis has to be translated into a model of underlying processing mechanisms (Leslie 1987; Leslie and Frith 1990). Our discussion in this chapter falls under the second approach.

We shall discuss the theory of mind hypothesis in relation to a particular model of the architecture of the cognitive system in development (Leslie 1987, 1988*a*; Leslie and Frith 1990; Leslie and Thaiss 1992). The model shares many of the general assumptions of current work in cognitive science and, in particular, in cognitive neuropsychology. Specifically, cognitive neuropsychology depends upon having a processing model of a given *normal* psychological function. Using such a model, one may identify theoretically a pattern of damage which could produce an observed pattern of impaired performance (see, for example Shallice 1988). Having such a cognitive model is also vital in allowing progress to be made on relating complex behaviour to neural systems (Frith *et al.* 1991; Johnson and Leslie, in preparation).

The cognitive neuropsychology of autism poses an additional challenge because autism is a developmental disorder. Consequently, the model of processing required has to account for normal development. By the same token, the model should allow for patterns of abnormal development to be specified. Thus the model makes explicit, within a single explanatory framework, the interrelation between normal and abnormal development. In Fig. 5.1 the entity to be modelled—i.e. the cognitive system—is illustrated as a

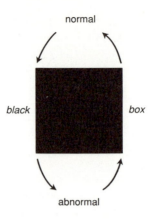

normal

black   *box*

abnormal

Fig. 5.1. 'Black box' model: theories of the internal architecture of the black box are required to account for both normal and abnormal development.

'black box'. The problem for the cognitive scientist is to develop theories concerning the internal structure or architecture of the black box. We believe that in understanding the development of knowledge, the scientist needs first to study the initial states of the architecture — the *core* that provides the specific engines of development.

We assume that the black box is to be understood as an arrangement of distinct functional components. Each of these components could, in principle and to a first approximation, be impaired independently. Now we can distinguish two possible senses of impairment — one in terms of surface behaviour, the other in terms of the underlying architecture of the black box. We know from several decades of cognitive research that the relationship between surface behaviour and underlying mechanisms is invariably complex. Indeed, part of the point in drawing this distinction is to highlight the possibility that similar patterns of surface behaviour may result from very different cognitive architectures. As long as one remains wedded to surface-behavioural definitions, whether of normal abilities or of syndromes of abnormality, progress will be limited. Mere description of surface behaviour must give way to the pursuit of underlying processing mechanisms.

For example, Leslie (1987) drew attention to the long and inevitably fruitless search for a behavioural definition of *pretending*; but, unlike others who had discussed this problem (for example Huttenlocher and Higgins 1978), Leslie developed a theoretical definition in terms of underlying processing mechanisms and their representations. This move had immediate consequences that went well beyond the surface behaviours associated with pretending. Specifically, the normal capacity for pretence could now be seen as a reflection of the normal capacity to employ and develop a theory of mind.

This, in turn, led to the hypothesis that the mechanisms constituting this capacity were impaired in autism, and to the attempts to test that hypothesis outlined below.

At present, the definition of autism is behavioural (see, for example, DSM IIIR (APA 1987)); eventually, however, this will give way to a definition which can play a systematic role in a causal theory of autism. In developing the cognitive level of such a definition, account will have to be taken of the complex relationship between surface behaviour and underlying mechanisms. Given that similar surface behaviour may have a variety of underlying sources, models of underlying impairment should account for overall patterns rather than local performances taken in isolation.

The neuropsychological literature abounds with examples of tasks where a patient can achieve an apparently normal level of performance by employing an abnormal strategy. For example, prosopagnosic patients recognize other people by clothing or accessories rather than by face (Ellis and Young 1989), dyslexic patients may use letter-by-letter reading (Shallice 1988), and amnesic patients may remember other people's names by means of mnemonic imagery (Wilson 1987). These patients can circumvent the impaired structure of the black box by using parts of its non-impaired structure. The result is an apparently normal performance on aspects of the relevant tasks.

Just as explicit models of the black box are essential to understanding impaired adult functioning, so they are also relevant to understanding impaired development. Abnormally developing children may, under some test conditions, show a similar pattern of performance to normally developing children, but do so for quite different reasons. For example, autistic children perform like normal three-year-olds on certain 'theory of mind' tasks. However, when we look at a broader range of performances, the profile shown by autistic children over a set of related tasks is quite different from that of both three- and four-year-olds. In fact, they exhibit an overall pattern that is never found in normal development. Constructing an explicit theoretical model of the internal architecture of the black box is vital to understanding such abnormal profiles.

Understanding normal development also requires a model of the black box. For example, how else can one understand the existence of U-shaped curves in development (Strauss 1982)? In these cases, a child shows similar performance on the same task at two distant points in development, separated by a period during which performance drops off. For example, Karmiloff-Smith (1988) has found that four- and eight-year-olds succeed, while six-year-olds fail, on a block-balancing task. But the similar surface behaviour of the youngest and oldest children has a dissimilar cause in terms of underlying processes. The four-year-olds use proprioceptive feedback to get the right answer; the six-year-olds use the geometrical centre of the

NORMAL

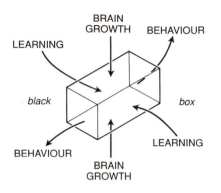

ABNORMAL

Fig. 5.2. Brain growth largely determines the core architecture of the black box, and constrains its powers of learning. Abnormal brain growth results in an abnormal architecture, which may then produce abnormalities affecting learning and behaviour.

object, and thus often fail; while the eight-year-olds rely on a naïve notion of the centre of gravity.

Another example of similar behaviours resulting from different underlying representations comes from the acquisition of irregular past-tense forms in English. The course of development from correct irregular form to incorrect regularized form and back again to correct irregular form is probably due to a shift from rote memorization to over-use of a rule to use of a rule limited by exceptions (e.g. Pinker and Prince 1988; see also Marcus *et al.* 1990).

A theory of normal development, then, has to be more than a description of changes in surface behaviour. Developmental theory must reveal the structure of knowledge and ability that is reflected in surface behaviour. It must also show how processes of brain-growth and learning create such knowledge and ability. In addressing these deeper questions, cognitive modelling of the internal structure of the black box will play an indispensable role. Likewise, theories of abnormal development will require an understanding of how abnormalities of brain-growth affect the cognitive structure of the black box. Figure 5.2 illustrates these notions. To date, little progress has been made on these questions. The scientific significance of the currently promising studies of autism will largely depend on the contribution they ultimately make to this general approach.

METAREPRESENTATION—A ROSE BY ANY OTHER NAME?

We turn now to consider the model of normal development that gave rise to the prediction that autistic children would be specifically impaired in 'theory of mind'. First, we reiterate some of the main concepts, in particular the notion of metarepresentation. We then discuss some subsequent interpretations of this idea which we feel are misguided or unfruitful. We then outline some recent findings and issues within the metarepresentational framework and the theory of the 'Theory of Mind Mechanism' (ToMM).

In recent studies of the child's 'theory of mind' a number of different ways of using the term 'metarepresentation' have grown up. Not surprisingly, this has caused a degree of confusion. In this section we shall attempt to clarify this terminological confusion and sharpen the important distinctions. The term was introduced into discussions of development by Leslie (Baron-Cohen *et al.* 1985; Leslie 1983, 1987) who borrowed the term from Pylyshyn (1978). Leslie gave the term 'metarepresentation' a technical sense in which it referred to a certain kind of *data structure* computed by our cognitive system. Leslie's notion was, from the outset, developed from the perspective of understanding the 'black box' as an information-processing device. Such a perspective does not restrict itself to characterizing what is consciously known, by the child or anyone else. It seeks instead to characterize the information-processing mechanisms which, on exposure to the environment, create or participate in creating conceptual knowledge, whether conscious or not.

The data structure called 'metarepresentation' was contrasted with another kind of data structure, which was dubbed 'primary representation'. A primary representation is illustrated for example by the kind of literal description of a situation that results from perception. The metarepresentation, by contrast, describes an agent's mental state—i.e. it provides a kind of 'agent-centred' description of a situation. It does this by describing a special kind of relation (an 'informational relation') between the agent and a 'situation'. To achieve this, the metarepresentation must have a number of components. The first of these components specifies who the agent is. The second component specifies the (informational) relationship between the agent and the following two components: an aspect of reality (described by a primary representation), and an imaginary situation (described by a 'decoupled' representation). Thus, primary representations constitute one component and decoupled representations another in a more complex relational structure, the whole of which is referred to by the term 'metarepresentation'. For example, **mother PRETENDS [of] the banana [that] 'it is a telephone'**. In this example, the agent (**mother**) is related to a proposition (**'it is a telephone'**) by way of holding an attitude (**PRETENDING**) that the proposition is true of an aspect of reality **(the banana)**.*

A metarepresentation, in the above sense of the term, is a data structure by means of which the child processes a certain type of information. This ability can be translated into a specific and limited understanding. It allows the child, under certain performance limitations, to represent particular attitudes (for example, PRETENDS) that agents can take to information. But, it would certainly be odd to suppose that processing metarepresentations somehow means that the child has a conscious theory of decoupling; and equally odd to suppose that, because a decoupled representation is, in effect, processed as if it were a copy or a report of a primary representation, it somehow implies that the child has a conscious theory that mental states are representations in the head. The theory of decoupling was invented by the theoretician, not by the child.

The ability to decouple is one component ability required to form and process metarepresentations. The ability to form and process metarepresentations is, in turn, an essential component of the ability to represent the propositional attitudes agents hold in relation to their descriptions of situations. A metarepresentation, in this sense, then, 'represents a representational (or informational) relation' — it represents the attitude an agent takes to a description of a particular aspect of reality. It seems to us that this constitutes a clear alternative to the notion that the pre-school child must consciously develop a general theory of representation and then apply it to understanding the mind.

Our assumption is, then, that the child is equipped to process decoupled representations in the right way (see Leslie 1987, 1988*a*, *b*; Leslie and Frith 1990, for ideas on the processing of decoupled representations). Decoupled representations form one component in a metarepresentation, and we likewise assume that the child is equipped to process metarepresentations. We believe that these mechanisms are subcomponents of the larger modular processing system we call 'ToMM'.

---

* This way of writing out a metarepresentation was introduced by Leslie (1988*a*) as a somewhat more elegant formulation than that found in Leslie (1987). Originally, the decoupling marks would have included in their scope an element, *the banana*, which was italicized to indicate that it corresponded to an aspect of reality. This then required a processor (dubbed the 'Interpreter') to find the corresponding object in primary representation and to anchor '*the banana*' to it. Leslie (1987) commented at the time that the italicized element acted like a variable. Leslie (1988*a*) then simplified the account by removing **the banana** from the decoupled expression, replacing it with an element that looks more like a variable ('**it**'), and displaying the primary representation of the real object within the metarepresentation as shown in the text. This makes the three-place relational structure of the metarepresentation more obvious. Leslie and Frith (1990) also note that this change renders the 'Interpreter' redundant, thus simplifying that part of the account which attempted to model the processing involved. Further questions can be raised about the two remaining parts of the 1987 processing model, which was geared to accounting for solitary pretence and does not easily generalize to understanding pretence in others or understanding propositional attitudes in general. However, we do not wish to pursue these questions here, since little of relevance hinges on the details of the 1987 'Decoupler' mechanism.

Misrepresenting the child's theory of mind

Unfortunately, the term metarepresentation has been taken up by others but interpreted in a radically different way — namely, from the perspective of the child's conscious knowledge. Perhaps this focus follows from an assumption, which we would reject, that the black box has but a single level of organization, corresponding to 'conscious knowledge'. In any case, this focus has given rise to the following, in our opinion, erroneous idea. Because mental states are, according to current scientific thinking, *representations*, the child's task must be to grasp consciously the fact that mental states are representations in the head. Only when she has accomplished this task are we allowed to speak of the child's theory of mind; only then are we allowed to speak of *meta*representation (conscious theorizing about representations). Consequently, the child's general theory of representation must be a central ingredient in the development of the child's theory of mind. This assumption has been made in various guises by most of the leading figures in the field (for example Flavell 1988; Forguson and Gopnik 1988; Gopnik and Slaughter 1991; Perner 1988; Wellman 1990; Wimmer and Hartl 1991; Zaitchik 1990). Foremost among these positions is one developed by Perner (1991) (see also Perner, this volume, Chapter 6). Perner argues that the pre-school child develops an explicit (conscious) theory of representational relations such as sense, reference, truth, and existence, and then applies this theory to understanding the mind. Furthermore, Perner insists that this is the only admissible sense of the word, metarepresentation: explicit knowledge about representation in general. Only by developing a general theory of representation and then applying it to mental states could the child or anyone, according to Perner, possibly understand false belief. Perner claims that a Representational Theory of Mind (RTM) is first constructed by the child at around four years of age, and constitutes a radical theory-change in the child's understanding of behaviour.

One of the unfortunate consequences of concentrating on the child's 'theory of representation' is that it shifts focus away from what we believe is central to the common-sense theory of mind, namely, the set of concepts exemplified by *pretends*, *believes*, and *wants* — i.e. the propositional attitudes. It seems to us that both adults and children tacitly hold the view that the behaviour of an agent is caused by the agent taking a given attitude to the truth of a given proposition about some aspect of the world. We shall sum this up in the phrase 'understanding agents and attitudes'.

The above attitude-based understanding is not enough, claims Perner, for understanding false belief. False-belief understanding, in Perner's view, requires that the child should project a representational *medium* into the mind of the agent — that the child should think of the agent as having, for example, a picture-in-the-head or a 'mental sand-box model' that represents reality for the agent. In our view, this proposed extra step for the child of having to project a representational medium into the agent's head is unwarranted, and we know of no evidence in its favour.

Notice that, in Perner's proposal above, the child will still have to understand the following two things. First, that any mental representation the agent might have will affect behaviour only by virtue of what the representation says. And what the representation 'says' is the meaning or proposition the representation expresses. One does not believe a picture but what the picture says. Second, the particular effect a given representation will have on behaviour will depend upon the particular attitude the agent takes to the proposition expressed by the representation. Suppose, for example, that John has in mind a representation of its raining outside. What effect will this have on John's behaviour? Well, that depends on whether John believes, pretends, or wants it to be raining outside. The predictions of John's behaviour that one's theory of mind makes will depend upon which attitude to which proposition one's theory attributes to John.

Going beyond the propositional attitude and projecting a representational medium into John's mind is not necessary for the everyday business of predicting and explaining behaviour. This basic work is carried out by the ability to grasp the role of propositional attitudes in the causation of behaviour — by understanding 'agents and attitudes'. It is the understanding of agents-with-attitudes that characterizes the intuitive common-sense theory of mind; it is this that emerges in the pre-school period; and it is this that is specifically impaired in autism. RTM, on the other hand, tries to explain what sort of thing propositional attitudes really are, rather as the atomic theory of matter explains what substances really are. Our suspicion is that such a view of mind probably does not emerge at all during the pre-school period. At any rate, passing standard false-belief tests at four years does not constitute evidence for a shift from a propositional attitude view to such a representational theory.

Why is an understanding of agents and attitudes not enough for Perner? It seems insufficient to Perner because he assumes an unnecessarily impoverished view of the information a propositional attitude can express. He assumes that attitude concepts could not possibly capture the fact that when Sally believes (wrongly) that the marble is in the basket, Sally's belief is nevertheless about a particular marble — the one sitting in the box. To understand this, Perner assumes, requires projecting a representational medium into Sally's mind. Unfortunately, he never makes clear why this is *required*, rather than being simply one way, among others, of capturing the fact that Sally's wrong belief is about a particular real marble.

One thing is clear. Whether conscious theorizing about representations is involved or not, to carry out the above analysis of believing requires the processing of a certain kind of data structure. This structure has to represent a relation (of *believing*) that holds between three things (like a mathematical function with three arguments). The first argument slot is filled by the agent (in this case, Sally); the second slot is filled by the aspect of reality that anchors the attitude (in this case, the marble); while the third slot is filled by the 'proposition' in relation to which the attitude is held (in this case, 'it is in the basket'). Whether or not this final slot is consciously theorized by the child to be a 'mental representation' is something that will require a lot of evidence, beyond merely showing that the child understands false beliefs. To our knowledge, no such evidence is currently available; and, as we shall see below in the next section, what evidence there is argues against.

We believe that the pre-school child has access to data structures of the above sort long before four years of age and success on standard false-belief tasks. We shall outline a model of the development of performance on belief tasks below. Here we wish to note that Leslie (1987) argued in some detail that exactly the above kind of data structure is available to the very young child to account for the normal capacity to pretend and understand pretence-in-others. Given that pretence involves pretending of particular objects — of *this* banana that it is a telephone (as opposed to pretending that bananas *in general* are telephones) — and that understanding pretence in someone else involves understanding which banana they are pretending is a telephone, it was clear that a three-place relational structure was required (Leslie 1987, pp. 414–20).

Thus the same analysis is required for understanding pretending as for understanding false belief. The difference between the two resides in the particular relation concepts involved, pretending versus believing; as we shall see below, important consequences follow from that conceptual difference. We believe that having access to such data structures, together with the system of inferences they support, constitutes a tacit and intuitive theory of mind, or, if you like, constitutes a tacit theory of the specific 'representational relations' (like *pretends*, *believes*, *wants*) that enter into the causation of agents' behaviour.

Before moving on, we want to end the terminological wrangle that has come to surround the term 'metarepresentation'. While we reject the strictures and admonitions put forward by Perner (for example, in Chapter 6, this volume), we do see an advantage to changing our own jargon slightly. We propose to replace our use of the term 'metarepresentation' with the term *M-representation*. We hope the reader will not enquire what the 'M' stands for. This term is so transparently a piece of jargon and so lacking in wider connotations that we trust we can be left to use the term as we wish. The meaning we give to it is the meaning we formerly gave to the term 'metarepresentation' — a three-argument, relational data structure computed by a processing mechanism whose job it is to relate agent's behaviour to agent's attitude.

M-REPRESENTATION AND THE ARCHITECTURE OF THE 'BLACK BOX'

The idea that autism involves an impairment in understanding agent's attitudes led to the prediction that autistic children will have specific problems in understanding other people's mental states, in particular, false belief. This prediction was confirmed (Baron–Cohen *et al.* 1985, 1986) and subsequently replicated using a variety of tasks and procedures (e.g.: Baron–Cohen 1989, 1991; Eisenmajer and Prior 1991; Leslie and Frith 1988; Leslie and Thaiss 1992; Perner *et al.* 1989; Roth and Leslie 1991; Russell *et al.* 1991; Sodian and Frith 1992; see Baron–Cohen, Chapter 4, this volume, for a review).

An account which tries to explain 'theory of mind' development in terms of the acquisition of a 'theory of representation' must either suppose that autistic children are specifically impaired in their understanding of representation or must seek an external explanation for their difficulties. This latter course is advocated by Perner (this volume, Chapter 6). It is easy enough to think up general impairments that might somehow or other impede development of a theory of mind; it is much more difficult, however, to account for the specific impairments in social cognition in autism while at the same time accounting for those abilities which are spared. One advantage of the view we advocate is that our model of the internal organization of the 'black box' can do this, accounting for the pattern of spared abilities as well as the disabilities in autism, while also at the same time providing an account of the specific basis for the normal development of a theory of mind.

According to our model, there is a domain-specific mechanism, which Leslie has dubbed the 'Theory of Mind Mechanism' (ToMM). This mechanism underlies our ability to conceive of our own and other people's mental states and to reason about behaviour in terms of such states. We believe that this mechanism, ToMM, comprises an inferential device, which infers 'states of mind' on the basis of behavioural events, together with the representational system we have called the 'M-representation'.

The M-representation is part of the normal capacity to conceive of mental states, such as pretence or belief, and to manage intentional communications with other agents. The autistic child's documented difficulties with pretend play, with true and false belief, and with communication can be explained on the assumption of an impaired ability to form and/or process this kind of representation. Such an impairment, however, will not fundamentally affect the autistic child's apprehension of physical artefacts, including *representational* artefacts such as photographs or maps.

The existence of the above spared ability has been confirmed by Leslie and Thaiss (1992; see also Leekam and Perner 1991; and Charman and Baron–Cohen 1992). Autistic children were given two different false-belief tasks along with two analogous 'false-photographs' tasks modelled on Zaitchik's

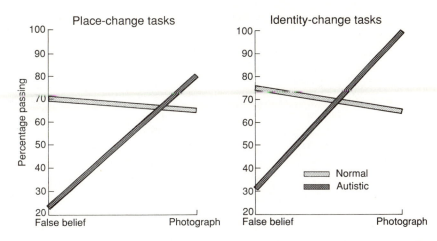

Fig. 5.3. Performance of normal four-year-olds and autistic children on two pairs of out-of-date representations tasks. One member of each pair is a 'standard' false-belief task, while the other is a false-photograph task (from Leslie and Thaiss 1992).

(1990) tasks. For example, the children were presented with a standard 'Sally and Anne' scenario in which an object is moved by Anne from where it was originally placed by Sally without Sally's seeing the change. The child is then asked where Sally thinks the object is. In the analogous photograph task, the child is shown Sally taking a polaroid photograph of the object in the first location, and placing the photograph face down on a table. The object is then moved to the second location and the child is asked where, in the photograph, the object is. In the first task, Sally's belief goes out of date, while in the second task Sally's photograph goes out of date. The autistic children, as expected, mostly failed the belief tasks, but were at or near ceiling on the photographs tasks — performing significantly *better* than the normal four-year-olds (see Fig. 5.3). A follow-up task, using a map that goes out of date, yielded similar results (see Fig. 5.4).

Because pictures and maps are representations but not agents, they cannot hold attitudes toward the situations they represent. A picture does not *believe* what it depicts! Because of this, ToMM and the M-representation are not engaged in processing pictures. Thus autistic children are quite capable of solving problems that involve out-of-date photographs and maps, even though the general problem-solving characteristics of an out-of-date photograph task are highly similar to the general characteristics of an out-of-date belief task. This striking result underscores the domain-specificity of autistic impairment in theory of mind, and highlights the usefulness of the ToMM model (Leslie and Thaiss 1992).

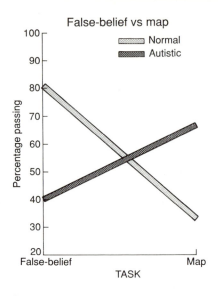

Fig. 5.4. Performance of normal four-year-olds and autistic children on an out-of-date belief and an out-of-date map task (from Leslie and Thaiss 1992).

FALSE BELIEF AND IMPAIRED 'EXECUTIVE FUNCTIONS'

The pattern of performance described above suggests minimally a dichotomous division of the internal structure of the 'black box' into a theory of mind-specific component, which consists of the processing and inferential mechanisms involved in the spontaneous attribution of mental states, (ToMM), and an independent component responsible for reasoning about the properties of representational artefacts.

The existence of other non-theory of mind factors is suggested by recent studies, in which typical 'frontal lobe' tests, such as the Wisconsin Card Sorting and the Tower of Hanoi, were administered to able autistic and learning-disabled children and adolescents (Ozonoff *et al.* 1991; see also Prior and Hoffman 1990). These studies showed that autistic children perform worse than controls on such tasks, and demonstrate the perseverative errors often shown by 'frontal' patients. This pattern of performance was interpreted in terms of an impaired 'executive function' that is responsible for inhibiting a previously established response pattern, and for the execution of planned action (Shallice 1988).

The above findings may reasonably lead to the suggestion that both the autistic and the normal young child's difficulties with false belief are due to the same impairment in 'frontal' functioning (see for example Ozonoff *et al.* (1991); Hughes and Russell (unpublished); and Harris (this volume,

Chapter 11). Russell *et al.* (1991) presented children with a competitive game in which they have to gain as many sweets or candies as possible by deceptively directing a competitor experimenter to an empty box. The children, but not the experimenter, can see which box is empty and which full by looking through small windows in the boxes. It was found that both autistic and three-year-old children persisted in pointing to the box containing the candy despite repeatedly losing it. Russell *et al.* (1991) suggest that both autistic and three-year-old normal children fail in false-belief tasks to inhibit their response to the immediate perceived reality. In both cases this might be due to the same 'executive functioning' difficulties — temporary in the one case, permanent in the other.

The above line of work makes an interesting and important contribution to the neuropsychology of autism. However, we think that the picture of underlying architecture it rests upon is too simple. We offer two reasons for thinking this. The first we have already outlined with the results of the photographs, maps, and drawings tasks (Charman and Baron–Cohen 1992; Leekam and Perner 1991; Leslie and Thaiss 1992). Autistic children are perfectly capable of 'inhibiting their response' or suppressing a 'salient perceived reality' when answering questions about a photograph or map.

When considering, then, the question of why autistic children fail false-belief tasks, we have to recognize that between false-belief tasks, on the one hand, and Tower of Hanoi and Wisconsin Card Sorting, on the other, both of which are failed by autistic children, rest the photographs and maps tasks, which they pass. Given that the photographs and maps tasks are far closer in problem-solving structure to false-belief tasks than either the Tower of Hanoi or Wisconsin Card Sorting are, this pattern of ability and disability implies that 'executive functioning' mechanisms are fractionated and various.

Here is our second ground for believing that a single 'executive functioning' account rests upon too simple a view of underlying architecture. Autistic children and normal three-year-olds do not fail false-belief tasks for the same reason. We described in our opening pages how important is the distinction between surface behaviour and underlying architecture, and how surface similarities can result from quite different underlying mechanisms. Leslie and Frith (1990) listed several ways in which autistic children differ from three-year-olds, despite the surface similarity they show in both failing false-belief tasks. Leslie and Thaiss (1992) discuss another in terms of the different pattern of errors three-year-olds and autistic children show on the 'Smarties' false-belief task. We saw above the excellent performance of autistic children on false photographs; most three-year-olds, by contrast, fail these tasks. Baron–Cohen (1991) has found an abnormal pattern of performance on so-called 'representational change' tasks, a pattern not found in normal development. To this list of dissimilarities we add another striking example provided by Roth and Leslie (1991).

Roth and Leslie modified a standard 'Sally and Anne' false-belief scenario in a number of ways which we hypothesized would engage the three-year-old's competence more directly. Chief among these modifications was a focus on a verbatim conversation between Sally and Anne. In essence, Sally returns to see that her chocolates are not where she left them, and asks Anne where they are. Anne has actually hidden the chocolates in a nearby box, but because she wants them for herself, she tells Sally that the dog took them and that they are now in the doghouse. The child is focused on this verbatim conversation and then asked 'Where does Anne think the chocolates are?' and 'Where does Sally think they are?' We performed an error analysis upon the answers to these questions given by three groups of children: five-year-olds, three-year-olds, and autistic adolescents.

Answers could fall into four categories: 'fully correct', in which the Speaker (Anne) is attributed a true belief, while the Listener (Sally) thinks what she has been deceptively told; 'attitude-based 1', in which the Speaker is attributed a belief based on what she said (not on what she should know), while the Listener thinks what she has been told; 'attitude-based 2', in which the Speaker is attributed a belief based on what she said, while the Listener is attributed a reality-based belief; and finally 'reality-based', in which both questions about Speaker and Listener are answered on the basis of reality. We expected that the five-year-olds would mostly give 'fully correct' answers, and thus demonstrate an adult-like understanding of the situation in which Anne deliberately lies to Sally. We did not expect the three-year-olds to have such a sophisticated grasp; but we nevertheless expected that they would have available a rudimentary concept of belief which they could demonstrate by attributing, either to Sally or to Anne, a belief that did not simply mirror reality. (A suitable control ensured that these children were not 'parroting' what Anne had said.) Finally, we hypothesized that the autistic group would not even show this level of competence, and would base their answers simply on reality. Table 5.1 summarizes the categories in our error analysis.

The results confirmed the above predictions. Five-year-olds apparently grasped Anne's deceptive intention, and were mostly 'fully correct'. The three-year-olds, consistent with their failing at 'standard' false-belief tasks, mostly did not understand the situation in this way and assumed that Anne (the Speaker) actually believed what she said. Most of the three-year-olds also said that Sally (the Listener) believed what she was told. In neither of the normal groups did any appreciable portion of the children give 'reality-based' responses (only a single child at each of the ages). Yet this was the dominant pattern of responding in the autistic group, suggesting that even a rudimentary concept of belief was not readily available. These results are visualized in Fig. 5.5.

The findings of Roth and Leslie (1991) show differences between all three groups, but with a major qualitative difference lying between the normal

Table 5.1. Attribution categories for error analysis in Roth and Leslie (1991)

Fully correct	=	Speaker thinks what is TRUE AND Listener thinks what is TOLD
Attitude-based 1	=	Speaker thinks what she SAID AND Listener thinks what is TOLD
Attitude-based 2	=	Speaker thinks what she SAID AND Listener thinks what is TRUE
Reality-based	=	Speaker: REALITY AND Listener: REALITY

children, who almost universally attribute attitudes, and the autistic adolescents, who typically do not. We suggest that this qualitative difference corresponds to the possession of an impaired or unimpaired ToMM.

How can we draw together the results from the extensive range of studies cited in this chapter into a single explanatory framework? The framework we want will account for *both* the normal capacity to acquire a theory of mind *and* the abnormal pattern of development found in autism. In the next section, we outline a theoretical model of the black box which begins to do this, drawing on some ideas of Roth (in preparation).

A NEUROPSYCHOLOGICAL MODEL

We shall summarize the set of phenomena that we believe a successful model will have to account for.

1. Mental-state notions are domain-specific and are tied to understanding agents. We need a cognitive mechanism which will spontaneously employ propositional-attitude concepts, and thus the M-representation. This mechanism will be an inferential engine capable of generating, in real time, analyses of behaviour relative to its theory of the attitudes. It should provide the pre-school child with an intuitive understanding of the mental states of agents, with minimum reliance on conscious problem-solving. Finally, it should have an architecture which allows the possibility of dissociable damage. These features point to something like the theory of ToMM.

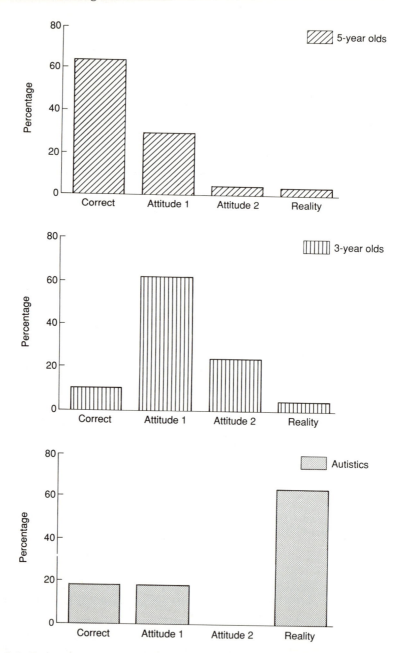

Fig. 5.5. Error analysis of responses of normal five- and normal three-year-olds and autistic adolescents on a non-standard false-belief task (after Roth and Leslie 1991).

2. The modular nature of ToMM allows it to be specifically damaged in autism in the following two senses: first, autistic children's theory of mind ability is impaired relative to their own verbal and non-verbal mental age and relative to other spared abilities; and second, such impairment is not the inevitable consequence of mental retardation or developmental handicap.

3. The normal three-year-old possesses an intact ToMM which allows her to deploy a theory of mind, including a concept of belief, despite performing differently from four- and five-year-olds (for example, failing standard false-belief tasks). Three-year-olds do succeed on some non-standard false-belief tasks (see for example Mitchell and Lacohée 1991; Roth and Leslie 1991; Wellman and Bartsch 1988; Zaitchik 1991) and are, under some circumstances, *more* willing to attribute a false belief than five-year-olds (Roth and Leslie 1991).

4. Autistic children are not simply 'stuck' at a three-year-old level of theory of mind development. On some (for example, the standard false-belief) tasks, autistic children perform like three-year-olds; but on other tasks they perform less competently than three-year-olds (see for example, Roth and Leslie 1991), and on yet others they show a different pattern of response (for example the 'Smarties' task—see Leslie and Thaiss 1992, and Baron–Cohen 1991). On structurally similar but non-theory of mind tasks, they can perform more competently than four-year-olds (see for instance Leslie and Thaiss 1992).

5. Normal four-year-olds perform reliably better on false-belief tasks than on analogous 'representation tasks' (Zaitchik 1990; Leslie and Thaiss 1992). Autistic children show the opposite pattern. Understanding representations as representations, therefore, is neither a necessary nor a sufficient condition for understanding belief. False belief tasks and representation tasks tap different mechanisms.

We believe that there is a domain-specific mechanism underlying normal theory of mind development. In addition to this specialized mechanism however, certain tasks, for example the standard false-belief tasks, require an additional co-operating mechanism. This additional non-specific mechanism is also required for certain non-theory of mind tasks, such as representation tasks. In the following we put forward some ideas on the nature of this additional factor, and on the developmental patterns it gives rise to.

To account for the above phenomena we propose a model with three distinct components, illustrated in Fig. 5.6. The first component, ToMM, is specific to theory of mind, and employs the M-representation. The second component embodies knowledge about representational artefacts, such as photographs and maps. We do not take a position on whether this second component constitutes another specialized mechanism or whether it is part of general encyclopaedic knowledge. The third component we have called

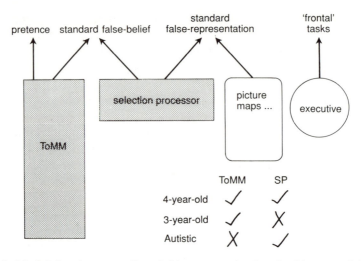

Fig. 5.6. Model showing some dissociable components involved in normal develop-ment. Normal four-year-olds possess both ToMM and the Selection Processor (SP) components, and can thus pass standard false-belief, photograph, and map tasks; three-year-olds possess ToMM but not yet SP, and can thus understand pretence and pass certain non-standard false-belief tasks; autistic subjects are impaired in ToMM but not in SP, allowing them to pass only false-photographs and maps tasks. An 'executive' component is also shown as independent of both ToMM and SP. This may be independently impaired in autism (from Leslie and Thaiss 1992).

the Selection Processor (SP). This last component co-operates with ToMM and with knowledge of representational artefacts to obtain the solution to particular classes of problem. SP is thus a more general mechanism than ToMM. However, we do not believe that SP is an entirely general mech-anism, because of the results on 'frontal tasks' in autism discussed in the previous section. So we have also shown an 'executive' component in Fig. 5.6 to make the point that we do not believe that SP should be identified with *general* executive functioning.

ToMM is required for the normal early capacity to pretend and to under-stand pretence-in-others and for attributing beliefs. To solve standard false-belief tasks, however, the co-operation of SP is required. SP is also required to solve standard false-photograph tasks. Both sets of tasks have a similar structure: in both, one state of affairs gives way to another state; an event related to the earlier state occurs (for example, the forming of a belief or the taking of a photograph), but does not occur again when the situation changes; the subject is then required to make an inference about the state of the belief or photograph; to do this, the subject must select, *for the purposes of inference*, her stored description of the earlier situation, resisting the intrusion of current reality. Three-year-olds do not fail standard false-belief tasks because they cannot remember the earlier situation — control questions show they do not have a memory problem as such. The SP performs the control function of selecting the appropriate counterfactual situation. Having selected a description of the appropriate state of affairs from memory, SP inputs this to a specialized inference device, to ToMM or, as the case might be, to the component dealing with representational artefacts.

The model postulates that SP becomes increasingly functional towards the fourth birthday. Before this time, the three-year-old can perform well on theory of mind tasks that do not require SP but instead exploit the on-board capabilities of ToMM. Among these on-board abilities is a capacity for counterfactual reasoning (for example, the ability to decouple) specialized for M-representing an agent's behaviour (Leslie 1987). ToMM succeeds to the extent that the imaginary or counterfactual content of an agent's mental state can be 'read off' behaviour as in pretence (Leslie 1988*a,b*) or in making an utterance (Roth and Leslie 1991). Other tasks reduce the need for SP by balancing the salience of the counterfactual and the factual situations (for example, where the factual situation is not definitely known, as in Wellman and Bartsch (1988) ('not own task') and in Zaitchik (1991). Intermediate levels of difficulty occur with scenarios where the agent's behaviour gives an immediate clue to her mental content (for example Bartsch and Wellman's (1989) explanation of action tasks) or in tasks where the job of SP is taken over by the experimenter (for example Wellman and Bartsch (1988) (inferred belief) or by a public representation (for example Mitchell and Lacohée 1991). Finally, standard false-belief tasks have a high degree of difficulty because the imaginary content of the mental state must be reconstructed purely on the basis of the agent's *exposure history*, without guidance from current behaviour and without compensation for a poorly developed SP.

SP is also required by 'standard' false-photograph tasks for the same reasons. This accounts for the correlation between successful performance on false-belief and false-photograph tasks in normal development, while allowing for false-belief competence to be ahead in the normal case (thanks to ToMM) and behind in autism (impaired ToMM).

Finally, autistic children (at least with verbal mental ages in excess of four years) possess SP intact, have developed knowledge of photographs and maps, but have an impaired specific component, ToMM. Thus, autistic children pass false-representation tasks but are impaired in their 'theory of mind'. We believe, for reasons given earlier, that performance on the Tower of Hanoi and WCS tasks should be explained by independent mechanisms. This neuropsychological model of development is summarized in Fig. 5.6.

THE IMPAIRMENT OF ToMM IN AUTISM: SOME ISSUES

The nature of the impairment

We have been arguing that there is a specific cognitive mechanism, ToMM, which drives a domain of normal development and which is impaired in autism. If this is correct then a number of questions arise having to do with the nature of this impairment at the cognitive level. Is ToMM simply absent, dysfunctional by degree, dysfunctional and delayed, or just delayed? in all or only some autistic children? could there be subcomponents of ToMM? only some of which are damaged?

Some of the above questions have already been touched upon. In what remains we want to make some brief remarks about some of the others. First, we consider the above model in the light of that group of autistic children who have been found to pass false-belief tasks. Frith *et al.* (1991) estimate this proportion to be around 25 per cent of higher-ability autistic children. Those who pass seem to require a minimum verbal mental age of about 5 years 6 months *together with* a minimum chronological age of around 11 years 6 months. Satisfying this joint mental and chronological age criterion does not guarantee success for the autistic adolescent — about half still fail. This pattern explains why sampling differences have led to some studies finding a mental-age correlation with passing (for example Eisenmajer and Prior 1991), others a chronological-age correlation (for example Leslie and Frith 1988), and yet others no reliable relationship with any background variables (for example Baron–Cohen *et al.* 1985).

In considering why some autistic children might pass, we should bear in mind the possibility that the autistic population, gathered together by applying clinical and fairly gross behavioural criteria, may be composed of neuropsychological subgroups. The more detailed our neuropsychological models become, the greater the opportunity to distinguish between finer and finer patterns of impairment. The logical conclusion of this process is single-case studies. Given the apparently varied biological aetiology for autism (Steffenberg and Gillberg 1991) the existence of subgroups would not be too surprising.

Neuropsychological subgroups might be related to degree of impairment in ToMM. Leslie and Frith (1990) listed three ways in which ToMM might be impaired in autism. There might be impairment in the capacity to deploy informational relations affecting the development of attitude concepts. There might be problems in forming or processing the decoupled component of the M-representation, affecting ToMM's on-board counterfactual ability. Or there might be difficulties with processing M-representations, for example with inferences. Any one of or any combination of these problems could account for what is currently known about autistic impairment in terms of theory of mind.*

As Eisenmajer and Prior (1991) point out, the above set of possibilities implies that ToMM may be impaired by degree. Such might be the case, if different subgroups were impaired in different combinations or if any particular impairment could be more or less severe. In this latter case, ToMM might still be functional in autism, but less so; development might yet proceed normally, but at a slower pace. Thus some autistic children may eventually develop a normal theory of mind. Set against this hypothesis of 'specific delay' (Baron–Cohen 1989), however, are the findings of abnormalities in autistic theory of mind performance discussed earlier (Baron–Cohen 1991; Leslie and Thaiss 1992; Roth and Leslie 1991).

One possibility that makes sense of why there should be a joint minimum verbal mental age and chronological age is that those autistic children who pass theory of mind tasks do so, not by virtue of a delayed but otherwise functional ToMM, but by employing compensatory strategies in the face of a quite dysfunctional ToMM (Frith *et al.* 1991). Such strategies are arrived at by exercising general reasoning abilities, and hence require a quite high level of such abilities, together with extensive practice and general knowledge, and hence do not appear before adolescence. Furthermore, there is no *guarantee* that such strategies will develop, even with high ability and an age of fourteen or fifteen years. All this contrasts with the fast, spontaneous, intuitive development of theory of mind in normal children during the pre-school years, when general knowledge and reasoning ability are limited.

One piece of evidence gives further support to the above possibility. Rivière and Castellanos (1988) found that passing standard false-belief tasks in autistic children was correlated with passing Piagetian seriation and conservation tasks. This is not, of course, true for the normal child, who fails seriation and conservation tasks until about six years or older. The authors suggest that the autistic child achieves success in false belief by using 'logical' or problem-solving skills that are not employed by, or available to, the normal pre-school child with his intuitive grasp of mental states.

What might such compensatory strategies look like? One possibility is that verbally able autistic children are eventually able to exploit the fact that

---

* Perner (this volume, chapter 6) claims that autistic children's success on false photographs and map tasks undermines the idea that they might be impaired in terms of the decoupling component of an M-representation. Perner, however, construes the notion of decoupling too broadly. According to the theory of ToMM, the M-representation is restricted to handling mental states (that is what the M-representation is specialized for). Perner dismisses this as '*ad hoc*'. However, a modular approach to cognitive architecture has been a central and principled feature of this research programme from its inception, hence for example the stress on modelling the specific innate basis of theory of mind and the importance of uncovering a specific deficit in autism. The study of autism is providing further evidence, it seems to us, of the importance of this way of thinking about the structure of cognition. Of course, the theory of the modularity of cognitive systems, though a *Zeitgeist* of cognitive science for more than a decade, might yet turn out to be false; but either way, one can hardly call it '*ad hoc*'.

verbal expressions lay out the structure of propositional attitudes (and other related notions) using a unique verb–argument structure where the object of the verb is another sentence. Learning will be very slow and uncertain because the autistic child lacks the intuitive grasp, which the young normal child has, of the underlying attitude concept encoded by the verb. The problem would then be to map such verbs on to something invariant across the non-linguistic contexts of utterance. Clearly, this invariant will be difficult to find without the availability of hypotheses framed in terms of attitude concepts. Perhaps the autistic child's understanding of representational artefacts comes in handy here. Perhaps the autistic adolescent can sometimes construct a representational theory of mind, exploiting representational and mechanistic analogies.

The notion of compensatory strategies may turn out to be an over-simplification by itself. The very early abnormalities affecting brain-growth, and producing a cascade of effects on development, very probably lead to abnormal arrangements for other parts of the architecture aside from ToMM (in addition to any other initial abnormalities). Some of these departures from the normal developmental trajectory may preclude the possibility that theory of mind development in autism could simply be delayed to adolescence, but otherwise be completely normal. We should expect that the cognitive architecture the autistic adolescent brings to bear on theory of mind problems will be different in a number of ways from that brought by the normal three-year old.

The origins of impairment

The assumption behind the ToMM model of autism is that biological damage, most probably in the pre-or perinatal period, affects the expression or growth of part of core cognitive architecture. Frith *et al.* (1991) have outlined how such damage might have a cascade of consequences affecting the cognitive, behavioural, and social levels. In this framework, ToMM plays a central explanatory role. Understanding the full range of signs and symptoms of this syndrome involves sorting out the basic neural impairments and the cognitive functions they directly affect from the further psychogenetic consequences of cognitive impairment. This is where a cognitive model of the architecture of the black box plays a vital role. However, a complete understanding of autism must include a neuroanatomical model that can be related to the computational properties of the circuits that implement the affected cognitive powers, for example ToMM (Frith and Frith 1991; Johnson and Leslie, in preparation). The biological basis of autism may itself involve a cascade of neurodevelopmental abnormalities. But each link in the chain must be directed toward explaining the specific pattern of spared cognitive abilities and cognitive disabilities found in autism rather than to generalities, such as 'social incompetence', 'affective impairment', or 'attentional problems'. This means putting forward models of cognitive architecture that are capable of accounting for specific patterns of normal development, as well as abnormalities.

The theory of ToMM aims to account for the specific innate basis of our capacity to acquire a theory of mind and for the impaired capacity found in autism. One reason for using scare quotes is to remind ourselves that the term 'theory of mind', (Baron–Cohen *et al*. 1985; Leslie 1987; Premack and Woodruff 1978) refers to our ability to make sense of behaviour in terms of mental states, and not to a concept of *mind per se*. We do not suppose that pre-school children explicitly theorize about *mind*! It makes much more sense, we believe, both from a cognitive and an evolutionary point of view, to suppose that theory of mind is one particular solution to the problem of understanding the behaviour of *agents*. ToMM, embodying this solution, allows the young child to attend to mental states and to their causal role in agents' behaviour. Thus, ToMM is one of the chief components which regulates attention to agents and, in communication, joint attention with agents (Leslie and Happé 1989).

In pursuing the above idea, we may ask: 'Is ToMM itself monolithic?' We doubt it. Leslie (1990, 1991*a*, in preparation; Johnson and Leslie, in preparation) has suggested a major division within ToMM into two subsystems. *System*$_1$ represents the behaviour of agents in terms of the *goal-directedness* of behaviour and the agent's *perceptual* contact with the environment. We believe that *system*$_1$ begins to develop in the first year of life, and is signalled by, for example, request-reaching and the following of eye-gaze. We are also inclined to think that *system*$_1$ is relatively intact in autism, certainly by the time these children are tested on 'theory of mind' tasks. This is indicated, for example, by their use of request-pointing, and their understanding of 'desire' and of the purpose of artefacts. The M-representation, on the other hand, is deployed by *system*$_2$ and begins to develop, slightly later than *system*$_1$, early in the second year, as signalled by the emergence of intentional communication (for example the comprehension of declarative pointing and showing), and then somewhat later by the emergence of pretence. We believe that it is *system*$_2$ which is specifically impaired in autism.

ToMM has been postulated to play a significant role in the control of attention to agents and of joint attention with agents (Leslie and Happé 1989). If ToMM is not monolithic, but comprises subsystems, then it would follow that attention to agents and joint attention would similarly not be monolithic, but fractionated, with a developmental course that reflected (in part) the development of ToMM. The development of joint-attention skills in autism, then, can be relevant to the study of ToMM. Could it be that impairment of agent-directed attention early in the first year of life gives rise to later impairment of ToMM?

There are some findings which indicate abnormalities in joint-attention skills fairly early on in autism, for example in autistic two- and three-year-olds (Baron–Cohen, this volume, Chapter 4; Mundy, *et al*. 1990). We cannot therefore rule out this possibility. However, there is a danger in arguing from impaired performance at two or three years backwards to impaired performance at, say, nine months. In normal development, attention to agents will depend upon a variety of controlling mechanisms which develop in systematic ways over the first two years. In autistic development, it is possible that the initial developmental trajectory of attention to agents is relatively normal until the second year of life, and increasingly abnormal thereafter. Such a model would be supported by evidence that social responding, and in particular, joint attention, is not notably abnormal during the first year of an autistic child's life.

There is, of course, a paucity of evidence on autistic infant development (the rarity of the condition all but precludes prospective studies). However, as Leslie and Frith (1990) noted, what little evidence there is suggests that non-severely retarded autistic infants do not show any glaringly obvious problems with social responding in their first year of life. A virtually unique prospective study by Knobloch and Pasamanick (1975) suggested that the social impairment in autism emerges later than the first year. Of 50 children referred to a paediatric service sometime before the age of three years because they showed a specific social impairment ('failure to regard people as persons'), none who had been referred during their first year turned out to be autistic on follow-up several years later. Indeed, these infants who were seen in their first year did not show social impairment later in life. However, 25 per cent of the children first seen at the age of two and showing social impairment, turned out, on follow-up, to be autistic. Of the older children, aged three years or more, this proportion rose to 80 per cent.

Since then two screening studies have supported this picture. Johnson *et al*. (1992) obtained routine infant screening records for a group of children subsequently diagnosed as autistic and compared them with records for a group of learning-disabled and a group of normal children. The learning-disabled group showed a sharp increase in abnormalities in motor skills, vision, hearing, and social behaviours at twelve months, while the autistic group were indistinguishable from the normal children in all these categories. At eighteen months, however, 57 per cent of the autistic group were reported as having problems in the social category as opposed to 26 per cent of the learning-disabled. The authors conclude that these data are consistent with a late-appearing social impairment in autism. An on-going study (Frith, pers. comm.) asks mothers of autistic children when they first became concerned about the social responsiveness of their child, and specifically whether they had any worries during the first year. Over two-thirds of the mothers report they did not. Baron–Cohen, Allen, and Gillberg (in press) have also found that lack of pretend-play and protodeclarative pointing in high-risk infants at eighteen months is predictive of autism at thirty months (see Baron–Cohen and Howlin, this volume, Chapter 21).

Perhaps most relevant is a prospective study of more than 1200 infants in which a standard questionnaire was given to mothers during routine health-screening (Lister-Brook, pers. comm.). The questionnaire focused upon sociability, gesture, and joint attention at twelve months. This was used to identify an 'at risk' group of infants. Follow-up studies to twelve years of age found 3 children among the original 1200 who were later diagnosed as suffering from 'classical autism' and a further 4 children as having 'autistic-like features'. Of these 4 'autistic features' children, 3 showed joint-attention behaviours at twelve months. Of the 3 classically autistic children, 2 were untestable psychometrically, and were estimated to have IQs, of less than 40. Given this level of retardation, it is not surprising that these two children failed to show joint-attention or pointing behaviours at fourteen months (along with 25 per cent of the whole sample). However, the other classically autistic child had a tested IQ of 61, had not been iden-tified as 'at risk', and had indeed showed both joint-attention and pointing behaviours at eleven months.

Although it is far too early to reach a firm conclusion on the question of whether attention to agents in non-severely retarded autistic babies is abnor-mal from a few months onwards, or whether it only becomes so later in development, what evidence there is fits with the idea that such problems tend to emerge during the second year. Of course, given our assumption that the primary biological damage occurs pre-or perinatally in autism, we might well expect some subtle problems from very early on that are detectable by expert observation (cf. Gillberg *et al.* 1990). But this is not what is demanded either by the classical affective account of autism (for example, Hobson, this volume, Chapter 10), or by a 'general failure to attend to agents' account: in either case, there should be major and obvious abnormalities in social responding from the outset. As far as we can tell, major obvious abnor-malities in social responding do not occur in non-severely retarded autistic babies until the second year of life.

CONCLUSION

We have argued that both normal and abnormal development need to be related to models of the core architecture of cognition. We reviewed some ideas about the structure of this core, specifically the mechanism we call ToMM, and related these ideas to normal development in the domain of social understanding — specifically, of 'theory of mind'. We assumed that neural structures implementing parts of the core cognitive architecture can be damaged biologically. In such cases, cognitive development will tend, in some respects, to be both delayed and abnormal, showing a profile of disabilities and spared abilities that reflects the abnormal cognitive architecture that has resulted developmentally from a damaged core. We argued that autistic development can (partly) be understood in terms of the consequences of a failure of ToMM, or a subsystem thereof, to emerge early and function well or at all. Throughout, our focus has been on what autism can teach us about mechanisms of normal development. Nevertheless, such an approach will inevitably have important implications for understanding autism from a more clinical perspective.

Acknowledgements

We are grateful to Franki Happé, Uta Frith, John Morton, Simon Baron-Cohen, and Helen Tager-Flusberg for very helpful comments on an earlier draft of this chapter.

REFERENCES

APA (American Psychiatric Association) (1987). *Diagnostic and statistical manual of mental disorders*, 3rd edn, revised. The American Psychiatric Association, Washington DC.

Baron–Cohen, S. (1989). The autistic child's theory of mind: a case of specific developmental delay. *Journal of Child Psychology and Psychiatry*, **30**, 285–97.

Baron–Cohen, S. (1991). The development of a theory of mind in autism: deviance and delay? *Psychiatric Clinics of North America*, **14**, 33–51.

Baron–Cohen, S., Allen, J., and Gillberg, C. Can autism be detected at 18 months? The needle, the haystack, and the CHAT. *British Journal of Psychiatry*. (In press.)

Baron–Cohen, S., Leslie, A.M., and Frith, U. (1985). Does the autistic child have a 'theory of mind'? *Cognition*, **21**, 37–46.

Baron–Cohen, S., Leslie, A.M., and Frith, U. (1986). Mechanical, behavioural and Intentional understanding of picture stories in autistic children. *British Journal of Developmental Psychology*, **4**, 113–25.

Charman, T., and Baron–Cohen, S. (1992). Understanding drawings and beliefs: a further test of the metarepresentation theory of autism (Research Note). *Journal of Child Psychology and Psychiatry*, **33**, 1105–12.

Eisenmajer, R., and Prior, M. (1991). Cognitive linguistic correlates of 'theory of mind' ability in autistic children. *British Journal of Developmental Psychology*, **9**, 351–64. Reprinted in *Perspectives on the child's theory of mind* (ed. G.E. Butterworth, P. Harris, A.M. Leslie and H.M. Wellman), Oxford University Press.

Ellis, H.D., and Young, A.W. (1989). Are faces special? In *Handbook of research on face processing* (ed. A.W. Young and H.D. Ellis). Elsevier, Amsterdam.

Flavell, J.H. (1988). The development of children's knowledge about the mind: from cognitive connections to mental representations. In *Developing theories of mind* (ed. J. W. Astington, P. L. Harris and D. R. Olson). Cambridge University Press, New York.

Fodor, J. A. (1983). *The modularity of mind*. MIT Press, Cambridge, Mass.

Forguson, L., and Gopnik, A. (1988). The ontogeny of common sense. In *Developing theories of mind* (ed. J. W. Astington, P. L. Harris and D. R. Olson). Cambridge University Press.

Frith, C. D. and Frith, U. (1991). Elective affinities in schizophrenia and childhood autism. In *Social psychiatry: theory, methodology and practice* (ed. P. E. Bebbington). Transaction Books, New Brunswick, NJ.

Frith, U. (1989). *Autism: explaining the enigma*. Blackwell, Oxford.

Frith, U., Morton, J., and Leslie, A. M. (1991). The cognitive basis of a biological disorder: autism. *Trends in Neurosciences*, **14**, 433–8.

Gillberg, C., Ehlers, S., Schaumann, H., Jakobsson, G., Dahlgren, S, O., and Lindblom, R., *et al.* (1990). Autism under age 3 years: a clinical study of 28 cases referred for autistic symptoms in infancy. *Journal of Child Psychology and Psychiatry*, **31**, 921–34.

Gopnik, A., and Slaughter, V. (1991). Young children's understanding of changes in their mental states. *Child Development*, **62**, 98–110.

Hughes, C., and Russell, J. (unpublished). Autistic children's difficulty with mental disengagement. MS, Department of Psychology, University of Cambridge.

Huttenlocher, J., and Higgins, E. T. (1978). Issues in the study of symbolic development. In *Minnesota Symposia on Child Psychology*, Vol. 11 (ed. W. Collins) Erlbaum, Hillsdale, NJ.

Johnson, M., and Leslie, A. M. (in preparation). A neural circuit analysis of the capacity to acquire a 'theory of mind'. MS. Carnegie–Mellon University.

Johnson, M., Siddons, F., Frith, U., and Morton, J. (1992). Can autism be predicted on the basis of infant screening tests? *Developmental Medicine and Child Neurology*, **34**, 316–20.

Karmiloff-Smith, A. (1988). The child is a theoretician, not an inductivist. *Mind and Language*, **3**, 183–95.

Knobloch, H., and Pasamanick, B. (1975). Some etiologic and prognostic factors in early infantile autism and psychosis. *Pediatrics*, **55**, 182–91.

Leekam, S., and Perner, J. (1991). Does the autistic child have a 'metarepresentational' deficit? *Cognition*, **40**, 203–18.

Leslie, A. M. (1983). Pretend play and representation in the second year of life. Paper presented to BPS Developmental Conference, Oxford.

Leslie, A. M. (1987). Pretense and representation: the origins of 'theory of mind'. *Psychological Review*, **94**, 412–26.

Leslie, A. M. (1988a). The necessity of illusion: perception and thought in infancy. In *Thought without language* (ed. L. Weiskrantz). Oxford University Press.

Leslie, A. M. (1988b). Some implications of pretence for mechanisms underlying the child's theory of mind. In *Developing theories of mind* (ed. J. Astington, P. Harris and D. Olson). Cambridge University Press.

Leslie, A. M. (1990), Domain specific mechanisms in theory of mind. Paper presented to International Conference on Domain Specificity in Cognition and Culture, October 1990, University of Michigan, Ann Arbor.

Leslie, A. M. (1991a). An information processing approach to the child's theory of mind. Paper presented to Society for Research in Child Development Biennial Conference, April 1991, Seattle, WA.

Leslie, A.M. (1991*b*). The theory of mind impairment in autism: evidence for a modular mechanism of development? In *Natural theories of mind*. (ed. A. Whiten). Blackwell, Oxford.

Leslie, A.M. (in preparation). The theory of ToMM. MS MRC Cognitive Development Unit, University of London.

Leslie, A.M., and Frith, U. (1988). Autistic children's understanding of seeing, knowing and believing. *British Journal of Developmental Psychology*, **6**, 315–24.

Leslie, A.M., and Frith, U. (1990). Prospects for a cognitive neuropsychology of autism: Hobson's choice. *Psychological Review*, **97**, 122–31.

Leslie, A.M., and Thaiss, L. (1992). Domain specificity in conceptual development: neuropsychological evidence from autism. *Cognition*, **43**, 225–51.

Marcus, G.F., Ullman, M., Pinker, S., Hollander, M., Rosen, T.J., and Xu, F. (1990) *Over-regularization*, Occasional Paper No. 41. Massachusetts Institute of Technology, Cambridge, Mass.

Mitchell, P., and Lacohée, H. (1991). Children's early understanding of false belief. *Cognition*, **39**, 107–27.

Mundy, P., Sigman, M., and Kasari, C. (1990). A longitudinal study of joint attention and language development in autistic children. *Journal of Autism and Developmental Disorders*, **20**, 115–28.

Ozonoff, S., Pennington, B.F., and Rogers, S.J. (1991). Executive function deficits in high-functioning autistic individuals: relationship to theory of mind. *Journal of Child Psychology and Psychiatry*, **32**, 1081–1106.

Perner, J. (1988). Developing semantics for theories of mind: from propositional attitudes to mental representation. In *Developing theories of mind* (ed. J. Astington, P.L. Harris and D. Olson). Cambridge University Press.

Perner, J. (1991). *Understanding the representational mind*. MIT Press, Cambridge, Mass.

Perner, J., Frith, U., Leslie, A.M., and Leekam, S.R. (1989). Exploration of the autistic child's theory of mind: knowledge, belief and communication. *Child Development*, **60**, 689–700.

Pinker, S., and Prince, A. (1988). On language and connectionism: analysis of a parallel distributed processing model of language acquisition. *Cognition*, **29**, 73–193.

Premack, D., and Woodruff, G. (1978). Does the chimpanzee have a theory of mind? *The Behavioral and Brain Sciences*, **4**, 515–26.

Prior, M.R., and Hoffman, W. (1990). Brief report: neuropsychological testing of autistic children through an exploration with frontal lobe tests. *Journal of Autism and Developmental Disorders*, **20**, 581–90.

Pylyshyn, Z.W. (1978). When is attribution of beliefs justified? *Behavioral and Brain Sciences*, **1**, 592–3.

Rivière, A., and Castellanos, J. (1988). Autismo y teoría de la mente. Paper presented to IV Congreso Nacional AETAPI, Cadiz.

Roth, D. (in preparation). Beliefs about false beliefs. MS, Tel Aviv University.

Roth, D., and Leslie, A.M. (1991). The recognition of attitude conveyed by utterance: a study of pre-school and autistic children. *British Journal of Develop-*

mental Psychology, **9**, 315–30. Reprinted in *Perspectives on the child's theory of mind* (ed. G. E. Butterworth, P. L. Harris, A. M. Leslie and H. M. Wellman). Oxford University Press, 1991.

Russell, J., Mauthner, N., Sharpe, S., and Tidswell, T. (1991). The 'windows task' as a measure of strategic deception in pre-schoolers and autistic subjects. *British Journal of Developmental Psychology*, **9**, 331–49. Reprinted in *Perspectives on the child's theory of mind* (ed. G. E. Butterworth, P. L. Harris, A. M. Leslie and H. M. Wellman). Oxford University Press, 1991.

Shallice, T. (1988). *From neuropsychology to mental structure.* Cambridge University Press.

Sodian, B., and Frith, U. (1992). Deception and sabotage in autistic, retarded and normal children. *Journal of Child Psychology and Psychiatry*, **33**, 591–605.

Steffenberg, S., and Gillberg, C. (1989). The aetiology of autism. In *Diagnosis and treatment of autism* (ed. C. Gillberg). Plenum, New York.

Strauss, S. (1982). *U-shaped behavioural growth.* Academic Press, New York.

Wellman, H. M. (1990). *The child's theory of mind.* MIT Press, Cambridge, Mass.

Wellman, H. M., and Bartsch, K. (1988). Young children's reasoning about beliefs. *Cognition*, **30**, 239–77.

Wilson, B. A. (1987). *Rehabilitation of memory.* Guilford Press, New York.

Wimmer, H., and Hartl, M. (1991). Against the Cartesian view of mind: young children's difficulty with own false beliefs. *British Journal of Developmental Psychology*, **9**, 125–38. Reprinted in *Perspectives on the Child's theory of mind* (ed. G. E. Butterworth, P. L. Harris, A. M. Leslie and H. M. Wellman). Oxford University Press, 1991.

Zaitchik, D. (1990). When representations conflict with reality: the pre-schooler's problem with false beliefs and 'false' photographs. *Cognition*, **35**, 41–68.

Zaitchik, D. (1991). Is only seeing really believing? Sources of true belief in the false belief task. *Cognitive Development*, **6**, 91–103.

6

The theory of mind deficit in autism: rethinking the metarepresentation theory

JOSEF PERNER

Before embarking on a rethink of the 'metarepresentation theory' I need first to outline this theory. This is not easy because the topic is fraught with terminological ambiguities arising from the everyday meaning of the word 'representation' and—as Fig. 6.1 illustrates—the changes in meaning effected by the prefix 'meta-'. In particular, the term 'metarepresentation' has taken on a different meaning in Leslie's (1987) hands than the meaning it had originally been given by Pylyshyn (1978).

The term 'metarepresentation' was first used in this context by Pylyshyn (1978, p. 593) in a comment on Premack and Woodruff's (1978) 'Does the chimpanzee have a theory of mind?' When discussing potential experimental demonstrations of the chimpanzee's ability to represent another organism's beliefs mentally, Pylyshyn spoke of a 'metarepresentational capacity' which requires 'the ability to represent the representational relation itself'. This is a very clear definition, as can be seen by closer inspection of the concept of representation.

REPRESENTATION

In cognitive science 'representation' is used in a wide sense, encompassing pictures, linguistic expressions, and even the mind. One common feature of all these *representational media* is that they are *about* something. For instance, the photo in your passport is *of* you (it tells us something *about* you). Similarly when I think of you my thoughts are *about* you. In all these cases we can discern three important parts: something [representational medium] which represents [representational relation] something else [representational content], summarized in the following schema:

medium ——[representational relation]——→ content

for example:

piece of paper ——[being a photo of]——→ you (as shown)

Fig. 6.1. Metaconfusions.

In its most common usage the word 'representation' refers to the *representational medium*, that is, when talking about your passport mugshot I mean by 'representation' the glossy piece in your passport, but not you [representational content] or the representational relationship.* Unfortunately, in the case of 'representation', natural use is not always so clear-cut. This is usually not detrimental to everyday use, since context disambiguates sufficiently. For theoretical discussion, however, one needs to be aware of these different uses to avoid potential confusion.

---

* At this point the cautious reader may feel tricked into a commitment bearing unwanted fruit later. The reader might object that a glossy piece of paper by itself is not a representation of anything. That is true, of course. That piece of paper would not be a representation of you if it did not stand in a representational relation to you. Nevertheless, the word 'representation' does not refer to the representational relationship, but to the medium that happens to bear such a relationship to you.

The same terminological problem occurs for simpler relations, like 'kicking'. If I kick you (stand in a kicking relation to you) then someone can refer to me as 'the kicker'. As in the case of the photo it is true to say that I would not be a kicker unless I bore such a relationship to you. Yet the word 'kicker' is used to refer to me (who happens to stand in such a relationship to you) rather than to the action of kicking. In this case the terminology seems perfectly clear. There are three different words: 'kicker', 'kicking', and 'kicked', and they are used to refer to three different conceptual elements involved in a kicking event, just as 'representation' ('signifier'), 'representing' ('signifying') and 'represented' ('signified') refer to the three conceptual elements when something represents (symbolizes) something else.

When representations of non-existing entities or events are involved, the word 'representation' can be loosely used to refer to what is represented (content). If one is not alerted to the fact that the word is used here in a different sense, an equivocation of representational medium and content results. This equivocation is particularly tempting in the case of something non-existent (for instance a *unicorn*), because the represented has no independent existence (by definition). So when asked what a picture of a unicorn represents, one might answer: 'nothing'. But that won't do, because if it represents nothing the picture would not be a representation at all. So it must represent something, namely a unicorn. One could ask another embarrassing question, namely where this thing represented is to be found. Now one might answer, 'Nowhere really, it's just in this picture, it's *just a representation.*' This, in fact, appeared to be the answer that David Hume gave to such questions about mental representations like ideas, thoughts, and the like, and for which he was reprimanded by his contemporary Thomas Reid (see Lehrer 1986).

We can use a typical philosopher' s ploy to make clear why Reid raised objections to equivocating the unicorn (represented) with its representation. If the two were the same then one could substitute one for the other in any sentence without changing the meaning or truth-value of that sentence. At first, this seems defensible when imagining a person pointing at a picture of a unicorn, saying, 'This is a unicorn.' True! If we now substitute 'representation of a unicorn' for 'unicorn', the resulting statement 'This is a representation of a unicorn' would still be true. However, this substitution is only apparently valid, since with the substitution a tacit shift in referent occurred from the content of the representation (unicorn) to the representational medium (picture) itself.* That two expressions are not, true synonyms becomes apparent in another round of substitution. The resulting statement, 'This is a representation of a representation of a unicorn' does not apply any more to the picture of a unicorn. We would expect to see something like a *photograph of a painting of a unicorn*, and not just a painting of a unicorn.

I have dwelled rather extensively on this ambiguity in the term 'representation' because it helps us understand some of the confusions in the use of the word 'metarepresentation'.

METAREPRESENTATION

In our use of the term 'metarepresentation' we should adhere to Pylyshyn's (1978, p. 593) implicit definition as *representing the representational relation itself*, because his definition is unambiguous, and he was one of the first

---

* This ambiguity in reference has so elegantly been illustrated by Magritte's drawing of a pipe with the words *Ce n'est pas une pipe* written underneath it.

to use this term in cognitive science. More loosely speaking, we can define metarepresentation as a *representation of a representation*, as long as it is clear that we always use 'representation' as unambiguously referring to representational medium, and not to representational content.

As a clear instance of metarepresentation we can refer to the above-mentioned photograph of a painting of a unicorn. It is a case of metarepresentation since it involves a representation [photo] of a representation [painting] of something [a unicorn]. More explicitly, it satisfies Pylyshyn's criterion because, by depicting a painting as a painting of a unicorn, the photo represents the fact that the painting stands in a *representational relation* to a unicorn.

Importantly, if in our definition 'representation of a representation' we did not adhere to the rule that 'representation' must be used to refer to a representational medium only, then all sorts of pictures would turn into metarepresentations. For instance, if we allowed the ambiguous use of 'representation' to refer to non-existent entities depicted in a picture, then the plain picture of a unicorn could be called a 'metarepresentation' on the grounds that it is a representation [medium] showing a unicorn which, because it doesn't exist, is a mere 'representation' (in the sense of representational content). By equivocating the two distinct senses of 'representational' (as medium and as content) the photo turns into a 'representation of a representation'. Clearly, a picture of a unicorn is not what one would ordinarily call a 'metarepresentation', and certainly not something that would qualify as such under Pylyshyn's original definition.

I put so much emphasis on pointing out these potential misapplications of the term 'metarepresentation' because—as I have elaborated elsewhere, for example in Perner (1991), Note 2.1 and p. 175—they have made their way in one guise or other into the developmental literature, and created avoidable confusion on the question of when children become able to metarepresent. And it helps being aware of these dangers when evaluating Leslie's (1987) analysis of pretend-play, and trying to understand his tendency to equate decoupled representations with metarepresentations.

DECOUPLING

Leslie (1987) made the valid point that a pretending child must have a representational mechanism that can do more than just form faithful reflections of external surroundings, that is, the infant needs more than this basic representational skill of 'representing aspects of the world in an accurate, faithful, and literal way' which Leslie (p. 414) calls a 'capacity for *primary representation*'. One can easily see why primary representations are not sufficient to sustain pretence.

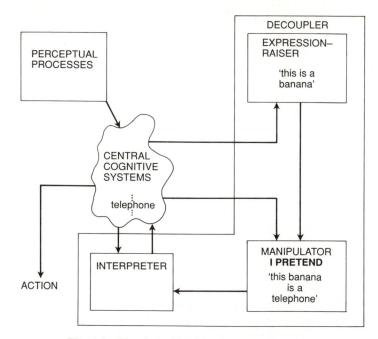

Fig. 6.2. The decoupler model of pretence.

The general assumption in cognitive psychology is that one cannot engage in meaningful action without an internal representation guiding such behaviour. So, for instance, a pretending child who knowingly engages in mock drinking behaviour with an empty cup must be entertaining a mental representation such as 'This cup is full' in addition to the primary representation 'This cup is empty', because the primary representation would not sustain drinking behaviour.

It is clear that 'This cup is full' cannot simply be added in the mind next to the primary representation, since it would result in what Leslie calls *representational abuse*. The child would be confused about whether the cup is full or empty. Drinking behaviour engendered by such confusion would not be pretence but a mere mistake. Hence, representations necessary to sustain pretence must be cordoned off (quarantined) from primary representations. This and some other necessary features for pretence, Leslie suggests, is achieved by a *decoupling* mechanism.

The decoupler in Fig. 6.2 (reprinted from Leslie 1987, Fig. 2) consists of three parts. The *expression raiser* copies primary representations into decoupling marks (which quarantine this new kind of representation off from the primary representations). The *manipulator* changes the copy of primary representations within decoupling marks, for example, from 'This cup is empty' to 'This cup is full', and represents that this changed representation is to be used for pretence purposes: *I pretend 'this cup is full'*. Finally the *interpreter* is necessary for co-ordinating the pretend representation with primary representations in order to ensure sensible execution of the pretend action.

I want to use this breakdown of the decoupler into three parts to ask what the predicted consequences would be if any one of these parts broke down, and then judge how these predictions might fit what we know about autistic children. However, before I engage in this discussion let me return to the terminological question of whether the representations created by the decoupler are 'metarepresentations' in Pylyshyn's sense.

DECOUPLED = METAREPRESENTATIONAL?

The first step in the decoupling process [expression raiser] produces copies of primary representations which are put inside decoupling marks (for which Leslie happened to use the inverted commas [",") conventionally used as quotation marks. To avoid confusion I will use ['] as quotation marks. The decoupling marks — Leslie claims — have three important consequences. I agree on two of them, but need to question the third.

(1) The quotation marks demarcate (*quarantine*) the copied representation from primary representations. This is obviously necessary for avoiding representational confusion when these copied representations are to be subsequently changed.

(2) The copied representations create problems of *logical opacity*. This follows from the fact that the decoupling marks separate the decoupled representation from the primary representation of reality. In other words the decoupling marks create different *contexts* which in turn create logical opacity — for example certain inferences and substitutions of expressions which are logically permissible within a single context become logically invalid across contexts. One should point out that this logical property is not created by decoupling alone. It is a much more pervasive phenomenon. For instance, the same problem occurs when temporal contexts are to be distinguished (see Perner (1988) pp. 159–61 and (1991) Chap. 3 for a more extensive treatment of this issue).

(3) Leslie also stipulates that decoupled representations be 'metarepresentations': 'Pretend representations . . . are in effect not representations of the world but representations of representations. For this reason I shall call them second order or, borrowing a term from Pylyshyn (1978), metarepresentations' (Leslie 1987, p. 417). Why should that be?

The main theoretical motivation for positing the decoupler was to explain how representational abuse could be avoided. This purpose is served by creating different contexts. But such contexts do not create metarepresentations. Why would one assume they did? The fact that pretend contexts create logical opacity cannot be a good reason, because — as was indicated above — temporal contexts also create opacity, but one would not want to speak of representations of past events as being metarepresentations of the current state of the world. The implications would be that events in the past represent present events — an absurd implication.

Pretend contexts differ from temporal contexts in that they are not created by past events in the world but by copying primary representations of the current world. However, this fact cannot be used to justify talk of 'metarepresentation', since copying does not create metarepresentations. If it did then any simple photograph, which is a copy of the negative on the film, would then be a metaphoto, or any message passed by several messengers would become a meta . . . metastatement — again, absurd consequences.

A third possible reason for thinking that decoupled representations are metarepresentations is indicated in the quote given above from Leslie himself. Since they are not *representations of the world* (which is true) they must be *representations of representations*. That conclusion is reminiscent of the equivocation of two different senses of 'representation' pointed out earlier in the example of a unicorn. Since the unicorn is non-existent it is called a 'mere representation' (in the sense of a mere representational content). By equivocating the two senses of 'representation' one could call a picture of a unicorn a 'metarepresentation', since the picture (medium) represents a mere representation [content]. If that is how Leslie used the term 'metarepresentation' in this context, then it does *not* conform to Pylyshyn's use (from which Leslie borrowed the term), and is unacceptable because it is based on an illicit equivocation of two different senses of the word 'representation'.

Finally, a fourth reason for thinking that Leslie's decoupler creates metarepresentations might be that the expression created by the manipulator, *I pretend 'this cup is full'* is referred to as a description of the child's own mental state: 'Thus, the representations underlying the child's pretense are equivalent to *internal reports* of the child's own mental state' (Leslie 1988, p. 28). One might or might not agree with this statement, depending on what is meant by 'mental'. The danger, though, is that 'mental' implies *mental representation*, in which case there would be metarepresentation. However, as I have argued so far, the above expression created by the manipulator does not represent a relationship to a representation, but to a state of affairs that happens not to be true. So, if one is content to mean by 'mental' a relationship to a non-factual state of affairs (for which there is some justification — see Perner (1991), Chapter 5), then one can agree that pretense is based on the child's report on her own *mental* state, as long as it is clear that that does not imply a *representational mental state*. But then, if it does not imply that the child is representing a representational state, again, there is no reason to speak of metarepresentation.

Since I cannot see which representation decoupled representations are supposed to represent, I can see no justification for calling them 'metarepresentations.' Leslie's theory that autistic children suffer from an impairment in their decoupling mechanism, therefore, would be better referred to as the 'decoupling [deficit] theory', rather than the 'metarepresentation [deficit] theory'. Later I will also explore the possibility that autistic children suffer from a genuine 'metarepresentation deficit' that is, that they cannot mentally represent that something is a representation. For now, I explore further the consequences of an inability to decouple, which would result in a much more profound deficit than an inability to understand representation (representing that something is a representation = metarepresentation).

ARE AUTISTIC CHILDREN DECOUPLING-IMPAIRED?

I take it that the objective is not to explain the autistic syndrome as a whole, but to explain what is specific to autism (see for example, Sigman *et al*. 1987, p. 104). The target group of interest for this purpose are those *intelligent* autistic children with an IQ of 70 or above and/or with a verbal mental age of at least three or four years. It is this subgroup of autistic children who are typically used in 'theory of mind' experiments. It has become clear that these children do not lack a theory of mind, simpliciter. Rather, their impairment is graded. One key finding is that only about 20–25 per cent of these children show understanding of false belief. So I take those intelligent autistic children who fail false belief as the target group whose problems the decoupling-deficit theory is supposed to explain.

To evaluate the decoupling-deficit theory in a systematic fashion it helps to distinguish different versions of the theory. The strong version would claim that autistic children lack a decoupler, and with it any one of its subcomponents. This version is too radical, since, as I will argue, autistic children do have some — if not all — of the subprocesses intact. A weaker 'medium version' of the theory would be that autistic children lack some — at least one — of the components of the decoupler. Finally, the 'weak version' assumes that all the component processes are available, but that the co-ordination of at least some of them is impossible.

Since the strong version falls with any weaker version, let me start with the medium version, going one by one through the functions served by Leslie's (1988, Fig. 2) decoupler, and showing that each function is operative in the target group of intelligent autistic children.

Expression raiser

The expression raiser is responsible for two distinct actions. It is supposed to provide copies of primary representations and to put these copies within decoupling marks [","]. Let me first consider the possibility that autistic children lack decoupling marks, and then the possibility that they cannot make copies of primary representations.

Leslie's main — and well-motivated — justification for decoupling marks is that they *quarantine* decoupled representations from primary representations, which is necessary for preventing representational confusion (abuse). As I have pointed out, such marks — or some other mechanism which serves such a quarantining function that puts representations into different contexts — are also needed for undecoupled representations, such as those that require temporal contexts. Informative statements about different times are, presumably, primary representations in Leslie's sense, since they represent accurately, faithfully, and literally how the world is and was. But there needs to be something like decoupling marks to separate what was from what is, and this separation creates all the problems of logical opacity — the alleged hallmark of pretence.

One early indicator of children's ability to remember a past state of affairs is their success on Piaget's (1937/1954) invisible displacement task or logical search task (Haake and Somerville 1985): bead disappears in experimenter's hand, hand goes under a cloth and re-emerges again. Child searches for the bead in the hand and finds it empty. By about eighteen months infants realize that the next best place to search is under the cloth. I have argued (Perner 1991, Chap. 3) that this requires representation of where the bead was before, namely inside the hand under the cloth. And the representation of this prior location needs to be separated by some context (decoupling) markers from the representation of where the child assumed the bead was (inside the hand outside the cloth) before the discovery that the hand was empty, if there is not to be total confusion about where the bead was at this point in time.*

If autistic children had no way of marking different temporal contexts (quarantining), then they could not solve invisible displacement tasks. However, several studies have found that they are not delayed on such object permanence tasks compared to children matched for mental age (Sigman and Ungerer 1981; Lancy and Goldstein 1982).

Another skill that requires 'decoupling' marks to separate different contexts (Perner 1991, Chap. 4), and which develops at the same age of about $1\frac{1}{2}$ years, is the ability to recognize oneself in the mirror or in photos (Lewis and Brooks–Gunn 1979). Decoupling is necessary in order to keep

---

* In fact, the assumption that these temporal context-markers fade quickly in the younger infant provides a good explanation for the so-called A-notB error (Perner 1991, pp. 236–8, Note 10.1)

the representation of oneself in the mirror (photo) separate from the representation of oneself in reality, or else one would be confused about where one really is: in front of the mirror or in the mirror. Again, autistic children of all shades seem capable of such self-recognition in mirrors (Neumann and Hill 1978; Sigman *et al.* 1987, p. 115). I take it that autistic children must have some mechanism that serves the function of decoupling marks to quarantine certain representations from representations of immediate reality.

The other function of the expression raiser is to form copies of primary representations. This function is not interesting in itself unless it is paired with the ability to then change these copies so as to represent something different from the world which gave rise to the primary representations. Changing copies of primary representations is one of the functions of the *manipulator*.

Manipulator

The manipulator — the second component in Leslie's decoupler — has two discernible functions. One is to manipulate decoupled representations, that is, to fiddle with the representations within decoupling marks — for example, to change it to 'this cup is full'. We could call this function: *manipulating the content* (of decoupled expressions). The other function is to put the decoupled expression into context, which in the case of pretend-play means marking it as a pretend context. In Leslie's scheme this is done by representing a relationship between the pretending person (the self or someone else) and the decoupled expression, for example: I pretend 'this cup is full'. Such constructions are commonly known as *propositional attitudes* because, in this case, the expression expresses the child's pretend attitude towards the proposition, 'This cup is full'.*

Again, let us check whether autistic children may lack a mechanism subserving these functions. If autistic children lacked the ability to manipulate decoupled expressions it would be impossible for them to think of hypothetical

---

* It is questionable whether Leslie would or should agree to calling this a propositional attitude. The term is appropriate if one assumes that decoupling marks just make explicit what our syntactic analysis assumes implicitly, namely that the component expression 'This cup is empty' is not a description of external reality. We know this even without any decoupling marks because the expression is syntactically embedded in the pretend sentence. If that is the only function of the quotation marks — and it is the only function explicitly motivated by Leslie's discussion about the need for avoiding representational abuse through quarantine — then we can speak of a propositional attitude.

However, the use of inverted commas as decoupling marks and the talk about being 'metarepresentational' suggests something different. Namely that 'This cup is full' does not express a proposition but expresses a (mis)-representation of the proposition 'This cup is empty'. In that case we are not faced with a propositional attitude but a *'representational attitude'*. However, if that is what Leslie had intended he would need to give a better justification — beyond the need for quarantine — for why infants should take such a sophisticated view of their pretend activity.

possibilities. Their thoughts would only be copies of external reality, and they would be incapable of reasoning about hypothetical physical events not directly perceived. They therefore could not have performed so well — even better than normal children — on the tasks used by Baron–Cohen, *et al.* (1986, Fig. 2: 'mechanical'), where they had to reason out the correct sequence of drawings depicting a physical event.*

The other function of the manipulator is to put the decoupled expression into context (for example, 'I pretend "this cup is full"'). This function in itself can be further divided into two distinguishable subfunctions. One of these is to provide the decoupled expression with a label, so that the system knows how and when to use the decoupled expression. For this purpose it would suffice to represent: For pretence: 'This cup is full'. In fact, for most forms of early pretence this would seem perfectly satisfactory, since there is no need to distinguish my pretend scenario from somebody else's. There is no need for that even in joint pretence, at which children from two years on are increasingly proficient (see for example, Dunn and Dale 1984), because both participants are co-operating within the same pretend scenario.

Concerning autistic children the question is whether they could be lacking this function. The answer must be 'no', since that labelling function is also required in the case of temporal contexts which are needed to understand invisible displacements or logical search tasks. And as we have seen, autistic children are not specifically impaired on these tasks. This labelling function is also required for keeping distinct the sensible and not so sensible

---

* This argument is based on the assumption that ordering pictures into the right sequence requires the trying out of possibilities as they come to mind and evaluating them against alternatives, for example: 'The rock drops into the pond and then goes halfway up the hill again' is a less plausible sequence than 'the rock rolls first halfway down the hill and then falls into the water.' Yet there is the possibility that autistic children may not have reasoned this way at all. They may have simply looked for a picture most similar to the one first presented (for example, everything is the same except that the circle [the rock] is slightly removed from its original location), then for the next most similar one, and so on. Judging from the three examples shown (Baron–Cohen *et al.* 1986, Fig. 1) this strategy would yield a perfect ordering for 'mechanical' problems, but a less perfect one for 'behavioural' (about 50 per cent success), and would give hardly any guidance for the 'intentional' problem. These percentages provide a pretty close match to autistic children's actual performance on these types of tasks. Hence this possibility cannot be ruled out.

Baron–Cohen (personal communication) argues that this is unlikely, since some autistic children gave 'physicalistic justifications' (Table 4) of their sequences. However, since these justifications were given after the series had been assembled, it could still be the case that they were descriptions of the final sequence or single pictures rather than reflections of having excluded implausible sequences. That is, these justifications were of the kind 'the rock hit the man', or 'the man is bleeding because the rock hit him', but in themselves do not give evidence for counterfactual reasoning about physical events of the kind: 'This picture could not have come last because then he would have started bleeding before the rock hit him.'

I just want to point out that in this case the interpretation of the picture-sequencing data given by the authors also becomes irrelevant, since children's success in ordering the mechanical and to some degree behavioural picture sequences would have nothing to do with the fact that these sequences involve purely mechanical or behavioural events.

mechanical-event sequences that one has to consider in the picture sequencing tasks by Baron–Cohen *et al.* (1986). And autistic children excel on these tasks.

The other subfunction of putting decoupled expressions into context is to *personalize* the context. This, of course, is not necessary for understanding time, and I can't see why it should be for most of children's pretence either.* However, it is clearly relevant for understanding false belief, since in that case it is crucial to understand that it is the other person's belief that the cup is full, when I know that it is empty. It is this function of the manipulator that requires the construction of *propositional attitudes*. Recently Leslie and Frith (1990) and Leslie and Thaiss (1992) have emphasized that it is this specific function of the decoupler which autistic children may be lacking.

In support of such a propositional-attitude deficit one could cite observations of autistic children's spontaneous language-production. They hardly ever talk about mental states (Tager–Flusberg 1989, this volume, Chapter 7), which are the most frequent cases where propositional-attitude constructions are used. However, even those children whose transcripts do not show any use of cognitive mental-state verbs like 'know', 'think', etc. give quite elaborate propositional-attitude constructions when talking about speech acts for example, Roger [4 years 4 months]: 'Jason, I told you to watch TV38'; 'Oh, you wanna tell the dentist how you cracked your tooth'; or Jack [at 7 years and 8 months]: 'Snoopy said she fuzzy' (Tager–Flusberg, e-mail message of 12 July 1991).

There is also increasing experimental evidence that children understand people being linked to propositions by perception and volition. One could cite evidence from Tan and Harris (1991) that autistic children understand quite well that when two people are facing each other with two objects between them then one of them sees object A in front of object B, while the other sees B in front of A. Here seeing must be understood as a relation to a proposition, i.e.: John sees that *A is in front of B*.†

---

* It is not necessary to personalize pretence for the typical joint pretence, since one and the same pretend scenario is shared among all participants. However, Norman Freeman (personal communication) reported that by three years children understand that different people can follow different scenarios. For instance, they understand that if Ann pretends that the ball of clay is an apple and Jack pretends it's a plum, then when in their joint pretence they turn the ball of clay into a pie, this pie will be an apple-pie for Ann but a plum-pie for Jack. To follow such 'parallel pretence' children must be able to personalize pretend scenarios: 'Ann pretends "this is an apple-pie"', while 'Jack pretends "this is a plum-pie"'.

† Unfortunately, these data need to be interpreted with caution. Despite the authors' claim that their task assesses Level 2 perspective-taking there is the possibility that it does not assess perspective-taking at all. Not implausibly, the propositional attitude construction, 'Which object would John say is in front?' may be glossed as the simpler question, 'Which object is (directly) in front of John?', and the whole task can be 'solved' by representing: 'A in front of John and B in front of Mary'. No propositional attitudes need be understood. This, by the way, could explain why performance on this alleged Level 2 task tended to be better than performance on the Level 1 task.

Other evidence against a propositional-attitude deficit might be the fact that autistic children answer propositional attitude questions about their past mental states with great accuracy, as for instance 'Which box did you want to open?' and 'Did you see him upside down or sitting up?' (Baron–Cohen, 1991*b*). And occasionally they use propositional-attitude constructions themselves in their talk about emotions, for example, 'She doesn't like to be alone' (Baron–Cohen 1991*a*, p. 19).

Interpreter

The third and last component in Leslie's decoupler serves to relate the product of the creative thinking efforts sustained by the expression raiser and manipulator back to reality. This, I guess, is necessary for all the tasks discussed above. For instance it is required in the picture-story tasks (Baron–Cohen *et al.* 1986). After the child has figured out — by manipulating her decoupled mental representations appropriately — which sequence is physically possible, she has to then be able to translate her reasoning result into the appropriate real-world action of putting the cards into the corresponding order. Since autistic children are extremely good at some of these tasks their impairment cannot be located in the interpreter — at least not at the present level of analysis.

The discussion so far has largely produced evidence that autistic children probably have all the individual functions served by Leslie's decoupler intact. In the course of the discussion these functions can be listed as follows:

Expression raiser
    Copying
    Quarantining (contextualization through decoupling marks)

Manipulator
    Manipulation of content (of decoupled expression)
    Labelling contexts
    Personalizing contexts (propositional attitudes)

Interpreter (not further investigated).

Since autistic children seem to have all these functions intact, we can reject the 'medium-strength version' of the decoupling-deficit theory of autism. This, however, still leaves the *weak version* of the theory as a viable possibility, namely that the deficit is one of co-ordinating different component functions of the decoupler. According to this version autistic children should fail on those tasks which require all (or at least more than one) of the component functions of the decoupler.

Coordinating decoupling components

Presumably the whole decoupling apparatus with all its functions is required in pretend play. After all, that is what the theory was designed to explain. Are autistic children capable of pretence or not?

Pretence. It is in fact one of the defining features of autism that there is an 'absence of imaginative activity, such as playacting of adult roles, fantasy characters, or animals; lack of interest in stories about imaginary events' (DSM IIIR (APA 1987) p. 39, B(3)). However, it is not clear whether autistic children are incapable of understanding pretence — as the decoupling-deficit theory suggests — or whether they are — for some unknown reason — reluctant to engage in such activity. Lewis and Boucher (1988) report the usual almost total lack of spontaneous pretence not only in autistic children but also in children with moderate learning difficulties. Furthermore, autistic children, when handed a car and some other toy and asked [elicited condition]: 'Show me what you can do with these?', were as good as normally developing younger children and children with moderate learning difficulty at producing what looks like genuine pretence — for example holding a straw upright, pushing car around it, and saying 'Lamppost'. Boucher and Lewis (1990), in response to Baron–Cohen's (1990) criticism of their study, point out that every one of their autistic children showed at least one such instance of pretence.

One simple interpretation of these data is that with age children become reluctant to engage in pretence unless explicitly asked to do so. Hence the normal children in the study, who were all younger than $5\frac{1}{2}$ years old, spontaneously engaged in such play, while the autistic children, who were between $6\frac{1}{2}$ and 16 years old, and those with learning difficulties, who were between about $5\frac{3}{4}$ and 11 years old, felt inhibited.

The autistic children's reluctance could be motivational or could be a cognitive inability. One way to help decide this issue is to assess the contingency between spontaneous pretence and children's ability to pass the false-belief task. If all the ones who pass false belief engage in spontaneous pretence, then one would be inclined to link the lack of spontaneity to a cognitive problem. Whereas if the children who pass false belief are particularly reluctant to engage in unsolicited pretence, then a motivational/embarrassment explanation seems to be called for.

Baron–Cohen (1991*b*) has relevant data, unfortunately not on spontaneous pretence but only on instructed pretence (for example: 'Can you pretend there is orange juice in this cup?'). Nevertheless, even with such explicit instructions many autistic children needed further prompts before they engaged in the appropriate pretend action. What is of interest here (Baron–Cohen, personal communication) is that none of the children who passed false belief (i.e., remembered their own prior false belief) needed any special prompting. Since the observed reluctance does not occur in the able but only in the cognitively more impaired children, we may tentatively conclude that the reluctance is due to cognitive impairment rather than to embarrassment.

However, the fact that all of those children who failed false belief could be brought to engage in pretence suggests that these children do not lack the decoupling mechanism as such, but rather lack the ease of normal children in employing this mechanism (see also Harris, this volume, chapter 11).*

Photo task. Another task on which autistic children do very well and which — to my mind — requires the larger part of all the functions of the decoupling mechanisms is Zaitchik's (1990) photo task. This is particularly interesting since Zaitchik modelled her task closely on Wimmer and Perner's (1983) false-belief task, with which autistic children have such notorious problems. In Leekam and Perner's (1991) version of Zaitchik's task a Polaroid picture is taken of a doll clad in blue. While the photograph develops, the doll's clothing is changed to green. Before the developed photo was shown to the children they were asked 'What colour is the doll in the picture?' The ability to answer this question correctly was compared to their ability to answer a very similar question about the false belief about the doll's colour of dress of a person who, like the camera, saw the doll dressed in blue, but missed the change to green. For normal three- and four-year-old children the two questions were about equally difficult; but autistic children showing the usual difficulty with belief had hardly any problems answering the question about the photograph. These results have been successfully replicated by Leslie and Thaiss (1992); and Charman and Baron–Cohen (1992) obtained comparable results by using drawings instead of photos.

Autistic children's ability to make correct predictions of how things are in the photo shows that their expression raiser must be intact. In Leslie's notation they must have formed a mental representation of the form: The photo shows 'the doll is wearing blue'. This cannot be a primary representation, since they have not yet looked at the photo. They must have copied the primary representation, 'The doll is wearing blue' that represented the situation at the time the photo was being taken. To safeguard against confusions with later changes in reality this copy has to be quarantined, i.e., put inside decoupling marks: '*The doll is wearing blue*'. These are the two functions the expression raiser is responsible for. Children also must have an — at least partially — functioning manipulator, as it is the function of the manipulator

---

* This kind of cognitive difficulty in switching from an established response pattern (treat empty cup as empty cup) to a novel response (treat empty cup as full) is reminiscent of the problem frontal lobe patients encounter in the Wisconsin Card Sorting Task (Kolb and Whishaw 1985, p. 430). In the extreme these patients can't switch strategies of their own accord even when realizing that the strategy is wrong. However they can do so when instructed to switch. In fact autistic adults and children tend to be impaired on that sort of task (Rumsey and Hamburger 1988; Ozonoff *et. al.* 1991).

to mark that the decoupled representation shows what is in the photo, i.e. 'The photo shows "the doll is wearing blue,"' in the same way as in pretend play, where the manipulator has to form the expression: 'I pretend "the doll is wearing blue."'

There are two interesting differences between the demands created by the photo task and those created by pretence. In the photo task the manipulator does not have to change the content of the copied primary representation, and the interpreter only needs to read off the mental representation of what is in the picture; whereas in pretence the interpreter has to ensure sensible pretend behaviour. Hence, one might conclude that we have narrowed down the autistic child's difficulty. It lies either in the manipulation of the content of decoupled representations or in the interpreter's ability to generate adequate behaviour based on decoupled representations.

Unfortunately, this cannot be the explanation for why children find false-belief tasks so difficult, because the false-belief task is exactly parallel to the photo task in these respects. And in any case, there is a question mark over whether autistic children are able to engage in pretence or not. So it seems more promising to narrow down the nature of the autistic deficit by looking at the differences between the belief and the photo task. However, the difference in performance on these two tasks is impossible to capture in terms of which components of Leslie's decoupler might be damaged.*

WHY ARE BELIEFS SO DIFFICULT, PHOTOS SO EASY?

I focus on two possible answers to the question of why false belief is so much more difficult to understand than the photo task. One explanation — favoured by Alan Leslie — is that autistic children's deficit is domain-specific, that is, the deficit affects only their understanding of mental states but not their understanding of other representations like photographs. Leslie (this volume, chapter 5) speaks of a ToMM, a mechanism that is specifically responsible for theory of mind.

Another possibility — one that I favour — is that there is a domain-general deficit in understanding the nature of representation, a deficit of something

---

* Autistic children's good performance on the photo task also poses a problem for those theorists who try to link these children's problem with the false-belief task to 'frontal lobe' problems of executive control (for example, Ozonoff *et al.* 1991; Hughes and Russell 1990). In particular, Hughes and Russell argue that autistic children are unable to point deceptively to the empty location in their pointing task (Russell *et al.* 1991) and in the false-belief task because they are unduly influenced by their immediate perceptual inputs (or presumably by what they know they would see if there were perceptual input). In our version of the false-belief task this would mean that they answer wrongly that Judy's friend thinks Judy is dressed in green because they see Judy standing there in green. But if attraction by the immediate perceptual input were the reason for their wrong response then they should be equally misled in the photo task. However, they aren't.

which is essential for understanding, for example, cases of misrepresentation. This understanding, one can show, is not required for correct performance in the photo task, but is essential for understanding false belief.

Domain-specificity

Leslie and Thaiss (1992) argued that the difficulty in the false-belief task is domain-specific. It is due to the fact that the belief is a relationship between an *agent* and a proposition. The same holds true for pretence (Leslie and Thaiss take it for granted that autistic children are incapable of pretend-play), but not for a photograph.

Domain-specificity can explain why autistic children have little problem with the photo task. It is simply that the alleged decoupling deficit does not apply to the domain of photographs. However, the reason why it does not apply to the photographs is not explained by the structure of the decoupler. It is an additional, *post hoc* stipulation extraneous to the basic assumptions of the decoupler.

Furthermore, the specific proposal that the difficulty with belief is due to an inability to deal with propositional attitudes and animate agents is unsatisfactory, because there is mounting evidence that autistic children can understand other kinds of propositional attitudes involving human *agents*, as has been outlined above in my discussion of the role of the *manipulator* in Leslie's decoupling mechanism. For instance, autistic children understand the relationship between a human agent and how that agent sees a scene (Tan and Harris 1991). They understand the link between desires and simple emotions (Baron–Cohen 1991*a*; Tan and Harris 1991); and they can remember their own views and desires (Baron–Cohen, 1991*b*).

Understanding representation

I have argued (Perner 1991) that understanding of false belief comes with a new view of representation that is acquired at about four years. At this age children become able to understand things like pictures and the mind *as* representations (i.e., to engage in metarepresentation in Pylyshyn's original sense of the term). Before that age children also have some understanding of representations, but they do not completely differentiate between the representational medium and its representational content — for example, the situation that is represented in a picture. They simply treat a picture in terms of the situation that is represented in the picture. For this reason I called them *situation theorists*. In contrast, I called the older children who can understand representations as representations (understand the picture as a medium which stands in a representational relation to the situation it depicts) *representation theorists*.

One should not think, however, that one kind of theory completely replaces the other one. Rather, both kind of views persist into adulthood. At heart we are situation theorists; but situation theorists, who, unlike the very young child, can take a representational view when pushed. This is evident from how we talk about pictures (and sometimes the mind). When exclaiming 'Look, there is Judy in the picture. She is wearing blue' we are talking as situation theorists. That is, we talk as if there were a situation in the picture involving Judy, who is wearing a blue dress.

Two important things are to be noticed in this kind of talk. With the word 'Judy' we make a direct reference to Judy being in the picture, and the expression 'in the picture' serves as a context-marker in much the same way that, in the statement 'Yesterday at your party Judy was wearing blue' the expression 'Yesterday at your party' marks a difference in the spatio-temporal context which differentiates it from that of current reality.

In contrast, as representation theorists we would talk about the picture as follows: 'Look, this picture shows Judy wearing blue.' Notice that in this statement we do not refer with 'Judy' to Judy in the picture, but to Judy in real life being somewhere else, and the expression 'this picture shows' does not mark a difference in context but refers to a representational relationship between the picture and what it represents.

The important point here is that a situation theory is perfectly sufficient for Zaitchik's photo task. All one needs to understand is that by getting Judy into the viewfinder and depressing the button Judy will be in the picture, and she'll be dressed there the same way as she was at the time the picture was taken, and she won't change her dress in the picture even when she does change it outside the picture. For understanding false belief, this way of looking at mental states is not adequate. For, conceiving of Judy's friend's belief as 'Judy being dressed in blue in her friend's mind' won't differentiate between 'the friend merely *thinking* of (imagining, dreaming) Judy being dressed in blue' and 'the friend *thinking that* (believing) Judy is dressed in blue' (Perner 1990; 1989). In other words, the friend's belief about what Judy is doing in the world cannot be conceived of as some mental situation in the friend's mind, because such a conception does not capture the important 'representational' nature of the belief by which it makes a claim about how Judy is outside the friend's mind. In stating a person's belief we are not just concerned about how that person's mind looks, but how that person *thinks the world is*. And to understand that the belief is *about* how the world is, a representational view of the mind is necessary. A situation theorist would miss the essential point, namely that aspect in which a belief (thinking that) differs from mere thoughts, images (thinking of) and the like.

Back to autism. My tentative proposal is that autistic children are limited to a 'situation theory'. In contrast to young children, however, they are more sophisticated situation theorists. They have learned about cameras and that the people in the picture will be doing the same things as when they were in the viewfinder and the picture was taken. Three-year-old children lack this knowledge. I speculated (Perner 1991, pp. 100–1) that young children simply have no specific expectation of what precisely should be happening in a photograph (what dress Judy is wearing) and so, desperate for an answer, they take what they see Judy doing outside the photo at the time as a clue (she is wearing green).

Recent pilot data provide some support for this view (Myers 1991). Children showed the typical problems with the question about a person in a drawing (Charman and Baron–Cohen 1992; Pollock 1990) when the draw- ing was at a distance, and their gaze tended to shift to the real person when that person's name was mentioned. When the drawing was held in front of them and they were made to look at the back of it when answering the question, even young three-year-olds gave mostly correct answers.

In summary, my proposal is that autistic children lack a proper concept of representation. They cannot represent that something is a representation, and hence one could say that they have a *metarepresentational deficit* in Pylyshyn's original sense of 'metarepresentational'. Such a proposal sits uncomfortably with Leslie's general claim that autism is due to a damaged neural mechanism which serves a particular representational function. For it would be unnatural to proclaim that there are specific neural mechanisms responsible for understanding specific concepts, such as the concept of representation.

In my final section I will now speculate about how such a conceptual deficit could arise in autistic children.

THEORY FORMATION: DEFECTIVE DATA-BASE

Why do autistic children have this particular metarepresentational diffi- culty? If one is to trust the experts (for example, Frith 1989) evidence seems overwhelming that autism has a genetic origin (Folstein and Rutter 1977). But what do the genes fail to specify? For those who believe in genetic deter- mination of individual concepts (for example, Piatelli-Palmarini 1986) it may be quite congenial to suggest autistic children congenitally lack specific concepts. However, this view seems difficult to sustain if one assumes concepts to be acquired.

Acquisition of concepts is widely held (Carey 1985; Keil 1989; Murphy and Medin 1985) to be an intrinsic part of the formation of relevant theories. In this case, the autistic child's failure to acquire a concept of representation is better viewed as the result of a congenital impairment in the relevant experi- ences (data-base) on which these theories are based. Let me speculate on one such possibility.

Perhaps there is a genetic defect in the neural mechanism for rapid shifts in attention (Courchesne, 1991). Such a deficit would affect the ease with which children can engage in typical human interaction, to which even neonates seem predisposed. Such interaction requires constant monitoring of what the other person is doing (mostly in the visual modality) in relation to what the child herself is doing (mostly in the proprioceptive modality). As a consequence, one of the earliest indication of autism is a difficulty in co-ordinating mutual attention (Curcio 1978; Mundy, *et al.* 1986; Baron–Cohen 1989*a*).

As a result, later achievements building on these skills are erratic. For instance, autistic children have some, but not a very solid, understanding of how seeing relates to knowing. Their performance on such tasks varies from very good (Leekam and Perner, unpublished data with a paradigm developed by Perner and Ogden 1988) to not so good (Leslie and Frith 1988; Perner *et al.* 1989), but is still better than their understanding of false belief. Similarly, most (if not all) of them can adjust their answers to questions to the question-asker's perceived knowledge. However, their use of this skill is very unreliable (Perner *et al.* 1989).

Similarly, one of the early signs of autism is the lack of spontaneous pretend-play. However, as the data of Lewis and Boucher (1988) indicate, autistic children are not incapable of engaging in pretence when such an activity is elicited or if they are instructed to pretend that something is the case (Baron–Cohen 1991*b*). Their lack of spontaneity in this respect may also originate in problems with rapid shifts in attention. Enactment of a pretend scenario takes place in the real world, so the competent pretender has constantly to monitor changes in the real world and in his internal pretend representation, so that his action is possible in the real world and, at the same time, the action stays true to the intended pretence. Again, if autistic children are only impaired in *rapid* shifts of attention, but can shift more slowly, then this will not make them incapable of pretence, but pretence will be a rather cumbersome enterprise for them.

Now, although the autistic child is not completely incapacitated in his understanding of either attentional/information-seeking processes or pretence, the impairment that there is will become forbidding when new concepts need to be formed which require a good understanding of both these areas. One example is the concept of a false belief by observing cases of erroneous behaviour as depicted in the false-belief stories used by Wimmer and Perner (1983).

In these situations a person gets information about the location of an object. Then the object's location is changed while the person is away, but the person is not informed about the change. Finally the person returns, and behaves as if the object were still in its old location. The 'situation theorist' can make sense of the fact that the person did attend to the object's first location, but not to the object's later location. He can also make sense of a person's acting as if the object were in a different location than it really is. But what is difficult to understand (Perner 1988: the puzzle of false belief) is why the person would look in the wrong place when she really wants the object. To understand this the child has to form a new concept — has to understand the person's belief where the object is a mental representation. With this concept the child can make sense of the observed sequence of events. The person's attending to the object's original location leads to a mental representation (belief) of where the object is. This representation stays unchanged as the object is moved without the person's knowing. However, being a representation of reality (not just a situation in the mind), this outdated information governs where the person will look for the object in the real world.

However, to hit on this new idea by observing relevant instances is, presumably, rather difficult. Yet the child will be greatly helped if she is fascinated by the relevant aspects. (They say that scientific discoveries are made by those who are well immersed in the problem domain, i.e., are aware of the relevant facts.) For then she will notice the fact that what the observed person attended to (the object's original location) coincides with the person's acting as if the object were still there.

So we see that if autistic children do not pay much attention to attentional facts and are relatively unfamiliar with as-if action (pretence), then their task of forming the concept of representation by working in this task domain seems remote. Though some (about 20–30 per cent of the very intelligent autistic children) do. For instance, they pass false belief tests (see, for example, Baron–Cohen *et al.* 1985; Perner *et al.* 1989), and they understand the difference between appearance and reality (Baron–Cohen 1989*a*), which are all hallmarks of having acquired the concept of representation. Most of these children, however, fail more complex versions involving second-order beliefs (Baron–Cohen, 1989*b*), though some eventually will understand even those (Happé 1991; Ozonoff *et al.* 1991).

Yet for those who do not make the step to understanding the mind as representational, many aspects of human interaction remain mysterious. For instance, although the relationship between desire and happiness can be understood by situation theorists, and many autistic children seem able to understand it (Baron–Cohen 1991*a*), I have argued that the understanding of conflicting desires, and hence the notion of competition, may require a representational understanding of mind (Perner *et al.* 1991). In fact, autistic children tend to find conflicting desires quite difficult (Harris and Muncer 1988; Harris 1991), and like three-year-old children (Gratch 1964), they don't show much sign of enjoyment or frustration when winning or losing in competitive games. This may also explain why autistic children's play consists of repetitious rituals, where the same person always 'wins' (Baron–Cohen 1992, personal communication).

Perhaps the most intriguing, but the least researched, consequence of autistic children's failure to understand the mind as we do may be a lack of insight into their own behaviour. For instance, I have speculated (Perner 1991, Chap. 9) that an understanding of the representational nature of their own mind gives children a better insight into their own volitional processes. And this insight enables them to gain better control over their desires and will. Hence a lack of such an insight might account for autistic children's tendency to have uncontrolled emotional outbursts (tantrums), and for a certain rigidity in coping with unusual changes to their routines. There is also evidence in the research by Ozonoff *et al.* (1991) that very able autistic adults who pass many of the belief tasks are still impaired on tasks requiring self-control and creative planning (see also Harris, this volume, Chapter 11). Furthermore, Frith and Frith (1991) have pointed out that there may be a link between the failure to understand the mind and volitional problems, which can explain not only the autistic syndrome but also aspects of schizophrenia.

In sum, I think the position that autistic children lack a 'theory of mind' because they genetically lack a computational module called a 'decoupler' has not kept up with the rapidly increasing evidence about autistic children's abilities in this area. An alternative account is that autistic children have a hereditary impairment affecting the personal data-base on which they would have to build a theory of mind. Depending on how severe this and related impairments are, they are able to reach a certain competence in understanding their own and other people's minds; but their understanding in this area always lags behind their other intellectual achievements.

Acknowledgements

I would like to thank Sue Leekam, Simon Baron–Cohen, Uta Frith, Helen Tager-Flusberg, and Paul Harris for helpful comments on earlier versions and for helping me to find my way to relevant parts of the literature on autism.

REFERENCES

APA (American Psychiatric Association) (1987). *DSM-III-R: Diagnostic and Statistical Manual of Mental disorders* (3rd edn, revised). The American Psychiatric Association, Washington DC.

Astington, J. W. (1990). Wishes and plans: children's understanding of intentional causation. Paper presented at the 20th Anniversary Symposium of the Jean Piaget Society, Philadelphia, PA, May 1990.

Baron–Cohen, S. (1989a). Perceptual role taking and protodeclarative pointing in autism. *British Journal of Developmental Psychology*, 7, 113–27.

Baron–Cohen, S. (1989a). Are autistic children behaviourists? An examination of their mental-physical and appearance-reality distinctions. *Journal of Autism and Developmental Disorders*, **19**, 579–600.

Baron-Cohen, S. (1989*b*). The autistic child's theory of mind: a case of specific developmental delay. *Journal of Child Psychology and Psychiatry*, **30**, 285–97.

Baron-Cohen, S. (1990). Instructed and elicited play in autism: a reply to Lewis and Boucher. *British Journal of Developmental Psychology*, **8**, 207.

Baron-Cohen, S. (1991*a*). Do people with autism understand what causes emotion? *Child Development*, **62**, 385–95.

Baron-Cohen, S. (1991*b*). The development of a theory of mind in autism: deviance and delay? *Psychiatric Clinics of North America*, **14**, 33–51.

Baron-Cohen, S. (1992). Out of sight or out of mind? Another look at deception in autism. *Journal of Child Psychology and Psychiatry,* **33**, 1141–55.

Baron-Cohen, S., Leslie, A.M., and Frith, U. (1985). Does the autistic child have a 'theory of mind'? *Cognition*, **21**, 37–46.

Baron-Cohen, S., Leslie, A.M., and Frith, U. (1986). Mechanical, behavioural and intentional understanding of picture stories in autistic children. *British Journal of Developmental Psychology*, **4**, 113–25.

Boucher, J., and Lewis, V. (1990). Guessing or creating? A reply to Baron-Cohen. *British Journal of Developmental Psychology*, **8**, 205–6.

Carey, S. (1985). *Conceptual change in childhood.* MIT Press, Cambridge, MA.

Charman, T., and Baron-Cohen, S. (1992). Understanding drawings and beliefs: a further test of the metarepresentation theory of autism. *Journal of Child Psychology and Psychiatry*, **33**, 1105–12.

Courchesne, E. (1991). Attention-shifting in autism. Paper presented at the Workshop on Theory of Mind and Autism, Seattle, April 1991.

Curcio, F. (1978). Sensorimotor functioning and communication in mute autistic children. *Journal of Autism and Childhood Schizophrenia*, **8**, 281–92.

Dretske, F. (1988). *Explaining behavior: reasons in a world of causes.* MIT Press, Cambridge, MA.

Dunn, J., and Dale, N. (1984). I a daddy: 2-year-olds' collaboration in joint pretend with sibling and with mother. In *Symbolic play.* (ed. I. Bretherton), Academic Press, New York.

Flavell, J.H., Flavell, E.R., and Green, F.L. (1983). Development of the appearance-reality distinction. *Cognitive Psychology*, **15**, 95–120.

Folstein, S., and Rutter, M. (1977). Infantile autism: a genetic study of 21 twin pairs. *Journal of Child Psychology and Psychiatry*, **18**, 297–321.

Frege, G. (1892/1960). On sense and reference. In *Philosophical writings of Gottlob Frege* (ed. P. Geach and M. Black), pp. 56–78. Basil Blackwell, Oxford.

Frith, U. (1989). *Autism: explaining the enigma.* Basil Blackwell, Oxford.

Frith, C.D., and Frith. U. (1991). Elective affinities in schizophrenia and childhood autism. In *Social psychiatry: theory, methodology and Practice* (ed P. Bebbington). Transaction Publications, New Brunswick, NJ.

Goodman, N. (1976). *Languages of art.* Hackett Publishing, Indianapolis, IN.

Gopnik, A., and Astington, J.W. (1988). Children's understanding of representational change and its relation to the understanding of false belief and the appearance-reality distinction. *Child Development*, **59**, 26–37.

Gratch, G. (1964). Response alternation in children: A developmental study of orientations to uncertainty. *Vita Humana*, **7**, 49–60.

Haake, R.J., and Somerville, S.C. (1985). Development of logical search skills in infancy. *Developmental Psychology*, **21**, 176–86.

Happé, F. (1991). Understanding of figurative language in autism. Paper presented at the Meeting of the Experimental Psychology Society, Sussex University, July 1991.

Harris, P.L. (1991). The work of the imagination. In *Natural theories of mind* (ed. A. Whiten). Basil Blackwell, Oxford.

Harris, P.L., and Muncer, A. (1988). Autistic children's understanding of beliefs and desires. Paper presented at the Annual Conference of the Developmental Section of the British Psychological Society, Coleg Harlech, Wales, September 1988.

Hughes, C.H., and Russell, J. (1990). One reason why deception is difficult for autistic subjects: failure to disengage from an object. Unpublished MS, Department of Experimental Psychology, University of Cambridge.

Keil, F.C. (1989). *Concepts, kinds, and cognitive development*. MIT Press, Cambridge, MA.

Kolb, B., and Whishaw, I.Q. (1985). *Fundamentals of human neuropsychology*. (2nd edn). Freeman, San Francisco.

Lancy, D.F., and Goldstein, G.I. (1982). The use of nonverbal Piagetian tasks to assess the cognitive development of autistic children. *Child Development*, **53**, 1233–41.

Leekam, S.R., and Perner, J. (1991). Does the autistic child have a 'metarepresentational' deficit? *Cognition*, **40**, 203–18.

Lehrer, K. (1986). Metamind: belief, consciousness and intentionality. In *Belief* (ed. R.J. Bogdan). Oxford University Press.

Leslie, A.M. (1987). Pretense and representation: the origins of 'theory of mind'. *Psychological Review*, **94**, 412–26.

Leslie, A.M. (1988). Some implications of pretense for mechanisms underlying the child's theory of mind. In *Developing theories of mind* (ed. J.W. Astington, P.L. Harris and D. Olson). Cambridge University Press, New York.

Leslie, A.M., and Frith, U. (1990). Prospects for a cognitive neuropsychology of autism: Hobson's choice. *Psychological Review*, **97**, 122–31.

Leslie, A.M., and Thaiss, L. (1992). Domain specificity in conceptual development: neuropsychological evidence from autism. *Cognition*, **43**, 225–51.

Lewis, V., and Boucher, J. (1988). Spontaneous, instructed and elicited play in relatively able autistic children. *British Journal of Developmental Psychology*, **6**, 325–39.

Lewis, M., and Brooks-Gunn, J. (1979). *Social cognition and the acquisition of self*. Plenum, New York.

Mundy, P., Sigman, M., Ungerer, J.A., and Sherman, T. (1986). Defining the social deficits of autism: the contribution of nonverbal communication measures. *Journal of Child Psychology and Psychiatry*, **27**, 657–69.

Murphy, G.L., and Medin, D. (1985). The role of theories in conceptual coherence. *Psychological Review*, **92**, 289–316.

Myers, D. (1991). Preschoolers' ability to understand the permanence of pictorial representation despite its conflict with the real world. Unpublished Third-Year Student Project, Laboratory of Experimental Psychology, University of Sussex.

Neumann, C.J., and Hill, S.D. (1978). Self-recognition and stimulus preference in autistic children. *Developmental Psychobiology*, **11**, 571–8.

Ozonoff, S., Pennington, B. F., and Rogers, S. J. (1991). Executive function deficits in high-functioning autistic children: relationship to theory of mind. *Journal of Child Psychology and Psychiatry*, **32**, 1081-1106.

Peerbhoy, D. (1990). Do children use intentionality in assessment of neutral and bad moral situations? Unpublished third-year experimental project, Laboratory of Experimental Psychology, University of Sussex.

Perner, J. (1988). Developing semantics for theories of mind: from propositional attitudes to mental representation. In *Developing theories of mind* (ed. J. W. Astington, P. L. Harris and D. R. Olson). Cambridge University Press, New York.

Perner, J. (1989). Is 'thinking' belief? Reply to Wellman and Bartsch. *Cognition*, **33**, 315-19.

Perner, J. (1990). On representing **that**: the asymmetry between belief and intention in children's theory of mind. In *Child's theories of mind* (ed. D. Frye and C. Moore). Erlbaum, Hillsdale, NJ.

Perner, J. (1991). *Understanding the representational mind*. MIT Press, Cambridge, MA.

Perner, J., and Ogden, J. (1988). Knowledge for hunger: children's problem of representation in imputing mental states. *Cognition*, **29**, 47-61.

Perner, J., Frith, U., Leslie, A. M., and Leekam, S. R. (1989). Exploration of the autistic child's theory of mind: knowledge, belief and communication. *Child Development*, **60**, 689-700.

Perner, J., Peerbhoy, D., and Lichterman, L. (1991). Objective desirability: bad outcomes. Conflicting desires and children's concept of competition. Paper presented at the Biennial Meeting of the Society for Research in Child Development, Seattle, WA, April, 1991.

Piaget, J. (1937/1954). *The construction of reality in the child*. Basic Books, New York.

Piatelli-Palmarini, (1986). In *Language learning and concept acquisition* (ed. W. Demopoulos and A. Marras). Ablex, Norwood, NJ.

Pollock, J. (1990). Children's understanding of 'false' drawings. Unpublished Manuscript, Department of Experimental Psychology, University of Oxford.

Premack, D., and Woodruff, G. (1978). Does the chimpanzee have a theory of mind? *The Behavioral and Brain Sciences*, **1**, 516-26.

Pylyshyn, Z. W. (1978). When is attribution of beliefs justified? The *Behavioral and Brain Sciences*, **1**, 592-3.

Rumsey, J. M., and Hamburger, S. D. (1988). Neuropsychological findings in high-functioning men with infantile autism, residual state. *Journal of Clinical and Experimental Neuropsychology*, **10**, 201-21.

Russell, J., Mauthner, N., Sharpe, S., and Tidswell, T. (1991). The 'windows task' as a measure of strategic deception in preschoolers and autistic subjects. *British Journal of Developmental Psychology*, **9**, 331-50.

Sigman, M., and Ungerer, J. A. (1981). Sensorimotor skills and language comprehension in autistic children. *Journal of Abnormal Child Psychology*, **9**, 149-65.

Sigman, M., Ungerer, J. A., Mundy, P., and Sherman, T. (1987) Cognition in autistic children. In *Handbook of autism and pervasive developmental disorders* (ed. D. J. Cohen, A. M. Donnellan and R. Paul). Wiley, New York.

Tager-Flusberg, H. (1989). An analysis of discourse ability and internal state lexicons in a longitudinal study of autistic children. Paper presented at the Biennial Meeting of the Society for Research in Child Development, Kansas City, MO, April, 1989.

Tan, J., and Harris, P. (1991). Autistic children understand seeing and wanting. *Development and Psychopathology*, **3**, 163–174.

Wimmer, H., and Perner, J. (1983). Beliefs about beliefs: representation and constraining function of wrong beliefs in young children's understanding of deception. *Cognition*, **13**, 103–28.

Yuill, N. (1984). Young children's coordination of motive and outcome in judgments of satisfaction and morality. *British Journal of Developmental Psychology*, **2**, 73–81.

Zaitchik, D. (1990). When representations conflict with reality: The preschooler's problem with false beliefs and 'false' photographs. *Cognition*, **35**, 41–68.

7

What language reveals about the understanding of minds in children with autism

HELEN TAGER-FLUSBERG

INTRODUCTION

Human language, like the communication systems of other species, has as one of its primary functions the sending and receiving of information. The well-known example of the complex bee dance (Von Frisch 1950) provides us with an interesting illustration of a referential communication system designed exclusively to inform other members of a hive where they might find nectar. From the beginning, children acquiring language can exchange information about many different topics, and are not simply limited to food. The power of human language, which lies in its syntax, is that it is unlimited in the range of information that it can transmit: it is creative, semantic, and can be used to talk about the past, the future, the hypothetical, and the non-existent.

Information about the world comes to us directly through perceptual sources, and indirectly (or via a symbolic, representational system) through language. Indeed, by the time children become proficient language-users they, like adults, have come to rely heavily on language (through conversation, narratives, books, and other media) as the major source of knowledge, including cultural knowledge. Arguably, much misinformation is also conveyed via language, and there are data to suggest that children come to understand false belief communicated verbally before they do so through misperception (see for example Zaitchik 1991).

Recent research in cognitive development has demonstrated the early development of knowledge about mental states (Astington *et al.* 1988; Perner 1991; Wellman 1990). There is a wealth of data to suggest that by the age of four the major constructs of the mind, such as desire, knowledge, and belief, are acquired and employed by children to explain action. Nevertheless, while the cognitive architecture of the mind (its representational form) is acquired early, the *contents* of the mind, especially more sophisticated understanding of social interaction, behaviour, and relationships, continue to develop throughout childhood (Shantz 1983). Language must play a significant role in the elaboration of this kind of knowledge. It is primarily through language that we come to know the contents of other people's minds — their thoughts, beliefs, feelings, and desires. We explain one another's motivations and actions, and in this way children come to acquire and elaborate their social knowledge.

It is, therefore, not surprising that language is the lens through which we typically view the child's developing theory of mind. We rely almost exclusively on children's verbal responses and explanations to infer their understanding of the theory of mind, as well as on their more spontaneous discussion of mental phenomena (cf. Bretherton and Beeghly 1982; Shatz *et al.* 1983).* In this chapter I focus on early language as the medium through which we might investigate the autistic child's impaired understanding of mental states. The chapter is divided into two parts: in the first, I outline a number of examples of language characteristics that are intimately related to the communication of mental states, and consider briefly what is known about their development in normal and autistic children. The second section of the chapter then focuses in greater detail on some features of autistic children's early language that reveal fundamental problems in certain aspects of their knowledge of theory of mind.

THE LANGUAGE OF MENTAL STATES

Several features of human language seem to have been designed especially to communicate about mental states. We can identify such features at the phonological, lexical, syntactic, and pragmatic levels of language, and together they confirm the view that language is the most significant source of information about feelings, desires, thoughts, fantasy, and so forth. Of course, in interpersonal interactions such linguistic information is complemented by non-linguistic communication in gesture, body posture, and especially facial expressions.

Lexical terms

At the most obvious level, the lexicons of languages are filled with many hundreds of words which represent all different kinds of mental states

* Although the emphasis in this chapter is on language as the major means by which we discern both normal and autistic children's theory of mind, clearly it is not the only means. Both Gómez *et al.* and Whiten in this volume highlight the significance and possibility of non-linguistic methods for the study of theory of mind in non-human primates. There is much to learn from their work, and the work of others in that field, about how to study the understanding of mental states without relying on language (see also recent research on deception in young children reviewed by Sodian and Frith, this volume, Chapter 8).

and mental-state categories. These terms fall into all the major open-class categories — nouns (for example, *idea, knowledge*), verbs (for example, *feel, want, hope, remember*), adjectives (for example, *happy, thoughtful*), and adverbs (for example, *sadly, wishfully*).

Differences in terms — their meanings and usage — among different languages reveal cultural variation in ways in which people may talk about and attach significance to such phenomena. For example, Lutz (1987) using an 'ethnotheoretical' approach, describes the unique aspects of emotion knowledge among the Ifaluk people of Micronesia. In contrast to our culture's emphasis on the private internal nature of emotion, the Ifaluk are reported to interpret emotions by referring to the situations and events in which they occur. This difference is reflected in their definitions of emotion terms, which typically include propositions referring to various situations associated with an emotion term, not to a mental state.

Children begin to acquire mental-state terms very early, before the age of two (Bretherton *et al.* 1981; Bretherton and Beeghly 1982). Although initially they use words like *want, know*, or *think* in idiomatic (for example 'I don't know') and conversational ways (Shatz *et al.*, 1983), by the time they are three years old children use a range of terms to refer to all categories of psychological states. Such uses have been taken as an early indicator of the child's developing theory of mind (Wellman 1990), especially as children begin to express their understanding of the contrast between mental states and objective reality (Shatz *et al.* 1983). Later on in this chapter I describe some data on the acquisition of mental-state terms by autistic children.

Prosodic features

The speech-sound system of language comprises segmental and non-segmental features. Segmental features include the basic speech sounds or phonemes of a language, while non-segmental features include the stress, pitch, timing, rhythm, and melodic tone of the sound signal (Crystal 1975). These prosodic characteristics are important for marking syntactic information and for communicating subtle word-meaning distinctions. More importantly from the perspective of this discussion, prosody is also intimately linked to the expression of emotions, mood, and attitudes (Frick 1985). We can identify speakers' affect and infer degree of social engagement from the prosodic contours of their speech. Thus prosody is used to convey information about broad categories of some basic emotional states. The speaker's attitude toward a listener, such as doubt, authority, and submission, is also conveyed through the melodic intonation of speech.

In addition, prosodic characteristics are used to communicate different kinds of messages. These aspects of the verbal signal are critical for distinguishing between literal and non-literal messages. For example, we use prosody to convey to a listener that what we say should be taken literally or as sarcasm, irony, hyperbole, or understatement (Winner 1988). The speaker communicates to the listener through prosody that the intended meaning of such an utterance is the opposite of its literal meaning. In this way, prosody is being used to manipulate the beliefs and interpretations of intentionality of the listener.

Research on very young normal children suggests that by the end of the second year they acquire the use of prosody to communicate social and affective information (Furrow 1984). Few studies have investigated children's ability to understand prosodic cues to make inferences about a speaker's mental states; however, there are some data suggesting that intonation cues tend to be ignored or are found difficult to interpret by children less than seven or eight years old (Ackerman 1983; Van Lancker *et al*. 1989; Winner *et al*. 1987). Although there is very little work that has been done in this area, the studies mentioned here suggest that while children can use intonation to express affective and related information quite early, comprehension of such information is a significantly later development.

From Kanner's earliest descriptions of autistic children's speech, their deficits in prosody, including lack of melodic tone, and atypical patterns of rhythm, stress, and pitch have been noted (Kanner 1946; Pronovost *et al*. 1966). These deficits appear to show the least improvement over time, persisting through adolescence and adulthood (Ornitz and Ritvo 1976; Simmons and Baltaxe 1975). No studies have examined in detail whether verbal autistic children do communicate affect or social meanings in their speech (Baltaxe and Simmons 1985); however, there is one study which suggests that they are impaired in their ability to interpret such information in the speech of others (Van Lancker *et al*. 1989). There is still a considerable amount of both theoretical and empirical work that needs to be done on prosody, especially in autism, as it could be especially revealing about the roots of theory of mind problems in this population.

Grammar

Many syntactic and morphological features of lanquages appear to be individually designed for conveying specific meanings and functions (Pinker and Bloom 1990). From the perspective of a belief–desire psychology, the most significant grammatical mechanisms are those which allow for the expression of propositional arguments embedded in other propositions: complementation and control. The complement construction is especially utilized for the expression of the propositional attitudes that are at the heart of a theory of mind. For example:

(1) John thinks that Bill will visit.

(2) Mary knows that Bill is sick.

(3) Stephen wishes that Bill would call.

In sentences (1)–(3) the clause containing some proposition about Bill is embedded in a matrix clause that has a mental-state verb which takes a clause as its object complement. This is the language of mental states, or propositional attitudes (cf. Leslie 1987).

Children begin to use primitive complement constructions as early as two, typically to talk about their own mental states (Limber 1973; Menyuk 1969). For example, Limber reports the following examples, which primarily involve verbs like *want* or *need*, and *see, watch,* or *look(it)*:

(4) I want Mommy do it.

(5) I see you sit down.

(6) I want go.

And by three they can express each set of propositional attitudes with a wide range of verbs, demonstrating their mastery of both the syntax and the conceptual developments associated with knowledge of mental states. Some examples, taken from Shatz *et al.* (1983) and Bretherton and Beeghly (1982), are provided in (7)–(9).

(7) Jim knows where it is.

(8) I think Mommy is beautiful.

(9) I didn't, 'cept I tricked you.

Syntax is not a linguistic domain that is specifically impaired in autism (Pierce and Bartolucci 1977; Tager-Flusberg 1981*a*), and one study that investigated grammatical development longitudinally (Tager-Flusberg *et al.* 1990) found that some autistic children do master complement constructions, so any difficulties they might have understanding mental states cannot be due simply to syntactic limitations, or limitations in the linguistic expression of propositional attitudes.

Pragmatics

Pragmatics is a component of language that encompasses a broad assortment of knowledge central to effective communication. This includes how to use language to convey different speech acts (requests, assertions, promises, etc.); the appropriate use of language in particular social contexts that may reflect different types of discourses (conversation, monologue, narrative, etc.); and structuring utterances to distinguish given/new information. A speaker must use pragmatic knowledge to organize the information to be communicated in the most useful manner. This entails taking into account what the speaker knows about the listener, including his or her current status, knowledge, feelings, and so forth. In turn, the listener must employ similar information in order to comprehend the speaker's intentions. From this description we see that effective communication is achieved when both speakers and listeners employ a theory of mind to structure the ongoing discourse (Sperber and Wilson 1986).

To illustrate these points, let us consider a couple of examples. In discourse, referents such as *book* or *chair* must be preceded by a determiner, for example an article: *a* or *the*. The choice of article depends on what the speaker thinks the listener knows about the referent in question. For example, when recounting an event, a new referent is introduced with the indefinite article (for example, 'a man'), but later mentions of the same referent will either be with an anaphoric substitute ('he'), or will have the noun preceded by a definite article ('the man'). Too many repetitions, however, of the full noun phrase rather than the pronoun give the discourse a pedantic or distinctly odd tone. In a different context, other considerations need to be kept in mind. If two people are engaged in conversation in a room with a single chair, it also sounds odd for one to say:

(10) Pass me *a* chair.

Mastery of article-use thus requires a speaker to track what the listener could be expected to know about a referent. These complex pragmatic characteristics of articles begin to develop during the pre-school years (Maratsos 1976), but are not fully grasped across a range of discourse contexts until middle childhood.

Another example that relates pragmatic knowledge to knowledge of mental states comes from narrative discourse (see Loveland and Tunali, this volume, Chapter 12), in which speakers must keep in mind what their listeners know about a topic in deciding what information to include in the narrative, what would be redundant, and how to organize a sequence of utterances to make the discourse as comprehensible as possible. Again, while children produce simple narratives when they are very young (Applebee 1978; Nelson 1986), pragmatic knowledge continues to develop throughout the middle childhood period.

For some time now, it has been recognized that language problems in autistic children and adults lie primarily in the pragmatic domain (Paul 1987; Tager-Flusberg 1981*b*; 1989*a*). Indeed, the pragmatic deficits of autistic people are one of the key features of the syndrome that are incorporated under the theory of mind hypothesis (Baron-Cohen 1988; Hobson, this volume; Chapter 10). Some examples of problems in autistic individuals include the absence or paucity of certain speech functions (Ball 1978; Mermelstein 1983; Wetherby 1986); difficulty communicating information

(Hurtig *et al*. 1982; Paul and Cohen 1984); and problems with speaker-listener relations (Baltaxe 1977). Adults with autism have also been shown to have difficulties responding appropriately to indirect requests (for example 'Can you colour the circle blue'), which they tend to interpret literally (Paul and Cohen 1985). The remainder of this chapter focuses more extensively on specific aspects of the language deficit in autism, including pragmatic impairments, that are considered to be closely linked to deficits in theory of mind (Baron-Cohen 1988; Frith 1989).

LANGUAGE AND THE THEORY OF MIND IN AUTISM

In the previous section, I discussed briefly several examples of language characteristics of autistic children that illustrate problems in their understanding of minds. In this part of the chapter I take up in greater detail some additional evidence for the idea that there is indeed an inextricable link between autistic children's problems with language and communication, and their theory of mind deficit, which is at the core of many of their social difficulties (Baron-Cohen 1988; Baron-Cohen *et al*. 1985; Frith 1989). The hypothesis that has guided much of the research that I have done in the past few years is that children with autism in the process of acquiring language, and even before this point, show quite specific, *selective* impairments in those areas that entail knowledge of some aspects of a theory of mind (Tager-Flusberg 1989*a*). One advantage of investigating autistic children's understanding of mind through language is that, unlike experimental tasks which are either passed or failed in an all-or-none fashion, data from naturalistic language are more graded or continuous. We are able to investigate *degree of impairment*, which ultimately must provide a more realistic picture of the deficits in this domain of functioning for autistic children.

Prelinguistic communication

It is generally recognized by researchers of normal language development that the onset of *intentional communication* is at about nine months of age. At this time infants begin to use gestures and sounds in intentional ways to communicate several sorts of messages (Bates *et al*. 1975; Foster 1990): social routines (such as hello and good-bye), protoimperatives (used to obtain a desired object or action), and protodeclaratives (used to indicate an object of interest). The most significant of these, from the point of view of autism, are protodeclaratives, which are found to be virtually absent (Curcio 1978). These pointing gestures are also part of what is included under the rubric of joint-attention behaviours (Baron-Cohen 1989*a*; Mundy and Sigman 1989; Mundy *et al*., this volume, Chapter 9).

There is some controversy about how to interpret the paucity of joint-attention behaviour in autistic children. While some have emphasized the affective components of joint attention (Hobson, this volume, Chapter 10; Mundy *et al.*, this volume, Chapter 9) others argue that it involves metare-presentational capacities, tying it developmentally to symbolic play and false belief (Baron-Cohen 1989*b*). The data presented in Kasari *et al.* (1990, see Mundy *et al.*, this volume, Chapter 9) suggest that the majority of joint-attention behaviours are accompanied by positive affect in normal and retarded infants, but positive affect is less frequent during joint attention episodes among autistic children. However, among all subjects joint attention was often not coupled with any affect expression; therefore it is difficult to interpret joint attention *exclusively* in terms of affect-exchange, as some researchers have done.

An alternative approach is to define joint attention* simply as a communicative act which conveys the meaning of a 'declarative' or *new information* (see also Leslie and Happe 1989; Sperber and Wilson 1986). When an infant or child engages her mother to observe an object that she is looking at or playing with, she is conveying some information to her mother. The information may be an emotional attitude toward an object, such as delight or interest, or the way the object appears, or how it moves, or that it is broken, and so forth. On this analysis, the joint attention deficit in autism involves impaired ability to communicate novel information, and may come from an impaired understanding of communication as a source of knowledge. Consistent with this interpretation are the findings from a number of recent studies which show correlations between joint-attention deficit and language development in children with autism (Loveland and Landry 1986; Mundy *et al.* 1990). It is tempting to consider of those autistic children who are capable of speech but fail to use it at a functional level that they simply never come to understand that the role of communication is to exchange information. This explanation also links early-appearing joint attention deficits to certain aspects of later pragmatic deficit in verbal autistic children, as we shall see in a later section.

Early language

Only a small proportion of autistic children develop some productive language, although typically, these verbal children have higher IQ levels, and a better prognosis than non-verbal autistic children (Rutter *et al.* 1967). Little

* My focus here is on the child's ability to direct gaze, using head turning, pointing, or showing gestures, not social referencing which is also considered sometimes a joint attention behavior. One could, however, offer a parallel analysis of social referencing which may involve the child looking to the mother in order to *receive* rather than transmit information about an object.

is known about the developmental picture of language-acquisition in this population, especially in terms of comparisons between autism, other developmentally delayed, and normal children. Recently, I conducted a longitudinal study which compared patterns and processes of language development in autistic and Down's syndrome children (Tager-Flusberg *et al.* 1990). The data discussed in this and later sections come from this study.

Six boys, all diagnosed with autism at a young age but who had begun to talk, were followed for between one and two years. A second group of six children with Down's syndrome, four boys and two girls, matched on productive language level and age to the autistic subjects, were included as a control group. The children were between three and six years old at the beginning of the study, and had begun to use combinatorial speech. They were visited in their homes every other month for about one hour. During this time the children engaged in different activities with their mothers, such as play, games, cooking, snacks, and so forth. Each visit was videotaped and audiotaped, and later verbatim transcripts, which included rich contextual information, were prepared in a form suitable for computer-assisted analysis. In all, there are about 60 hours of language data from the autistic children, and 50 hours from the Down's syndrome children, which constitute a very rich and unique data-base.

Surprisingly, at the earliest stages of language development, when mean length of utterance (MLU, a well-known measure of language level: Brown 1973) was between about 1.0 and 2.0, the autistic and Down's syndrome children's language samples were strikingly similar. Comparisons of their grammatical knowledge, lexicon size and contents, form-class distribution (Tager-Flusberg *et al.* 1990), and even turn-taking abilities (Tager-Flusberg and Anderson 1991) revealed no differences between the groups. All the children, autistic and Down's syndrome, spoke about people and objects, naming them, discussing their relationships, and their actions. In the following examples taken from two autistic boys we see how they speak about themselves and others as intentional agents of actions:

(11) Daddy work.

(12) Rick go bathroom.

(13) Open it.

The groups were not indistinguishable: the autistic children all had unusual prosodic patterns, and tended to be somewhat more imitative (echolalic) than the Down's syndrome children. Both groups in fact imitated significantly more than normal children, yet for all children imitation served the same functions: to serve conversational ends rather than as a means for acquiring grammar (Tager-Flusberg and Calkins 1990). There was one aspect of the autistic children's language that appeared distinctly different, even at the earliest stages. The autistic children all made pronoun-reversal errors, referring to themselves as 'you' and their mothers as 'I' or 'me'. None of the Down's syndrome children made this error. In fact, at this early stage of language development the majority of the autistic children's pronouns were correct; only about 10 per cent of the first- and second-person references involved errors like the following (taken from the same two boys):

(14) You want candy.

(15) I write.

These errors were three times more frequent when the children's MLUs reached between 2.0 and 4.0, and after that they declined sharply, and eventually, for two of the autistic children, disappeared (Tager–Flusberg 1991).

In addition to pronoun-reversing errors, autistic children during this period asked a variety of questions, some of which also illustrated a unique kind of form–function error, as illustrated in the following examples:

(16) Did we say bye?

(17) Want cracker?

These utterances are clearly marked as questions by rising intonation, yet from the context in which they were spoken and the choice of verb, it is clear that they should have been made as statements (Tager–Flusberg 1989*b*).

I suggest that both kinds of unusual error patterns have the same source: the autistic child who confuses pronouns such as I/you, and questions/statements, has not yet completely figured out the speaker/listener discourse roles (cf. Baltaxe 1977). As such, an error comes from the child as speaker, producing an utterance that would have been spoken by the listener — in these instances, the mother (cf. Charney 1980). Autistic children's confusion about discourse roles comes from their impaired understanding that individuals have different conceptual perspectives; that individuals may perceive, interpret, remember, and value the same situation in different ways (see also Hobson 1990). Loveland (1984) demonstrated that normally developing children do not begin to use pronouns until they are able to solve spatial perspective problems. Experimental studies with autistic subjects have shown that they generally have no problems with spatial perspective-taking (Baron-Cohen 1989*a*; Hobson 1984; Leslie and Frith 1988); however the children in these studies were considerably older than the young children who are still making pronoun errors in the natural-language data. This discrepancy can be interpreted in two ways. It could be that conceptual perspective-taking, as the basis for understanding discourse roles, is more complex than simple visual perspective-taking; thus although autistic children can pass visual perspective tasks, they still have difficulties with discourse roles. Alternatively, autistic children who are still confusing pronouns might also make errors on a visual perspective task — the critical experiment which includes both kinds of problems has yet to be done. The language data do suggest that higher functioning, more verbal, children do eventually come to understand discourse roles, and these may be the kinds of children who pass the perspective-taking tasks in the studies conducted by Hobson (1984) and Baron-Cohen (1989*a*).

One interesting aspect of this interpretation is that pronoun errors and question–statement errors are directly linked to problems in early knowledge of people with minds (presumably for both autistic and other children who make these errors). Yet because some autistic children eventually come to understand discourse roles, at least as evidenced by correct pronoun-usage, this aspect of mental state knowledge can later be acquired, albeit quite belatedly, by these children. The advantage of the longitudinal language data is that they allow us to see that some early problems in understanding aspects of mind can later be solved by at least some children. This, however, is not to say that *all* aspects of a theory of mind may ultimately develop in any person with autism (cf. Baron–Cohen 1989*c*, this volume, Chapter 4).

Talking about mental states

One central question that can be asked about autistic children is whether they talk at all about mental states as normal children do (Bretherton and Beeghly 1982; Shatz *et al.* 1983), or whether their language is limited to talk about behaviour or action. The spontaneous speech data from the longitudinal study of language-acquisition in autistic and Down's syndrome children were analysed to try to provide some answers about mental-state language in both groups of subjects (Tager-Flusberg 1992).

The transcripts were searched for lexical terms falling into four major categories of psychological states: perception, desire, emotion, and cognition. Table 7.1 lists the words that were found for each of these categories in the samples from both groups of children, and, as can be seen, there was considerable overlap. Shatz *et al.* (1983) point out that many uses of psychological-state terms may not in fact refer to actual mental states. For example children will often use *I want* as a way of saying *give me*. In our analyses of the subjects' transcripts, we therefore used context and other clues to pick out only those utterances in which the child was actually referring to a mental state of desire, emotion, or cognition. For perception, this kind of functional analysis distinguished between talk about perception and calls for joint attention (for example, *Look at this*!).

The major findings from this study revealed that while autistic children were no different from the Down's syndrome children in their talk about perception (across all five senses), and mental states of desire and emotion, they had significantly fewer examples of calls for joint attention, and

Table 7.1 Psychological-state terms used by autistic and Down's syndrome subjects

Desire	care, want, wish[†]
Perception	
Vision	look, see, watch
Hearing	hear, listen, loud,[†] noise
Touch	cold, dry,[†] feel,[*] hard, hot, hurt, messy,[*] ouch, soak,[†] touch, wet, yucky[*]
Smell	smell
Taste	taste,[†] sour,[†] yucky[†]
Emotion	
Behaviour	cry, hug, kiss, laugh,[†] scream,[†] smile
Emotion	angry,[*] bad,[†] better,[*] calm, fun, good,[*] happy, hate, like, love, mad,[*] sad, scare, surprise,[*] upset,[†] worry[†]
Cognition	believe,[*] dream, figure,[*] forget,[*] guess,[*], idea,[*] know, make believe,[†] mean,[*] pretend, remember, think, trick,[*] understand, wonder

[*] Down's syndrome children only
[†] Autistic children only

references to cognitive mental states (see Fig. 7.1). The significant differences between the groups were not surprising. The data on the lack of calls for attention support findings from other studies of non-verbal joint-attention behaviour. The results suggest that problems with joint attention persist even in verbal autistic children. In addition, the relative lack of language referring to cognitive or epistemic states fits well with the theory-of-mind hypothesis. If autistic children do not understand minds, they are unlikely to talk about their contents. Incidentally, the paucity of talk about cognition was not due to the lack of appropriate syntax, since four of the six autistic children did use complement constructions with other verbs (for example, 'Alison said "See you tomorrow"'). In contrast, all four of the Down's syndrome children whose grammars included complementation did also provide examples of cognitive mental-state reference.

Perhaps the more interesting findings from this study are that these autistic children *do* talk about desire and emotion. They even demonstrated some understanding of the causal links between emotions and the situations which cause them, or their consequences. In a simple sense, then, verbal autistic children are not strict behaviourists. They have acquired some of the components of a belief–desire psychology, which they use to explain action. These findings from naturalistic language data complement the findings from some recent experimental studies by Baron-Cohen (1991) and Tan and Harris (1991).

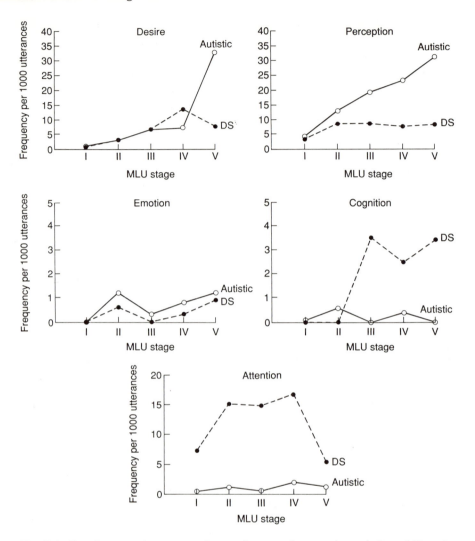

Fig. 7.1. Developmental patterns of mental-state references in autistic and Down's syndrome children.

According to Wellman (1990; this volume, Chapter 2), normal children develop a desire psychology by the age of about two; significantly earlier than they understand belief. The autistic children in this study demonstrate the acquisition of some aspects of an early understanding of mind, and fit the general developmental picture. How then might we interpret the relative absence of epistemic state terms? It may be that most autistic children never even come to a basic understanding of belief, so they do not talk about it or related cognitive states. These, then, may be the children who also fail on tasks that tap false-belief understanding (cf. for example Baron-Cohen *et al.* 1985). Alternatively, it may well be that some children, even the ones from our longitudinal study, do eventually come to talk about some epistemic states. Even if they do acquire these mental-state terms, they might fail to demonstrate the conceptual understanding of belief (as is true for normal three-year-olds: Wellman 1990). Thus some children with autism, while comprehending in a limited sense the meanings of terms like *think, know*, or *believe*, may still fail the false-belief task. It may even be the case that for some children, experience in using these cognitive-state terms and the syntax of embeddings more generally are instrumental in promoting the conceptual developments that underlie the acquisition of a belief-based psychology.

Conversations with autistic children

Even older high-functioning verbal people with autism are difficult to talk to. Conversations are limited to a small number of topics, often revolving around those few which are the special interest of the autistic person. This difficulty autistic children have generating a wide range of topics may parallel their deficit in spontaneously generating symbolic play sequences (Baron-Cohen 1987; Lewis and Boucher 1988). Even when the autistic person does engage in conversation the non-autistic listener has a hard time participating in the discourse or learning much from his or her conversational partner (Paul 1987). In this final section I take up the issue of what kinds of impairments autistic children show in their conversational skills, focusing on the ability to maintain an ongoing *topic* of conversation.

Bloom and her colleagues found that as young normally developing children became more advanced linguistically, they were more able to respond to their mothers' utterances in a contingent, or topically-related way (Bloom *et al.* 1976). The most significant change in their conversational competence at later stages of language development was the increased ability to add new relevant information to the ongoing topic of discourse. In order to investigate these changes in autistic children, we employed the coding scheme devised by Bloom *et al.* (1976) to the language samples collected in the longitudinal study (Tager-Flusberg and Anderson 1991).

Four samples spanning a twelve-month period were coded from each child in the study. Only adjacent utterances from the child (those following directly after a mother's utterance) were selected for analysis. They were coded as contingent (related to the topic of the mother's prior utterance) or non-contingent (not topically related). Contingent utterances were then divided further according to the means used to maintain topic. If the child simply imitated, provided a routine or yes/no response, or recoded the previous utterance, these were classified as not adding new information. Child

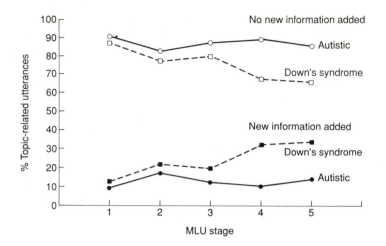

Fig. 7.2. Distribution of contingent utterances that do and do not add new information.

utterances which expanded on the mother's prior utterance, perhaps adding a new topic, or opposing something the mother said, were classified as adding new information.

The main findings from this study were that at the early stages, when MLU was between 1.0 and about 3.0, there were few differences between the autistic and the Down's syndrome children. They were all relatively more contingent than non-contingent in their conversations. As MLU grew longer than 3.0, the groups began to diverge. The Down's syndrome children, like the normal children in Bloom and colleagues' study, became more contingent at this stage, primarily by increasing the proportion of responses that added new information, as can be seen in Fig. 7.2. Thus, as their linguistic abilities advanced, so too did their conversational abilities: the ability to use longer and syntactically more complex sentences allowed the children to converse in a more interesting way, telling their mothers things they may not have known. In contrast, the autistic children showed no change in the proportion of contingent utterances, or in the proportion of utterances with new information. Even though the autistic children showed linguistic advances, there were no changes in their conversational competence.

In a related study we coded the functions of the wh-questions asked by the autistic and Down's syndrome children in the same set of transcripts. The main difference between the groups was that the autistic children asked significantly fewer questions that would elicit new information from their mothers. The groups were not different in the proportion of other types of questions (for example, asking permission, requests for clarification) that they addressed to their mothers (Tager-Flusberg 1989*b*).

Unlike normal or Down's syndrome children, autistic children do not seem to develop the understanding that conversations ought to entail the exchange of information. This appears to be at the heart of what makes communication with autistic people so difficult (Paul and Cohen 1984). This deficit in communication in linguistically-advanced autistic children relates back to the problems that prelinguistic autistic children have with early aspects of communication — the paucity of protodeclaratives. From the onset of intentional communication through the acquisition of a formal linguistic system, autistic children demonstrate a fundamental impairment in their understanding that communication and language exist for the exchange of information or knowledge. Because understanding where information comes from is at the core of understanding belief, it is no wonder that autistic children show impairments in their acquisition of a theory of mind.

SUMMARY

Throughout this chapter I have argued, from several different perspectives, that language is inextricably tied to our understanding of minds. The specialized and multifaceted forms and functions of human language appear to have been designed for communicating with other people about mental states. One of the primary functions of language, to serve as a major source of knowledge, is impaired in autistic children even in the prelinguistic period. It is this impairment which links deficits in joint attention, later problems with communication, and the understanding of belief.

The child's developing understanding of minds is quite remarkable; so too is her or his acquisition of language. It is not surprising that these twin accomplishments of the pre-school period are what make the normal child a full-fledged member of society. Until we know how to change the conceptual worlds of autistic people, they will be on the fringes of that society, because the social world will remain incomprehensible to them.

Acknowledgement

Support for the preparation of this chapter was provided by a grant from the National Institute of Deafness and Other Communication Disorders (RO1 DC 01234).

REFERENCES

Ackerman, B. (1983). Form and function in children's understanding of ironic utterances. *Journal of Experimental Child Psychology*, **31**, 487–507.

Applebee, A. (1978). *The child's conception of story: ages two to seventeen.* University of Chicago Press.

Astington, J. W., Harris, P. L., and Olson, D. R. (1988). *Developing theories of mind.* Cambridge University Press.

Ball, J. (1978). *A pragmatic analysis of autistic children's language with respect to aphasic and normal language development.* Unpublished dissertation, Melbourne University.

Baltaxe, C. A. M. (1977). Pragmatic deficits in the language of autistic adolescents. *Journal of Pediatric Psychology*, **2**, 176–180.

Baltaxe, C. A. M., and Simmons, J. Q. (1985). Prosodic development in normal and autistic children. In *Communication problems in autism*, (ed. E. Schopler and G. Mesibov) Plenum, New York.

Baron-Cohen, S. (1987). Autism and symbolic play. *British Journal of Developmental Psychology*, **15**, 139–48.

Baron-Cohen, S. (1988). Social and pragmatic deficits in autism: cognitive or affective? *Journal of Autism and Developmental Disorders*, **18**, 379–402.

Baron-Cohen, S. (1989a). Perceptual role-taking and protodeclarative pointing in autism. *British Journal of Developmental Psychology*, **7**, 113–27.

Baron-Cohen, S. (1989b). Joint attention deficits in autism: towards a cognitive analysis. *Development and Psychopathology*, **1**, 185–189.

Baron-Cohen, S. (1989c). The autistic child's theory of mind: a case of specific developmental delay. *Journal of Child Psychology and Psychiatry*, **30**, 285–98.

Baron-Cohen, S. (1991). Do people with autism understand what causes emotion? *Child Development*, **62**, 385–95.

Baron-Cohen, S., Leslie, A. M., and Frith, U. (1985). Does the autistic child have a 'theory of mind?' *Cognition*, **21**, 37–46.

Bates, E., Camaioni, L., and Volterra, V. (1975). The acquisition of performatives prior to speech. *Merrill–Palmer Quarterly*, **21**, 205–26.

Bloom, L., Rocissano, L., and Hood, L. (1976). Adult–child discourse: developmental interaction between information processing and linguistic knowledge. *Cognitive Psychology*, **8**, 521–52.

Bretherton, I., and Beeghly, M. (1982). Talking about internal states: the acquisition of an explicit theory of mind. *Developmental Psychology*, **18**, 906–21.

Bretherton, I., McNew, S., and Beeghly-Smith, M. (1981). Early person knowledge as expressed in gestural and verbal communication: when do infants acquire a 'theory of mind'? In *Infant social cognition* (ed. M. E. Lamb and J. R. Sherrod). Erlbaum, Hillsdale, NJ.

Brown, R. (1973). *A first language.* Harvard University Press, Cambridge, MA.

Charney, R. (1980). Speech roles and the development of personal pronouns. *Journal of Child Language*, **7**, 509–28.

Crystal, D. (1975). *The English tone of voice.* Edward Arnold, London.

Curcio, F. (1978). Sensorimotor functioning and communication in mute autistic children. *Journal of Autism and Childhood Schizophrenia*, **8**, 281–92.

Foster, S. H. (1990). *The communicative competence of young children.* Longman, London.

Frick, R.W. (1985). Communicating emotion: the role of prosodic features. *Psychological Bulletin*, **97**, 412–29.

Frith, U. (1989). *Autism: explaining the enigma*. Basil Blackwell, Oxford.

Furrow, D. (1984). Young children's use of prosody. *Journal of Child Language*, **11**, 203–13.

Hobson, R.P. (1990). On the origins of self and the case of autism. *Development and Psychopathology*, **2**, 163–81.

Hurtig, R., Ensrud, S., and Tomblin, J.B. (1982). The communicative function of question production in autistic children. *Journal of Autism and Developmental Disorders*, **12**, 57–69.

Kanner, L. (1946). Irrelevant and metaphorical language in early infantile autism. *American Journal of Psychiatry*, **103**, 242–6.

Kasari, C., Sigman, M., Mundy, P., and Yirmiya, N. (1990). Affective sharing in the context of joint attention interactions of normal, autistic and mentally retarded children. *Journal of Autism and Developmental Disorders*, **20**, 87–100.

Leslie, A.M., and Happé, F. (1989). Autism and ostensive communication: The relevance of metarepresentation. *Development and Psychopathology*, **1**, 205–12.

Lewis, V., and Boucher, J. (1988). Spontaneous, instructed and elicited play in relatively able autistic children. *British Journal of Developmental Psychology*, **6**, 315–24.

Limber, J. (1973). The genesis of complex sentences. In *Cognitive development and the acquisition of language* (ed. T.E. Moore). Academic Press, New York.

Loveland, K.A. (1984). Learning about points of view: spatial perspective and the acquisition of I/you. *Journal of Child Language*, **11**, 535–56.

Loveland, K.A., and Landry, S. (1986). Joint attention and language in autism and developmental language delay. *Journal of Autism and Developmental Disorders*, **16**, 335–49.

Lutz, C. (1987). Goals, events and understanding in Ifaluk emotion theory. In *Cultural models in language and thought* (ed. D. Holland and N. Quinn) Cambridge University Press.

Maratsos, M. (1976). *The use of definite and indefinite reference in young children*. Cambridge University Press.

Menyuk, P. (1969). *Sentences children use*. MIT Press, Cambridge MA.

Mermelstein, R. (1983). The relationship between syntactic and pragmatic development in autistic, retarded, and normal children. Paper presented at the Eighth Annual Boston University Conference on Language Development, Boston, MA.

Mundy, P., and Sigman, M. (1989). The theoretical implications of joint-attention deficits in autism. *Development and Psychopathology*, **1**, 173–83.

Mundy, P., Sigman, M., and Kasari, C. (1990). A longitudinal study of joint attention and language development in autistic children. *Journal of Autism and Developmental Disorders*, **20**, 115–23.

Nelson, K. (ed.) (1986). *Event knowledge: structure and function in development*. Erlbaum, Hillsdale, NJ.

Ornitz E.M., and Ritvo, E.R. (1976). The syndrome of autism: a critical review. *American Journal of Psychiatry*, **133**, 609–22.

Paul, R. (1987). Communication. In *Handbook of autism and pervasive developmental disorders* (ed. D.J. Cohen and A.M. Donellan). Wiley, New York.

Paul, R., and Cohen, D.J. (1984). Responses to contingent queries in adults with mental retardation and pervasive developmental disorders. *Applied Psycholinguistics*, **5**, 349–57.

Paul, R. and Cohen, D. J. (1985). Comprehension of indirect requests in adults with mental retardation and pervasive developmental disorders. *Journal of Speech and Hearing Research*, **28**, 475–479.

Perner, J. (1991). *Understanding the representational mind*. MIT Press/Bradford, Cambridge, MA.

Pierce, S., and Bartolucci, G. (1977). A syntactic investigation of verbal autistic, mentally retarded and normal children. *Journal of Autism and Childhood Schizophrenia*, **7**, 121–34.

Pinker, S., and Bloom, P. (1990). Natural language and natural selection. *Behavioral and Brain Sciences*, **13**, 707–84.

Pronovost, W., Wakstein, M., and Wakstein, P. (1966). A longitudinal study of the speech behavior and language comprehension of fourteen children diagnosed atypical or autistic. *Exceptional Children*, **33**, 19–26.

Rutter, M., Greenfield, D., and Lockyer L. (1967). A five-to fifteen-year follow-up study of infantile psychosis: II Social and behavioral outcome. *British Journal of Psychiatry*, **113**, 1183–99.

Shantz, C. (1983). Social cognition. In *Handbook of child psychology: cognitive development* (ed. J. H. Flavell and E. M. Markman), Vol. 3 (ed. P. H. Mussen). Wiley, New York.

Shatz, M., Wellman, H., and Silber, S. (1983). The acquisition of mental verbs: a systematic investigation of first references to mental state. *Cognition*, **14**, 301–21.

Simmons, J. Q., and Baltaxe, C. A. M. (1975). Language patterns of adolescent autistics. *Journal of Autism and Childhood Schizophrenia*, **5**, 333–51.

Sperber, D., and Wilson, D. (1986). *Relevance: communication and cognition*. Harvard University Press, Cambridge, MA.

Tager-Flusberg, H. (1981*a*). Sentence comprehension in autistic children. *Applied Psycholinguistics*, **2**, 5–24.

Tager-Flusberg, H. (1981*b*). On the nature of linguistic functioning in early infantile autism. *Journal of Autism and Developmental Disorders*, **11**, 45–56.

Tager-Flusberg, H. (1989*a*). A psycholinguistic perspective on language development in the autistic child. In *Autism: new directions in diagnosis, nature, and treatment* (ed. G. Dawson). Guilford, New York.

Tager-Flusberg, H. (1989*b*). *The development of questions in autistic and Down syndrome children*. Gatlinburg Conference on Research and Theory in Mental Retardation. Gatlinburg, TN.

Tager-Flusberg, H. (1991). A longitudinal analysis of the acquisition of personal pronouns in autistic children. Unpublished MS, University of Massachusetts.

Tager-Flusberg, H. (1992). Autistic children's talk about psychological states: deficits in the early acquisition of a theory of mind. *Child Development*, **63**, 161–72.

Tager-Flusberg, H., and Anderson, M. (1991). The development of contingent discourse ability in autistic children. *Journal of Child Psychology and Psychiatry*, **32**, 1123–34.

Tager-Flusberg, H., and Calkins, S. (1990). Does imitation facilitate the acquisition of grammar? Evidence from autistic, Down syndrome and normal children. *Journal of Child Language*, **17**, 591–606.

Tager-Flusberg, H., Calkins, S., Nolin, T., Baumberger, T., Anderson, M., and Chadwick-Dias, A. (1990). A longitudinal study of language acquisition in autistic and Down syndrome children. *Journal of Autism and Developmental Disorders*, **20**, 1–21.

Tan, J., and Harris, P.L. (1991). Autistic children understand seeing and wanting. *Development and Psychopathology*, **3**, 163–74.

Van Lancker, D., Cornelius, C., and Kreiman, J. (1989). Recognition of emotional-prosodic meanings in speech by autistic, schizophrenic, and normal children. *Developmental Neuropsychology*, **5**, 207–26.

Von Frisch, K. (1950). *Bees, their vision, chemical senses and language*. Cornell University Press, Ithaca, NY.

Wellman, H. (1990). *A child's theory of mind*. MIT Press/Bradford, Cambridge, MA.

Wetherby, A. (1986). Ontogeny of communication functions in autism. *Journal of Autism and Developmental Disorders*, **16**, 295–316.

Winner, E. (1988). *The point of words*. Harvard University Press, Cambridge, MA.

Winner, E., Windmueller, G., Rosenblatt, E., Bosco, L., and Best, E. (1987). Making sense of literal and nonliteral falsehood. *Metaphor and Symbolic Processes*, **2**, 13–32.

Zaitchik, D. (1991). Is seeing really only believing? *Cognitive Development*, **6**, 91–103.

8

The theory of mind deficit in autism: evidence from deception

BEATE SODIAN AND UTA FRITH

DECEPTION—REVISITED

In 1978 Premack and Woodruff raised the question 'does the chimpanzee have a theory of mind?' They argued that acts of deception would be a good indicator of the presence of a theory of mind in the deceiver, on the grounds that deception involves the manipulation of others' behaviour by influencing their beliefs about reality. Woodruff and Premack (1979) trained chimpanzees to mislead an opponent by way of deceptive pointing. After several months of training, some of these animals pointed consistently to an empty container when a hostile trainer was present; that is, they showed behaviour that had a deceptive *effect*. Does this impressive achievement indicate theory of mind? Not necessarily. In order to infer the presence of a theory of mind it is necessary to show that the chimpanzees had deceptive *intent*, i.e. that they pointed to an empty container with the intention *to make an opponent believe* that there was food in the container. Such a conclusion is difficult to draw on the basis of data such as those of Woodruff and Premack. It is possible to argue that the chimpanzees may have learned to 'mindlessly' employ a behavioural routine without conceptualizing their victim's beliefs.

In the debate about the significance of the findings of Woodruff and Premack, Dennett (1978), Bennett (1978) and Harman (1978) outlined how conclusive evidence could be gathered on an individual's ability to represent others' beliefs: the subject is aware that he or she and another person observe a certain state of affairs x. Then, in the absence of the other person, the subject witnesses an unexpected change in the state of affairs from x to y. The subject now knows that y is the case. Does the subject understand that the other person still believes that x is the case?

Wimmer and Perner (1983) introduced this 'unexpected change' paradigm to the study of the child's theory of mind. The results of their initial and many subsequent studies employing variations of this paradigm are the main source of evidence for a 'cognitive watershed' in children's understanding of the mind around the age of about four years. Children above the age of four years understand that the other person believes to be true what they know to be false, whereas younger children indicate that the other person will believe what they know to be true. Thus, children below the age of about four years seem to lack the understanding of belief that is central to our common-sense interpretation of human action.

Baron-Cohen *et al.* (1985) proposed that it is precisely the lack of this interpretative framework, i.e., the failure to develop a 'theory of mind', that underlies the severe cognitive, social, and communicative impairments observed in autistic individuals. Autistic children's performance in the unexpected-change paradigm confirmed this prediction: autistic children with a mental age well above four failed the false-belief task that was passed by normal and Down's syndrome children of lower mental age. Since then considerable evidence has been accumulated that is consistent with the assumption that autistic persons fail to develop an informational-representational conception of the mind.

Four-year-old children's *success* in the unexpected-change paradigm provides good evidence for their understanding of belief; it is hard to see how they could correctly assess another person's wrong belief about a state of affairs *and* infer how that person's wrong belief will constrain his or her future actions (Wimmer and Perner 1983) without a genuine understanding of belief. However, it has been debated whether three-year-old normal children's and autistic children's *failure* on the false-belief task is equally compelling evidence for the absence of a concept of belief. Perner *et al.* (1987), who systematically ruled out a series of alternative explanations for three-year-olds' poor performance on the false-belief task, concluded that these children's failure to attribute false beliefs to others does indicate a conceptual deficit. However, critics argued that the task of attributing beliefs to story figures (or even to real people) may in itself be too computationally complex or not motivating enough for very young children. These children may nevertheless display a genuine understanding of belief when given an opportunity to *manipulate* other persons' beliefs in highly-motivating contexts (Chandler *et al.* 1989). Thus, a decade after Woodruff and Premack's paper on deception in chimpanzees, cognitive developmentalists became interested in the emergence of deceptive action in children.

The interest has focused almost exclusively on the question of whether three-year-old children are capable of intentional deception, with proponents of an early-onset view arguing that children as young as $2\frac{1}{2}$ years can intentionally deceive, and proponents of a late-onset view arguing that on the rare occasions when children younger than four years show behaviours that have deceptive effects, they do not do so with deceptive intent. It may seem futile to try to determine whether children are capable of deception at the ages of $2\frac{1}{2}$, 3 or $3\frac{1}{2}$ years. Note, however, that the issue here is not whether belief understanding emerges at the age of 3 or $3\frac{1}{2}$, but whether or not there is conceptual change in the young child's developing theory of mind. If it could be shown that, contrary to previous findings, children possess a genuine understanding of belief at well below the age of four years, this would support the view that the common-sense psychological theory that we use in explaining and predicting human behaviour emerges at a very early age. If, on the other hand, the empirical findings indicate a specific deficit in three-year-olds' (and younger children's) understanding of belief, this would support the view that a representational–informational conception of the mind is acquired relatively late and replaces an earlier, non-representational view of the mind (Perner 1991).

CAN THREE YEAR OLD CHILDREN DECEIVE?

Surprisingly little cognitive developmental research has directly addressed the development of deception in children. Studies of children's performance in strategic hiding games that require the flexible use of deceptive strategies showed that only around the age of five years do children start to manipulate their opponents' intentional states deliberately (DeVries 1970; Gratch 1964; Shultz and Cloghesy 1981). However, success on these tasks requires a number of skills besides the ability to manipulate others' beliefs intentionally. Proponents of an early-onset view (for example, Dunn 1991) argue that anecdotes and observational data indicate that children start to lie and deceive at a much earlier age (although they may initially not be very good at it). As with anecdotes on deception in non-human primates (see Premack 1988 for a critique; also see Whiten, this volume, Chapter 17) it is often difficult to decide whether such early lies are acts of genuine deception, i.e. whether the child understands how her utterance will influence the listener's beliefs.

Take, for example, the case of Piaget's daughter Jacqueline, who, at the age of eighteen months, managed to be taken out of her playpen by deceptively indicating that she had to go to the bathroom (Piaget 1954). This anecdote resembles the one about Ashley's dog (reported by Dennett 1978), who made her master get up from her (the dog's) favourite chair by scratching the door and giving Ashley the impression that she had decided to go out. Both Jacqueline's and the dog's ploys *worked* by deception. However, both of them may have simply used a behavioural routine to manipulate the others' behaviour in the desired way without conceptualizing the effect of their strategy on their victim's belief. Thus, an important criterion for distinguishing genuine from apparent acts of deception is that the critical act is not *known* to elicit the opponent's action in the current setting (Perner 1991, Chapter 8).

In an early, conceptually sophisticated, analysis of diary data Clara and William Stern distinguished between genuine lies, i.e. false utterances that were made with the intention to deceive, and various forms of mistakes, fantasies, and 'pseudo'-lies (Stern and Stern 1909). Interestingly, the age at which 'true-lies' were first observed was four years. However, pseudo-lies, such as 'no' responses when accused of a forbidden action, were observed much earlier. The reason for classifying them as pseudo-lies was that Stern and Stern regarded them as affective responses rather than intentionally false statements of fact (the young child denies a misdeed to escape blame or because he does not wish it to be remembered). Anecdotal examples, for instance 'I didn't break the lamp and I won't do it again' (Vasek 1986) abound to support the view that young children may know some effective strategies for manipulating others' behaviour (for example, for avoiding reprimand) without understanding their effect on others' beliefs.

This should be kept in mind when interpreting findings from experimental studies that three-year-olds and even younger children can be brought to employ strategies that have deceptive effects. Chandler *et al.* (1989) report the use of 'deceptive' strategies in children as young as $2\frac{1}{2}$ in the following task. A treasure was hidden by a puppet, which left visible tracks to the hiding-place (one of four boxes). The child's task was to make it difficult for another person to find the treasure. Many $2\frac{1}{2}$- and 3-year-olds used the obvious strategy of wiping out the tracks. Sodian *et al.* Experiment 1 (1991) replicated the finding that such young children could be brought to wipe out tell-tale tracks, but only after massive prompting: 'What about the tracks. Can you do something to the tracks so that [name] won't find the ball'?. Three-year-olds' attention had to be drawn explicitly to the tracks before they did anything to them; whereas four-year-olds spontaneously eliminated them without specific prompting.

Chandler *et al.* claim that three-year-olds and even younger children in their experiment wiped out tracks with deceptive intent. However, they could just have enjoyed the activity of wiping out tracks, as our colleague Paul Luetkenhaus, expert on toddlers' intentional action, suggested when he first heard about Chandler and colleagues' findings: 'Give a two-year-old a sponge and she'll wipe out anything.' To test this interpretation we added a second condition to Chandler *et al.*'s task (Sodian *et al.* 1991, Expt. 2). In this control condition the objective was to make it easy for a friendly person to find the treasure. In the other condition, the child was instructed to make it difficult for the other person, because the opponent would keep the treasure if he found it. In both conditions children were given the option of either wiping out the tracks or reinforcing their clarity by adding an extra line.

Less than a third of the three-year-olds whom we tested applied the strategies of reinforcing and wiping out tracks in the correct selective way, whereas almost all four-year-olds did. Moreover, while 6 (of 20) three-year-olds correctly wiped out for the opponent and reinforced for the co-operator, 3 three-year-olds showed exactly the opposite pattern. Thus, three-year-olds did not seem to choose their strategy with the objective of the game in mind.

Chandler *et al.* also report that three-year-old children laid false tracks to empty locations, although only after some prompting, and that they pointed deceptively when questioned. We found very little evidence of spontaneous laying of false tracks in children younger than four but we were fairly successful in getting children to accept this technique as 'a good idea' when we modelled it for them. However, only 3 (of 14) three-year-olds used the technique for the competitor only; the others either used it for both cooperator and competitor or for the cooperator only. In contrast, 10 (of 16) four-year-olds showed the correct selective use. Similarly, while most four-year-olds pointed to an empty location when questioned by the competitor, only very few three-year-olds did so spontaneously. When 'deceptive' pointing was observed in the younger children, it occurred only after they were prohibited by the experimenter from uncovering the treasure to the competitor.

Thus, the results of our study indicate that three-year-old children can be led to produce deceptive ploys, but that they show no clear understanding of their effect. The finding that children younger than four used deceptive and informative strategies indiscriminately, whether asked to mislead a competitor or to inform a collaborator, seriously undermines the view that these young children possess a genuine understanding of belief.

Hala *et al.* (1990) have gathered more evidence in support of their position that three-year-olds show a genuine understanding of deception. They report that three-year-olds are just as good as four-year-olds at selecting an appropriate response according to whether they are asked to help or hinder an experimenter in finding a treasure. At first sight, their procedure appears to be similar to the one used in Sodian *et al.* (Expt 2); however, closer scrutiny reveals that a confounded assessment of children's selectivity was created: when children were asked to deceive the experimenter, a track to the hiding-place was already in place. In contrast, when children were asked to help the experimenter, no track was in place. This difference may have biased children to produce 'false tracks' more often in the deception condition. Although they could obviously lay new trails in either condition, the option of laying a true trail had been pre-empted in the deception condition, but not in the co-operative condition. Moreover, children were systematically warned against providing accurate information in the deception condition only: throughout this procedure '. . . subjects were cautioned not to blurt out the true hiding location' (Hala *et al.* 1990, p. 87).

Let us now review the evidence for a late-onset view. Support for the assumption that three-year-olds do not possess a genuine understanding of deception comes from studies showing their pervasive failure to engage in acts of deception. Several studies have used variations of Woodruff and

Premack's paradigm, and found drastic differences between three- and four-year-olds in the use of a deceptive pointing strategy (LaFreniere 1988; Russell *et al.* 1991; Sodian 1991). Russell *et al.* (1991) demonstrated three-year-olds' failure particularly impressively. In the first phase of their experiment, children did not know in which of two boxes a piece of chocolate was hidden, but were asked to point to one box. A confederate of the experimenter looked inside the box which the child indicated. If the chocolate happened to be in that box, she kept it, otherwise the child could keep it. This was repeated for 15 trials to make sure that the children understood the effect their pointing had. Then the original boxes were replaced by boxes with windows facing the child so that the child could see where the chocolate was. Children now had a chance to deliberately point to the empty box in order to mislead the experimenter. Three-year-olds did not do so spontaneously, and failed to do so on the following 20 trials in spite of their growing frustration at losing to the experimenter. In contrast 10 (of 16) four-year-olds pointed to the empty box on the first trial, and all except for one did so at some later point.

Another study showed that three-year-olds advised a story character to point truthfully even when this story figure explicitly said that she did not want the competitor to find the reward, and that three-year-olds gave a truthful description of a state of affairs ('the box is open', i.e. the opponent can get the sweet) even when a false description that would be instrumental in keeping the opponent away from the reward was modelled for them ('do you want to say the box is locked?') (Sodian 1991).

While these tasks may seem a little contrived, Joan Peskin (1992) found that three-year-olds also fail to misinform in a situation of high affective involvement that resembles real-life conflict situations between children. In her experiment, three- to five-year-old children faced the problem of an opponent's desiring always exactly the same object that the child herself preferred. Since the opponent got the first choice, the only way of securing the desired object for herself was for the child to tell him that she wanted a different one from the one she in fact preferred. More than half the four-year-olds and about 80 per cent of the five-year-olds used this strategy. In contrast, less than 20 per cent of the three-year-olds did so. While four-year-olds' performance improved over a series of trials, three-year-olds persistently told the truth despite mounting frustration.

All these findings are consistent with the assumption that children below the age of about four years do not possess a concept of belief. However, they are also consistent with a number of other explanations. It is possible, for instance, that three-year-olds differ from four-year-olds in their understanding of the competitive nature of deception tasks — i.e., that they do not understand that the competitor's gain is their loss, or that they are simply not motivated enough to try to win in such situations. This explanation does of course not account for their failure on the 'standard' false-belief attribution task, but it has been argued that the hypothetical story format of these tasks may make them unnecessarily difficult for young children. Another possibility is that three-year-olds differ from older children in their ability to inhibit a response towards a salient object (see Russell *et al.* for such a proposal). They may be 'irresistibly drawn' to point towards the location where the object actually is (or to describe truthfully the object they have in mind), even though they may understand that they have the possibility to manipulate another person's belief. Thus, their problem on both false-belief attribution and deception tasks may be a lack of executive control of behaviour rather than a conceptual deficit in the representation of belief.

Therefore, demonstrations of three-year-olds' failure on deception tasks are not sufficient to support the assumption of a *specific* conceptual deficit. Rather, it is necessary to show that these children succeed on control tasks that pose similar requirements to the deception tasks except that they do not involve the manipulation of *beliefs*. Two studies have employed this method, and gathered some first evidence for the specificity of the deficit.

Peskin (1992) showed that, while almost all three-year-olds failed to misinform the opponent, they knew how to exclude him physically from the game. *Physical manipulation* was contrasted systematically with *belief manipulation* in a study of deception and sabotage (Sodian 1991, Exp. 3). The rationale for this study was as follows: if young children's failure on deception tasks is specific to the manipulation of beliefs, then they should be able to hinder an opponent from attaining a desired object by blocking his physical access to this object, while failing to do so by manipulating his beliefs. A good way to hinder someone from taking something out of a box is to lock the box. Only if there is no lock available is it necessary to resort to such sophisticated tactics as deception. Of course, young children's willingness to lock something from a competitor is only informative if it can be shown that they do not use this strategy indiscriminately. Thus, it is necessary to show that they lock the box for an opponent but leave it open for a friendly person.

Three-year-olds' performance on this task indicates that they can discriminate quite easily between co-operator and opponent in this situation. Over 70 per cent locked the box for the opponent and left it open for the cooperator on the first pair of trials, and only 2(of 26) children failed to do so on the second trial pair. In contrast, when there was no lock available, only 26 per cent of these three- to four-year-old children lied to the opponent ('the box is locked'), when asked whether they wanted to (truthfully) say that it was open or whether they wanted to say it was locked, even though it was pointed out to them that the opponent could not see that the box was unlocked, and that being lazy, he wouldn't bother coming all the way to the candybox if he thought it wasn't worth the effort anyway.

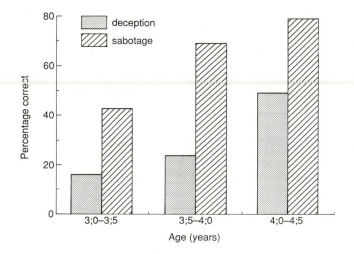

Fig. 8.1. Deception and sabotage in normal children.

In a second task, physical manipulation was contrasted with deceptive pointing. In the deceptive pointing condition the child was asked to which box (the full or the empty one) she wanted to point for the cooperator or the competitor, respectively. In the parallel sabotage task, she was given the choice of which box she wanted to lock. In the competitive condition, the decision to lock the full box is fairly straightforward. In the cooperative condition, it would appear reasonable to leave both boxes open. However, if one is required to lock one of the boxes, then it only makes sense to lock the empty one so that the cooperator (who gives the child the Smartie when he can find it) does not open this one and leave, disappointedly. Note that to pass this task, the child has to operate on the empty box, just as she is required to point to the empty box in the deception task. Therefore, if it can be shown that children lock the empty box in the cooperative sabotage condition while they fail to point to the empty box in the deception condition, this provides evidence against the view that their failure to point deceptively is due to a problem in inhibiting a response towards a location where a salient object is hidden.

The results support the view that children's failure on the deceptive pointing task is, in fact, specific to deception. Figure 8.1 shows that even in the youngest age-group (3;0 to 3;6-year-olds) about half the children performed correctly on *both* sabotage trials, i.e. locked the full box to hinder the competitor from gaining the reward, and the empty box to help the cooperator attain the reward, while less than 20 per cent correctly pointed to the empty box in the deception condition. As expected, the complex sabotage task did pose some difficulty for 3- to $4\frac{1}{2}$-year-old children. Only about 60 per cent correctly distinguished between cooperator and competitor on the first trial pair (all mistakes were mistakes of 'overprevention', i.e. the full box was locked in both conditions). However, as Fig. 8.1 shows, performance on this complex sabotage task was substantially better at all ages than performance on deceptive pointing. This finding (a significant difference between performance on complex sabotage and on deceptive pointing) was replicated for normal three-year-olds by Sodian and Frith (1992).

These results support the assumption of a specific conceptual deficit in three-year-old children. Results of the sabotage tasks indicate that three-year-olds certainly understand that they can hinder an opponent from winning a reward in an experimental situation, and that they are quite motivated to do so. In fact, many three-year-olds spontaneously tried to hinder the opponent from opening the box by physical means even when there was no lock available (i.e., they truthfully indicated where the treasure was hidden, but tried to hold the box shut with their hands, to remove it from the table, etc.). The finding that even young three-year-olds are better at employing a quite complex behavioural manipulation strategy that involves a counter-intuitive action of operating on an empty location than at using a fairly straightforward belief-manipulation strategy gives additional support to the view that they have acquired some expertise in the manipulation of others' behaviour, but fail to understand that they can manipulate others' beliefs.

In summary, it appears that proponents of an early-competence view have as yet failed to demonstrate that children younger than four can engage in acts of genuine deception. The conceptual deficit view, on the other hand, is supported by the finding of sharp developmental progression between the ages of three and four years in various deception tasks, and by evidence for the specificity of three-year-olds' difficulty on these tasks.

The reason to review research on deception in normal children in some detail in a book on autism is that the argument that was developed here is relevant to the evaluation of the metarepresentational theory of autism. In the past five years it has been well established that autistic children fail in a wide range of tasks that tap their ability to represent mental states, and specifically their understanding of epistemic states. It has also been established that this failure is not due to mental retardation in general, but is fairly specific to autism, since control groups of mentally retarded (usually Down's syndrome) children of equivalent mental age perform on these tasks like normal pre-school children. However, many of these findings are not only compatible with a specific metarepresentational account of autism, but also with the assumption of a more general social or communicative impairment.

To demonstrate the specificity of the presumed metarepresentational deficit in autistic children we therefore need to show that these children pass control tasks that are similar to the metarepresentational tasks in all important aspects except that they do not require the representation of *mental* states. Baron-Cohen, (this volume, Chapter 4) provides a review of other studies pursuing such a strategy. In studying deception and sabotage in autistic children, we explore whether autistic children's difficulty with false belief is specific to belief-representation, or whether it is due to a more general lack of social understanding or interest.

AUTISTIC CHILDREN'S FAILURE TO DECEIVE:
HOW PROFOUND AND HOW SPECIFIC IS THE DEFICIT?

While normal children's parents often recall their pre-school children's first transparently obvious attempts to deceive them, autistic children's parents notice the striking absence of such behaviours in their children. Sometimes, parents report their relief at observing the first very clumsy 'lie' in their intellectually able autistic adolescent. One mother who was very concerned about her sixteen-year-old autistic son's behaviour problems (attacking her physically, throwing food and furniture), was even more worried about his social naïvety, characterizing him as 'too good,totally honest,and unable to lie'. She reported, for instance, that her son would stick to a given instruction such as 'go to bed at 10 p.m.' whether or not she was there to enforce it.

Systematic investigation of autistic children's ability to lie and deceive has started only very recently; to date, three studies have been conducted, which all found evidence for a severe impairment of belief-manipulation skills in autistic children (Russell *et al.* 1991; Baron-Cohen, 1992; Sodian and Frith, 1992). Russell *et al.* report that autistic children with a verbal mental age of well above four years persistently failed in their 'windows' task — that is, they truthfully pointed to the location where the reward was hidden, and continued to do so on 20 trials despite persistent frustration, whereas a control group of Down's syndrome children easily passed this task, just like normal four-year-olds.

What is the source of this difficulty? Do autistic children fail to understand that they could make the opponent falsely believe that the reward is hidden in the empty box, or is their failure due to an inability to inhibit a behavioural response, as Russell *et al.* suggest? Do autistic children have the prerequisites for understanding such competitive games at all? Do they understand what the opponent's goals are, and do they understand these goals as opposed to their own goals in the game? and, if they understand that the opponent will take a reward away from them, do they care?

We addressed these issues in our study of deception and sabotage in autistic children (Sodian and Frith, 1992). If we want to argue that autistic children's difficulty with deception is due to a specific failure to acquire a concept of belief, we have to rule out the possibility that they fail deception tasks because of a general inability to form 'anti-plans', i.e. plans about how to prevent a competitor from executing his plan and reaching his goal. The 'sabotage' tasks that were described earlier require such anti-planning; but there the anti-plan does not involve a manipulation of the opponent's belief. If autistic children passed these tasks while failing the parallel deception tasks, this would comprise evidence for a specific difficulty with belief-manipulation. We used the two deception (lying and deceptive pointing) and sabotage (locking a box in simple and complex tasks) tasks that we described above. We chose these sabotage tasks as a fairly difficult test for autistic children. As these children often show stereotyped behaviours, would they lock a box in an experimental task indiscriminately for cooperator and an opponent? Success in discriminating between these two conditions would appear to be quite convincing evidence for autistic children's task understanding.

A second aim of our study was to investigate whether some improvement in the autistic children's understanding of belief occurs with ability and experience. Previous studies of false-belief understanding do not suggest that this is the case, although there has been little systematic exploration of developmental progress in autistic children's understanding of belief. However, it seems likely that very high-functioning autistic adolescents begin to use simple deceptive strategies at some point in development. We therefore included a group of very able autistic adolescents in our study (verbal MA 7–12 years, age equivalent). Table 8.1 provides an overview of the experimental and control groups studied.

Whether deceptive strategies are used with deceptive intent or whether they are used as behavioural manipulation strategies that merely have deceptive effect is hard to determine. This is the lesson from Premack and Woodruff's original studies. We attempted to gather some evidence on very able autistic individuals' understanding of deception by asking them to explain a doll's deceptive actions. We also assessed performance on a traditional false-belief attribution task to determine to what extent our deception tasks and the 'standard' false-belief task tap the same abilities.

As Fig. 8.2 and 8.3 show, autistic children's performance on our 'lying' task and the parallel sabotage task ('do you want to say the box is locked or do you want to say it is open?' vs. 'do you want to lock the box or do you want to leave it open?') very clearly supports the assumption of a profound and specific deficit in autistic children's ability to manipulate belief.

Figure 8.2 shows that the 'lying' task was easily passed by almost all normal four-year-olds and by all retarded children with a mental age of about five years. In contrast, almost all autistic children with a mental age between about four and five years failed this task, and even in the group with a verbal mental age between seven and twelve only 60 per cent succeeded.

Table 8.1 Subjects grouped according to diagnosis and verbal mental age*

MA band	Autistic	Mentally retarded	Normal
very low		$n = 11$ CA 10;0–15:0 mean 11;6 MA lower than 4;0 (below test norms)	$n = 19$ CA 3;1–4:0 mean 3;6 (younger group)
low	$n = 8$ CA 6;0–17;9 mean 12;7 MA 4;0–5;5 mean 5:0	$n = 9$ CA 10;0–0–14;1 mean 12;5 MA 4;3–5;3 mean 4;6	$n = 20$ CA 4;1–5;2 mean 4;7 (older group)
middle	$n = 6$ CA 8;9–19;2 mean 14;8 MA 5;6–6;3 mean 5;11	$n = 9$ CA 10;7–16;7 mean 13;2 MA 5;6–7;0 mean 6;3	
high	$n = 5$ CA 12;3–16;6 mean 14;5 MA 7;6–12;0 mean 8;9		

* Verbal MA was assessed by the British Picture Vocabulary Test (BPVS) and the German version of the WPPSI Vocabulary subtest (HAWIVA).
Reproduced by kind permission from Sodian and Frith (1992).

Performance on the 'lying' task was highly correlated with performance on the 'standard' false-belief task, independently of mental age.

The results of the parallel 'sabotage' task (see Fig. 8.3) showed that autistic children happily prevented a competitor from getting a sweet by locking the box in which it was hidden. More importantly, they did not apply this strategy indiscriminately, but left the box open for a cooperative partner, just as normal and mentally retarded children of equivalent mental age did. This is remarkable, as one might have expected autistic children to show stereotyped behaviour (i.e., to lock the box regardless of experimental condition). Thus, when physical means were available, autistic children would help the cooperator and hinder the competitor. They failed in this objective *only* when physical means were not available.

One might argue that the language demands of the 'lying' task made this task more difficult than the parallel 'sabotage' task, regardless of its requirements for mental-state manipulation. Children had to understand

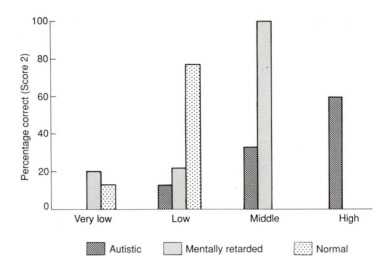

Fig. 8.2. One-box task: deception (lying).

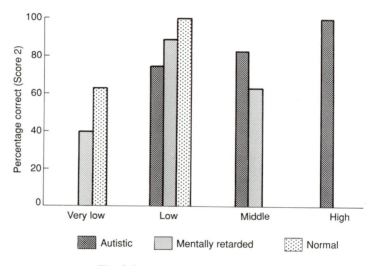

Fig. 8.3. One-box task: sabotage.

fairly elaborate instructions, such as 'Do you want to say it is locked or do you want to say it is open?' While we cannot discount this possibility from our study, a recent experiment posing minimal language demands also attests to autistic children's specific difficulty with deception. Baron-Cohen (1992) explored autistic children's performance in a naturalistic guessing game, the

Penny Hiding Game that has been previously employed with normal children (Gratch 1964; DeVries 1970). This game is a two-person game in which the subject is actively involved, either as a guesser or as a hider. In order to win, the hider minimally has to ensure that the guesser does not SEE the penny. Baron-Cohen calls this '*object occlusion*'. However, keeping the penny out of sight is not sufficient. The game is also about preventing the guesser from getting access to INFORMATION about where the penny might be. Baron-Cohen calls this '*information occlusion*'. The simplest way to ensure information occlusion is to keep the penny behind one's back while transferring (or pretending to transfer) it from one hand to the other, and then to keep it in a closed fist while inviting the guess. It is also essential not to say anything, and to keep the other hand closed, too. A more advanced method of information occlusion is misinformation, i.e. deliberately attempting to mislead the opponent into thinking the penny is in the empty hand. Baron-Cohen's results showed that autistic children enjoyed the game as one of object occlusion, but the majority failed to show even simple information occlusion, while normal four-year-olds and mentally retarded children of comparable mental age proficiently employed information-occlusion strategies.

A charming anecdote of a normally developing boy, aged 2;10 years, illustrates these ideas in another context. The boy, Michael, had been given a pair of attractive crocodile-shaped scissors and, understandably, wanted to take them with him when he went to bed. Equally understandably, his mother did not want him to have the scissors in the bed. Therefore, little Michael hit on the idea of wrapping the scissors in his blanket, 'hiding them from Mummy', as he told his father. When the mother came in and asked him 'What's in your blanket?' he told her 'Scissors hiding'.

Consistent with the results of Russell *et al.* (1991) we also found that most autistic children with a mental age of about four to seven years failed to employ a deceptive pointing strategy (pointing to an empty box) to mislead an opponent, although their failure on this task was not as pervasive as on the 'one box' deception task and not as pervasive as we expected from the findings of Russell *et al.* on a very similar task. This is illustrated in Fig. 8.4.

As Fig. 8.5 shows, on the parallel (complex) sabotage task, almost all autistic children correctly locked the full box when it was the competitor's turn to come and try to get the Smartie. As we outlined above, to make this 'sabotage' task parallel to the deceptive pointing task, it was necessary to introduce a rather complicated cooperative condition: to help the cooperator to find the Smartie right away, it was necessary to lock the empty box and leave the full box open. There was a tendency in autistic children to perform better on the sabotage than the deception tasks (when the full range of scores was taken into account), but the difference failed to reach significance. Thus, there was no unequivocal evidence that autistic children

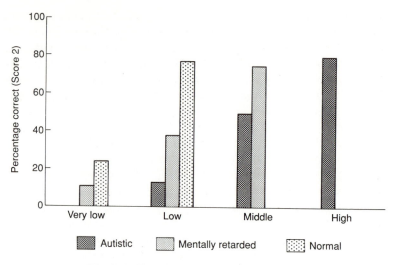

Fig. 8.4. Two-boxes task: deceptive pointing.

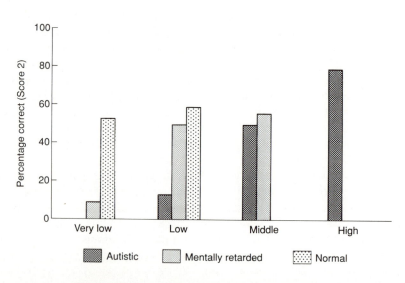

Fig. 8.5. Two-boxes task: complex sabotage.

can employ such *sophisticated* behavioural manipulation strategies while they still fail extremely simple belief-manipulation tasks.

A possible explanation for this finding is that both the cooperative part of the sabotage task and the competitive part of the deception task required children to operate upon an empty box, and thus to inhibit a response directed towards a location where a salient object was hidden. Thus, autistic children's difficulty with the deceptive pointing task may not be specific to belief-manipulation but may reflect a more general problem in the executive control of behaviour, as Russell *et al.* suggest. Autistic children may be 'irresistibly' drawn towards the location where a salient object really is, unable to inhibit 'truthful' pointing or a verbal description of a state of affairs that corresponds to reality. This proposal clearly warrants further exploration (see also Harris, Chapter 11, this volume). However, our results indicated that both handicapped groups had a specific difficulty in the complex sabotage task that could not be accounted for by their problems in belief-attribution or belief-manipulation.

It should also be noted that the proposal of a failure in executive control of behaviour is inconsistent with some recent evidence on autistic children's ability to understand non-mental representations. Leekam and Perner (1991), as well as Leslie and Thaiss (1992) found that autistic children passed a task developed by Zaitchik (1990) that is exactly parallel to the false-belief task but involves understanding of *non-mental* representations ('false' photographs where a photographic representation conflicts with the real world). If these children's failure on the standard false-belief task had been due to an inability to inhibit a response to a salient object, then they should have reacted in the same way on the photographs task. Thus, there is converging evidence from recent studies on autistic children's theory of mind that their deficit is in fact specific to understanding mental representations (see Chapter 6, by Perner, this volume).

We included a heterogeneous population of mentally retarded children in our study, and found that in this group there was significant delay in the acquisition of a theory of mind not only relative to chronological age but also relative to mental age. The tasks that were mastered by normal four-year-olds were generally only mastered by retarded children who were in their teens, and had a mental age of above five. It is possible, therefore, that not only autistic children but also some subgroups of mentally retarded children acquire a concept of belief with a marked delay. Nevertheless, there were still significant differences between mentally retarded and autistic children on the standard belief-attribution task when mental age was taken into account (Sodian and Frith, 1992). Note that the rationale for including mentally retarded children in studies of autistic children's cognitive deficits is not to show that the presumed deficit (or delay) occurs *only* in autistic children and in no other clinical group. Rather, what has to be shown is that the presumed deficit is not simply a general characteristic of retardation. The clear-cut difference we found between autistic and retarded children on the lying task indicates that autistic children have specific difficulties in manipulating beliefs that cannot be attributed to retardation *per se*.

The inclusion of a subgroup of highly able autistic children in our study made it possible to address the question of whether there is developmental progress in autistic children's acquisition of a concept of belief. Do autistic adolescents eventually come to understand belief? We found that there was developmental progress with mental age on the deceptive pointing task, with about 80 per cent of the highest mental-age group (seven to twelve years) succeeding. However, just as in Premack and Woodruff's (1978) original studies, caution is warranted in interpreting this finding as evidence for genuine belief understanding. After all, behaviours that have the effect of deceiving an opponent can be employed without actual deceptive intention on the part of the child. Nevertheless, there were four autistic children who showed perfect performance on both the lying and the deceptive pointing tasks, and also passed the belief-attribution task. These are the most likely candidates to have acquired the ability to understand and manipulate beliefs. Since these individuals were all over twelve years old, and since nobody has yet demonstated the presence of this ability in a younger autistic child, we assume that they would have developed such understanding with a marked delay.

The question still remains what kind of ability these individuals have acquired. Autistic children who passed false-belief tasks were investigated further by Baron-Cohen (1989). He found that these subjects, who were all over twelve years old and very able, still failed a more complex belief task (John thinks that Mary thinks . . .). This was in contrast to Down's syndrome controls. Happé (1991) also investigated able autistic subjects with a battery of theory of mind tasks, acted out by either actors or puppets. The tasks she used included those involving false-belief attribution and deception. Interestingly, when these two classes of tasks were compared it turned out that the autistic group as a whole seemed to find the deception tasks harder than the belief-attribution tasks. A similar effect was found in a story-comprehension task, where again stories involving complex deception proved particularly puzzling to able autistic individuals. In contrast, there was no task effect for young normal controls (who were between seven and eight years old). This suggests that the autistic subjects may have applied task-specific strategies rather than bringing to bear a general 'stance' towards mental-state concepts. Very few of this select minority of able autistic individuals tested by Happé passed complex (second-order) false-belief tasks, and fewer still a corresponding deception task which involved double bluff—where you have to tell the truth when a lie is expected, in order to deceive your opponent. That some autistic subjects could solve such a task is remarkable. However, it cannot be ruled out that they worked out the solution by some surface-level strategy.

It would clearly be important to investigate the successful individuals further, and to establish if there is evidence for a theory of mind in their everyday behaviour or whether they have developed relatively sophisticated behavioural strategies to compensate for their deficit. This question has implications for efforts to train social understanding in autistic children. Parents and teachers in fact try to teach autistic children to 'understand' jokes and lies. As Bok (1978) has shown in her monograph on the moral choices underlying all types of lies, we must assume that deception is employed very frequently in all normal interpersonal relationships. It would be extremely interesting to see whether teaching efforts merely parallel those of Premack and Woodruff's training of chimpanzees, or whether such efforts can eventually lead to the acquisition of a concept of belief that can be applied flexibly to a variety of situations and that can give rise to an awareness of the moral issues inherent in deception. It is tantalizing that clear-cut evidence for the employment of mere behavioural routines versus deeply understood concepts still remains elusive.

Acknowledgement

This chapter was written whilst the first author was a Post Doctoral Fellow at the Department of Brain and Cognitive Sciences, Massachussetts Institute of Technology, supported by a grant from the Deutsche Forschungsgemeinschaft.

REFERENCES

Baron-Cohen, S. (1989). The autistic child's theory of mind: a case of specific developmental delay. *Journal of Child Psychology and Psychiatry*, **30**, 285-98.
Baron-Cohen, S. (1992). Out of sight or out of mind? Another look at deception in autism. *Journal of Child Psychology and Psychiatry*, **33**, 1141-55.
Baron-Cohen, S., Leslie, A.M., and Frith, U. (1985). Does the autistic child have a 'theory of mind'? *Cognition*, **21**, 37-46.
Baron-Cohen, S., Leslie, A.M., and Frith, U. (1986). Mechanical, behavioural, and Intentional understanding of picture stories in autistic children. *British Journal of Developmental Psychology*, **4**, 113-25.
Bennett, J. (1978). Some remarks about Concepts. *Behavioral and Brain Sciences*, **1**, 557-60.
Bok, S. (1978). *Lying: moral choice in public and private life*. Harvester Press, Hassocks, Sussex.
Chandler, M., Fritz, A.S., and Hala, S. (1989). Small scale deceit: deception as a marker of 2-, 3-, and 4-year-olds' early theories of mind. *Child Development*, **60**, 1263-77.
Dennett, D. (1978). Beliefs about beliefs. *Behavioral and Brain Sciences*, **4**, 568-70.
De Vries, R. (1970) The development of role-taking as reflected by behaviour of bright, average, and retarded children in a social guessing game. *Child Development*, **41**, 759-70.
Dunn, J. (1991). Understanding others: evidence from naturalistic studies of children. In *Natural theories of mind* (ed. A. Whiten). Blackwell, Oxford.

Frith, U. (1989). *Autism: explaining the enigma*. Blackwell, Oxford.

Gratch, G. (1964). Response alteration in children: a developmental study of orientations to uncertainty. *Vita Humana*, **7**, 49–60.

Hala, S., Chandler, M., and Fritz, A. S (1991). Fledgling theories of mind: deception as a marker of 3-year-olds' understanding of false belief. *Child Development*, **62**, 83–97.

Happé, F. G. E. (1991). *Theory of mind and communication in autism*. Unpublished Ph.D. thesis. London University.

Harman, G. (1978). Studying the chimpanzee's theory of mind. *Behavioural and Brain Sciences*, **1**, 591.

LaFreniere, P. J. (1988). The ontogeny of tactical deception in humans. In *Machiavellian intelligence. Social expertise and the evolution of intellect in monkeys, apes, and humans* (ed. R. W. Byrne and A. Whiten). Oxford University Press.

Leekam, S. and Perner, J. (1991). Does the autistic child have a metarepresentational deficit? *Cognition*, **40**, 203–18.

Leslie, A. M., and Frith, U. (1987). Metarepresentation and autism: how not to lose one's marbles. *Cognition*, **27**, 291–4.

Leslie, A. M., and Frith, U. (1990). Prospects for a cognitive neuropsychology of autism: Hobson's choice. *Psychological Review*, **97**, 122–31.

Leslie, A. M., and Thaiss, L. (1992). Domain-specificity and conceptual development: neuropsychological evidence from autism. *Cognition*, **43**, 225–51.

Perner, J. (1991). *Understanding the representational mind*. MIT Press, Cambridge, Mass.

Perner, J., Leekam, S. R., and Wimmer, H. (1987). Three-year-olds' difficulty with false belief: the case for a conceptual deficit. *British Journal of Developmental Psychology*, **5**, 125–37.

Perner, J., Frith, U., Leslie, A. M., and Leekam, S. R. (1989) Exploration of the autistic child's theory of mind: knowledge, belief, and communication. *Child Development*, **60**, 689–700.

Peskin, J. (1992). Ruse and representations: on children's ability to conceal information. *Developmental Psychology*, **28**, 84–9.

Piaget, J. (1954). *The construction of reality in the child*. Basic Books, New York.

Premack, D. (1988). 'Does the chimpanzee have a theory of mind?' revisited. In *Machiavellian intelligence: social expertise and the evolution of intellect in monkeys, apes, and humans* (ed. R. W. Byrne and A. Whiten) Oxford University Press.

Premack, D. and Woodruff, G. (1978). Does the chimpanzee have a theory of mind? *Behavioral and Brain Sciences*, **1**, 515–26.

Russell, J., Mauthner, N., Sharpe, S., and Tidswell, T. (1991). The 'windows task' as a measure of strategic deception in preschoolers and autistic subjects. *British Journal of Developmental Psychology*, **9**, 133–49.

Shultz, T. R., and Cloghesy, K. (1981). Development of recursive awareness of intention. *Developmental Psychology*, **17**, 465–71.

Sodian, B. (1991). The development of deception in young children. *British Journal of Developmental Psychology*, **9**, 173–88.

Sodian, B. and Frith, U. (1992). Deception and sabotage in autistic, retarded and normal children. *Journal of Child Psychology and Psychiatry*, **33**, 591–605.

Sodian, B., Taylor, C., Harris, P. L., and Perner, J. (1991). Early deception and the child's theory of mind: false trails and genuine markers. *Child Development*, **62**, 468–83.

Stern, C and Stern, W. (1909). *Erinnerung, Aussage and Luege in der ersten Kindheit*. Barth, Leipzig.

Vasek, M.E. (1986). Lying as a skill: the development of deception in children. In *Deception: perspectives on human and non-human deceit* (ed. R. W. Mitchell and N.S. Thompson). State University of New York Press, New York.

Wimmer, H. and Perner, J. (1983). Beliefs about beliefs: representation and constraining function of wrong beliefs in young children's understanding of deception. *Cognition*, **13**, 103–28.

Woodruff, G. and Premack, D. (1978) Does the chimpanzee have a theory of mind? *Behavioral and Brain Sciences*, **4**, 515–26.

Woodruff, G. and Premack, D. (1979). Intentional communication in the chimpanzee: the development of deception. *Cognition*, **7**, 333–62.

Zaitchik, D. (1990). When representations conflict with reality: the preschooler's problem with false beliefs and 'false' photographs. *Cognition*, **35**, 41–68.

Part III The theory of mind hypothesis of autism: critical perspectives.

9

The theory of mind and joint-attention deficits in autism

PETER MUNDY, MARIAN SIGMAN, AND
CONNIE KASARI

The recent use of false-belief tasks and related research paradigms (Wellman 1988; Wimmer and Perner 1983) has made an invaluable contribution to the study of autism. With few exceptions (Oswald and Ollendick 1989) there is a wealth of elegant experimental data to support the hypothesis that people with autism often display difficulty on these paradigms (Baron–Cohen *et al.* 1985; Leslie and Frith 1988; Perner *et al.* 1989). The information gleaned from this type of research has led to the articulation of the 'theory of mind' model of autistic psychopathology (Baron–Cohen 1988; Leslie 1988; Frith 1989). This model has provided an important framework that has facilitated hypothesis-testing and debate pertaining to the nature and treatment of people with autism.

In brief, this model suggests that autistic individuals perform relatively poorly on false-belief tasks because they do not adequately develop the cognitive capacity to represent the internal beliefs, feelings, and thoughts of others. In turn, the inability to represent the covert cognitive and affective processes of others is thought to make a significant contribution to the social deficits that are intrinsic to this disorder. Currently, though, it is not clear whether the theory of mind model is explanatory with respect to autistic individuals at different levels of development. In particular, several researchers have begun to debate issues concerning the application of the theory of mind model to the social deficits that arise very early in the development of the autistic child, such as deficits in non-verbal joint-attention skills (Baron–Cohen 1989*a*; Hobson 1989*a*, Harris 1989; Leslie and Happé 1989; Mundy and Sigman 1989*a*, *b*; Rogers and Pennington, 1991). Let us briefly review this debate.

THE JOINT-ATTENTION DEBATE

Bruner and Sherwood (1983) described joint-attention behaviours as a class of prelinguistic social communication skills that involve the use of gestures to share attention *vis-à-vis* objects or events (for example, showing toys). Deficits in joint-attention behaviours are manifest in autistic children with mental ages well below thirty months (Mundy *et al.* 1990). Moreover, in normal development this class of behaviours typically begins to emerge in the six- to twelve-month age-period (Bakeman and Adamson 1982; Hannan 1987; Leung and Rheingold 1981). On the other hand, explicit knowledge of other people's minds may only be evident after a child has achieved a mental age of thirty to thirty-six months (Lewis and Osborne 1990; Wellman 1988). Thus, joint-attention (JA) and theory of mind (TOM) skills appear to emerge during different periods of development, with JA skills emerging considerably earlier than codified TOM skills.

This observation led us to question whether the later emerging, and presumably relatively advanced, cognitive processes involved in TOM may validly be used to explain earlier emerging problems in the acquisition of JA (Mundy and Sigman 1989*a*, *b*). It may well be that joint-attention behaviours involve simpler cognitive processes than those involved in repre- senting the internal states of others. Nevertheless, these early JA cogni- tive processes may involve developmental precursors of those involved in TOM (Baron–Cohen 1989a; Gómez *et al.*, this volume, Chapter 18; Mundy and Sigman 1989*a*, *b*). Baron–Cohen (1989*a*), Harris (1989), and Leslie and Happé (1989) all note that the early emergence of JA skills does not, in and of itself, preclude the possibility that JA skills may be a very early index of the development of the capacity to represent some aspects of the internal psychological processes of others.

Understanding which, if either, of these alternatives best describes the relation between JA and TOM disturbance may have implications both for an understanding of the course and nature of autistic psychopathology and for an understanding of the course of early, normal social cognitive development. However, for lack of empirical data, it is currently difficult to make a compelling case for either alternative (Leslie and Happé 1989). Therefore, one goal of this chapter is to present data from a preliminary study that attempts to forge an empirical link between JA skill deficits and TOM disturbance. In this study of very young autistic children we examine the relation between JA skills and the capacity for pretence as indexed by pretend-play performance. Pretend-play skills as a measure of the capacity for pretence hypothetically reflect an arena of cognitive development essential to TOM acquisition (Leslie 1987, 1988; Wellman, this volume, Chapter 2). Thus, the study of the relation between JA and symbolic skills development may provide one approach to examining alter- native models of the relation between JA and TOM deficits in autistic children.

A second question emerging in the joint-attention debate was whether the TOM model is broad enough to account for what appears to be an affective contribution to JA deficits among autistic children (Mundy and Sigman 1989*a*, *b*). The current TOM model emphasizes a singularly cognitive explanation of autistic social pathology. However, others have suggested that an affective disturbance may be primary to the nature of autism (Hobson 1989*a*, *h*; Kanner 1943). Moreover, autistic JA skill deficits appear to be associated with a disturbance in the conveyance of affect in conjunction with the use of referential gestures (Kasari, Sigman, Mundy, and Yirmiya 1990). The apparent involvement of affect in autistic JA deficits may have implications for current theory.

Psychological theories of autism have tended to emphasize either cognitive or affective processes as central to autistic psychopathology (Hobson 1989*b*; Leslie and Happé 1989). The phenomenon of JA deficits, however, fosters the consideration of how *both* cognitive and affective disturbances may contribute to early-emerging autistic social deficits. Our recent theoretical efforts have followed this path, and have attempted to articulate a model that describes the interaction of affect and cognition in the development of JA and subsequent social cognitive deficits among autistic children. A consideration of this model will constitute the final section of this chapter. However, before this discussion and the presentation of data on JA and pretend-play development, it may first be useful to review the research that suggests that JA skill deficits are an important characteristic of the early development of autistic children.

JOINT-ATTENTION DEFICITS IN AUTISTIC CHILDREN

It was Frank Curcio (1978) who initially published the observation that autistic children tended to display far fewer joint-attention gestures than other types of non-verbal acts, such as requesting gestures. The latter involve the use of eye-contact and/or gestures for an instrumental or imperative purpose. Hence, these gestures are used to gain another person's aid in obtaining objects or events. Examples include pointing to a toy that has just been removed from reach, or giving a sealed container to someone to elicit aid in opening the container. These behaviours involve what have been referred to as *triadic* exchanges, in that they demand that the child and a second person share or co-ordinate attention *vis-à-vis* some third object or event (Bakeman and Adamson 1984). Joint-attention behaviours are also characterized by triadic exchanges. However, they are not characterized by an instrumental function. Rather, they serve a declarative purpose. They involve the use of gestures and eye-contact to share an awareness of, or the experience of an object or event. Examples here include a child's use of eye-contact and pointing to an object that is within the reach of the child, or showing a toy to a care-giver.

Subsequent to Curcio's seminal observation, research has demonstrated that autistic children display more robust deficits on measures of joint attention than measures of requesting or other types of non-verbal social-communication skills when compared to developmentally matched controls with mental retardation, language delays, or normal development (Baron–Cohen 1989b; Loveland and Landry 1986; Mundy *et al.* 1986; Mundy *et al.* 1990; Sigman *et al.* 1986; Wetherby and Prutting 1984). In our own research we have demonstrated that these deficits are apparent in comparisons of children with verbal age approximations below eighteen months and chronological ages below forty-eight months (Mundy *et al.* 1990). Thus, a specific deficit in the emergence of joint-attention skills, as opposed to other types of non-verbal communication skills, appears to characterize the development of very young autistic children.

Among young autistic children joint-attention deficits are apparent on a variety of behaviours. Not only do they show deficits on pointing and showing gestures, but also on following line-of-regard trials accompanied by a point* and on simple referential looking (Loveland and Landry 1986; Mundy *et al.* 1986). The former refer to the capacity of the child to follow or understand the direction of someone's gaze, while the latter refers to the tendency to look back and forth between a person and an interesting event such as an active mechanical toy. By contrast, autistic children display relative facility with a variety of non-verbal requesting skills such as pointing to out-of-reach objects (Baron–Cohen 1989b; Mundy *et al.* 1986), giving objects, and making eye-contact after a toy has been moved out of reach or an actived toy has ceased action (Mundy *et al.* 1986).

It is important to note that the specificity of autistic joint-attention deficits is quite high. In our research, deficits on referential looking alone have correctly discriminated 94 per cent of a sample of low-functioning autistic children from a sample of children with mental retardation matched on a mean mental age of 26 months (Mundy *et al.* 1986). Using different methods Lewy and Dawson (personal communication) found that measures of gestural joint attention correctly classified approximately 90 per cent of their young autistic sample relative to controls.

---

* Line-of-regard tasks may be sensitive to developmental shifts in autistic children. In older autistic children line-of-regard tasks do not appear to be problematic (see, for example, Baron–Cohent 1989b). However, there is evidence of a deficit in following direction of gaze in younger autistic children, even when the direction of gaze is emphasized with pointing (Mundy *et al.* 1986). This is noteworthy because line-of-regard skills may be spatial in nature, and may not require the cognitive skills that are central to TOM (Baron–Cohen, 1989a; Leslie and Happé 1989). If so, early line-of-regard difficulty among young autistic children may suggest that, initially, autism is characterized by more basic cognitive deficits than those involved in TOM. This is an important issue for further research.

Although deficits in these skills are displayed by most young autistic children, individual differences in joint-attention skills are also displayed by autistic children, and these differences are important. Current research suggests that joint-attention skill development is both a concurrent and predictive correlate of individual differences in language-acquisition among autistic children (Loveland and Landry 1986; Mundy *et al.* 1987; Mundy *et al.* 1990). Moreover, recent research has demonstrated that differences in social context or social stimulation may have an effect on the expression of joint-attention skills in autistic children (Lewy and Dawson 1991). These data suggest that joint-attention skills may be an important domain for early intervention efforts. Furthermore, the data from group comparisons and research on individual differences also suggest that joint-attention skill deficits are an important marker of the social-communicative developmental disturbance that characterizes the early presentation of the syndrome of autism.

What then is the nature of this early-emerging developmental disturbance? An answer to this question may hinge on understanding how non-verbal joint-attention skills and requesting skills differ. Since relatively uneven development across these behaviour domains is a characteristic of autism, the comparative task analysis of joint-attention and requesting skills has become an important issue of theoretical concern in the study of autism (Baron–Cohen 1989*b*; Gomez *et al.*, this volume, Chapter 18; Mundy *et al.* 1986).

DIFFERENCES BETWEEN NON-VERBAL JOINT ATTENTION AND REQUESTING

Mundy *et al.* (1986) considered the possibility that joint-attention and requesting skills might differ with respect to three factors: attention demands, cognitive demands, and affective processes. As was stated above, the attention demands of joint-attention skills appear to be triadic. That is, they demand the flexible and smooth transition between attention to self, attention to some object or event, and attention to another person (Bakeman and Adamson 1984). Hence, in referential looking, children direct attention to an interesting object/event, shift attention to another person, and then shift attention again to reorient themselves to the object/event.

It may be that autistic children have difficulty with JA skills because they are not proficient with a flexible, triadic attention-deployment (Mundy *et al.* 1986). However, requesting skills also appear to require flexible, triadic attention-deployment. For example, when a toy is placed out of reach an autistic child may be observed to fixate the toy, reach to the toy, shift attention to another person while maintaining the reach, and then shift attention again in order to reorient back to the toy. Thus, on the surface, the complexity of the attention demands of JA and requesting skills seem to be similar.

This, of course, does not put the attention-deficit hypothesis to rest. There is evidence that brain-stem systems involved in the mediation of attention-control may play a role in the pathogenesis of autism (Courchesne 1991). Moreover, older autistic individuals do seem to have difficulty in the flexible shifting of attention on problem-solving tasks (Ozonoff *et al.* 1991). Nevertheless, a detailed model of how attention-control processes may explain differences in the development of JA and requesting skills among autistic children has not been presented. This remains a challenge to the field. Alternatively, current theory clearly suggests that these skills may differ with regard to both affective (Bruner 1981; Kasari *et al.* 1990; Mundy and Sigman 1989c) and cognitive processes (Baron–Cohen 1989a; Mundy *et al.* 1986). Let us first consider the possible affective differences in these skills.

Affect and joint-attention

The emergence of non-verbal communication begins with caregiver–child interactions that are not focused on objects or events. Rather, the initial phase of early communication development (0–5 months) involves sharing affective states through the conveyance of affective signals between the care giver and the child (Adamson and Bakeman 1982; Trevarthen 1979; Werner and Kaplan 1963). Joint-attention behaviours, rather than requesting behaviours, may be the more direct descendants of these early affective interactions because a primary purpose of joint-attention behaviours appears to be to share the experience of an event or object with others (Baron–Cohen 1989a; Kasari *et al.* 1990; Mundy *et al.* 1986; Mundy and Sigman 1989c). That is, the declarative or experience-sharing function of joint-attention behaviours may involve conveyance of affect to a greater degree than is involved in the instrumental function of requesting behaviour. Bruner (1981) noted this difference when he suggested that affect or some mood-marking procedure may distinguish joint-attention from requesting gestures.

This hypothesis has recently been supported in a study of the degree to which joint-attention and requesting behaviours are associated with displays of facial affect in young autistic, mentally retarded, and normal children (Kasari *et al.* 1990). The results of this study indicated that, among children with autism, the display of JA behaviours was typically accompanied by *neutral* affect. In contrast, children with mental retardation or normal development most often displayed *positive* affect to another person in conjunction with joint-attention behaviours. Moreover, within the normal

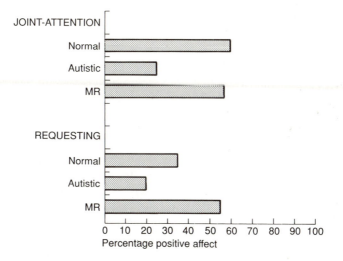

Fig. 9.1. Percentage of positive affect displayed during joint-attention and requesting behaviours by children with autism, mental retardation, or normal development.

sample JA behaviours were associated with the display of positive affect significantly more often than were gestural requests (see Fig. 9.1)

This difference in affect display between categories of non-verbal communication cannot be attributed to one specific type of JA behaviour, but rather holds true for comparisons of a variety of joint-attention and requesting behaviours. In a recent study, the frequency and duration of four requesting and four JA behaviours were rated for a sample of 32 normal infants with a mean chronological age of twenty months (Mundy *et al.* 1992). An independent coder rated how long the infants expressed facial affect to another person while engaged in each type of non-verbal behaviour. The results indicated that more positive affect was displayed in conjunction with each of four types of JA behaviour relative to the set of four requesting behaviours. On average positive affect was displayed for over 50 per cent of the duration of the display of all types of joint-attention behaviours (range = 56 per cent to 70 per cent). In contrast, the mean duration of positive affect displayed during requests did not exceed 36 per cent for any type of behaviour (range = 18 per cent to 36 per cent).

These data confirm that JA behaviours and requesting behaviours may be distinguished on the basis of affect. Joint attention not only involves the coordination of attention *vis-à-vis* some object or event, but also involves the conveyance of affect. The latter presumably reflects some aspect of the child's response to the object or event. Within this interpretative frame, these data are consistent with the hypothesis that joint-attention deficits may involve a disturbance in affective as well as in cognitive development in autistic children (Kasari *et al.* 1990; Mundy *et al.* 1986; Mundy and Sigman 1989*c*).

Joint attention and symbolic skill development

Although JA and requesting may differ on the basis of affect, theory also suggests a cognitive interpretation of the differences between these early forms of social communication.

Intentional non-verbal communication behaviours have been considered by some to be indicative of rudimentary knowledge about others' intentions and perceptions (Bretherton *et al.* 1981; Poulin-Dubois and Shultz 1988; Leslie and Happé 1989). However, in discussing the social cognitive demands of non-verbal communication, few have considered joint-attention and requesting skills as distinct categories of behaviour. Indeed, it has been suggested that *both* requesting skills and joint-attention skills implicitly reflect knowledge of others. For example, Poulin-Dubois and Shultz (1988, p. 113), state: 'A widely accepted hypothesis regarding the emergence of knowledge about others' intentions proposes that infants manifest this ability when they develop intentional communication for instrumental purposes.' If instrumental non-verbal acts (i.e., requesting acts) involve knowledge of others to the same degree as joint-attention behaviours, how could the theory of mind model explain the finding that people with autism tend to display deficits on the latter but not the former?

To deal with this question it may be useful to borrow from Werner and Kaplan (1963), and hypothesize that these skill domains reflect different aspects of social cognition. That is, while JA skills may involve understanding others as *agents of contemplation*, requesting skills may only involve understanding others as *agents of action* (Mundy *et al.* 1986; also see Baron–Cohen 1989*b*). In understanding others as agents of action, requesting skills may only require representation of previously experienced and generalized action sequences (for instance, 'other retrieves object'). This is akin to what Jean Mandler (1988) has described as sensorimotor or procedural representations of motor acts and perceptual knowledge. According to Mandler this representational system develops in infancy, but is relatively automatic, and does not necessarily involve conscious monitoring while employed.

Mandler (1988) has also described another type of representational system that may develop in infancy. This type of representation is consciously attended to, and is more symbolic than sensorimotor representation. Mandler (1988, p. 116) calls this 'conceptual' or 'declarative representation', and describes this as 'potentially expressible (declarable) knowledge is symbolic knowledge: it has the status of a comment on something that has been experienced . . . it is also illustrated by conscious awareness of what it is you are currently perceiving. If you are conscious of what you are seeing you are engaged in conceptual thought' (p. 116).

Joint-attention skills are often called proto-declarative skills, because they appear to involve a comment on the child's experience. As such, they may involve what Mandler has described as symbolic, conceptual representation to a greater degree than non-verbal requesting skills. The symbolic, consciously monitored representational process described by Mandler would appear to require something similar to the cognitive decoupling process that Leslie (1987, 1988) proposes is involved in the process of metarepresentation. The compromised capacity to acquire higher-order representations or metarepresentations via decoupling has been proposed as the foundation for false-belief and theory of mind deficits displayed by autistic children (Leslie 1988; this volume, Chapter 5). In this view, then early joint-attention skills may, to a greater degree than requesting skills, reflect an early manifestation of the symbolic (or metarepresentational) cognitive skills that potentiate the child's development of a theory of mind (cf. Baron–Cohen 1989a; Leslie and Happé 1989).

This hypothesis suggests a possible empirical link between joint-attention skills and a proposed cognitive substrate of theory of mind development. This link may be open to empirical inquiry through examining the relations between joint attention and an index of the emergence of metarepresentational skill, such as symbolic-play development.

JOINT-ATTENTION AND PLAY DEVELOPMENT

Symbolic play skills are thought to provide a marker of the acquisition of the capacity for pretence. This, in turn is assumed to index the development of skills that are either indicative of the emergence of the TOM module (Leslie 1987, 1988) or indicative of the acquisition of cognitive skills that may be necessary but not sufficient for the emergence of a codified theory of mind (Wellman, this volume, Chapter 2). In either case, symbolic-pretence skill development is considered to be a major milestone in the development of TOM. Furthermore, consistent with the theory of mind model of autism, young autistic children display marked deficits on measures of pretence play (Baron–Cohen 1987; Riquet *et al.* 1982; Sigman and Ungerer 1984; Wing *et al.* 1977). However, neither the relative timing of the emergence of play and joint-attention deficits in autistic children, nor the relations between these areas of deficit are well understood.

It may be that joint-attention skill and symbolic play skill deficits reflect similar cognitive processes that emerge at the same developmental level among autistic children (Harris 1989; Baron–Cohen 1989a). In this case JA and symbolic-play skills may be expected to emerge in a simultaneous pattern with deficits in both areas, emerging at approximately the same time in the development of the autistic child. Alternatively, joint-attention skills may reflect earlier, more basic social cognitive skills than do symbolic-play skills (Mundy and Sigman 1989*a*; Wellman, this volume, Chapter 2). In this case, JA and symbolic-play skills may be expected to emerge in a sequential pattern, with deficits in the former being apparent before deficits in the latter as autistic children develop (Mundy and Sigman 1989*a*).

We have examined these alternative hypotheses in a preliminary longitudinal study of a sample of 15 autistic children and a language-matched sample of mentally retarded children. These samples were assessed longitudinally over a 13-month interval, and have been described in detail elsewhere (Mundy *et al.* 1990).

In brief, these young children with autism or mental retardation were under five years of age and were matched on receptive and expressive language-age estimates (means = 12.5 months and 13.6 months respectively). At the initial and follow-up assessments these children were administered measures of non-verbal joint-attention and requesting skills contained within the Early Social-Communication Scales (Seibert *et al.* 1982). The children were also administered a play assessment (Ungerer and Sigman 1981) that provided frequency measures of both functional and symbolic play. The former involves behaviours that imply recognition of the social conventions of object-use (for example, holding a toy telephone receiver to an ear) and theoretically index simple, first-order representational skill, but not the second-order representational skills involved in pretence acts (Leslie 1987; Baron–Cohen 1987). Symbolic play involves pretence acts such as substitution play or the use of one object as if it were a different object (for example, using a sponge on a spoon as if it were food), and doll-as-agent play, wherein a doll is imbued with the ability to perform acts as an independent agent, such as 'driving' a toy truck or 'waving' good-bye. The definitions of these symbolic play acts satisfy the relatively rigorous criteria used in defining pretend-play indices of metarepresentational capacities (Baron–Cohen 1987).

The data from the group comparisons in this study provided two notable findings (see Table 9.1). First, these data were consistent with the sequential model of joint-attention and symbolic-play development (Mundy and Sigman 1989*a*). Joint-attention deficits were manifested by the autistic sample in the first testing session, and this was before before the emergence of either functional or symbolic-play deficits (see Table 9.1). The autistic children also displayed significant deficits on joint-attention skills at the later assessment. However, consistent with previous findings, the autistic children did not display a significant deficit on requesting skills at either the first or the second assessment.

Table 9.1 Mean frequency of joint attention and play acts for the two groups

	Language-matched MR		Autistic	
	Test 1	Test 2	Test 1	Test 2
Joint attention	18.1*	24.7*	9.0	12.2
Request	17.0	21.6	14.7	17.2
Functional play	9.3	14.5*	9.3	10.5
Symbolic play	0.8	2.3	0.4	1.3

* $p < .05$, significantly different from the autistic group.

A second important result of this study was that functional-play deficits were observed on the second assessment, before the clear appearance of symbolic-play deficits in this sample of young autistic children (see Table 9.1). These data were consistent with earlier, limited evidence of a functional-play deficit among young autistic children (Mundy *et al.* 1986). However, they contrast with those from studies of older children, which suggest that autistic children display relatively poor development on measures of symbolic–pretence-play skill, but not on measures of simpler first-order representational skill, such as functional play skills (Baron–Cohen 1987; Frith 1989). Like the age differences in data on line-of-regard trials (see footnote on p. 184), these findings suggest an important principle. *The type or level of cognitive or social impairment that is characteristic of autistic children may change over time and development* (Mundy and Sigman 1989a; Rutter and Garmezy 1983; Sigman 1989). In this case, there appears to be a limited period of time when play deficits among young autistic children may reflect a developmental delay of first-order representational process, before second-order, symbolic process deficits are apparent.

How are the sequential developmental phenomena described in this study to be interpreted? In our view, the disparity between the emergence of joint-attention and play-skill deficits is consistent with the notion that these deficits may reflect different developmental levels of cognitive impairment. The foregoing data suggest that autistic JA disturbance appears to emerge during, or even before a period of development in autistic children that is also characterized by a disturbance in functional play skills. Also, as has been noted earlier, some evidence of line-of-regard disturbance is evident in the early development of autistic children (Mundy *et al.* 1986). Hypothetically, the data on early functional play and line-of-regard deficits suggest a period of *presymbolic* cognitive disturbance in young autistic children. Given their developmental contiguity with these deficits, the early emergence of JA deficits may also initially reflect a presymbolic area of developmental disturbance in autistic children. Such a hypothesis is consistent with other work on the development and meaning of non-verbal communication development (Sugarman 1984; Wellman, this volume, Chapter 2).

If this hypothesis is true, then the foregoing account of the cognitive differences between joint attention and requesting would appear to be less than valid. Yet, whether viewed from the perspective of Mandler or others, the notion that these two types of skills differ in cognitive demands seems so intuitively plausible that this hypothesis is difficult to relinquish (see also Gómez *et al.*, this volume, Chapter 18). It may not be surprising, then, that when the correlational data in this study were examined, subtle differences in the cognitive correlates of these types of skills appeared.

Across the two samples there was a consistent difference in the profile of correlations between play skills and JA versus requesting skills. This profile difference was most pronounced in the MR sample. In the initial testing period functional play but not symbolic play was significantly correlated with both joint-atttention and requesting skills in the MR sample ($r = 0.48, p < 0.05; r = 0.54, p < 0.025$ in one-tailed analyses respectively). On follow-up, symbolic play but not functional play was significantly correlated with joint-attention skills ($r = 0.58$, $p < 0.025$, one-tailed analysis). Neither type of play was correlated with requesting skills on follow-up.

These data raise the possibility that there may be a developmental transition in the factors that mediate joint-attention performance. Early on, both joint-attention development and requesting may be mediated by basic cognitive factors involved in functional-play development. This may involve first-order representation, or what Mandler (1988) refers to as 'sensorimotor representation'. Moreover, this association may be relatively more powerful for the simpler, early-emerging joint-attention behaviours such as referential looking. However, with development and the advent of more sophisticated joint-attention skills such as pointing and showing, the more advanced cognitive processes reflected in symbolic play may become involved (cf. Gómez *et al.*, this volume, Chapter 18).

In addition, predictive correlations in this study also supported the notion of some type of association between play and joint-attention development. Initial joint-attention-skill scores, but not requesting-skill scores, were a significant predictor of symbolic-play development at follow-up in the MR sample ($r = 0.46$, $p < 0.05$; $r = 0.08$ in one-tailed analyses respectively) and approached significance in its predictive association with symbolic play in the autistic sample ($r = 0.43$, $p < 0.10$; $r = 0.25$ in one-tailed analyses respectively). Thus joint-attention skills appear to precede the emergence of symbolic-play skills, and may also reflect factors associated with individual differences in the subsequent development of symbolic-play skills. Of course, JA skills also appear to precede and predict subsequent individual differences in language skills, another arena of symbolic development (Mundy *et al.* 1990*a*).

What factors may be involved in this type of association? Is the possible relation between joint-attention and symbolic-play development best explained in terms of a single cognitive process or in terms of the interaction of multiple processes? These questions will be addressed in the final section of this chapter.

THE DEVELOPMENT OF JOINT-ATTENTION AND SYMBOLIC SKILLS

Holding in abeyance a variety of methodological issues, such as small sample sizes and resulting limits in power,* let us assume that the data from this preliminary study provide a valid reflection of phenomena associated with the development of joint-attention, requesting, functional-play, and symbolic-play skills. These data suggest that, early in development, joint-attention skills, requesting skills, and functional-play skills may share a common and presumably presymbolic source of variance. Subsequently, variance in joint-attention skills may be associated with individual differences in later-emerging symbolic-play skills; but requesting skills do not appear to be predictive of symbolic skills. How is this pattern of data to be interpreted?

One obvious hypothesis is that, as with functional play, early joint-attention skills involve basic representational skills. This representational component may give rise to the predictive relation of joint-attention skills to the more elaborate type of representational skill involved in symbolic play. However, this hypothesis does not account for the data on requesting skill. These skills were also correlated with functional-play performance; hence they too may involve basic representational skill. However, requesting skill did not predict symbolic-play development, and requesting skills are not a focus of autistic developmental disturbance, as joint-attention skills are. Thus, it may require something more than recourse to the hypothetical involvement of basic representational skills to explain joint-attention development and its possible relation to emerging symbolic-play skills.

Perhaps a purely cognitive explanation is not sufficient to the task of explaining the data on autistic joint-attention development presented in this chapter. Indeed, some would argue that reductionist attempts to explain a

---

* The spectre of low power due to small sample sizes raises the possibility that these data may provide an imprecise reflection of the true relations between non-verbal communication and pretence-play development. This may well be true, and we are in the process of collecting longitudinal data from larger samples of children with autism and mental retardation to address this and other issues.

fundamental disturbance in social development solely in terms of deficits in cognitive processes are inconsistent with research that suggests that significant aspects of social development may be explained in terms of biological systems that are specific to social behaviour (Brothers 1990; Fein *et al.* 1986; Panksepp and Sahley 1987; Walters *et al.* 1990). In accord with these arguments let us elaborate upon a model that considers the affective component of joint-attention skills, as well as cognitive factors, in explaining autistic deficits in joint-attention skills and allied phenomena (Mundy and Sigman 1989*b*). As all models are based on general assumptions, let us be explicit about some of ours.

Our current model assumes that the early stages of autism are characterized by at least two paths of developmental disorder. One is a basic disturbance in *arousal self-regulation* that is associated with atypical affective responses to stimulation (Dawson and Lewy 1989). The other is a disruption specific to the cognitive development of representational skills. One early manifestation of the latter may be a delay in the development of functional-play skills. Subsequently this disturbance may be manifest in deficits in a variety of skills, including skills in the use of concrete operations (Yirmiya, Sigman, Kasari, and Mundy 1992).

Neither of these disturbances, in and of themselves, may constitute an absolute or core deficit of autism (Ungerer 1989). Thus, at any one time, the measurement of one or another domain may yield evidence of deficits in some autistic children, but not all autistic children. Furthermore, the relative impairment in each domain may give rise to broad individual differences in the population of autistic individuals. That is, some autistic children may be relatively more impaired in affective rather than representational processes, and vice versa. However, the interaction or synergism of these two disturbances is assumed to contribute to the aberrant course of development in all autistic children.

More explicitly, this model assumes that a disturbance of arousal self-regulation gives rise to atypical experience with *affect* in social interactions (Dawson and Lewy 1989). Combined with compromised representational-skill development, the young child with autism has difficulty developing an understanding of the social-signal value of affect and the related capacity to share experience with others. Operationally, one of the earliest effects of this combined affective–cognitive disturbance is deficient joint-attention skill development. However, other early-emerging non-verbal skills that do not involve both representational and affective processes to a great a degree, such as non-verbal requesting, are relatively unimpaired.

This model also assumes that deficits in joint-attention skills reflect a disturbance in a fundamental developmental process that may contribute to subsequent disturbance in symbolic skill development. With regard to this assumption the model incorporates Werner and Kaplan's (1963) hypothesis that the ability of an infant to share experience with a caregiver with regard to an object of reference is an important experiential context that stimulates symbolic development.

When Werner and Kaplan raised this provocative hypothesis, they were not very explicit with regard to the mechanism by which early states of shared experience might contribute to symbolic development. However, one aspect of this mechanism may be suggested by *skill theory*, and the proposition that the development of specific social and cognitive skills is tied to the availability of specific experiences and, therein, specific information about the world (Fischer and Pipp 1984). Accordingly, just as specific types of experience gained in object interaction may be important to nonsocial aspects of early cognitive skill development (object permanence, understanding spatial relations, etc.), it may also be that the exchange of affect between the young child and others provides a unique source of information that plays an important early role in fostering the development of specific aspects of social cognition (Malatesta and Izard 1984).

In the early experiences of the infant, affect may well be a unique source of information about the covert aspects of others' behaviour. However, for affect to provide useful information, it must become interpretable in some sense for the non-verbal infant. How, then, does affective information become interpretable or take on meaning for the child? We would suggest that in joint-attention and affective sharing contexts such as referential looking, the child is afforded an opportunity to *compare* proprioceptive affective information elicited by an external referent with the perceived affective information emanating from others relative to the same referent. This opportunity to compare the emotion of the self with affect of another may be one of the first in the ecology of the young child that provides the right combination of information for the cognitive system of the child to begin to develop symbolic representational schemes for the covert psychological activity of others.

This developmental process may unfold in the following sequence. In the first five months of life the infant gains considerable practice with the simultaneous experience of proprioceptive and exteroceptive affective information in face-to-face interaction (Trevarthen 1979). During this period the infant's mnemonic and related representational systems also become more sophisticated. In particular, we assume that advances in these cognitive skills play a role in the reported increase in three-to five-month-old infants' ability to compare and abstract the invariant information presented by different exemplars of facial and vocal affect (see for example Schwartz *et al.* 1985; Walker-Andrews and Grolnick 1983).

Admittedly, this type of research has dealt only with the infants' ability to abstract invariance across the affect displayed by others or analogues of others (such as pictures of faces). However, on the basis of this information it does not seem implausible to suggest that by about five to six months infants can not only analyse and compare the affective displays of others, but also begin to analyse and compare their own proprioceptive affective information with the affective information provided by others, especially when events generate these simultaneously. Hence we are speculating that, by about six months of age, infants generate rudimentary representations that enable them to compare their own affective experience (for example, non-verbal feeling state; pattern of facial muscle movement) with the affective display of another person while engaged in shared face-to-face affective exchanges (see also Meltzoff and Gopnik, this volume, Chapter 16).

One outcome of this process of relating representations of the affect of self and other may be the developing understanding of *social action schemes* such as { I Smile [then] Other Smiles }. The development of this type of social action scheme may contribute to the development of triadic non-verbal communication skills in the six- to nine-month developmental period. To reiterate, communication is said to be triadic because the infant begins to use direction of gaze, eye-contact and gestures to coordinate attention with others relative to an external referent object or event (Bakeman and Adamson 1984). With regard to basic joint-attention behaviour such as referential looking we have suggested that, when an object or event elicits an affective response from an infant, the tendency to look to another person in reference to this object/event is initially enhanced by the earlier-acquired expectation that { I Smile [Then] Other Smiles } (Mundy and Sigman 1989*b*).

Repeated experience with affective exchange in reference to an object or event may, in turn, give rise to further developments in social cognition. In the context of these repeated experiences the infant may begin to develop a social understanding of the means–ends relations as they pertain to external causality and emotion. In other words, because there is a temporal association between the emotion experienced by the child and the perceived emotion displayed by another person *that corresponds to an external event or object*, this type of situation may be a prepotent stimulus for the child to learn that environmental stimuli can elicit similar emotional reactions from themselves and others. A corollary assumption is that the repeated exposure to the simultaneous experience of self-affect and other-affect as related to a third object contributes to the infants' increased sophistication in the cognitive interpretation of the affect of others.

Here we assume that hierarchical integration is a basic process of cognitive development (Sternberg 1984). Accordingly, some time after six months of age the infants may begin to integrate their representations of the affect of the self with their representations of the affect of others. This integration yields a higher-order representation of the affect of others, which may be symbolic in nature.

In this integrative process the representation of the child's affective state may be 'decoupled' (Leslie 1987), and become associated with representations of the facial affect of others. That is, in the context of joint attention the maturing infant perceives the affective signals of another, converts these to a mental representation, and then tags or associates this with a representation of the affect simultaneously being expressed by the child. With repeated experience this cognitive tagging process becomes automatic. Consequently the infant gradually becomes adept at understanding the affective displays of others in terms of associated representations of the infant's own affect. For example, when viewing a smile the young infant's cognitive system may call up the representation of [Self-Smile] and tag it to the representation of [Other-Smile], thus understanding this affective display in terms of the associated affective experiences of the infant. In some sense, then, the infant begins to be able to interpret the affect of others 'as if it were my own'.

It is also important to emphasize that the infant in this model is increasingly able to call up a representation of self-affect that is stored in memory and tag this to the perceived affect of others to facilitate interpretation *even when the infant is not engaged in a similar affective display or experience*. Thus, this type of integrated self-other affective representation may be used to signify affective information that is not immediately perceptually available to the infant. As a signifier, this type of representational scheme takes on the characteristic of a symbolic thought (Piaget 1962). Moreover, in this process, the affect of others takes on an enhanced meaning for the child, because the representation of the child's affect (for example [Smile]) may be far richer than a simple scheme for facial or vocal affect. It may include representations of the actual feeling state or affective tone that the child has experienced in association with specific affective displays. To a degree, then, in this process the affect of others may come to be interpretable within the context of representations of the child's own feeling states. According to this model, then, joint-attention skills need not be considered to be an early manifestation of symbolic or metarepresentational process, at least not in the early stages of joint-attention skill-development. Rather, joint-attention skills presumably reflect part of a developmental process of integrating 'self-affect' with 'other-affect' that initially involves first-order representational skill applied to the realm of affective experience, and ultimately contributes to the development of symbolic representations in the service of interpreting the affect of others.

CAVEATS AND COMMENTS

This model, of course, is circumscribed. It attempts only to organize our thinking with respect to one possible path of association between joint-attention skills and the subsequent development of symbolic skills. It makes no claim with respect to the general mechanisms of symbolic development, as does the seminal model of Leslie (1987). Indeed the model described above dovetails with this model of symbolic development, as it incorporates the operation of integrating first-order representations to generate higher-order representations, or 'metarepresentations' of affect. It follows, then, that our musings are not inconsistent with the theory of mind model of autistic psychopathology, except in suggesting that the factors that initially yield joint-attention deficits may not involve symbolic processes *per se*. However, the disturbance of joint-attention skill development may contribute to deficit in the development of specific types of symbolic schemes that provide a foundation for the early development of social cognition.

We hope that it is also obvious that aspects of this model are quite sympathetic to the many ideas espoused by Hobson (1989*a,b*, 1990). However, Hobson's model emphasizes the possible negative sequelae of an autistic deficit in the *perception* of affect. As hopefully is clear at this point, we would argue that a disturbance in the *expression* of affect may be as problematic for autistic children, because this could undermine the early stages of self–other affective comparison. Errors in this process could be expected to contribute to affect-perception difficulties early in the life of autistic children. Moreover, since it may be argued that there is evidence of both an expressive affective disturbance and a representational skill deficit early in the development of autistic children (Mundy and Sigman 1989*a*), our current thinking has been drawn to an attempt to understand the potential unfolding consequences of the interaction of disorders in both affective and cognitive processes.

Beyond the alternative cognitive and affective models described by Baron–Cohen, Hobson, Leslie, and Frith, other models of autistic developmental pathology may be pertinent in the explication of joint-attention deficits. Independently of our own efforts, Rogers and Pennington (1991) have ascribed the core deficit of autism to a disturbance in the comparison of self- and other-representations. In their cogent paper they take a primarily neurocognitive perspective, and suggest that this type of disturbance is mediated most strongly by a specific deficit in representational mapping as regulated by the prefontal cortex. Alternatively, as noted earlier, it may still be argued that a deficit in the types of comparative processes that have been described in this chapter has at its core a disturbance in the control of attention (Courchesne 1991).

A definitive description of the processes that lead to joint-attention deficits in autistic children is beyond the scope of our current knowledge. The description of this process remains one of many important items on the agenda for future research on autism. The fundamental point of the last section of this chapter, however, has not been to explain these deficits, but rather to suggest that the outcome of the disruption of joint-attention skill-development may be a disturbance in the development of the capacity to compare the affect of the self with the affect of others. The pervasive disturbance of this process may be one fundamental aspect of the developmental disturbance that contributes to social-behavioural and social-cognitive deficits in autistic children. If the independent arrival of researchers at a similar hypothesis is an indicator of the potential validity of the hypothesis, then the related work of Hobson (1990), Rogers and Pennington (1991), and Meltzoff and Gopnik (Chapter 16, this volume) emphasizes the need to consider further the role of deficits in the cognitive capacity to compare the affect of self and that of others in autistic psychopathology.

Finally, we acknowledge that, although both affect and cognition have roles in our current model, we still emphasize cognitive processes (representation, comparison, abstraction of invariance, etc.). However, we do not want to suggest that the capacity to share affect is largely a cognitive process (cf. Hobson 1989*a*). This model does not attempt to describe the mechanisms of shared affect. For example, this model does not address the issue of whether or not infants are inherently motivated to exchange affect with others. Rather, this model simply attempts to describe how the process of sharing affect may be fodder for the cognitive system that ultimately yields incipient elements of social cognition and perhaps a theory of mind.

CONCLUSION AND SUMMARY

Clinical and theoretical imperatives focus the intellectual energies of many of us on the fundamental aspects of human nature that are concealed within the puzzle of autism. The false-belief paradigm and the theory of mind model have provided the field with a powerful lens through which to examine the nature of autism. Indeed, this volume is a testament to the heuristic value of the seminal contribution of the London group (Baron–Cohen *et al.* 1985). This model has also provided a touchstone for others in the development of alternative affective models (Hobson 1989*b*), as well as alternative cognitive models (Rogers and Pennington, 1991), of autistic psychopathology.

The theory of mind model has been extremely valuable in our own efforts to plumb the meaning of joint-attention deficits in young children with autism. In this chapter we have attempted to review and elaborate on this work. Extant research, and new data from a preliminary study, suggest that joint-attention deficits in children with autism: (1) involve an affective component; and (2) are apparent earlier in development than are deficits in symbolic/pretence-play skills; but also that (3) JA and symbolic-play skills may share a common source of variance. This last point provides a link between joint-attention deficits and the theory of mind model. In attempting to understand this link we have proposed that joint-attention skill development reflects the capacity of the child to share and compare affective experiences with others *vis-à-vis* some third referent. We have also suggested that interaction between the developing representational system of the child and the capacity to share affective states with others contributes to the development of symbolic schemes concerning the internal experience of others. Thus, joint-attention skills may be a behavioural reflection of a process of integration of affect and cognition that contributes to the child's early steps toward developing a theory of mind. In the case of children with autism, deficits in joint-attention skills may reflect early developmental missteps in this process.

Acknowledgement

Preparation of this manuscript was supported by NINCDS grant NS 25243 and NICHD grant HD 17662.

REFERENCES

Adamson, L., and Bakeman, R. (1982). Affectivity and reference: concepts, methods, and techniques in the study of communication development of 6-to-18-month-old infants. In *Emotion and early interaction* (ed. T. Field and A. Fogel). Erlbaum, Hillsdale, NJ.

Bakeman, R., and Adamson, L.B. (1984). Coordinating attention to people and objects in mother–infant and peer–infant interaction. *Child Development*, **55**, 1278–89.

Baron–Cohen, S. (1987). Autism and symbolic play. *British Journal of Developmental Psychology*, **5**, 139–48.

Baron–Cohen, S. (1988). Social and pragmatic deficits in autism: cognitive or affective? *Journal of Autism and Developmental Disorders*, **21**, 37–46.

Baron–Cohen, S. (1989*a*). Joint-attention deficits in autism: towards a cognitive analysis. *Development and Psychopathology*, **1**, 185–9.

Baron–Cohen, S. (1989*b*). Perceptual role-taking and protodeclarative pointing in autism. *British Journal of Developmental Psychology*, **7**, 113–27.

Baron–Cohen, S., Leslie, A.M., and Frith, U. (1985). Does the autistic child have a 'theory of mind'? *Cognition*, **21**, 37–46.

Bretherton, I., McNew, S., and Beegly-Smith, M. (1981). Early person knowledge as expressed in gestural and verbal communication: when do infants acquire a 'theory of mind'? In *Infant social cognition* (ed. M.E. Lamb and L.R. Sherrod). Erlbaum, Hillsdale, NJ.

Brothers, L. (1990). The social brain: a project for integrating primate behavior and neurophysiology in a new domain. *Concepts in Neuroscience*, **1**, 27–51.

Bruner, J. (1981). Learning how to do things with words. In *Human growth and development* (ed. J. Bruner and A. Garton). Oxford University Press.

Courchesne, E. (1991). Attention-shifting in autism. Paper presented at the Workshop on Autism and Theory of Mind. Seattle, WA.

Curcio, F. (1978). Sensorimotor functioning and communication in mute autistic children. *Journal of Autism and Childhood Schizophrenia*, **8**, 282-92.

Dawson, G., and Lewy, A. (1989). Arousal, attention, and the socioemotional impairments of individuals with autism. In *Autism, nature, diagnosis, and treatment* (ed. E. Dawson). Guilford Press, New York.

Fein, D., Pennington, B., Markowitz, P., Braverman, M., and Waterhouse, L. (1986). Toward a neuropsychological model of infantile autism: are the social deficits primary? *Journal of the Academy of Child and Adolescent Psychiatry*, **25**, 198-212.

Fischer, K., and Pipp, S. (1984). Process in cognitive development: optimal level and skill acquisition. In *Mechanisms of cognitive development* (ed. R. Sternberg). Freeman, New York.

Frith, U. (1989). A new look at language and communication in autism. *British Journal of Disorders of Communication*, **24**, 123-50.

Hannan, T. (1987). A cross-sequential assessment of the occurrences of pointing in 3- to 12-month-old human infants. *Infant Behavior and Development*, **10**, 11-22.

Harris, P. (1989). The autistic child's impaired conception of mind. *Development and Psychopathology*, **1**, 191-6.

Hobson, R. P. (1989*a*). On sharing experiences. *Development and Psychopathology*, **1**, 197-204.

Hobson, R. P. (1989*b*). Beyond cognition: a theory of autism. In *Autism: nature, diagnosis and treatment* (ed. G. Dawson). Guilford Press, New York.

Hobson, R. P. (1990). On acquiring knowledge about people and the capacity to pretend. Response to Leslie (1987). *Psychological Review*, **97**, 114-21.

Kanner, L. (1943). Autistic disturbance of affective contact. *Nervous Child*, **2**, 217-50.

Kasari, C., Sigman, M., Mundy, P., and Yirmiya, N. (1990). Affective sharing in the context of joint attention interactions of normal, autistic and mentally retarded children. *Journal of Autism and Developmental Disorders*, **20**, 87-100.

Leslie, A. (1987). Pretense and representation: the origins of 'theory of mind'. *Psychological Review*, **94**, 412-26.

Leslie, A. (1988). Some implications of pretense for mechanisms underlying the child's theory of mind. In *Developing theories of mind* (ed. J. Astington, P. Harris and D. Olson). Cambridge University Press.

Leslie, A., and Frith, U. (1988). Autistic children's understanding of seeing, knowing and believing. *British Journal of Developmental Psychology*, **6**, 315-24.

Leslie, A., and Happé, F. (1989). Autism and ostensive communication: the relevance of metarepresentation. *Development and Psychopathology*, **19**, 205-12.

Leung, H., and Rheingold, J. (1981). Development of pointing as a social gesture. *Developmental Psychology*, **17**, 215-20.

Lewis, C., and Osborne A. (1990). Three-year-olds' problems with false belief: conceptual deficit or conceptual artifact. *Child Development*, **61**, 1514-19.

Lewy, A., and Dawson, G. (1991). Effects of social stimulation on joint attention skills in young autistic children. Paper presented at the Biennial Meeting of the Society for Research in Child Development, Seattle, Washington.

Loveland, K., and Landry, S. (1986). Joint attention and language in autism and developmental language delay. *Journal of Autism and Developmental Disorders*, **16**, 335–49.

McDonald, M. A., Mundy, P., Kasari, C., and Sigman, M. (1989). Psychometric scatter in retarded, autistic preschoolers as measured by the Cattell. *Journal of Child Psychology and Psychiatry*, **30**, 599–604.

Malatesta, C., and Izard, C. (1984). The ontogenesis of human social signals: from biological imperative to symbolic utilization. In *The psychobiology of affective development* (ed. N. Fox and R. Davidson). Erlbaum, Hillsdale, NJ.

Mandler, J. (1988). How to build a better baby: on the development of an accessible representational system. *Cognitive Development*, **3**, 113–36.

Mundy, P., and Sigman, M. (1989*a*). The theoretical implications of joint attention deficits in autism. *Development and Psychopathology*, **1**, 173–83.

Mundy, P., and Sigman, M. (1989*b*). Second thoughts on the nature of autism. *Development and Psychopathology*, **1**, 213–17.

Mundy, P., and Sigman, M. (1989*c*). Specifying the nature of social impairment in autism. In *Autism: new perspectives on diagnosis, nature, and treatment* (ed. G. Dawson). Guilford Press, New York.

Mundy, P., Sigman, M., Ungerer, J. A., and Sherman, T. (1986). Defining the social deficits in autism: the contribution of non-verbal communication measures. *Journal of Child Psychology and Psychiatry*, **27**, 657–69.

Mundy, P., Sigman, M., Ungerer, J. A., and Sherman, T. (1987). Nonverbal communication and play correlates of language development in autistic children. *Journal of Autism and Developmental Disorders*, **17**, 349–64.

Mundy, P., Sigman, M., and Kasari, C. (1990*a*). A longitudinal study of joint attention and language development in autistic children. *Journal of Autism and Developmental Disorders*, **20**, 115–23.

Mundy, P., Kasari, C., and Sigman, M. (1992). Nonverbal communication, affective sharing, and intersubjectivity. *Infant Behavior and Development*, **15**, 377–82.

Oswald, D., and Ollendick, T. (1989). Role taking and social competence in autism and mental retardation. *Journal of Autism and Developmental Disorders*, **19**, 45–62.

Ozonoff, S., Pennington, B., and Rogers, S. (1991). Executive function deficits in high-functioning autistic children: relationship to theory of mind. *Journal of Child Psychology and Psychiatry*, **32**, 1081–105.

Perner, J., Frith, U., Leslie, A. M., and Leekam, S. R. (1989). Exploration of the autistic child's theory of mind: knowledge, belief, and communication. *Child Development*, **60**, 689–700.

Piaget, J. (1962 [1951]). *Play, dreams and imitation in childhood*. Norton, New York.

Poulin-Dubois, D., and Shultz, T. R. (1988). The development of the understanding of human behavior: from agency to intentionality. In *Developing theories of mind* (ed. J. Astington, P. Harris and D. Olson). Cambridge University Press, New York.

Riquet, C., Taylor, N., Benaroya, S., and Klein, L. (1981). Symbolic play in Autistic, Down's, and Normal children of equivalent mental age. *Journal of Autism and Developmental Disorders*, **11**, 439–48.

Rogers, S., and Pennington, B. (1991). A theoretical approach to the deficits in infantile autism. *Development and Psychopathology*, **2**, 137–62.

Rutter, M., and Garmezy, N. (1983). Developmental psychopathology. In *Handbook of child psychology*, Vol. 4, (ed. E. Hetherington). Wiley, New York.

Schwartz, G., Izard, C., and Ansul, S. (1985). The five-month-old's ability to discriminate facial expressions of emotion. *Infant Behavior and Development*, **8**, 65–77.

Seibert, J.M., Hogan, A.E., and Mundy, P. (1982). Assessing interactional competencies: the Early Social-Communication Scales. *Infant Mental Health Journal*, **3**, 244–59.

Sigman, M., and Mundy, P. (1987). Symbolic processes in atypical children. In *Symbolic development in atypical children* (ed. D. Cicchetti and M. Beeghly). Jossey-Bass, San Francisco, California.

Sigman, M., and Ungerer, J.A. (1984). Cognitive and language skills in autistic, mentally retarded, and normal children. *Developmental Psychology*, **20**, 293–302.

Sigman, M., Mundy, P., Sherman, T., and Ungerer, J.A. (1986). Social interactions of autistic, mentally retarded and normal children with their caregivers. *Journal of Child Psychology and Psychiatry*, **27**, 647–56.

Sugarman, S. (1984). The development of preverbal communication. In *The acquisition of communicative competence* (ed. R.L. Schiefelbusch and J. Pickar). University Park Press, Baltimore.

Trevarthen, C. (1979). Communication and cooperation in early infancy: a description of primary subjectivity. In *Before speech: the beginning of interpersonal communication* (ed. M. Bullova). Cambridge University Press, New York.

Ungerer, J. (1989). The early development of autistic children: implications for defining primary deficits. In *Autism: new perspectives on diagnosis, nature, and treatment* (ed. G. Dawson). Guilford Press, New York.

Ungerer, J., and Sigman, M. (1981). Symbolic play and language comprehension in autistic children. *Journal of the American Academy of Child Psychiatry*, **20**, 318–37.

Walker-Andrews, A., and Grolnick, W. (1983). Discrimination of vocal expressions by young infants. *Infant Behavior and Development*, **6**, 491–8.

Walters, A., Barrett, R., and Feinstein, C. (1990). Social relatedness and autism: current research, issues and directions. *Research in Developmental Disabilities*, **11**, 303–3216.

Wellman, H. (1988). First steps in the child's theorizing about the mind. In *Developing theories of mind* (ed. J. Astington, P. Harris and D. Olson). Cambridge University Press.

Werner, H. and Kaplan, S. (1963). *Symbol formation*. Wiley, New York.

Wetherby, A.M., and Prutting, C.A. (1984). Profiles of communicative and cognitive–social abilities in autistic children. *Journal of Speech and Hearing Research*, **27**, 367–77.

Wimmer, H., and Perner, J. (1983). Beliefs about beliefs: representation and constraining function of of wrong beliefs in young children's understanding of deception. *Cognition*, **21**, 103–28.

Wing, L., Gould, J., Yeates, S.R., and Brierly, L.M. (1977). Symbolic play in severely mentally retarded and autistic children. *Journal of Child Psychology and Psychiatry*, **18**, 167–78.

Yirmiya, N., Sigman, M., Kasari, C., and Mundy, P. (1992). Empathy and cognition in high-functioning children with autism. *Child Development*, **63**, 150–60.

10

Understanding persons:
the role of affect

PETER HOBSON

INTRODUCTION

It was in 1943 that Kanner first drew attention to the strange and probably unique quality of 'aloneness' manifested by children with early childhood autism. Autistic individuals relate to other people, and especially though not exclusively to other *people*, in a way that is markedly abnormal. One can feel as well as see that this is so, and Kanner was so struck by the fact that he identified the children's inability to make 'affective contact' with others as their cardinal abnormality. He also suggested that the study of autistic children might help to further our ideas about 'the constitutional components of emotional reactivity' in non-autistic people (Kanner 1943, p. 250).

It is perhaps surprising that almost four decades were to pass before experimental psychologists made a concerted effort to investigate autistic individuals' understanding of other people, and specifically their understanding of the minds of others. After all, there was plenty of observational evidence to suggest that autistic children not only have problems in relating socially, but also find other people very difficult to fathom. A central concern of this chapter is the nature of the connection between autistic people's biologically based impairments in social–affective relatedness, and their limited intellectual (cognitive) grasp of the nature of other persons *as* persons with their own mental life. This topic in developmental psychopathology is twinned with another in normal psychology, and following Kanner's (1943) recommendation I shall attempt to explore how the phenomena of autism might shed light on the rôle of a normal infant's capacity for patterned affective relatedness with others, in determining the course of early psychological development.

In fact, as the present volume testifies, a number of partly separate lines of research have converged to establish the topic of autistic children's 'understanding of minds' as a central issue for developmental research. I shall select three for mention. Towards the end of the 1970s, workers with an interest in psycholinguistics (for example Baltaxe 1977; Fay and Schuler 1980; Menyuk 1978; Tager-Flusberg 1981) were giving increasing prominence to autistic children's impairments in the 'pragmatics' of language comprehension and use, that is, their deficient understanding of the ways in which language is modified according to the social context in which it occurs. Autistic children frequently overlook or mistake the manifold respects in which language needs to be adapted and understood according to the psychological perspectives and communicative intents of speakers and listeners. Thus it came to be appreciated how, to a significant degree, autistic children's characteristic language abnormalities might reflect social incomprehension rather than more restrictively 'linguistic' deficits. The relative contribution of different kinds of impediment to language development, and the extent and mode of influence of more broadly 'social' factors in this regard, have continued to be important issues for research and debate (see for example Baron–Cohen 1988; Frith 1989*a*; Goodman 1989; Hobson 1989*a*; Tager-Flusberg 1989; Wetherby 1986).

At about the same time as this shift in emphasis was gathering momentum, clinical experience led me to propose that autistic children fail to acquire a fully developed concept of 'persons' with minds (Hobson 1982). My initial programme of research was designed to investigate the source of this conceptual impairment. The hypothesis was that it arose through the children's lack of basic perceptual–affective abilities and propensities that are required for a person to engage in 'personal relatedness' with others. The approach was one that drew upon the tradition of genetic epistemology. I had become convinced that in order to acquire knowledge of the nature of persons with minds, an individual needs to have experience of reciprocal, affectively-patterned relations with others; and in order to become engaged in personal relations, an individual needs to perceive and react to the bodily appearances, expressions, and actions of others with what philosophers have called 'natural' reactions involving feelings (Hamlyn 1974; Hampshire 1976; Peters 1974). Or to put this another way, infants and young children must have biologically-given 'prewired' capacities for direct perception of, and empathic responsiveness to, the bodily-expressed attitudes of other people, for the development of intersubjective understanding to begin at all. Correspondingly, if autistic children were to lack such prerequisites for personal relatedness and interpersonal awareness, this might account for their relative failure to understand the minds of others—a failure that itself has drastic implications for more developed forms of communication and thought. The major part of this chapter will be devoted to refining and developing this position, and to reviewing the evidence for its relevance in the case of autism.

A third approach of the early 1980s, pioneered by Baron–Cohen, Leslie, and Frith (for example, Baron–Cohen *et al.* 1985, 1986), arose through cross-fertilization between empirical and theoretical work on social cognition in normal young children and research in the field of autism. The hypothesis adopted by these workers is that autistic children lack some form of cognitive ability—according to Leslie (1987), the ability to form representations of other people's representations of the world—that is necessary for a child to acquire a 'theory of mind'. The principal focus of the early experiments was upon autistic children's difficulties in understanding the nature of people's beliefs, and especially the existence and implications of false beliefs, but the scope of this work has diversified considerably (see for example Baron–Cohen 1989a, b; Leslie and Frith 1988; Perner *et al.* 1989; and other chapters in this volume). This research programme has provided the principal source of evidence that autistic children do indeed face problems in appreciating the nature of such mental states as belief in themselves and others.

Now I have already stated that the above three avenues of research are convergent. In view of the important points of divergence that rightly continue to excite controversy among the advocates of different theoretical perspectives on these matters, I think this point needs to be emphasized. What the approaches have in common is a commitment to the view that autistic individuals' deficient understanding of people's minds is intrinsically related to a range of concurrent and/or secondary handicaps that are intellectual as well as social in nature. In this important sense, they are *all* concerned with cognitive development. They agree that in order to understand why the syndrome of autism is characterized by a cluster of seemingly diverse disabilities, and specifically a typical profile of impairments in social relations, in language, and in creative, flexible symbolic thinking and play, we need to appreciate the nature and implications of autistic children's limitations in social understanding (see for example Baron–Cohen 1988, 1990; Frith 1989a,b; Hobson 1989a, 1990a, 1992; Leslie and Frith 1990; Loveland 1991).

What of the matters over which there is disagreement? I shall not survey these now, since I hope a number of the most important issues will emerge as the chapter proceeds. By way of clearing the ground, however, I would highlight three points.

Firstly, there is contention over the way one should characterize both the development of 'mind' and the acquisition of knowledge or 'theory' about the mind (Samet, this volume, Chapter 19). For example, Leslie (1987) has attempted to capture the essence of mental functioning by identifying 'cognitive computations' in a deliberately abstract model of mental processes that posits the existence of 'representations' and operations on representations. He explains the early growth of interpersonal understanding by the 'switching on' of novel (meta)representational capacities that underpin concepts of 'mind'. I shall be adopting a contrasting approach that emphasizes the importance of an infant's innately configured and perceptually-anchored interpersonal *relations* in providing foundations for the older child's concepts of persons with bodies and minds. In this regard, Baron–Cohen's recent (1990) formulation that autistic children have a *'cognitive'* deficit of 'mind-blindness' is likely to muddy the conceptual waters. The reason is that it conflates two very different levels of interpersonal understanding – in my own terms, the conceptual ('cognitive') and the perceptual–relational – that need to be distinguished and accorded appropriate places in the developmental picture. I shall suggest how autistic children's limited concepts of mind (or failures to perform appropriate 'computations over representations') may result from, but not equate with, perceptual–relational deficits. Such relational deficits may sometimes involve a combination of partial mind-blindness and partial mind-deafness as facets of the central impairment in interpersonal coordination and inter-subjective contact (Hobson 1991*a*). Thus I think that the metaphor of mind-blindness is apt for many autistic children (and indeed I have emphasized how congenital blindness itself may lead to autistic-like clinical phenomena: Hobson 1990*a*, 1991*b*), but this is precisely because it draws attention to the possible significance of impaired perceptual and especially perceptual–affective capacities which tend to be overlooked when 'cognition' is abstracted from its proper developmental context.

My second, related point is that there is implicit if not explicit disagreement over the conceptual categories we should employ in our psychological theories of early development. The tripartite division of mental phenomena into cognitive, conative, and affective psychological functions has been popular in continental Europe since the seventeenth century (O'Neil 1968); but is this conceptual apartheid as appropriate when applied to infant psychology as it is when applied to adults? From the uncontested fact that autistic children have cognitive deficits, it has been implied (for instance by Rutter 1983, and by Baron–Cohen 1989*c*) that it is more plausible or fitting to seek the source of the children's social impairment in the 'cognitive' rather than the 'affective' domain. It seems to have been difficult to entertain the very idea that affective deficits *could* lead to specific cognitive deficits. Although I believe that (within this restricted frame of reference) they can and do, I want to adopt a more radical alternative position by changing the conceptual framework within which this debate takes place. I believe that what autistic individuals relatively lack are certain specific forms of 'affective–cognitive–conative' *relatedness* that normally occur from infancy onwards (see also Hobson 1990*b*). In due course I shall attempt to specify what this means.

The above two sets of issues introduce a third: what 'unit of study' is most appropriate for research on autism, if this involves pathology of 'infant-level' developmental processes? Specifically, is the appropriate unit of study the individual child's 'thoughts', 'beliefs', or 'feelings' considered in isolation, or is there a need to be concerned not only with the individual's own configurations of behaviour-cum-experience, but also with the nature of the transactions that occur *between* people?

It is surely sensible to embark on research into autism by seeking whatever is specific and universal to autistic individuals. Indeed, one characteristic that might turn out to be universal and relatively specific is a certain kind of impairment in understanding other people's minds. This is why I have emphasized autistic children's deficient concept of persons, and why Baron-Cohen, Leslie, and Frith have suggested that they lack a theory of mind. If this level of description captures what is homogeneous across autistic individuals, does it also constitute a basic level of psychological explanation? In one respect, as I have indicated, there is agreement that such lack of understanding *is* basic for explaining a range of autistic children's other psychological deficits. On the other hand, there is disagreement over how much it explains, and whether there are more fundamental levels of explanation involving developmentally antecedent psychological causes of this 'cognitive' disorder. In particular, is autistic children's deficit in conceptual understanding a necessary and sufficient explanation for their impaired interpersonal relations, or is it not *only* the source of a number of social–communicative difficulties but also the *outcome* of disorder in yet more basic processes of interpersonal relatedness?

At some point or other, psychological explanation reaches a bedrock beyond which further explanation has to be in non-psychological (for example, in physiological or neuroanatomical) terms. Have we reached such a point in saying that autistic children do not have an understanding of persons with minds? If, as Leslie (1987) suggests, the ability to understand minds is determined by the capacity for metarepresentation; if the cognitive capacity to 'metarepresent' the representations of others arises *sui generis* on an innate basis; and if this cognitive capacity is what autistic children innately lack, then the answer is probably 'Yes'.

I think the answer is 'No'. I believe there are antecedent psychological causes for autistic children's social-conceptual impairments. Moreover, to return to the issue of which 'unit of study' is appropriate here, I suggest that at *this* level of explanation we need to characterize such 'causes' by what is abnormal about autistic children's deficient *inter*personal experience. In one sense the focus remains on the individual child, but in another it shifts to what happens or fails to happen *between* people. I suggest that the manifest heterogeneity in the physical and psychological factors predisposing to autism yields homogeneity in so far as these diverse conditions disrupt the patterning of what transpires between autistic individuals and others (see also Hobson 1991*a*; Trevarthen 1989). In other words, autism may reveal the importance of what Vygotsky called the 'interpsychological category' of mental functioning (see Wertsch 1985), or what Trevarthen (1979) has named 'intersubjectivity', for the individual infant's psychological development. It is to this issue that I now turn.

THE NATURE OF INTERPERSONAL UNDERSTANDING

In what sense is an individual's understanding of minds *inter*personal? The most obvious sense is that the individual conceives 'mindfulness' to be a property of herself as well as an attribute of others. She knows that she herself has (first-person) experiences, and ascribes a similar quality of subjective life to other 'selves' (third-person attributions). The characteristic of having a mind is one conferred across persons.

This might seem a trivial use of the adjective 'interpersonal', but it is not. For it highlights the problem of how one individual arrives at a concept of a self who is *both* comparable to *and* differentiated from other selves who share the vital characteristic of having minds. How does a child become aware that she is like others in having a mind, yet different from others in so far as she has her own feelings, thoughts, beliefs and so on, and other people have theirs? Could it be that an infant first conceptualizes her own mental life, and by 'transferring' or 'projecting' the concepts acquired through her own private experience, arrives at the notion or 'theory' that other people have minds as well (see for example Harris 1989)? Elsewhere I have presented arguments to suggest that no, this cannot be the infant's route to an initial understanding *that* there are minds (Hobson 1990*c*, 1991*c*), even though rôle-taking of this kind undoubtedly takes place after infancy. There are two species of problem here: firstly, problems in explaining how a child could acquire the capacity to reflect on her own mental states earlier than, and as a precondition for, the ascription of minds to others; and secondly, problems that have to do with the basis or justification for applying analogy 'from her own case' if she does not *already* grasp that other people are fitting subjects for analogy in having a mind like she does.

The problems are overcome by recognizing that we need to begin by characterizing the earliest forms of psychological connectedness that exist between an infant and others, and then to account for the processes of differentiation that allow the individual to conceptualize herself and others as separate centres of psychological attitudes and experiences, that is, as separate 'selves'. This is the perspective adopted by Werner and Kaplan (1963) in their description of the 'distancing or polarization' that occurs between infant and care-taker in the 'primordial sharing situation', a development that 'emerges slowly in ontogenesis from early forms of interaction which have the character of "sharing" experiences *with* the Other rather than of "communicating" messages *to* the Other' (p. 42). The child's capacity for reflective self-awareness emerges as and when the child can take the rôle of the other (Mead 1934), and this can only occur once she has already recognized the other to be a person with whom she can identify, and thereby adopt an attitude to herself as a source of attitudes.

If this account is correct, it raises the following questions (amongst others):

(a) How are the most primitive forms of psychological connectedness or 'intersubjectivity' effected?

(b) How does the infant come to differentiate herself from others in such a way as to conceive of herself and others as distinct persons with minds? Or to put this another way, how does she grasp the nature of psychological attitudes in herself and others? As Strawson (1962) has emphasized, it is essential to the concept of a mental state that it can be ascribed to a person to whom it belongs, so that to understand minds is also to understand the nature of selves who have minds.

These two questions may be subsumed under a third: What are the conditions that allow for the possibility of knowledge of persons?

Shortly I shall argue that it is in addressing these questions that we discern the rôle of 'affect' in providing an essential foundation for interpersonal understanding. For now, I need to clarify the nature of psychological attitudes, and to indicate something of the *kind* of concept that the concept of 'persons' is.

Many but not all psychological attitudes are experienced. There is something 'that it is like' to feel happy or sad, to think of roses or to anticipate finishing this chapter (Nagel 1979). Many if not all psychological attitudes also have directedness or 'aboutness': one can feel sad *about* the death of a relative or think *that* an argument is valid. As Brentano (1973 [1874]) indicated, there is a sense in which the objects of mental attitudes have 'intentional inexistence', for one may have feelings or beliefs about objects or events that do not actually exist. On the other hand, there is another sense in which much of our mental life has reference to something 'out there'. As Bechtel (1988) emphasizes, the intentionality of mental states is just their ability to be *about* events in the world.

However, psychological attitudes are not simply or always modes of subjective experience. They also involve actions and other bodily expressions. Many such actions and postures have directedness towards specific objects and events in the environment. Indeed, animals as well as humans can be seen to be enjoying this meal or fearing that intruder. On the other hand, not all psychological attitudes find such direct and immediate expression in behaviour: you cannot (necessarily) know what I choose to conceal, nor can you (necessarily) tell that I am thinking of making a cup of tea. These observations raise the possibility that there is a developmental story to tell about the emancipation of 'mental' capacities of mind from behavioural expressions of mind, and a corresponding story about a young child's progressive understanding of (relatively) 'bodily inexpressive' mental functioning.

A central point here is that the directedness as well as the quality of mental attitudes may be manifest in behaviour. The subtlety of this issue for development in infancy may be illustrated by the equivocation in the following quotation from Werner and Kaplan (1963, p. 77): 'Eventually, a point is reached where the infant organism "refers" to, or better, is directed toward, distant objects with his entire body: he strains towards objects and at the same time emits "call-sounds", which are part and parcel of the bodily directedness.' Even though the infant is not meaning to 'refer', there is a way in which her body expresses a psychological attitude that has reference to specific aspects of the world. We can perceive that it is so. When can infants perceive something similar in the behaviour of others?

Before I address this question, I want to draw together much that has gone before by considering how the concept of a 'person' is more primitive than the concept of a 'mind' or 'body'. This philosophical issue has developmental implications. Earlier we enquired whether or to what degree development proceeds from individual self-consciousness to recognition of other minds, or progresses instead from the starting-point of intersubjectivity to later self- and other-awareness. At the present juncture, the question is whether a conception of human bodies precedes, and is distinct from, awareness of human minds, or whether there is something more primitive, a concept of bodily-cum-mental 'persons', relative to which these other concepts are derivative. Strawson (1962, p. 135–7) is forthright on this matter: '. . . the concept of a person is the concept of a type of entity such that *both* predicates ascribing states of consciousness *and* predicates ascribing corporeal characteristics, a physical situation, etc., are equally applicable to a single individual of that single type . . . The concept of a person is logically prior to that of an individual consciousness. The concept of a person is not to be analysed as that of an animated body or of an embodied anima.'

I shall argue that just as the concept of a bodily-cum-mental person is primitive, so too there are forms of *perception* in which to perceive bodies is also to perceive aspects of 'mind'. This perspective also returns us to the question of how first- and third-person attributions of mental life are coordinated. As Strawson (1962) suggests, it is essential to the character of the predicates we apply to persons that they have both first- and third-person ascriptive uses, and that they are both self-ascribable other than on the basis of observation by the subject who has them, and other-ascribable on the basis of behavioural criteria. To learn their use is to learn both aspects of their use. The question is therefore how the coordination of subjective experience in the 'self' and observation of behaviour in the 'other' is achieved.

I need to make a final point, for which I have not space to argue (see Hamlyn 1974; Hobson 1991*b*). This is that experience of affectively-patterned personal relatedness is *constitutive* of the concept of persons. If one had no experience of personal relatedness with others, one would not be able to grasp the kind of 'thing' a person is. The reason is that in order to be said to know something we must at least understand what it is for something to be the thing in question, and a part of that understanding is constituted by the relations in which the things stand to us (Hamlyn 1974). It is only through engagement in personal relations which involve feelings that one enjoys a 'form of life' in which the concept of 'persons' with their own value and specifically 'personal' attributes can get a purchase.

THE ROLE OF AFFECT

There is no precise meaning to the term 'affect'. It is easy enough to list principal members of the class of 'universal' emotions, those that have relatively constant expressions across cultures and can be recognized by all peoples of the world (Izard 1977; Ekman 1984); but there are less easily specified patterns and intensities of bodily expressive feelings, what Stern (1985) calls 'vitality affects', that would certainly qualify in this category. Partly in order to simplify my account, I shall concentrate on the former group of 'categorical affects' such as happiness, sadness, anger, fear, and surprise. Thus I shall begin by working within the traditional conceptual framework by focusing on the category of 'affect', and then stretch and refashion the boundaries of the 'affective' domain as I proceed. Let me attempt to summarize the properties of 'affective' psychological function that I believe to be especially important for the genesis of interpersonal understanding:

1. Each of the fundamental emotions is associated with a predisposition to more or less universal forms of experience *and* facial, vocal, and possibly gestural expression (Darwin (1965) [1872]; Izard 1977; Ekman 1984; Eibl-Eibesfeldt 1972). In addition, there are innately determined propensities to perceive appropriate emotional 'meanings' in the physiognomic and other bodily expressive actions of others (Walker-Andrews 1988). There are various meanings to the word 'meanings' here (see, for example, Hobson 1992; Walker-Andrews 1988), but these include the ways in which patterns of bodily–vocal expressiveness are coordinated with the 'action tendencies' of the organism (Frijda 1986). To be afraid is to be inclined to flee; to be angry is to be inclined to attack. As the philosopher Hampshire (1976) expresses it: 'In direct dealings with men, and outside the context of fiction, we perceive, and react to, the physiognomy of persons almost as immediately as to the full behaviour of which the facial expression is the residue. The expression, considered by itself, is as much a sign, or even a part, of incipient behaviour as it is a sign of inner feeling' (p. 78).

2. Correspondingly, there is a basic human capacity for 'direct perception' of feelings *in* the bodily expressions of others. This has been emphasized by a number of philosophers (for example Merleau-Ponty 1964; Scheler 1954; Wittgenstein 1980). As Woodruff Smith (1989) puts it: 'I not only see "her", I also see that "she is feeling sad" . . . Such an experience we feel intuitively is a direct awareness of the other's grief . . . And such is our acquaintance with others' (p. 134).

3. Emotions are interpersonally transmissible. My hunch is that it is partly this property that leads us to identify an 'affective' domain in the first place. To quote another philosopher, Peters (1974, p. 39): 'We are able to understand the expressions of the mental states of others because of the psychological law that expressions have the power, under normal conditions, to evoke corresponding experiences in the minds of observers.' In the 'affective' sphere, patterns of biologically determined bodily inter-co-ordination between people are also patterns of emotional coordination between people. To perceive a happy face as a happy face *is* to be inclined to respond with corresponding feelings and behaviour. To perceive an angry gesture towards oneself *is* to be inclined to react with feelings such as fear. Similar transpersonal influences may also be discerned in the domain of vitality affects, as on the ubiquitous occasions of 'affect attunement' (Stern 1985). The critical point is that what we call 'affect' is largely concerned with modes of *inter*personal perception and responsiveness that result in experiential as well as behavioural coordination between and across individuals. So-called 'affect' is the original means to and mode of 'primary intersubjectivity' (Trevarthen 1979).

4. Emotions have directedness. We feel things 'about' events that occur around us. Moreover, this directedness of a person's emotional attitudes is frequently manifest to others in the directedness of the person's behaviour.

5. Early affective attitudes amount to primitive forms of 'judgement'. It is not the case that 'cognitions' come first and 'emotions' later. Rather, as Plutchik (1984, p. 208) expresses it: 'Evaluations are a part of the total process that involves an organism interacting with its environment in biologically adaptive ways.' If one adopts the cognitive/affective dichotomy, it is *as* true that early cognition depends upon affect as it is true that affect depends on cognition. Better not to accept the dichotomy: 'affective relatedness' is the primitive concept here.

6. Signal-dependent affective 'communication' is a feature of the social life of higher animals. It is a principal mode of inter-individual coordination, but such coordination may also have reference to the shared environment: many animals can read the significance of other individuals' affectively expressive behaviour as this 'refers' to happenings in the surroundings. Indeed Gouzoules *et al.* (1985) trace how from affective signalling systems,

certain species of monkey seem to have evolved affect-independent communicative vocalizations that designate external referents. Thus there is an evolutionary heritage to the perception as well as to the expression of affective forms of inter-individual relating, 'sharing', and referring.

I trust it is now self-evident how intimately the properties of what one might call 'affective relatedness' are bound up with the existence of inter-subjective connectedness. One organism can perceive and relate to the 'body' of another with highly configured patterns of feeling and action that correspond with the feelings and actions of the other. One organism can also perceive something of the outer-directedness of the psychological attitudes of another in the other's behaviour. It is not the case that to begin with, behaviour is perceived in a cool, detached way, and is only subsequently interpreted as 'mental'; nor does the idea of subjective mental life 'behind' behaviour come from nowhere.

Yet it remains to trace the way in which self and other become differentiated, that is, the way in which an individual comes to understand self and other as separate persons with their own 'first-person' psychological orientations to the world. It is also necessary to account for a child's growing appreciation of the manner in which a person represents (and potentially misrepresents) aspects of the world by subsuming these under particular descriptions 'for' that person. Here it may be most helpful to consider development in infancy, and to focus upon the infant's perception of the 'directedness' of affective attitudes. Two-month-olds may react to the bodily-expressed feelings of care-takers with corresponding or coordinated actions and emotional expressions of their own (Haviland and Lelwica 1987). Even at such an early stage, the directedness as well as the timing of the care giver's behaviour towards or away from the infant can drastically influence such interpersonal-affective exchanges (Murray and Trevarthen 1985). By the end of the first year of life, infants may be observed to respond not only to a caregiver's relatedness towards themselves, but also to the caregiver's attitude towards environmental objects and events, a phenomenon called 'social referencing' (Campos and Stenberg 1981; Feinman 1982; Walden and Ogan 1988). For example, an infant of twelve months who is confronted by a strange situation that causes anxiety may appraise and react to the caregiver's own affective appraisal of the situation; or again, the infant may look towards the focus of the care-taker's fearful attentiveness, seemingly to discover what is disturbing her (Klinnert *et al.* 1983; Sorce *et al.* 1985).

It is important to observe that, even in such situations of 'social referencing', the infant need not yet be conceptualizing the attitude of the care-taker as a 'propositional attitude', that is, conceiving the care-taker to be feeling *that* the situation is safe, dangerous, or whatever. But at the same time, in its manner of relating to the care-taker's ways of relating to the situation, the infant demonstrates an awareness that the other person is a 'thing' with attitudes to an environment that is *also* the infant's own environment. The attitudes perceived in the other person are affective in nature, and they can confer new affective meanings on the things related-to. For instance, an infant may be reassured about the nature of a strange object by the caregiver's smiling at it. Note how the *self-same* object begins with one 'meaning' for the infant—it is a source of anxiety—but through the infant's own attitude to another person's object-directed 'meaning-conferring' attitude, it comes to have an altered significance. To the infant, the world is becoming one in which different people can find different 'meanings' in a single situation or event (Hobson 1990a; Mundy *et al.* this volume, Chapter 9). I believe that this triangulation involving self, other, and object-of-attitude is vitally important as the context within which a child comes to distinguish 'thought' from 'thing' (or initially, 'attitude' from 'thing'), and grasps how it is possible to confer new subjective meanings on reality-as-given, as in creative symbolic play.

At the same time as recognizing the objects of another person's affective attitudes, therefore, the infant is now relating to the other person as a source of attitudes. The ten-month-old infant will request things, point things out to, or show things to, the other, often looking back and forth between the object and the other's eyes; and she may even tease the other (Bretherton and Bates 1979; Bretherton *et al.* 1981; Harding and Golinkoff 1979; Trevarthen and Hubley 1978; Reddy 1991; and see Wellman, this volume, Chapter 2). Once again, all one requires to explain this is that the infant should adopt a particular kind of (emotionally involved) stance towards the 'behaviour' that expresses the affective engagement of the other. And this is precisely the point. To begin with, the infant's apprehension of the psychological attitudes of others is *neither* a non-mentalistic behavioural appraisal, *nor* does it yet entail a conceptual grasp of 'representing minds'. Rather, the nature of the child's affective engagement with the other person's affective attitudes (both child-related and environment-related) is the matrix from which more conceptual forms of self–other connectedness and differentiation emerge (Hobson 1989b). For example, a child's growing capacity to engage with another person's affective attitudes towards herself is what lifts primitive forms of self-experience into reflective self-awareness. She can now adopt an attitude to her own attitudes and to her own 'self' as a source of attitudes, feelings, and 'points of view'. Along with this comes a deeper understanding of the nature of other 'selves' as both like oneself, but distinct and 'other' (Hobson 1990b).

As a result of these developments, the child comes to acquire the correlative concepts of an 'objective' world and of 'subjective' psychological orientations to that world. Elsewhere I have attempted to trace how the development of a child's *cognitive* capacities for creative symbolic play and context-sensitive thought and language may have an intimate connection with this accomplishment (Hobson 1989*a*, 1990*a*; also Mundy *et al.*, this volume, Chapter 9).

THE CASE OF AUTISM

If the foregoing account is broadly correct, then it becomes plausible to suggest that early and severe impairments in interpersonal-affective relatedness might constrain a child's understanding of persons with minds, and through this the child's capacity for creative symbolic play and for contextually appropriate forms of language and thinking. My hypothesis is that the syndrome of autism provides evidence for the validity as well as the plausibility of such a developmental scheme. For the present purposes, I shall focus on the early parts of this story. I shall make the simplifying assumption that autistic children *do* have deficits in understanding the minds of others, and review some of the evidence that impairment in the children's affective relatedness with others is integral to such deficits. Of course it would be stronger to test the claim that the impairment in relatedness *causes* the conceptual difficulties, and indeed I think that very young and very retarded autistic children's abnormal social relatedness illustrates how this is developmentally prior to, and very likely causative of, the later conceptual problems. However, if one is testing a population of children who have concurrent disabilities in different domains (in this case, in personal relations and in understanding people), it is often difficult to determine the mode of association between the various facets of disorder, and in particular to establish whether one causes or is caused by another. Therefore I shall merely attempt to indicate some lines of evidence that illustrate how autistic children do have a relatively specific impairment in the recognition and understanding of emotion. I shall not attempt a full survey of the relevant literature, since I have presented comprehensive reviews elsewhere (Hobson 1991*d*, 1992).

The first kind of evidence is clinical. As Kanner (1943) and others have described, autistic children tend to give other people the experience that they are being treated as pieces of furniture rather than as persons. One can *feel* the lack of intersubjective affective contact with autistic individuals. I believe that this is an 'objective' clinical fact, but one that is very hard to establish by empirical measures. Certainly, the 'measuring instrument' for such a phenomenon needs to be another human being, which simply makes the point that *inter*subjectivity is in question.

The second kind of evidence is experimental. I have emphasized that at one level, autistic disturbances of affective contact are interpersonal.

It follows that the particular predisposing conditions for such disturbances *in the child* need not be homogeneous. For example, congenitally blind children may be predisposed to 'autism' in part because they lack visually-derived sources of information about the nature of other people's relatedness, including affective relatedness, both towards themselves and towards a visually-shared world.* On the other hand, I have adopted the hypothesis implicit in Kanner's (1943) writing that many autistic children lack the constitutional components of affect perception and responsivity that are necessary foundations for the capacity for personal relatedness with others, and thus for understanding 'other minds'. The following experiments have a bearing on this hypothesis.

The first experiment I shall describe is one designed to test the hypothesis that autistic children are abnormal in their relative lack of attentiveness to facial expressions of emotion (Weeks and Hobson 1987). We pairwise matched 15 autistic and 15 non-autistic retarded children and young adolescents for chronological age, sex, and performance on three subtests of the Verbal Scale of the WISC-R. Given that autistic children's 'verbal' abilities are among the most impaired of their intellectual capacities, these represent stringent matching procedures for investigating deficits specific to autism. Subjects were given a task of sorting photographs to 'go with' one or other of a pair of target photographs showing the head and shoulders of individuals who differed in three, two, or one of the following respects: sex, age, facial expression of emotion, and the type of hat they were wearing. The principal finding was that the majority of non-autistic children sorted according to people's facial expression ('happy' versus sad or 'non-happy') before they sorted according to type of hat (floppy versus woollen); but most autistic children gave priority to sorting by type of hat. Moreover, when in the course of the experiment the number of contrasting features in the target photographs was progressively reduced, all 15 non-autistic sooner or later sorted by emotional expression without being told to do so, but only 6 of the 15 autistic children did this ($p = 0.0003$, Fisher's exact test). Finally, 5 of the 15 autistic children, but none of the 15 non-autistic children, failed to sort consistently by facial expression when given explicit instructions to do so. Only after we had completed this experiment did we discover

---

* I have argued this case in detail elsewhere: Hobson 1990*a*, 1991*b*. My colleagues Rachel Brown and Maggie Minter and I have been collecting data that seems to corroborate case-reports suggesting an unusually high prevalence of echolalia, personal pronoun difficulties, deficits in creative symbolic play, and problems with 'sharing' amongst congenitally blind children. The central theoretical notion is that vision is specially important for enabling sighted children to see (literally) the ways in which different people have differentiated but complementary psychological orientations towards given objects and events. Congenitally blind children might be delayed in acquiring the insight into subjective versus objective reality that such experience affords; and those with additional impediments to intersubjective personal relations may be especially inclined to develop the full syndrome of 'autism'.

that Jennings (1973) had conducted a similar but not identical unpublished study which yielded results that were comparable to our own.

In a second set of studies my colleagues and I examined subjects' understanding of the 'meaning' of bodily expressions of emotion, by testing whether they could identify how the different expressions of given emotions are coordinated with each other and with appropriate kinds of situational event. For example, are autistic children and young adults comparable to intellectually matched non-autistic retarded or normal subjects in their ability to recognize how a scared face 'goes with' a quaking bodily gesture, a frightened vocalization, and a fear-inducing event? We have applied several somewhat differing methodological approaches to this question (Hobson 1986a, b; Hobson et al. 1988b), and I shall refer to only one of these. For the experiment in question, autistic and non-autistic retarded children and adolescents were individually matched according to age and verbal ability. They were asked to select which one of six photographed, standardized facial expressions of emotion (Ekman and Friesen 1975) was the one to 'go with' each of six successive emotionally expressive vocalizations (both non-verbal and verbal) recorded on audiotape. There was a control task in which subjects selected photographs of non-emotional objects or events for corresponding audiotaped sounds. For example, we presented photographs and sounds of six kinds of bird, six kinds of electrical appliance, six kinds of 'walking', and so on. The results were that, relative to non-autistic retarded control subjects, the autistic subjects performed significantly less well on the emotion tasks than on the non-emotion tasks. In a subsequent experiment (Hobson et al. 1989), subjects were asked to give free-response labels for a subset of these photographs and sounds presented separately, and again the autistic subjects were selectively poor at naming feelings *vis-à-vis* non-personal objects. Supportive and complementary findings have been reported by Tantam et al. (1989), Macdonald et al. (1989), van Lancker et al. (1989), and Ozonoff et al. (1991).

In a third study we focused upon autistic individuals' capacity to discriminate among standardized photographs of happy, unhappy, angry, and afraid faces (Hobson et al. 1988a). Once again autistic and non-autistic retarded subjects were matched for age and verbal ability. The 'emotions' task was to match the same expressions displayed by different people (i.e. to match the faces for emotion across changes in identity), and the control task was to match the faces for identity across changes in emotional expression. A match-to-sample procedure was employed, first using full faces, then faces with blanked-out mouths, and then faces with blank mouths and foreheads. The materials were designed so that even the last set of photographs retained some 'feel' of the emotions in the faces. The results included a significant second-order interaction of diagnosis by condition (emotion, identity) by form of face (full-face; blank-mouth;

blank-mouth-and-forehead): whereas on the identities task, the performance of the two groups showed a similar steady decline as the photographs became increasingly blanked-out, on the emotions task the performance of autistic subjects worsened more abruptly than that of control subjects as cues to emotion were progressively reduced. The autistic subjects seemed relatively unable to use the 'feel' in the faces to guide performance. Not only this, but correlations between individual subjects' scores on the identity and emotion tasks were higher for autistic than for non-autistic subjects, suggesting that autistic subjects might have been sorting the expressive faces by non-emotional perceptual strategies.

A fourth study was designed to examine autistic subjects' understanding of emotion-related concepts (Hobson and Lee 1990). We took two groups of autistic and non-autistic retarded adolescents and young adults who were pairwise matched for age and *overall* scores on the British Picture Vocabulary Scale (BPVS; Dunn *et al.* 1982), a British version of the Peabody Picture Vocabulary Test (Dunn 1965). In this test, individuals are presented with a series of plates in which drawings are arranged in groups of four. Subjects are given instructions such as 'Point to . . . dentist', or 'Show me . . . surprise', and they respond by indicating the appropriate picture. We then asked independent raters to select the items of the BPVS that were 'emotion-related', which as it turned out included word–picture combinations in which the words to be judged were 'delighted', 'disagreement', 'greeting', and 'snarling', as well as more obvious 'emotional' words such as 'horror' and 'surprise'. Thus we were able to compare subjects' scores on these items with their scores on 'non-emotion' items of the BPVS that are equally difficult for normal children. As expected, non-autistic retarded subjects achieved similar scores on the selected emotion-related and emotion-unrelated items, but autistic subjects with almost identical overall BPVS scores were specifically impaired on the emotion-related *vis-à-vis* the emotion-unrelated items. Moreover, the results could not be attributed to the abstract nature of such concepts, since autistic and non-autistic subjects were equally able to judge non-emotion-related abstract words *vis-à-vis* equally difficult 'concrete' words. Once again, therefore, this time at a conceptual level, significant group differences between autistic and non-autistic subjects were demonstrable specifically in the realm of emotional understanding.

Despite the evidence that has accumulated from the aforementioned studies, there are workers who are understandably sceptical about the existence of autism-specific emotion-recognition deficits. The principal source of doubt is that there are several published studies, including some of our own, in which verbal MA-matched autistic and non-autistic subjects have been found *not* to differ significantly on certain tests of emotion recognition or understanding (for example Baron–Cohen 1991; Braverman *et al.* 1989; Hertzig *et al.* 1989; Ozonoff *et al.* 1990; Prior *et al.* 1990). There are numerous methodological considerations here (see Hobson 1991*d*), but I would stress two points. Firstly, there are a number of studies in which verbal MA-matched autistic and non-autistic subjects *have* differed in levels and/or profiles of performance on such tasks (see above). Secondly, there is reason to believe that on tasks in which autistic as well as non-autistic children's 'verbal ability' seems to correlate with 'emotion-recognition ability', subjects' performance has *not* been determined by linguistic task-related difficulties. It would be curious to dismiss the poor performance of autistic subjects on tests of 'emotion-recognition', simply on the grounds that often this is no worse than their performance on tests of language ability! Indeed, the possibility arises that autistic children's severe problems with language could stem in part from their lack of experience of a shared, 'co-referenced' world – a deficit that might be caused by impediments to affectively patterned intersubjective communication and understanding. In keeping with this suggestion, Mundy *et al.* (1990) reported that within their sample of young autistic children, language development was predicted by a measure of gestural non-verbal attention one year previously, but not by measures of initial language score or IQ.

In addition, there are complementary avenues of experimental research that are highly suggestive of marked abnormality in autistic children's manner of 'affective' relating to others. In this regard, I should stress that perception of affective meanings in other people entails the capacity for appropriately patterned emotional responsiveness to the bodily 'expressions' of others (so that in fact I would agree with Mundy *et al.*, this volume, Chapter 9, that autistic children's problems with the experience and expression of affect may be part and parcel of their difficulty in perceiving as well as conceiving of subjective emotional life). More specifically, autistic children's emotional expressiveness is often muted or idiosyncratic, and frequently uncoordinated with the expressive behaviour of others (Dawson *et al.* 1990; Hertzig *et al.* 1989; Kasari *et al.* 1990; Macdonald *et al.* 1989; Ricks 1975; Snow *et al.* 1987; Yirmiya *et al.* 1989). So, too, young autistic children differ from MA-matched non-autistic retarded and developmentally language-delayed children in exhibiting few indicating gestures such as pointing to or showing objects to others, 'infantile-level' gestures which have the aim of sharing experiences (Curcio 1978; Landry and Loveland 1988; Mundy *et al.* 1986; Sigman *et al.* 1986; Mundy *et al.*, this volume, Chapter 9). There is also provisional evidence that, although autistic children can understand something of other people's visuo-spatial perspectives (Baron–Cohen 1989*a*; Hobson 1984; Leslie and Frith 1988), an ability that might be derived through processes that are relatively independent of the children's experiences of intersubjectivity, they may well be abnormal in their propensity to engage in 'affective' modes of social referencing (Mundy *et al.* 1986; Sigman and Mundy 1989).

In summary, then, there is substantial clinical and experimental evidence to suggest that many autistic children may lack what Kanner (1943) called 'the constitutional components of emotional reactivity' that are necessary for normal interpersonal relatedness. It remains possible, at least for more advanced autistic subjects, that 'emotion-recognition deficits' are secondary to other kinds of conceptual or linguistic deficit; it is even conceivable that autistic children's difficulties in establishing 'affective contact' with others through non-verbal communication are the result of essentially non-affective (? 'cognitive') deficits. Personally, I do not find the first of these suggestions very persuasive, nor the second very plausible. On the other hand, I have tried to emphasize that even if all autistic children have a critical impairment in 'affective relatedness', they do not all need to share the same underlying component deficits – and of course there *will be* 'lower-level' explanations, perhaps even psychological explanations, for the impairments in intersubjective coordination. Indeed, it may be important to distinguish two rather different facets of 'autistic' social impairment. One has to do with abnormality in the kinds of one-to-one interpersonal relatedness and communicative exchange that Trevarthen (1979) has called 'primary intersubjectivity'. The other has more to do with the child's incapacity to perceive another person's affective relatedness towards a shared world, that is, an inability to discern the directedness as well as qualities of a person's psychological attitudes towards objects 'out there' – a problem that may prove to be a source of 'autistic-like' features in congenitally blind children.

Although there are many questions that are yet to be settled, there is clearly a case for arguing that the cause of autistic individuals' deficient understanding of minds is their limited experience of affectively patterned personal relatedness. In many cases there appears to be biologically-based impairments in the autistic child's capacities for perception of, and normally patterned responsiveness to, the affective expressions and behaviour of other persons. If this is so, then autism provides a dramatic and tragic illustration of the rôle of affect, or better, the role of affectively patterned personal relatedness, in the development of a child's understanding of the nature of persons with minds.

CONCLUSIONS

I have tried to indicate how a person's understanding of the minds of other people and herself is grounded in forms of *inter*personal relatedness that are 'affective' in nature. It is because of the very intimate connection between bodies and minds as aspects of 'the person', and because of the corresponding intimacy between the perceptible expressiveness of a person's 'body' and the expressiveness of *certain* aspects of that person's 'mind', that truly interpersonal understanding can develop at all. There are innately determined sensibilities through which one individual can 'read' and respond affectively to another individual's affective attitudes, not only towards herself but also towards elements in the outside world. This fact is of great importance for the way in which different people's first-person subjective experiences and 'selves' are both linked to and differentiated from each other (also Barresi and Moore 1990).

The perspective from autism promises to teach us much about these matters. We may have been slow to appreciate the full significance of Kanner's (1943) suggestion that (most) autistic children have 'come into the world with innate inability to form the usual, biologically provided affective contact with people' (p. 250). This proposal might explain far more than we have so far realized about the psychological basis for the syndrome of 'autism', and so too it may reveal the profound importance of personal–affective relatedness for the course of normal child development.

Acknowledgement

I am greatly indebted to my colleague Anthony Lee for his long-standing collaboration and for his help in the preparation of this chapter.

REFERENCES

Baltaxe, C. A. M. (1977). Pragmatic deficits in the language of autistic adolescents. *Journal of Pediatric Psychology*, **2**, 176–80.

Baron–Cohen, S. (1988). Social and pragmatic deficits in autism: cognitive or affective? *Journal of Autism and Developmental Disorders*, **18**, 379–402.

Baron–Cohen, S. (1989a). Perceptual role-taking and protodeclarative pointing in autism. *British Journal of Developmental Psychology*, **7**, 113–27.

Baron–Cohen, S. (1989b). Are autistic children 'behaviorists'? An examination of their mental–physical and appearance–reality distinctions. *Journal of Autism and Developmental Disorders*, **19**, 579–600.

Baron–Cohen, S. (1989c). The autistic child's theory of mind: a case of specific developmental delay. *Journal of Child Psychology and Psychiatry*, **30**, 285–97.

Baron–Cohen, S. (1990). Autism: a specific cognitive disorder of 'mind-blindness'. *International Review of Psychiatry*, **2**, 81–90.

Baron–Cohen, S. (1991). Do people with autism understand what causes emotion? *Child Development*, **62**, 385–95.

Baron–Cohen, S., Leslie, A. M., and Frith, U. (1985). Does the autistic child have a 'theory of mind'? *Cognition*, **21**, 37–46.

Baron–Cohen, S., Leslie, A. M., and Frith, U. (1986). Mechanical, bchavioural and Intentional understanding of picture stories in autistic children. *British Journal of Developmental Psychology*, **4**, 113–25.

Barresi, J. and Moore, C., (1990). A theory of human social understanding. Submitted for publication.

Bechtel, W. (1988). *Philosophy of mind: an overview for cognitive science*. Erlbaum, Hillsdale, NJ.

Braverman, M., Fein, D., Lucci, D., and Waterhouse, L. (1989). Affect comprehension in children with pervasive developmental disorders. *Journal of Autism and Developmental Disorders*, **19**, 301–16.

Brentano, F. (1973[1874]). *Psychology from an empirical standpoint* (trans. A.C. Rancurello, D.B. Terrell and L.L. McAlister). Routledge & Kegan Paul, London.

Bretherton, I., and Bates, E. (1979). The emergence of intentional communication. In *Social interaction and communication during infancy*. New directions for child development, No. 4, (ed. C. Uzgiris). Jossey-Bass, San Francisco.

Bretherton, I., McNew, S., and Beeghly-Smith, M. (1981). Early person knowledge as expressed in gestural and verbal communication: when do infants acquire a 'Theory of mind'? In *Infant social cognition* (ed. M.E. Lamb and L.R. Sherrod). Erlbaum, Hillsdale, NJ.

Campos, J.J., and Sternberg, C.R. (1981). Perception appraisal and emotion: the onset of social referencing. In *Infant social cognition* (ed. M.E. Lamb and L.R. Sherrod), Erlbaum, Hillsdale, NJ.

Curcio, F. (1978). Sensorimotor functioning and communication in mute autistic children. *Journal of Autism and Childhood Schizophrenia*, **8**, 281–92.

Darwin, C. (1965 [1872]). *The expression of emotions in man and animals*. University of Chicago Press.

Dawson, G., Hill, D., Spencer, A., Galpert, L., and Watson, L. (1990). Affective exchanges between young autistic children and their mothers. *Journal of Abnormal Child Psychology*, **18**, 335–45.

Dunn, L.M. (1965). *Expanded manual for the Peabody Picture Vocabulary Test*. American Guidance Service, Circle Pines, Minnesota.

Dunn, L.M., Dunn, L.M., and Whetton, C. (1982). *British Picture Vocabulary Scale*. NFER–Nelson, Winsdor.

Eibl-Eibesfeldt, I. (1972). Similarities and differences between cultures in expressive movements. In *Non-verbal communication* (ed. R.A. Hinde), pp. 297–312. Cambridge University Press.

Ekman. P. (1984). Expression and the nature of emotion. In *Approaches to emotion* (ed. K.R. Scherer and P. Ekman). Erlbaum, Hillsdale, NJ.

Ekman, P., and Friesen, W.V. (1975). *Unmasking the face. A guide to recognizing emotions from facial cues*. Prentice-Hall, Englewood Cliffs, NJ.

Fay, W.H., and Schuler, A.L. (1980). *Emerging language in autistic children*. Edward Arnold, London.

Feinman, S. (1982). Social referencing in infancy. *Merrill-Palmer Quarterly*, **28**, 445–70.

Frijda, N.H. (1986). *The emotions*. Cambridge University Press.

Frith, U. (1989*a*). A new look at language and communication in autism. *British Journal of Disorders of Communication*, **24**, 123–50.

Frith, U. (1989*b*). *Autism: explaining the enigma*. Blackwell, Oxford.

Goodman, R. (1989). Infantile autism: a syndrome of multiple primary deficits? *Journal of Autism and Developmental Disorders*, **19**, 409–24.

Gouzoules, H., Gouzoules, S., and Marler, P. (1985). External reference and affective signaling in mammalian vocal communication. In *The development of expressive behavior*, (ed. G. Zivin). Academic Press, New York.

Hamlyn, D. W. (1974). Person-perception and our understanding of others. In *Understanding other persons* (ed. T. Mischel), pp. 1–36. Blackwell, Oxford.

Hampshire, S. (1976). Feeling and expression. In *The philosophy of mind* (ed. J. Glover). (Original work published 1960.)

Harding, C. G., and Golinkoff, R. M. (1979). The origins of intentional vocalizations in prelinguistic infants. *Child Development*, **50**, 33–40.

Harris, P. L. (1989). *Children and emotion*. Blackwell, Oxford.

Haviland, J. M., and Lelwica, M. (1987). The induced affect response: 10-week-old infants' responses to three emotion expressions. *Development Psychology*, **23**, 97–104.

Hertzig, M. E., Snow, M. E., and Sherman, M. (1989). Affect and cognition in autism. *Journal of the American Academy of Child Psychiatry*, **28**, 195–9.

Hobson, R. P. (1982). The autistic child's concept of persons. In *Proceedings of the 1981 International Conference on Autism, Boston, USA* (ed. D. Park), pp. 97–102. National Society for Children and Adults with Autism, Washington DC.

Hobson, R. P. (1984). Early childhood autism and the question of egocentrism. *Journal of Autism and Developmental Disorders*, **14**, 85–104.

Hobson, R. P. (1986*a*). The autistic child's appraisal of expressions of emotion. *Journal of Child Psychology and Psychiatry*, **27**, 321–42.

Hobson, R. P. (1986*b*). The autistic child's appraisal of expressions of emotion: a further study. *Journal of Child Psychology and Psychiatry*, **27**, 671–80.

Hobson, R. P. (1989*a*). Beyond cognition: a theory of autism. In Autism: nature, diagnosis, and treatment (ed. G. Dawson). Guilford, New York.

Hobson, R. P. (1989*b*). On sharing experiences. *Development and Psychopathology*, **1**, 197–203.

Hobson, R. P. (1990*a*). On acquiring knowledge about people and the capacity to pretend: response to Leslie. *Psychological Review*, **97**, 114–21.

Hobson, R. P. (1990*b*) On the origins of self and the case of autism. *Development and Psychopathology*, **2**, 163–81.

Hobson, R. P. (1990*c*). Concerning knowledge of mental states. *British Journal of Medical Psychology*, **63**, 199–213.

Hobson, R. P. (1991*a*). What is autism? In *Psychiatric clinics of North America*, **14**, pp. 1–17.

Hobson, R. P. (1991*b*). Through feeling and sight to self and symbol. In *Ecological and interpersonal knowledge of the self* (ed. U. Neisser). Cambridge University Press, New York.

Hobson, R. P. (1991*c*). Against the theory of 'Theory of Mind'. *British Journal of Developmental Psychology*, **9**, 33–51.

Hobson, R. P. (1991*d*). Methodological issues for experiments on autistic individuals' perception and understanding of emotion. *Journal of Child Psychology and Psychiatry*, **32**, 1135–1158.

Hobson, R. P. (1992). Social perception in high-level autism. In *High-functioning individuals with autism*, (ed. E. Schopler and G. Mesibov). Plenum, New York.

Hobson, R. P., and Lee, A. (1989). Emotion-related and abstract concepts in autistic people: evidence from the British Picture Vocabulary Scale. *Journal of Autism and Developmental Disorders*, **19**, 601–23.

Hobson, R.P., Ouston, J., and Lee, A. (1988*a*). What's in a face? The case of autism. *British Journal of Psychology*, **79**, 441–53.

Hobson, R.P., Ouston, J., and Lee, A, (1988*b*). Emotion recognition in autism: coordinating faces and voices. *Psychological Medicine*, **18**, 911–23.

Hobson, R.P., Ouston, J., and Lee, A. (1989). Naming emotion in faces and voices: abilities and disabilities in autism and mental retardation. *British Journal of Developmental Psychology*, **7**, 237–50.

Izard, C.E. (1977). *Human emotions*. Plenum, New York.

Jennings, W.B. (1973). *A study of the preference for affective cues in autistic children*. Unpublished Ph.D. thesis, Memphis State University.

Kanner, L. (1943). Autistic disturbances of affective contact. *Nervous Child*, **2**, 217–50.

Kasari, C., Sigman, M., Mundy, P., and Yirmiya, N. (1990). Affective sharing in the context of joint attention interactions of normal, autistic, and mentally retarded children. *Journal of Autism and Developmental Disorders*, **20**, 87–100.

Klinnert, M.D., Campos, J.J., Sorce, J.F., Emde, R.N., and Svejda, M. (1983). Emotions as behavior regulators: social referencing in infancy. In *Emotion: theory, research and experience*, Vol. 2: *Emotions in early development* (ed. R. Plutchik and H. Kellerman). Academic Press, New York.

Landry, S.H., and Loveland, K.A. (1988). Communication behaviors in autism and developmental language delay. *Journal of Child Psychology and Psychiatry*, **29**, 621–34.

Leslie, A.M. (1987). Pretense and representation: the origins of 'theory of mind' *Psychological Review*, **94**, 412–26.

Leslie, A.M., and Frith, U. (1988). Autistic children's understanding of seeing, knowing and believing. *British Journal of Developmental Psychology*, **6**, 315–24.

Leslie, A.M., and Frith, U. (1990). Prospects for a cognitive neuropsychology of autism: Hobson's choice. *Psychological Review*, **97**, 122–31.

Loveland, K.A. (1991). Autism, affordances and the self. In *Ecological and interpersonal knowledge of the self* (ed. U. Neisser). Cambridge University Press, New York.

Macdonald, H., Rutter, M., Howlin, P., Rios, P., LeCouteur, A., Evered, C., and Folstein, S. (1989). Recognition and expression of emotional cues by autistic and normal adults. *Journal of Child Psychology and Psychiatry*, **30**, 865–77.

Mead, G.H. (1934). *Mind, self, and society* (ed. C.W. Morris). University of Chicago Press.

Menyuk, P. (1978). Language: what's wrong and why? In *Autism: a reappraisal of concepts and treatment* (ed. M. Rutter and E. Schopler). Plenum, New York.

Merleau-Ponty, M. (1964). The child's relations with others (trans. W. Cobb). In *The primacy of perception*, pp. 96–155. Northwestern University Press.

Mundy, P., Sigman, M., Ungerer, J., and Sherman, T. (1986). Defining the social deficits of autism: the contribution of non-verbal communication measures. *Journal of Child Psychology and Psychiatry*, **27**, 657–69.

Mundy, P., Sigman, M., and Kasari, C. (1990). A longitudinal study of joint attention and language development in autistic children. *Journal of Autism and Developmental Disorders*, **20**, 115–28.

Murray, L., and Trevarthen, C. (1985). Emotional regulation of interactions between two-month-olds and their mothers. In *Social perception in infants* (ed. T.M. Field and N.A. Fox). Ablex, Norwood, NJ.

Nagel, T. (1979) (originally 1974). What is it like to be a bat? In idem, *Mortal questions*. Cambridge University Press.

O'Neil, W.M. (1968). *The beginnings of modern psychology*. Penguin, Harmondsworth, Middlesex.

Ozonoff, S., Pennington, B.F., and Rogers, S.J. (1990). Are there specific emotion perception deficits in young autistic children? *Journal of Child Psychology and Psychiatry*, **31**, 343–61.

Ozonoff, S., Pennington, B.F., and Rogers, S.J. (1991). Executive function deficits in high-functioning autistic individuals: relationship to theory of mind. *Journal of Child Psychology and Psychiatry*, **32**, 1081–105

Perner, J., Frith, U., Leslie, A.M., and Leekam, S.R. (1989). Exploration of the autistic child's theory of mind: knowledge, belief and communication. *Child Development*, **60**, 689–700.

Peters, R.S. (1974) Personal understanding and personal relationships. In *Understanding other persons* (ed. T. Mischel). Blackwell, Oxford.

Plutchik, R. (1984). Emotions: a general psychoevolutionary theory. In *Approaches to emotion* (ed. K.R. Scherer and P. Ekman). Erlbaum, Hillsdale, N.J.

Prior, M.R., Dahlstrom, B., and Squires, T.-L. (1990). Autistic children's knowledge of thinking and feeling states in other people. *Journal of Child Psychology and Psychiatry*, **31**, 587–601.

Reddy, V. (1991). Playing with others' expectations: teasing and mucking about in the first year. In *Natural theories of mind* (ed. A. Whiten). Blackwell, Oxford.

Ricks, D.M. (1975). Vocal communication in pre-verbal normal and autistic children. In *Language, cognitive deficits, and retardation* (ed. N. O'Connor). Butterworth, London.

Rutter, M. (1983). Cognitive deficits in the pathogenesis of autism. *Journal of Child Psychology and Psychiatry*, **24**, 513–31.

Scheler, M. (1954). *The nature of sympathy* (trans. P. Heath). Routledge & Kegan Paul, London.

Sigman, M., and Mundy, P. (1989). Social attachments in autistic children. *Journal of the American Academy of Child and Adolescent Psychiatry*, **28**, 74–81.

Sigman, M., Mundy, P., Sherman, T., and Ungerer, J. (1986). Social interactions of autistic, mentally retarded and normal children and their caregivers. *Journal of Child Psychology and Psychiatry*, **27**, 647–56.

Snow, M.E., Hertzig, M.E., and Shapiro, T. (1987). Expression of emotion in young autistic children. *Journal of the American Academy of Child and Adolescent Psychiatry*, **26**, 836–8.

Sorce, J.F., Emde, R.N., Campos, J., and Klinnert, M.D. (1985). Maternal emotional signaling: its effect on the visual cliff behavior of 1-year-olds. *Developmental Psychology*, **21**, 195–200.

Stern, D.N. (1985). *The interpersonal world of the infant*. Basic Books, New York.

Strawson, P.F. (1962) (originally 1958). Persons. In *The philosophy of mind* (ed. V.C. Chappel). Prentice-Hall, Englewood Cliffs, NJ.

Tager-Flusberg, H. (1981). On the nature of linguistic functioning in early infantile autism. *Journal of Autism and Developmental Disorders*, **11**, 45–56.

Tager-Flusberg, H. (1989). A psycholinguistic perspective on language development in the autistic child. In *Autism: nature, diagnosis, and treatment* (ed. G. Dawson). Guilford, New York.

Tantam, D., Monaghan, L., Nicholson, H., and Stirling, J. (1989). Autistic children's ability to interpret faces: a research note. *Journal of Child Psychology and Psychiatry*, **30**, 623–30.

Trevarthen, C. (1979). Communication and cooperation in early infancy: a description of primary intersubjectivity. In *Before speech* (ed. M. Bullowa). Cambridge University Press.

Trevarthen, C. (1989). The relation of autism to normal socio-cultural development: the case for a primary disorder in regulation of cognitive growth by emotions. Published in French 'Les relations entre autisme et développement socioculturel normal: arguments en faveur d'un trouble primaire de la régulation du développement cognitif par les emotions'. In *Autisme et troubles du développement global de l'enfant* (ed. G. Lelord, J. P. Muk and M. Petit). Expansion Scientifique Française, Paris.

Trevarthen, C., and Hubley, P. (1978). Secondary intersubjectivity: confidence, confiding and acts of meaning in the first year. In *Action, gesture and symbol: the emergence of language* (ed. A. Lock). Academic Press, London.

van Lancker, D., Cornelius, C., and Kreiman, J. (1989). Recognition of emotional-prosodic meanings in speech by autistic, schizophrenic, and normal children. *Developmental Neuropsychology*, **5**, 207–26.

Walden, T. A., and Ogan, T. A. (1988). The development of social referencing. *Child Development*, **59**, 1230–40.

Walker-Andrews, A. S. (1988). Infants' perception of the affordances of expressive behaviors. In *Advances in infancy research*, Vol. 5, (ed. C. Rovee-Collier). Ablex, Norwood, NJ.

Weeks, S. J., and Hobson, R. P. (1987). The salience of facial expression for autistic children. *Journal of Child Psychology and Psychiatry*, **28**, 137–52.

Werner, H., and Kaplan, B. (1963). *Symbol formation*. Wiley, New York.

Wertsch, J. V. (1985). *Vygotsky and the social formation of mind*. Harvard University Press, Cambridge, Mass.

Wetherby, A. M. (1986). Ontogeny of communicative functions in autism. *Journal of Autism and Developmental Disorders*, **16**, 295–316.

Wittgenstein, L. (1980). In *Remarks on the philosophy of psychology*, Vol. 2, trans. C. G. Luckhardt and M. A. E. Aue (ed. G. H. von Wright and H. Nyman). Blackwell, Oxford.

Woodruff Smith, D. (1989). *The circle of acquaintance*. Kluwer Academic, Dordrecht.

Yirmiya, N., Kasari, C., Sigman, M., and Mundy, P. (1989). Facial expressions of affect in autistic, mentally retarded and normal children. *Journal of Child Psychology and Psychiatry*, **30**, 725–35.

11

Pretending and planning

PAUL HARRIS

Kanner's original description of autistic children mentions their impoverished play (Kanner 1943). Subsequent epidemiological studies have confirmed that an imaginative poverty is indeed typical of the autistic syndrome. In this chapter, I take a closer look at this imaginative poverty. My aim is to show that it illustrates the central problem of autism.

Let us begin with the epidemiological evidence. In a thorough study of children in Camberwell, Wing (1978) identified all children who had been designated as severely mentally retarded or had shown some aspects of the autistic syndrome — difficulties in social relationships or an insistence on routine. This large group of children ($n = 158$) was divided into two sub-groups. One subgroup ($n = 84$) was composed of those children who exhibited both of Kanner's main symptoms of autism to some degree: a lack of affective contact with other people and a resistance to change, as shown by repetitive or stereotyped activities. The other subgroup ($n = 74$) showed none of the major signs of autism: they did take pleasure in social contact, at least within the bounds set by their intellectual handicap. Wing went on to look at the incidence of other aspects of the autistic syndrome in each subgroup; in particular, symbolic play was evaluated.

Each child was classified as showing one of three levels of symbolic play: 'no symbolic play' meant that the child did not use toys or materials to represent real objects or situations; 'stereotyped symbolic play' meant that the child did engage in pretend-play, but in a narrow, repetitive fashion, without regard to suggestions from other children; finally, 'true symbolic play' meant that the child added new themes and greater complexity to his or her play in the course of development.

The findings concerning symbolic play were very clear: most of the children in the sociable subgroup (77 per cent) showed true symbolic play, whereas the vast majority of the children in the non-sociable subgroup (99 per cent) did not. The asymmetry became even clearer if attention was confined to those children with a verbal and non-verbal mental age of 20 months or more. (This is a reasonable strategy because pretend play is quite limited even in normal children until that age.) All the children in the sociable subgroup ($n = 57$) exhibited true symbolic play, whereas only one of the children in the non-sociable subgroup ($n = 31$) did so. Thus, a tight link was found between the major autistic symptoms as described by Kanner and a deficit in symbolic play.

Closer inspection of the Camberwell data suggests the existence of two different developmental trajectories for the two subgroups. First, we may consider children who had developed few or no symbolic skills: children with a verbal age below 20 months and little or no symbolic play, again dividing them into sociable and non-sociable subgroups (Wing 1978, Table 5). At this developmental level, almost all the non-sociable children are either 'aloof' or 'passive' in their social interaction (Wing and Gould 1979, Table V). Among children who had developed more advanced symbolic skills, the two subgroups show a different pattern of progress. Sociable children who have more advanced language also display true symbolic play. By contrast, non-sociable children who have more advanced language rarely display true symbolic play: the vast majority only engage in stereotypic pretend-play (Wing 1978, Table 5). Admittedly, 'aloof' social behaviour has mostly disappeared among these non-sociable children; but they remain either 'passive' or 'odd' (Wing and Gould 1979, Table V).

In sum, this major survey produced two clear results. First, among children with social difficulties, deficits in pretend-play are almost ubiquitous. Second, these two problems remain linked even though development may bring about a transformation in the way that they manifest themselves. Aloof behaviour may disappear, but the children remain passive or odd. Pretend-play may emerge, but it remains stereotypic.

Wing's study included children with social abnormalities, but not all of them had received a formal diagnosis of autism. Is there the same link between social abnormalities and impoverished play among children who have received a formal diagnosis of autism? Sigman and Ungerer (1984), Baron–Cohen (1987), and Lewis and Boucher (1988) filmed autistic children engaged in free play with a variety of toys. In all three studies, it was found that the autistic children were less likely to engage in pretend-play with the toys compared with retarded or normal children of the same verbal ability.* Moreover, these group differences were quite stable across different age-groups. Looking at the three studies combined the children ranged from three to fifteen years of age.

In the next section, I outline three different conceptions of the link between deficits in social interaction and in pretend-play.

---

* The three studies did not find exactly the same type of group difference. Pretend-play can be divided into 'functional' play (e.g. playing with a toy car as a car), and more elaborate 'symbolic' play, in which pretend entities (or properties) are invented — for example, the car is filled with make-believe petrol. Baron–Cohen found group differences for symbolic play, Lewis and Boucher for functional play, and Sigman and Ungerer for both types. Nevertheless, a finding common to all three studies was that autistic children produced virtually no symbolic play in this free-play set-up.

A MOTIVATIONAL DEFICIT?

Pretend-play, especially pretend-play with toys and dolls, often turns on the recreation of a simple human drama: the doll wants a drink or a bath. Autistic children might have no interest in re-enacting these little scenes, not because they cannot imagine them, but because they have less emotional involvement with people. In much the same way, some of us are not interested in reading science fiction; we can imagine life on a distant planet, but have no emotional investment in such an alien setting. This motivational hypothesis implies that autistic children's alleged limitations in pretend-play are a consequence of their relative indifference to social contact. Indeed, according to this hypothesis, their imaginative poverty is more apparent than real. If they were observed in a domain that interested them, their imagination should be normal.

At first glance, this motivational hypothesis has support from some experimental studies which show that autistic children engage in more pretence if they are encouraged to do so rather than left to their own devices. Instead of simply allowing children to play freely with a variety of toys (as described above), children have been given either a general prompt or a more specific prompt by an adult. For example, in the case of a general prompt, the experimenter might present the child with selected props (for example a toy car and a box) and make an orienting remark such as: 'What can these do? Show me what you can do with these' (Lewis and Boucher 1988). In the case of a specific prompt, the experimenter might provide more structure, by modelling a make-believe action with some props (Riguet *et al.* 1981), or giving the child suitable props together with a definite verbal instruction, for example: 'Show me how the doll eats her dinner' (Lewis and Boucher 1988; Sigman and Ungerer 1984).

All of these studies have reported that prompting, be it general or specific, elicits more pretend-play from autistic children. Thus, observation of free play alone may lead to a lower estimate of their capacity for pretence. A similar conclusion was reached by Gould (1986), who assessed autistic children's free play and also their play in a structured test. She noted that the autistic children often received a lower assessment in the free-play context.

It would not be surprising if normal or retarded children were also helped by prompting. Hence we can ask whether prompting attenuates or magnifies the differences between autistic children and relevant control groups. Here the evidence is more equivocal. Group differences were just as marked after prompting in two studies (Riguet *et al.* 1981; Sigman and Ungerer 1984). On the other hand, they were eliminated in two other studies (Gould 1986; Lewis and Boucher 1988).

In summary, studies that have used prompting or structure provide some support for the motivational hypothesis: autistic children do engage in more pretence if they are prompted rather than left to their own devices. Indeed, prompting sometimes eliminates group differences.

However, the motivational hypothesis makes other predictions that find little support. If autistic children do not engage in pretend-play because they are indifferent to its human content, we should still expect them to engage in other imaginative activities in a normal way: they should be interested by physics, chess, or science fiction. Indeed, autistic children do sometimes develop elaborate hobbies that seem deviant because they lack any obvious human content. For example, they study calendars or time-tables. However, clinical descriptions of such hobbies suggest that they are unusual not simply for their content. The autistic person appears to be interested in these systems precisely for their cyclicity and orderliness. There is no indication that they are explored in a playful or creative fashion. In short, such preoccupations appear to be higher-level manifestations of the rigidity and inflexibility that Wing and Gould found in the make-believe play of children with autistic symptoms.

Thus, although, the motivational hypothesis implies that the apparent imaginative poverty of autistic children is confined to a well-demarcated content area, namely the area of human interaction, the evidence suggests that their difficulties are more pervasive. Even in a domain that preoccupies them, autistic people show few signs of inventiveness. At the same time, the motivational hypothesis has pointed up the possibility — which has some support — that autistic children are particularly unimaginative when left to their own devices. They can become more imaginative upon instruction. We shall take up this point again later.

UNDERSTANDING MENTAL STATES

Recent research with autistic children, as discussed throughout this book, has suggested that they have special difficulties in understanding the mental states of other people. Accordingly, they might fail to engage with their toys and dolls not because they are indifferent to other people, but because they lack some basic ability for understanding psychological experience. Just as they fail to understand the mental states of a real person in everyday life, or the mental states of the protagonist in a story that they hear, so too, they fail to impute certain mental states to the dolls that they play with. To the extent that pretend-play often calls for the attribution of psychological states to dolls, it will necessarily be impoverished. According to this hypothesis, the child's social and imaginative difficulties are simply different manifestations of the same problem: an inability to understand the nature of certain mental states.

This hypothesis makes several testable predictions. First, it implies that certain aspects of autistic children's pretend-play should be intact. They should be able to engage in object-substitution, because this does not call for the attribution of a psychological state. And again, they should be able to pretend that a doll can walk or eat or sleep, because this again does not call for the attribution of a psychological state.

There is some evidence in favour of each of these predictions. Lewis and Boucher (1988), as was noted earlier, gave children prompts to engage in make-believe. Many of these prompts required that children should produce an object-substitution — for example, should treat a blue serviette as a swimming-pool, or cotton wool as snow. In addition, some prompts called for the child to treat a doll as an agent who could eat or swim or walk. In response to such prompts, the autistic children were just as competent as the control groups.

Other evidence also shows that autistic children can manage certain simple forms of pretence. Consider the studies of Hobson (1984), Tan and Harris (1991), and Baron–Cohen *et al.* (1985), in which autistic children were tested for their understanding of perception and belief. If we set to one side for a moment the main aim of these experiments, it is interesting to note that they employed dolls in order to enact a simple scene. In Hobson's study children were asked to place a doll where it could not be seen by one or two other dolls giving chase. In the study by Tan and Harris (1991) two dolls could see different targets, or saw them from a different vantage-point. In the study by Baron–Cohen and his colleagues, one doll deposited a marble, and a second doll moved it in the absence of the first.

In none of these studies did the authors report that the autistic children had difficulty in pretending that the dolls were in certain key respects like people. In the experiments of Hobson and Tan and Harris, the autistic children were remarkably accurate in assessing what the dolls could 'see'. In the study by Baron–Cohen and his colleagues, the autistic children correctly remembered where the first doll had placed the marble, and, although they mostly ignored her subsequent false belief about its location, they did appear to understand that she would go to retrieve it. These results reinforce the claim that autistic children can pretend that a doll is capable of various simple sensorimotor actions, such as walking or running from one place to another, or holding, placing, and retrieving an object; they can also pretend that a doll can have certain simple sensations, notably seeing another doll or object.

In summary, there are indications that autistic children's difficulties in pretend-play are confined to certain kinds of make-believe and not others. They can treat one object as another object, and they can treat a doll as if she were a person who talks, acts, and even sees. On the other hand, they have difficulty in imagining a belief that they do not share, whether that belief is to be attributed to a doll or another person. This selective pattern of difficulty provides further ammunition against the motivational hypothesis discussed earlier. If autistic children are disinclined to engage in doll-play because they have little emotional involvement with people, they should be poor at attributing activities such as walking and eating to a doll, as well as psychological states such as belief.

However, the proposal that autistic children have a selective difficulty in attributing psychological states, particularly the mental state of belief, enjoys at once an advantage and a disadvantage. On the one hand, it is commendably specific in the deficits that it predicts. On the other hand, the autistic syndrome encompasses symptoms that appear to fall outside its purview. One of the standard symptoms of autism is an *insistence on sameness* or the preservation of routine. In the same vein, Wing's (1978) description of the limited pretend-play of the more advanced non-sociable children in her survey indicates that their play, almost without exception, remained stereotypic and routinized. It is not obvious how the inflexibility of autistic children, whether in play or in everyday life, can be linked to their alleged lack of psychological insight. Admittedly, one might speculate that autistic children's routines are an attempt to reduce the anxiety they feel when faced with the apparent unpredictability of other people's action (cf. Baron–Cohen 1989*a*). However, there is as yet no evidence that such routines are triggered by such socially-induced anxiety. In the next section, we turn to a third possible interpretation of the autistic child's imaginative poverty, which conceptualizes their inflexibility as a central cognitive characteristic, rather than as a symptom of social anxiety.

INTERNAL VERSUS EXTERNAL EXECUTIVE CONTROL

The basic claim to be developed is that autistic children are deficient in planning their actions and responses. A broad distinction will be made between actions that can be guided by standard or habitual schemas evoked by the current context, and those that must be guided by a plan specially formulated for the current task. A great deal of human action requires a marriage of both forms of control; but certain forms of action require a shift in the balance between the two forms. Specifically, some actions require that a familiar or habitual schema, normally evoked by the current context, should be set aside so that guidance can be achieved instead by a temporary, internally-formulated plan. Autistic children, it will be claimed, have special difficulties in over-riding external or habitual control in this autonomous, planful fashion.

Observation of the development of pretend-play in normal children reveals a steady decrease in reliance on the familiar, habitual schemas that are evoked by the current context, and a steady increase in internal control.

At first, the child repeats familiar actions outside their normal context (Nicolich 1977; Piaget 1951). For example, instead of sitting on a cushion, the child lies down on it even though it is not bedtime, and pretends to go to sleep. Thus, from the very beginning of pretending, there is a limited ability to set aside the schema normally elicited by the current context and to carry out a different pretend action. In the course of development, there is an increasing detachment from the external context combined with an increasing capacity for the internal creation of a complex script. For example, the child no longer needs partial reminders (for example, a cushion that resembles a pillow) in order to execute a pretend action. The child can pretend to go to sleep using a cloth or soft toy as a substitute for the pillow. In addition, the child can execute a longer sequence of pretend actions: the doll is made to stir its make-believe tea and then to drink it.

Surveying the development of pretend-play, we can see that the control of behaviour shifts its locus: whereas reality-adapted problem-solving is often guided by the current external context, pretend-play receives minimal support from that context — increasingly complex plans are formulated and executed with limited contextual support. Indeed, the action schemas that would normally be cued by the immediate context must be set aside to make way for an internally-formulated plan. Thus, depending on the play theme that is currently active, props have a new identity or property conferred on them, and imaginary objects are brought into existence despite their objective absence.

An explanation of the autistic child's impoverished pretend-play, therefore, is that the ability to achieve this shift in the locus of executive control is distorted or grossly delayed. As compared to a normal child or even a retarded child of the same mental age, the autistic child is more reliant on the schemas evoked by the current context, and has great difficulty in deliberately guiding his or her behaviour or responses according to an internally conceived plan that over-rides those schemas.

The evaluation of this claim will be divided into three parts: experimental work on planning and flexibility; the pattern of competence and deficit shown by autistic children in pretend-play; and finally, autistic children's problems in understanding psychological states. Having reviewed these three sources of evidence, I shall compare my hypothesis with other discussions of executive planning by autistic children.

Planning and flexibility

The most direct evidence on autistic children's capacity for planning and flexibility comes not from work on pretend-play, but from work on problem-solving. Prior and Hoffman (1990) gave children a maze-learning task (Milner 1965), and the Wisconsin card-sorting (WCS) task. As compared with control groups matched for chronological and mental age, the autistic children were worse on both. Similar findings were obtained by Ozonoff, Pennington and Rogers (1991*a*). They administered the WCS task and the Tower of Hanoi problem to autistic and learning-disabled children and adolescents. Again, the autistic subjects performed worse on each task. In both studies, there was clear evidence of perseveration. When error scores on the card-sorting test were examined, the groups differed on perseverative errors only, with the autistic children repeating a strategy that was no longer correct much more often than controls. For example, they might classify a card in terms of its colour if that had been the previously correct criterion for sorting, rather than in terms of a newly-imposed criterion, such as shape.

Ozonoff and her colleagues found that the Tower of Hanoi problem had considerable discriminative power: it successfully classified 80 per cent of the children in both groups. Accordingly, it is worth describing this task in more detail. The child is given a board with three vertical pegs. Three circular disks of different sizes (each with a hole in their centre) can be stacked vertically on the pegs. When the disks are stacked on the same peg from the largest up to the smallest they form a tower that constitutes the goal state; at the outset of the task, the three disks are distributed among different pegs. The task is to move the disks from one peg to another in order to build the tower on a designated peg. In moving the disks, the child must respect the rule that a larger disk cannot be placed on top of a smaller one, and must try to achieve the goal state with as few moves as possible. Good performance requires an ability to think ahead—to select moves in terms of the consequences that they will force in the future. For example, if the smallest disk is moved directly to the goal peg, it may need to be moved away again later because of the rule forbidding any larger disk to be placed on top of it. Selection of the next move must be based not on immediate progress toward the goal but on the advantages and disadvantages of various hypothetical intermediate arrangements that might be created, including those that do not immediately bring a disk closer to the goal peg. The implication, therefore, is that autistic children have difficulty in guiding their current behaviour in terms of this non-existent but foreseeable context,* particularly when a more direct approach to the goal appears to be available. This deficiency in planning is, of course, what would be predicted by the current hypothesis.

Supportive findings have also been obtained by Russell and his colleagues

---

* Autistic children can solve object-permanence problems and simple visual perspective-taking tasks. Hence, they can imagine an existing, but invisible object or situation. The deficit that I have identified should arise only when they must hold in mind a non-existent situation, especially when it calls for a different response from the one suggested by the current situation.

(Hughes and Russell 1991; Russell *et al.* 1991). The so-called 'windows' task that they administered is especially informative because it is so simple. On each trial, children were shown two boxes; each box had a window that made its contents visible. One box was clearly empty whereas the other contained a sweet. To win the sweet, children had to indicate the *empty* box. (Some were asked to do this verbally, others to point; the results were similar in each case.) If instead they indicated the box with the sweet in it, they lost it. Autistic children had tremendous difficulty in coping with this paradox. They repeatedly indicated the box containing the sweet, even though they thereby lost the sweet each time. Indeed, some of the autistic subjects made this mistake on 20 trials in succession, whereas none of the mentally handicapped controls did so.

This task illustrates especially clearly the executive difficulties of autistic children. They must suppress the familiar and habitual schema of indicating the object that they want. Instead, they must guide their response in terms of the temporary and unfamiliar plan of pointing at the empty box in order to obtain what they want.*

The development of pretend-play

In the preceding sections, three different conclusions were established with respect to pretend-play by autistic children. First, given the opportunity to play freely with toys, they are less likely to play with them in a functionally appropriate manner (for example, lifting the receiver of a toy telephone, or pushing a toy car into a garage), and they almost never produce more advanced forms of make-believe such as object-substitutions, the activation of a doll as an agent, or the creation of imaginary entities (Baron–Cohen 1987; Lewis and Boucher 1988; Ungerer and Sigman 1981).

Second, autistic children are, none the less, capable of genuine make-believe if they are prompted. When given some props and a general prompt

---

*Baron–Cohen (personal communication) points out that the Tower of Hanoi task and Russell's 'windows' task appear to rely on a competitive spirit: the aim of the tasks is to make as few errors as possible and/or gain as many sweets as possible. Autistic children might do poorly, not because they are unable to plan, but because they are not competitive. However, the majority of autistic children in the windows task failed by pointing to the chocolate on all 20 trials (Russell *et al.* 1991, Appendix B). Such systematic failure implies that they wanted to gain the sweet, without knowing how to do so. By contrast, a lack of competitive spirit should lead to a few initially correct responses, followed by random responding or withdrawal from the task. No autistic child showed this pattern.

Moreover, work in progress (Hughes and Russell 1991) shows that autistic children do poorly on planning tasks in which the element of competition appears to be non-existent. For example, in one task a marble must be retrieved by means of a detour reach instead of a standard line-of-sight reach.

Finally, a lack of competitiveness should lead autistic children to do poorly on the WCS as a whole. Yet their difficulties were confined to perseveration. They successfully identified as many categories as the control group.

(for example 'What can these do?'), they can create object-substitutions or treat a doll as an active agent (Lewis and Boucher 1988).* They also understand when an adult treats a doll as someone who can walk or search or see (Baron–Cohen *et al*. 1985; Hobson 1984; Tan and Harris 1991). Ungerer and Sigman (1981) found that autistic children failed to produce any symbolic play before prompting (for example, such things as an object-substitution, activating a doll or creating an imaginary entity); whereas the majority did so after prompting (Ungerer and Sigman 1981, Table 4). More generally, it is clear that autistic children benefit from adult guidance (Gould 1986; Lewis and Boucher 1988; Riguet *et al*. 1981), even if it does not invariably eliminate group differences.

Finally, the wide-ranging epidemiological survey of Wing and Gould (Wing 1978; Wing and Gould 1979) shows that despite this competence for make-believe, especially in structured settings, play remains inflexible and repetitive. This conclusion holds not only for children with a formal diagnosis of autism, but also for children with autistic symptoms. An illustration of this kind of stereotypy might be helpful. D. is an autistic child aged 8 years. She is obviously capable of pretend-play, but it is repetitive. For example, one of her favourite routines is pretending to be a waitress. At meal-times, she goes around the family circle asking each member what he or she would like to drink. Having received an order, she briefly mimes writing it down, or tells her customer that it is not available. The circle completed, she leaves the table to get the order. This role-play remains repetitive and fragile. First, she repeats the pretend script often; second the repetition rarely incorporates new variations — the same drinks are typically available at the same prices; third, the pretence is not sustained — her customers are lucky to get the drink they ordered, or any drink at all. Once D. has left the table, the script often collapses unless there is further prompting from her customers.

The hypothesis that autistic children suffer from a deficiency in internal

---

* The interpretation of their own 1988 findings by Lewis and Boucher has been questioned. Baron–Cohen (1990) points out that whereas genuine pretence calls for the creation of imaginary identities or properties, children may have simply guessed what to do with the props and carried out a simple sensorimotor action. For example, the experimenter handed children a car and a box saying: 'What can these do? Show me what you can do with these.' A child who put the car in the box might have been pretending that the box was a 'garage'. Equally, the child might simply have guessed at what to do with the two objects; after all, the box could not be put inside the car. In their reply, however, Boucher and Lewis (1990) note that each autistic child provided at least one example of creating imaginary identities or properties. For example, one child who was handed the car and the box, said 'garage', put the box on the floor with its flaps oriented appropriately, opened a flap, put the car inside, and closed the flap. Another child, handed a car and some string, placed one end of the string to the back end of the car, saying 'get petrol'. Such examples appear to show that the autistic children were genuinely pretending, rather than mechanically acting on the props, even if they did so at the behest of the experimenter.

relative to external, contextual control handles these various observations quite well. First, the hypothesis implies that autistic children will have difficulty in spontaneously generating and imposing their own make-believe conceptions on a given context. For example, they will not spontaneously animate a doll or pretend that a cup has tea in it. Yet it allows for the possibility that when autistic children are given an external prompt to pretend — say, they are handed a car and a piece of string — and verbally encouraged to show what might be done with the props, then they should be able to generate a pretence. In such cases, the locus of control has partly shifted back to the external contextual frame created by the adult, which includes the particular props that have been handed to the child and the verbal cue. Finally, the hypothesis necessarily implies that pretend-play among autistic children will remain repetitive and inflexible: even if it is fairly elaborate (as in the case of D.) it remains a well-rehearsed script that is prompted by a recognizable context (potential 'customers' seated around a table); it will not take on a momentum of its own, such that new props and variations become incorporated in a consequential fashion.

Understanding psychological states

Recent research with autistic children has provided convincing evidence that they have difficulty in understanding certain mental states. The planning hypothesis provides also provides a satisfactory account of these difficulties.

At first, it appeared that autistic children might altogether lack any conception of mind; but the evidence now points to a narrower problem. Summarizing the main findings, autistic children have no obvious difficulty (relative to controls matched for mental age) in understanding that people may vary in their perceptual experience (Baron–Cohen 1989b; Hobson 1984; Tan and Harris 1991) or in simple desires and preferences (Baron–Cohen 1991; Harris 1990). On the other hand, many (but not all) autistic children fail to understand the impact of beliefs, notably false beliefs, on a person's actions (Baron–Cohen *et al.* 1985) or emotions (Baron–Cohen 1991). Finally, almost all autistic children and adolescents fail to understand the inipact of second-order beliefs, i.e. beliefs about beliefs (Baron–Cohen 1989c; Ozonoff *et al.*, 1991a).

The acquisition of language by autistic children provides striking corroboration of this alleged pattern of competence and deficiency (Tager-Flusberg 1992; and Chapter 7, this volume). Relative to Down's syndrome controls, autistic children are unimpaired in their talk about perception and desire, but make markedly fewer references to cognitive states, including belief.

Why should autistic children succeed in understanding some psychological states but not others? Most approaches to psychological understanding both in normal children and in autistic children have claimed that children are either acquiring or failing to acquire a theory of mind. Hence development is described in terms of the acquisition of particular theoretical insights (cf. Perner 1991; Wellman 1990). In contrast, I have argued that progress is achieved by means of an increasingly complex process of *simulation* or imagination (Harris 1989, 1990). An obvious extension of this claim is that autistic children, by virtue of their deficits in planning, have difficulty in carrying out such a simulation process.

Two different levels of simulation were described by Harris (1990). At the simpler level, the child holds the current, actual situation constant, but imagines a different intentional stance toward some aspect of that current reality. At the more complex level, the child does not hold the current situation constant, but instead imagines a hypothetical situation that conflicts with what is currently known to be the case, and then proceeds to imagine a different intentional stance toward that hypothetical reality. The distinction between these two levels differentiates quite well between the tasks that autistic children usually pass and those that they fail.

Scrutiny of the tasks that autistic children pass shows that they do not require the child to set aside the current situation and to imagine instead a conflicting hypothetical situation. The child must simply acknowledge that people can see or not see, want or not want an object in the immediate context. The tasks do not, therefore, present the child with a conflict between a prediction derived from an imagined, hypothetical state of affairs and a prediction derived from the actual state of affairs. By contrast, that conflict is apparent in the tasks that autistic children fail. For example, in the simpler first-order belief task, the child knows that currently a particular box is empty, but to predict another person's false belief, the child must imagine the alternative, conflicting state of affairs that the other person takes to be true, for example that the box still contains chocolate. The second-order belief problem is more demanding still. It requires that the child imagine the content of the false belief that person A attributes to person B, even though that content conflicts not only with current reality but with person B's actual belief.

A similar analysis can be advanced for other theory of mind tasks that give autistic children difficulty: the differentiation between mental and physical entities and between appearance and reality (Baron–Cohen 1989*d*). In the mental–physical task, children are asked to make predictions about two story characters, one who has physical contact with an actual object (for example, is given a cookie), and one who is conjuring up a mental object (for example, is thinking or dreaming about a cookie). To understand such stories and make accurate predictions about the two characters, children must conceptualize and honour the distinction between having an object and merely thinking about but not having an object. In the appearance–reality task, the child must conceptualize and honour the opposition between a hypothetical identity (for example, the rock-like appearance of a proxy object) and an actual identity (for example, the fact that the proxy object is actually a sponge).

Thus, according to this analysis, theory of mind tasks, pretend-play, and planning tasks can be given a similar type of explanation. In each case, the child must envisage a non-existent hypothetical state of affairs and make a response or prediction that is tailored to that hypothetical state of affairs. That response will typically run counter to the response or prediction suggested by the actual state of affairs. To the extent that autistic children have difficulty in shifting the locus of control from the current, known context to a hypothetical, and conflicting context, they will exhibit similar difficulties on all three tasks.

ALTERNATIVE CONCEPTIONS OF EXECUTIVE FUNCTIONING

In this section, I consider in more detail whether or not these difficulties can be traced to a common underlying cause.

Theory of mind difficulties as a primary cause

One possible causal hypothesis is that the autistic child's primary problem is best tapped by theory of mind tasks, particularly those tasks which call for self-monitoring. That difficulty, in turn, leads to impoverished pretence and poor performance on tasks such as the Tower of Hanoi and the WCS task.

The argument might run as follows:[*] autistic children have a fundamental problem in representing their own mental states and those of other people, as manifested, for example, by their difficulty with the false-belief task. The ability to represent one's current mental state is a critical component of plan-based action, because it allows a hypothetical action to be held in mind, and evaluated in the face of competing alternatives.

Accordingly, the autistic child faced with the Tower of Hanoi can imagine a possible move, but having imagined that move cannot represent the fact that he or she is currently reflecting on a possible move. Unable to encode the imagined alternative as 'a move that I am currently considering' and to hold it in mind, the imagined move exerts no influence on the child's overt action; instead, the child carries out whatever action schema is cued by the immediate context. In line with this speculative hypothesis, informal observation of autistic children confirms that they rarely do make announcements

---

[*] I am grateful to Simon Baron–Cohen for suggesting this possible line of explanation to me.

such as: 'I've got an *idea* . . .'. Notice that this speculation can also be extended to children's pretend-play: although they are capable of pretend-play, they are unable to identify and hold in mind their own ideas for a pretend sequence. Hence, their pretence is impulsive and brief. It can only be sustained if it is routinized and stereotypic, rather than guided by a newly-formulated plan.

In sum, a plausible case can be made for the hypothesis that autistic children's planning and pretence limitations are a *result* of their lack of psychological insight. There is, however, evidence against this hypothesis. Ozonoff *et al.* (1991*b*) report that Asperger's syndrome children show no lack of psychological insight on standard theory of mind tasks. Yet they do have planning difficulties on the Tower of Hanoi and the WCS task. The clear implication is that their planning difficulties cannot be traced back to an inability to represent their own ideas. *

It is possible, of course, to respond to this evidence by arguing that the planning difficulties of autistic children and Asperger's children have a different origin: a lack of psychological insight in the case of autistic children, and an unknown but different cause in the case of Asperger's children. Such an argument is weak, however, because it is tantamount to the claim that there is a fundamental difference in the aetiology of two syndromes. Yet in many respects the two syndromes are hard to distinguish, and some writers have argued for a continuum rather than a sharp dissociation.

More generally, the findings of Ozonoff *et al.* (1991*b*) show that theory of mind difficulties are unlikely to be the 'primary' or underlying cause of the various problems exhibited by autistic children. They are not found among children with Asperger's syndrome, and by implication they are probably not responsible for the planning difficulties that are common to both autistic and Asperger's children.

Executive control deficits as a primary cause

An alternative causal analysis is to single out tasks such as the 'windows' task, or the Tower of Hanoi, to claim that they tap a primary deficit in planning, and to trace the co-occurring problems on theory of mind tasks and on pretence back to that planning deficit. Hughes and Russell (1991) advance an argument of this type. They point out that if autistic children

---

* Josef Perner points out that even though Asperger's children might (with difficulty) gain the psychological insight that is a prerequisite to successful planning, that insight might not be sufficient to ensure successful planning. This defence of the hypothesis that theory of mind difficulties could cause planning problems implies that Asperger's children have a less robust or efficient psychological understanding than normals despite their success on standard theory of mind tasks. A more searching assessment of their competence at theory of mind tasks might uncover such a limitation.

have problems in planning, then their own cognitive processes will be deviant, and that may distort their experience and conception of how the mind works. Consider once again the Tower of Hanoi problem. We might expect a normal subject to imagine a possible move, and then, holding the outcome of that hypothetical move in mind, to imagine the next move, and so forth. Only when a legitimate set of moves leading to the goal state has been identified will the subject actually make a move. At any point in this planning process, the subject could be interrupted and asked to report the possible move or sequence of moves currently being evaluated. In this respect, the Tower of Hanoi problem highlights an everyday phenomenon: we often think prior to acting, we are aware of what we are thinking, and we can if necessary report the content of such mental planning.

Suppose, however, that this planning process cannot be sustained by autistic children. As they attempt to imagine a possible move, it is cut short by their tendency to make a move in terms of the configuration with which they are immediately confronted. As a result, their introspective life will be different from that of a normal child. There is little mental planning to introspect about. Hughes and Russell conclude, therefore, that autistic children will lack the kind of mental experiences upon which to build a so-called theory of mind. Some support for this interesting specu-lation is provided by Baron–Cohen (1989*d*) and Ozonoff *et al.* (1991*a*). When autistic children were asked what the brain does, they were likely to answer in terms of behavioural rather than cognitive activities, or not to reply at all.

None the less, the evidence gathered by Ozonoff *et al.* (1991*b*) also poses a problem for the claim that planning difficulties lead to a lack of psychological insight. Recall that children with Asperger's syndrome performed poorly on planning tasks such as the Tower of Hanoi; these children still passed the theory of mind tasks. It is conceivable, of course, that they solved the theory of mind tasks in a non-standard fashion, with little recourse to self-monitoring; but, for the moment, it appears that a partial dissociation between performance on theory of mind tasks and tasks such as the Tower of Hanoi or WCS is possible.

Multiple forms of planning

So far, we have examined two proposals that seek to identify a single primary cause for the trio of difficulties identified earlier. An alternative possibility is that there may be no single cause, but rather a conjunction of potentially dissociable causes.

Consider, once again, the claim that I advanced earlier: autistic children exhibit a trio of difficulties which can all be conceptualized as different manifestations of an underlying deficiency in internal relative to externally-guided regulation. Although there might be a single indivisible mechanism responsible for all such forms of regulation, it is also possible that the regulatory system contains sub-areas of specialization which concentrate on different types of planning task. For example, one area might be specialized for pretence, and another for spatial tasks like the Tower of Hanoi.

Ozonoff *et al.* (1991*a*) have underlined the likelihood of such specialization within the regulatory system.* They point out that a variety of planning difficulties are linked with damage to the prefrontal cortex, and, following Goldman-Rakic (1987), they suggest that different areas of prefrontal cortex might be specialized for different types of planning problem depending on the type of representation required for task solution. For example, a theory of mind task might require the representation of another's belief, whereas a Tower of Hanoi problem might require the representation of a hypothetical spatial array. To the extent that prefrontal cortex is divided into specialized sub-areas it should be possible to find subjects who perform poorly on tasks such as the Tower of Hanoi, while succeeding on the false-belief task, or vice versa. Evidence for such a double dissociation would be very important, because it would show that whatever causes difficulties with tasks such as the Tower of Hanoi need not cause problems on theory of mind tasks, and vice versa.

As noted earlier, recent evidence suggests at least a partial dissociation. Asperger's children perform poorly on the Tower of Hanoi and the WCS, but perform at the same level as controls on theory of mind tasks. Moreover, preliminary results gathered by Lofts (1991) show that patients with frontal-lobe damage, who typically perform badly on planning tasks, are none the less able to solve the false-belief task.

In summary, evidence from clinical groups, and from physiological analysis (Goldman-Rakic 1987), shows that the regulatory system does exhibit areas of specialization. We might speculate that children with autism and with Asperger's syndrome differ in the areas of specialization that are deficient.

CONCLUSIONS

Recent research with autistic children has focused on their problems in understanding mental states. The robust findings that have emerged in that domain have tended to draw attention away from other familiar aspects of the syndrome: the deficits in pretend-play, and the absence of flexibility. In this chapter, I have attempted to redress that balance. The main claim

---

* I am very grateful to Sally Ozonoff for drawing my attention to the implications of her findings with Asperger's children.

is that autistic children find it difficult to regulate their behaviour and their expectations in terms of a hypothetical context that runs counter to the current, immediately available context. Instead, they tend to fall back on habitual schemas prompted by the immediate situation. This hypothesis predicts three deficits: a deficit in unprompted pretend-play, a deficit in problem-solving that calls for advance planning, and a deficit in understanding mental states that are directed at a hypothetical situation that runs counter to current, known reality. Robust evidence exists for all three deficits. Nevertheless, although the three deficits can be given a common psychological analysis, we should be cautious in assuming that they each derive from a common neuropsychological cause. Research aimed at discovering dissociations among the three deficits will be especially informative in this respect.

Acknowledgement

I am very grateful to Simon Baron–Cohen, Sally Ozonoff, and Joseph Perner for their helpful comments on an earlier draft of this chapter.

REFERENCES

Baron–Cohen, S. (1987). Autism and symbolic play. *British Journal of Developmental Psychology*, **5**, 139–48.

Baron–Cohen, S. (1989*a*). Do autistic children have obsessions and compulsions? *British Journal of Clinical Psychology*, **28**, 193–200.

Baron–Cohen, S. (1989*b*). Perceptual role-taking and protodecalarative pointing in autism. *British Journal of Developmental Psychology*, **7**, 113–27.

Baron–Cohen, S. (1989*c*). The autistic child's theory of mind: a case of specific developmental delay. *Journal of Child Psychology and Psychiatry*, **30**, 285–97.

Baron–Cohen, S. (1989*d*). Are autistic children 'Behaviourists'? An examination of their mental–physical and appearance–reality distinctions. *Journal of Autism and Developmental Disorders*, **19**, 579–99.

Baron–Cohen, S. (1990). Instructed and elicited play in autism: A reply to Lewis and Boucher. *British Journal of Developmental Psychology*, **8**, 207.

Baron–Cohen, S. (1991). Do people with autism understand what causes emotion? *Child Development*, **62**, 385–95.

Baron–Cohen, S., Leslie, A.M., and Frith, U. (1985). Does the autistic child have a theory of mind? *Cognition*, **21**, 37–46.

Boucher, J., and Lewis, V. (1990). Guessing or creating? A reply to Baron–Cohen. *British Journal of Developmental Psychology*, **8**, 205–6.

Goldman-Rakic, P.S. (1987). Circuitry of primate prefrontal cortex and regulation of behavior by representational memory. In *Handbook of physiology: the nervous system* (ed. V.B. Mouncastle, F. Plum and S.R. Geiger). American Physiological Society, Bethesda, MD.

Gould, J. (1986). The Lowe and Costello symbolic play test in socially impaired children. *Journal of Autism and Developmental Disorders*, **16**, 199-213.

Harris, P.L. (1989). *Children and emotion: the development of psychological understanding*. Blackwell, Oxford.

Harris, P.L. (1990). The work of the imagination. In *The emergence of mindreading* (ed. A. Whiten). Blackwell, Oxford.

Hobson, R.P. (1984). Early childhood autism and the question of egocentrism. *Journal of Autism and Developmental Disorders*, **14**, 85-104.

Hughes, C. and Russell, J. (1991). Unpublished data. Department of Experimental Psychology, Cambridge University.

Kanner, L. (1943). Autistic disturbances of affective contact. *Nervous Child*, **2**, 217-50.

Lewis, V. and Boucher, J. (1988). Spontaneous, instructed and elicited play in relatively able autistic children. *British Journal of Developmental Psychology*, **6**. 315-24.

Lofts, A. (1991). Unpublished M.Sc. project, Experimental Psychology, Sussex University.

Milner, B. (1965). Visually-guided maze learning in man: Effects of bilateral hippocampal, bilateral frontal, and unilateral cerebral lesions. *Neuropsychologia*, **3**. 317-38.

Nicolich, L. (1977). Beyond sensorimotor intelligence: assessment of symbolic maturity through analysis of pretend play. *Merrill-Palmer Quarterly*, **23**, 89-99.

Ozonoff, S., Pennington, B.F., and Rogers, S.J. (1991*a*) Executive function deficits in high-functioning autistic individuals: relationship to theory of mind. *Journal of Child Psychology and Psychiatry*, **32**, 1081-106.

Ozonoff, S., Rogers, S.J., and Pennington, B.F. (1991*b*) Asperger's syndrome: evidence of an empirical distinction from high-functioning autism. *Journal of Child Psychology and Psychiatry*, **32**, 1107-22.

Perner, J. (1991). *Understanding the representational mind*. MIT Press, Cambridge, Mass.

Piaget, J. (1951). *Play, dreams and imitation*. Heinemann, London.

Prior, M., and Hoffman, W. (1990). Brief report: neuropsychological testing of autistic children through an exploration with frontal lobe tests. *Journal of Autism and Developmental Disorders*, **20**, 581-90.

Riguet, C.B., Taylor, N.D., Benaroya, S., and Klein, L.S. (1981). Symbolic play in autistic, Down's, and normal children of equivalent mental age. *Journal of Autism and Developmental Disorders*, **11**, 439-48.

Russell, J., Mauthner, N., Sharpe, S., and Tidswell, T. (1991). The 'windows' task as a measure of strategic deception in preschoolers and autistic subjects. *British Journal of Developmental Psychology*, **9**, 331-49.

Sigman, M., and Ungerer, J.A. (1984). Cognitive and language skills in autistic, mentally retarded and normal children. *Developmental Psychology*, **20**, 293-302.

Tager-Flusberg, H. (1992). Autistic children's talk about psychological states: deficits in the early acquisition of a theory of mind. *Child Development*, **63**, 161-72.

Tan, J., and Harris, P. L. (1991). Autistic children understand seeing and wanting. *Development and Psychopathology*, **3**, 163–74.

Wellman, H. M. (1990). *The child's theory of mind*. MIT Press, Cambridge, Mass.

Wing, L. (1978). Social, behavioral and cognitive characteristics: an epidemiological approach. In *Autism: a reappraisal of concepts and treatment* (ed. M. Rutter and E. Schopler). Plenum, London.

Wing, L., and Gould, J. (1979). Severe impairments and associated abnormalities in children: epidemiology and classification. *Journal of Autism and Developmental Disorders*, **9**, 11–29.

12

Narrative language in autism and the theory of mind hypothesis: a wider perspective

KATHERINE LOVELAND AND BELGIN TUNALI

Understanding of other people, awareness of their knowledge, beliefs, and affective states, is essential to normal human communication (Bates 1976; Rommetveit 1974). The 'theory of mind' model asserts that these inter-personal factors in communication rest upon an ability to develop a mental representation for the contents of other people's minds, or *metarepresentation* (Leslie 1987). People with autism have been hypothesized to have difficulty developing metarepresentational ability (Baron–Cohen *et al*. 1985).

Although the metarepresentation explanation can be criticized on theoretical grounds (Hobson 1989*a,b*, this volume, Chapter 10; Loveland 1991; Loveland and Tunali 1991; Mundy and Sigman 1989; Mundy, Sigman, and Kasari, this volume, Chapter 9), it seems clear that autistic people do have problems in this area: that is, they make errors when asked to predict what other people know, believe, or feel (Baron–Cohen *et al*. 1985; Perner *et al*. 1989). What are the implications of such a deficit for language-use? In this chapter we discuss the implications of the autistic difficulty in understanding other persons for a specific kind of language-use: narrative language.

Narrative language consists of extended, organized discourse by a speaker or writer. It has been widely studied in young normally-developing children (see, for example, Applebee 1978; Stein and Trabasso 1981; Pellegrini and Galda 1990; Lucariello 1990; Kontos *et al*. 1986), usually in the form of stories. However, it can be studied only in a few people with autism; most simply do not have sufficient language. Accordingly, there are also few studies that are about narratives by autistic people.

Baron–Cohen *et al*. (1986), as part of a study of autistic children's theory of mind, had subjects tell the stories depicted in sets of ordered pictures. Although they did not examine the structure, accuracy, or pragmatic aspects of the narratives, they did find that autistic children were less likely than other children (including more severely impaired children with Down's syndrome) to talk about characters' mental states. Similarly, Tager-Flusberg (1989) reported that, in extended transcripts of spontaneous discourse, autistic children rarely talked about mental states, and seemed to have difficulty relating events to emotions. Tager-Flusberg and Quill (1987) had autistic subjects tell stories about a set of pictures. They examined structural characteristics of the stories, and found that, compared to stories by normal comparison subjects, stories by autistic children were shorter, less complex, and contained more errors in grammar and word-choice; Scopinsky (cited in Bruner and Feldman, this volume, Chapter 13) also found stories by autistic subjects were shorter and grammatically simplified in comparison with stories by normal subjects of similar age.

Other studies have found pragmatic deficits (see Tager-Flusberg, this volume) in the narrative language of autistic persons. Baltaxe and Simmons (1977) analysed the bedtime monologues of an autistic girl. They found that these reflected only a hearer's perspective, in contrast to the bedtime monologues of normal children, which characteristically imitate a dialogue between two persons (speaker and hearer). This difference may suggest a difficulty managing changes in point of view, as in the well-known problems of autistic persons in using I/you pronouns (Loveland and Landry 1986). Similarly, Baltaxe (1977) found that the conversational discourse of high-functioning autistic adolescents reflected numerous pragmatic deficiencies, including confusions in speaker/hearer point of view, rudeness, difficulty determining what is relevant, and other problems that suggest a failure to appreciate the listener's needs; Bruner and Feldman (this volume, Chapter 13) found that autistic subjects are able to take turns and respond in a conversational context, but are unable to extend the conversation by adding new, relevant information to previous comments.

Loveland *et al.* (1990*a*) studied pragmatic aspects of story re-telling (from a puppet show or videotape) in subjects with autism or Down's syndrome. Both groups produced narratives that were deficient in grammar, organization, and other structural characteristics. However, autistic subjects also produced considerable bizarre and irrelevant speech, and they tended to accompany their re-tellings with uninformative gestures, which resembled a moving puppet, but did not add to the story. Moreover, some autistic subjects told stories that indicated they understood the puppets mainly as moving objects, rather than as characters in a story. These subjects seemed to be deficient not only in awareness of the listener's needs, but also in a cultural perspective underlying the telling of stories (viz, the knowledge of what a story *is*) (cf. Heath 1986; Bruner and Feldman, this volume, Chapter 13). Similarly, Bruner and Feldman (this volume, Chapter 13) report studies in which autistic subjects used fewer pragmatic markers (time, place, etc.) than other subjects when telling a story, and also seemed to lack a grasp of how to *narrate*, as opposed to merely *describe*, a series of events.

It is difficult to draw conclusions from the existing studies relating to narrative language in autism, because of the variety of methods, purposes, and data examined. Thus, there is a need for a framework in which to examine autistic narratives.

KINDS OF NARRATIVES

In our view, the narrative should not be considered only as a text or a sample of speech collected and analysed for its content. Rather, we treat the various types of narratives as kinds of *communicative acts* (Searle 1969). This viewpoint permits us to examine not only the content of narratives, but how they function for the speaker and listener. Because the speaker and listener are continually engaged in both social and informational exchange, the narrative must take place within the context of their relationship. This relationship entails some degree of *intersubjectivity* (Rommetveit 1974), such that the speaker and listener share not only a common language code, but also knowledge of the topic discussed, of the world at large, of social and cultural conventions for communication, and of each other as persons and as individuals. Because narratives are an extended form of discourse that ordinarily proceeds without conversational interruption (Roth 1986), they have usually been considered apart from the social and cultural context in which they occur (though not always: cf. Heath 1986; Pellegrini and Galda 1990).

In this chapter we also construe the term 'narrative' more broadly than is usually done. Most studies of narrative lanquage have focused on story-telling. However, there are several other forms of extended, organized discourse that have structural or functional characteristics in common with story-narratives. We find it useful to compare these to story-narratives in terms of their implications for persons with autism.

Based on these theoretical considerations, we propose the following categories as a basis for analysis and further research:

1. *Story-narratives*. This is the most frequently studied category, and is what is most often meant by 'narrative'. A story-narrative is expected to be an organized series of causally-related event-descriptions that deal with some topic or lead to some point. Ordinarily, stories are supposed to have a beginning, a middle, and an end, recognizable characters, and some plot conflict that is resolved. Of course, not all attempts at story-telling meet these criteria. Studies of young children's stories show that they begin by stringing together sets of statements that lack causal relations (Applebee 1978: 'heaps' and 'sequences'). Only after several years' practice do children begin to produce adult-like narratives.

Stories may be either fictional or anecdotal; the requirements to produce original fictional narratives are different from the requirements for producing an original anecdote. Whereas the anecdote concerns what *was*, the fiction concerns what *might be*. Fiction requires the deliberate manufacturing of events and/or characters (imagination). Both fictions and anecdotes require an understanding of event causality (Kemper and Edwards 1986) and of linguistic tools for describing events (Duchan 1986). In both cases the narrator must select and organize information to be presented to the listener. However, in the anecdote, the information is already given, although the teller must determine which items are important ('. . . At the zoo we saw the giraffes. Then we got some hot dogs and ate them.'). In the case of fiction, the characters and/or the events involving them must be entirely constructed by the teller.

Stories, whether fictional or anecdotal, may be either original, or re-tellings of stories originated by others. In the case of re-tellings, much of the distinction between fiction and anecdote disappears, since the teller need not construct the fiction. An intermediate case occurs when the teller makes up a fiction concerning an established character or event ('Tell me a story about Robin Hood . . .'). Jokes are an example of re-tellings that must be executed within stylistic constraints to be effective; they are easily ruined by someone who 'gets the timing wrong' or otherwise distorts the parameters necessary for the effective telling of a particular joke.

Stories may also be spontaneous or elicited. Truly spontaneous stories are produced at the pleasure of the teller. Elicited stories are produced on demand, most commonly in response to something such as a picture that serves as a focus. Many studies of narrative story-telling have been done in this way (for example, Tager-Flusberg and Quill, 1987; Baron–Cohen, *et al.* 1986). This technique has the virtue of insuring some degree of uniformity of content among the narratives of different tellers, enabling the experimenter to make comparisons more directly. However, it does not reveal whether the subject can make up a story 'out of whole cloth'.

2. *Script narratives*. These are narrative accounts of generalizations about events: the way things 'usually happen' (Nelson 1986; Fivush and Slackman 1986). An example would be, 'Every morning when I get up, I take a shower and brush my teeth. Then I dress and go downstairs to eat breakfast . . .'. Although there is nothing to prevent a person from saying things like this spontaneously, it would be difficult to collect many of these from the spontaneous speech of any group, because they are rather uncommon. Usually, then, they are elicited.

The script narrative bears an interesting relationship to the anecdote. Like the anecdote, it is based on real events of a personal nature. Unlike the anecdote, however, it represents information about the common structure of many events recurrent over time, rather than the specifics of one series of events in particular. Like story-based narratives, the script narrative should be organized and should have a recognizable topic.

3. *Informative/didactic narratives.* These are narratives that are produced in order to convey specific information to someone. A spontaneous example might be giving detailed instructions ('I want you to go to the hall closet, and open the large wooden chest at the back under the Christmas wrappings. Inside, under some sweaters you will find your winter coat. Look in the pockets for your brother's brown gloves with the leather patches.') An elicited example might be giving directions to someone when asked. Another type of informative narrative is the speech, sermon, or lecture. Like story-based narrative speech, informative narratives are expected to be organized and to lead to some point (although, of course, not all of them do). However, unlike stories, they do not usually consist of causally-related event descriptions.

4. *Recitations/performances.* These occur when an individual recites learned narrative speech, for instance, the Pledge of Allegiance, the alphabet, a TV commercial, Bible verses, or other material that is well-learned at the time it is uttered. Such recitations and performances are common. However, their appropriateness is closely linked to the specific social and cultural context in which they occur. For example, reciting the alphabet may be appropriate in the classroom, but not in church. The recitation or performance differs from the story-based narrative, in that it need not consist of an organized series of event-descriptions. Moreover, its content need not be understood by the narrator for the recitation or performance to be successful. Thus, recitations and performances are distinctly different in both content and function from story-based narratives.

THE EFFECTS OF AUTISM ON NARRATIVES

On the basis of what is known about the specific difficulties of persons with autism in communication and social behaviour, we can make predictions about the difficulties posed for them by different types of narrative language. In doing so, we present examples of narrative language in autism that have not been previously presented.

1. *Recitations and performances.* The least difficult should be recitations and performances, because these are memorized verbatim and do not necessarily require understanding of content. Moreover, they do not require the speaker to select or organize the material. However, the speaker must choose when and where they will be produced; the same performance that is entirely appropriate or permissible in one setting might be quite inappropriate in another.

Some people with autism are adept at producing recitations and performances. There is nothing inherently abnormal about having them in one's repertoire, since many non-disabled children and most adults have some (such as the Pledge of Allegiance, the Lord's Prayer, etc.). However, some autistic people have well-learned narratives of this nature that are different from those of normal people, not only in that they are produced at inappropriate times and places, but in that they are of unusual, idiosyncratic content, and thus may or may not have, for the speaker, the meaning they hold for other people. An example would be the child who memorizes favourite commercials and repeats them at odd moments. Such speech is often labelled 'delayed echolalia'; but that does not mean it is necessarily empty or meaningless for the speaker. Like immediate echolalia, delayed echoing can function meaningfully for the autistic speaker, on a social/pragmatic level as well as a semantic level of communication (McEvoy *et al.* 1988).

A good example of someone with a repertoire of performances was our patient R.K., a male child with autism and moderate mental retardation. His animated renditions of nursery songs and commercials (aged eight) and his impersonations of people seen on television (aged eleven) all had a highly stereotyped, overlearned quality. Nevertheless, he received a great deal of social reinforcement for them, and he produced them often and on demand. For R.K., these performances served as a social outlet, through which this very disabled child could receive attention and approval. Interestingly, the performances were much more extended than any original utterances he was able to produce.

2. *Re-tellings of stories or other material.* Re-tellings should be somewhat more difficult for the person with autism than recitations and performances, even though the material to be narrated is not original. For re-tellings, in addition to memory demands, there is ordinarily a need to make references clear, to interpret the meaning of the material, and convey it to the listener clearly. These task demands can be highly challenging for the person with autism who is poor at anticipating and accommodating to the listener's needs (Baltaxe 1977; Loveland *et al.* 1989; Loveland *et al.* 1990*a*).

The difficulties presented by re-telling a story are illustrated by the following example from M.R., a sixteen-year-old, high-functioning male with autism who was a participant in our study of narrative story-telling (Loveland *et al.* 1990*a*). About an hour after a viewing a videotaped skit, the story of which he had been asked to narrate for a listener, he spontaneously re-told the story to his mother during a free interaction session. Mother had not seen the videotaped skit, which was about a thief who tries to steal money from an office but is driven off by a secretary wielding an umbrella. Mother's responses are in parentheses:

I saw that there was a kid stealing someone else's wallet.

(There was?)

Yeah, and they, she had an umbrella.

(Was it make-believe or did it really happen?)

It really happened. Why do you have to hit an umbrella you take the money?

(What?)

Why do you have to hit the umbrella?

(Who hit an umbrella?)

That was the lady did.

(Who had the umbrella?)

The robber . . . the robber was taking the money.

(From who?)

From the secretary.

(And what did the secretary do?)

They hit umbrella.

(She hit an umbrella?)

Yeah, she had to hit into a kid . . . a thief.

(Oh, a thief. Did the secretary have an umbrella?)

Yeah, she did.

(Okay. Did she hit the robber?)

Yeah, she did.

(With her umbrella?)

Yeah.

(Okay.)

Why did they have to hit him?

(Why do you think?)

So someone won't take the money away.

(Yeah.)

Did you find R. [M.'s surname] in the phone-book?

This attempt at a narrative quickly breaks down because the listener does not understand. M.'s mother asks increasingly specific questions over the course of the interaction, ending in yes/no questions, as she begins to see what he is describing. His meaning is obscured by distortions of grammar ('they hit an umbrella') and a failure to specify the referents of pronouns, among other problems. His assertion that the story 'really happened' is also remarkable, since it seems to betray a lack of awareness that this was a fiction. He asks several times (apparently) why the thief was hit with the umbrella, although he later supplies the obvious answer that it was to prevent the money from being taken. It is hard to know the source of his confusion about this event, although it may indicate a failure to understand characters' motivations. At the end, M. abruptly changes topics to ask about whether his mother saw their surname in the phone book they examined earlier; the phone book is a special interest for M., although his parents try to discourage him from it. No attempt is made to shift topics gracefully or with regard to the listener's interests.

3. *Anecdotes and informative/didactic narratives.* These may be more difficult for autistic persons than performances or re-tellings, because the speaker must convey information that was gained through experience (for example, telling what happened this morning when you went to see the doctor; explaining how to play a game). Both require that the speaker should determine what information it is important to present, organize the information in a coherent way, select verbal means to convey the information so that it will be understood, and accommodate to the needs of the listener. This is in contrast to re-tellings, in which the content and organization are already given for the speaker.

The following is a spontaneous, original anecdote, produced by M. R. several minutes after being told that someone else's wallet was stolen, as part of an experiment involving ability to respond appropriately to another person's distress (Loveland and Tunali 1991).

You know, my dad's car got tooken off. The names of his car got tooken off. And then the car . . . they glue on. He had to have the police come. The wire got cut down. His car, he had to go to the car shop. It's fixed now. It was on Labor Day. I left my bike in; my brother's bike I left it in.

This anecdote is remarkable in a number of respects. First of all, it is not at all clear exactly what happened to M's father's car (was it stolen, or only damaged, or both?). The pertinent information simply is not given. Instead, M. displays word-finding problems ('the names of his car' = licence plates? model name? other?), poor use of anaphora ('they glue on'?), poor grammar ('got tooken off'), and lack of transitions to clarify relationships among parts of the narrative ('I left my bike in . . .': how is this related to the rest of the story?). The lack of cohesive devices, in particular, contributes to incoherence.

It is unusual to encounter written narratives by autistic people. Written narratives differ from spoken ones in that the speaker (i.e. the writer) does not have the opportunity to interact directly with the listener (reader). Thus, the writer must anticipate the reader's needs with greater precision. Volkmar and Cohen (1985) reported a narrative written by Tony W., a young adult with a history of autism. The authors did not discuss the narrative in terms of its linguistic or pragmatic aspects, but focused instead on its implications for diagnosis and prognosis. Tony W. wrote an original statement describing his experiences as a child and adolescent with autism. It is a vivid depiction of his feelings and thoughts during this period. Interestingly, although Tony W. wrote in some detail about his own feelings and thoughts, he did not mention those of anyone else. His narrative is largely interpretable to the reader, but it is also deficient in grammar, organization, word choice, spelling, and punctuation. A sequence of events is presented, but transitions are not clearly marked; instead, loosely connected ideas are run together in long paragraphs. Lack of transitions, unclear pronoun references, and poor grasp of idiomatic usage make the statement difficult to follow.

Informative/didactic narratives by autistic people are also rare. In a recent study of referential communication in autism and Down's syndrome (Loveland *et al.* 1989), we asked autistic subjects to explain to a naive listener how to play a simple board game. The following is a didactic narrative by M.D., a high-functioning 27-year-old male with autism. The listener's comments, which were limited to very general prompts during this part of the observation, are in parentheses:

You have these animals. If you get these and then you can get on this one [points]. Then you can keep this animal here [points]. If you get on this one [points], the animal, you get to keep the animal. You do here or here [points], you have to go back. And here you get a shortcut [points]. . . . How old are you?

(Tell me more about this game.)

You just . . . whenever you land right here [points], you get it.

Of the ten possible essential pieces of information about the game, M.D. conveys only three: that landing on certain squares means the player gets a toy animal, or has to go back, or takes a shortcut. He leaves out important information such as the use of the spinner and the game pieces, where to start and finish the game, the fact that players take turns, and so on. At the end of his first attempt, he changes the topic abruptly ('How old are you?'). After further prompting, he is unable to give any new information, although he does return to the topic. It is worth noting that when much more specific prompting was introduced (for instance, 'Tell me what this is for?', while holding the spinner) M.D. was able to supply the remaining information.

Another participant in the same study, E.R., produced a didactic narrative that was also impoverished in content, but was also different in some respects. E.R. was a high-functioning 22-year-old male with autism:

First I put the moose on here [places animal on board], and then the goat on here, and then the camel over here, and then the kangaroo over here, and then the bear over here, and then the alligator over here, and then the giraffe over here, and then the hippo over here, and then the elephant over here, and then the jaguar over here, and then the elk over here, and finally the lion over here. It's easy.

(Tell me some more about this game.)

I'll take the yellow one. Play the game.

(What else?)

When we're winning you play the game. But if you lose, you're out of the game [throws arms up]. I take the yellow one. Take the green one. You go over here [traces path]. You see, I go first [points to self].

E. R.'s first attempt to explain the game is remarkable for its perseveration. He simply places every animal on the board in order, narrating his actions as he goes. When prompted further, he produces a little more information: that game pieces must be selected and that players take turns. He also inserts material that is only marginally relevant, and is certainly not informative ('When we're winning you play the game . . .'). Like subject M. D., however, E. R. gave most of the remaining information when specific prompts were supplied.

M. D. and E. R. were among the most able autistic subjects in our study, and their didactic narratives were consequently among the best. Even so, they had marked difficulty selecting, organizing, and presenting to someone else information that they knew. They also tended to include material that strayed from the topic or was uninformative or repetitive. Their narratives contrasted with those of the Down's syndrome subjects, which were much more informative, even with little prompting (Loveland *et al.* 1989).

4. *Script narratives.* Script narratives may be harder still for the person with autism, because they involve generalization over many events and selection of the most relevant information. They may be particularly vulnerable to distortion by the autistic person's selective, idiosyncratic, and poorly acculturated view of the world (Loveland, 1991; Loveland and Tunali 1991; Bruner and Feldman, this volume, Chapter 13).

Though there are many studies of script narratives in young normal children, we know of none in persons with autism. The following narrative, by one of our clinic patients, was collected in the attempt to elicit a script narrative. S. W. is a ten-year-old child with Pervasive Developmental Disorder whose intellectual functioning is in the Borderline range. He has an obsessive interest in hotels, and has memorized the locations of all the hotels of several hotel chains in major Texas cities and along interstate highways. His therapist (BT) asked him to describe a 'typical' holiday routine for his family. Instead, S. W. began by describing what he planned to do *that* Christmas, and quickly changed to discussion of his favourite topic, hotels.

(What do you usually do for Christmas?)

This year I'm gonna go to my cousin's. His name is Warren . . . We're going to go up there after Christmas and Granny and I will take a special vacation. We're gonna go to a hotel in Beaumont and we're gonna spend the night. . . . Me and my granny . . . I love doing that! . . . I always love doing that stuff . . . going to travel . . . That's great! That's my favorite thing! . . . I love travel! I say 'I want to travel' but I can't do it every time [raises his hands in the air]. Granny has saved up two 20s for the Best Western. We like this! It's called the Jefferson Inn. It's in Beaumont . . . remember, the Gulf coast! The Hotel 6 there compared to the one in Baytown . . . It's *ugly*! [grimaces].

(It is?)

Yeah! . . . It's yellow-painted . . . but the one compared to Baytown . . . it's much, much better . . . that used to called . . . [grimaces] . . . it's . . . much, much better.

This attempt to elicit a script narrative did not succeed. S.W. talked instead about what he planned to do at the present time, not what he usually did. He then switched to a different topic entirely, abandoning the question asked by his therapist. It may be that S.W. is unable to generalize across events in the way the construction of a script narrative requires. It may also be that S.W. lacks awareness of the kind of narrative structure that is conventional for the type of answer he was called upon to give. The extended discourse he did produce displays the unusual and idiosyncratic view S.W. has of what is important and interesting: his idea of travel is to visit and compare different hotels.

5. *Original story narratives*. Original story-telling should be most difficult for the person with autism, because it involves a great deal of choice with few structuring limits. It requires a knowledge of cultural conventions for content and style, the ability to construct and coherently connect meaningful, causally-related, but non-factual, events, the ability to organize these events into a recognizable story (for example, there is a beginning, a middle, and an end, not just a description), as well as a grasp of necessary language tools.

As of the time of writing, we have not seen any examples of truly original story-telling by autistic persons, i.e., specimens not elicited by pictures or other props. However, problems similar to those we predict for original stories are evident in the following spontaneous re-telling by T.B. (aged ten) during one of his individual therapy sessions. T.B. has a diagnosis of Pervasive Developmental Disorder, and is very high-functioning (Full-scale IQ 128). T.B. has a large number of tapes of 'Tom and Jerry' cartoons which he watches for several hours every day. He has numbered each cartoon (a total of 250) and memorized the cartoon numbers and titles, and the content of each. The following narrative is from a session in which he decided to share with his therapist some of his favourites. The therapist's responses are in parentheses:

This one is called 'Flirty Birdy . . . Number 92, tape 3. This is where Tom tries to make a sandwich out of Jerry and this big bird tries to take it from Tom and then the only way Tom can get Jerry back from the bird . . . [laughs] . . . Do you know what happens then? [sounds excited]

(Tell me.)

Tom acts like Toots . . . That's how it ends [laughs].

(Who is Toots?)

Toots is the girl cat . . . but that doesn't mean Toots is in this one. Tom just acts like Toots . . . yeah . . . no, he acts like a girl bird.

(What does Tom do?)

I don't know . . . Tom doesn't eat Jerry. He just tries to help some baby birds. Tom does all the stuff for baby birds. [Begins to look upset.] That's the end of this one. I'll tell you about the next one now.

The opening of this narrative effectively introduces a topic, beginning with a title and a statement of the essential conflicts involved in the plot. However, the narrative breaks down when T.B. tries to explain how the conflicts are resolved. He first provides a resolution ('Tom acts like Toots . . . That's how it ends.') that is unclear to the listener, although it is completely satisfactory to him. When she questions him further, he elaborates but reveals some uncertainty in his own understanding of the events in the cartoon (did Tom behave like a female cat or a female bird?). It remains unclear, at that point, how the character's actions resolved the plot. When the listener asks for more information, T.B. becomes frustrated and changes the subject.

The following narrative is from the next therapy session with T.B. and the same therapist.

This is called 'Jerry's Diary' . . . Tape 3, number 105. This is where . . . this is first Tom tries to be mean to Jerry because Jerry came out of his mouse-hole . . . then he listened to the radio and said . . . the radio said to Be Nice To Animals week . . . and then Jerry, I mean Tom did a lot of stuff to Jerry. You'd think that he was being nice to Jerry and guess what . . . ?

(Tell me.)

Tom was reading Jerry's diary and all this funny stuff happened [laughs] . . .

(What kind of funny stuff?)

It had parts of a cartoon I haven't seen . . . I mean it's not on any of our tapes . . . Here are parts of 'Yankee Doodle Mouse', which is number 37. This also has parts of a cartoon called 'Serenade' . . . That's the one I have put on a separate tape . . .

(I see . . . tell me what happened.)

And then Jerry, no . . . Tom tries to give Jerry a pie so Jerry can eat it . . . he throws it in Jerry's face [laughs]' . . . Yeah, that's how it ends. It is funny [laughs], isn't it?

This narrative also begins well, with an introduction of the title and topic. However, here the basic plot conflict is not clearly identified, although some events are mentioned. These events are strung together but are not linked into a coherent sequence that makes sense for the listener. Nevertheless, T.B. finds the story both interesting and amusing. At several places he laughs and talks about how funny the story is. Even though he has not provided enough information for the listener to share the joke, he clearly expects her to find it as funny as he does. An ending event for the story is mentioned (Tom hits Jerry with a pie), but it is not explained how this might resolve the plot.

Both these narratives reflect problems similar to those observed in earlier examples, such as lack of organization and clarity, poor grammar, intrusions, and incoherence. These problems persist, even though the speaker in this case is both intellectually and linguistically far more able. This individual is well-acquainted with the idea of a *story* and with some of the conventional elements of the story: beginning, middle, and end. To some extent he succeeds in conveying these elements. However, in both narratives he displays a marked failure to anticipate the listener's needs for information. He also tends to describe events without conveying their relative importance or marking their relationships; something about what Bruner and Feldman (this volume, Chapter 13) call *the act of storytelling* seems to be missing. T.B.'s narratives also suggest that he may himself have an impoverished understanding of the content of the cartoons he is trying to describe; when asked to explain or elaborate, he does not really succeed. His explanations suggest he does not fully understand characters' thoughts or motivations; rather, he simply reports their actions. Further, T.B.'s interest in the Tom and Jerry cartoons assumes a larger significance for him, as part of an elaborate private fantasy system. Not only has he memorized each of the cartoons, he has made it clear that he expects the therapist to share his interest and understanding of them, and to have memorized them as well. This entirely unrealistic expectation appears symptomatic of a larger difficulty in appreciating how meaning is shared.

SUMMARY

On the basis of the foregoing analysis and the little evidence available, we can make some predictions about what future studies of narrative language in autism might find. Autistic narratives will reflect, first of all, disordered *language*, (see Tager-Flusberg, this volume, Chapter 7) such as difficulties in grammar, word-finding, and semantics, and a failure to use language tools such as cohesive devices to mark organization. They will also tend to differ in *content* from normal persons' narratives, including more bizarre, irrelevant, or inappropriate material that may reflect an idiosyncratic world-view. The narratives will also include pragmatic errors reflecting a poor understanding of the *listener's knowledge-state* (as for example when the speaker refers to things or persons unknown to the listener without giving adequate explanation) and of the *listener's affective state* (for example inattention to gestural and facial feedback from the listener). Similarly, we expect them to reflect poor awareness of the *thoughts of characters* in the narrative, and poor understanding of *characters' affects and motivations*. Finally, we would expect the narratives of autistic people to reflect a poor appreciation of the *social and cultural context* in which narration is taking place (for example, failure to understand what makes a joke funny, or how fiction differs from non-fiction).

WHICH PROBLEMS ARE SPECIFIC TO AUTISM?

It is apparent that we would expect autistic persons' narratives to be deficient in several important areas: language itself, social/pragmatic constraints, awareness of the listener's needs, and understanding of the thoughts, feelings, and motivations of story characters, and of the social and cultural context of narration. It is tempting to conclude that these deficiencies reflect directly upon the central deficits of autism. However, some of these characteristics also show up in the narratives of school-aged children with language-learning disabilities (LD).

Although LD and non-disabled (ND) peers have similar ability to recall the order and structure of presented stories (see for example Weaver and Dickinson 1982; Loveland *et al.* 1990*b*), LD children do recall less information (Graybeal 1981; Hansen 1978; Loveland *et al.* 1990*b*), and they may have more trouble comprehending and drawing inferences from the stories (Oakhill 1984). Immature narrative styles in LD children have also been reported by several authors (for example Westby *et al.* 1984; Feagans and Short 1984; Liles 1985), who have found 'descriptive' rather than storytelling responses to pictures, reduced complexity, and poorer use of cohesive devices.

LD children asked to produce an original story also have difficulty. LD children's stories are reported to be similar to ND peers' stories in terms of story grammar characteristics, but less informative about characters and settings (Roth 1986). They are also less explicit about the middle part of the story, in which the character(s) attempt to respond to the central conflict of the story, suggesting that LD students are poorer at determining what the listener must know in order to understand the story.

LD children may also have trouble identifying and portraying the feelings and motivations of characters in a story. Westby (1985) argued that some LD children, particularly younger ones, have difficulty understanding feelings, and that somewhat older LD children have trouble understanding how feelings arise as a result of events. Similarly, other investigators have found that LD children are less able to interpret accurately the emotions and gestures of others (Bryan 1978; Gerber and Zinkgraf 1982; Saloner and Gettinger 1985; Vogel 1974, 1975). In some cases, difficulty interpreting affect may be related to specific patterns of underlying neuropsychological deficiency (Ozols and Rourke 1985; Loveland *et al.* 1990*b*), such as that associated with Arithmetic Disability.

LD children seem to be poorer communicators than ND peers when narrating a story or teaching, not only on the linguistic level (i.e. within the narrative itself), but also from the standpoint of social awareness. Feagans (1982) found that when an adult listener pretended not to understand the child's re-telling of a story, LD subjects were less able to rephrase the information for the listener. Feagans and Short (1986) had similar findings in a referential communication task. Other authors have found LD children to be poorer at providing and revising information for a listener in conversation as well (Bryan and Pflaum 1978; Donahue *et al.* 1980; Noel 1980; Spekman 1981).

Taken together, these findings suggest that people with autism are not alone in having difficulty with aspects of narratives such as effective use of language, understanding and portraying characters' affect and motivations, and awareness of the listener's needs. One conclusion to be drawn from this evidence is that difficulty understanding other persons, at least as it affects communication, does not appear to be limited to autism. In fact, this narrative deficiency may simply reflect developmental delays, since young children and mentally retarded persons are reported to have most of the same difficulties understanding others. Thus, although some aspects of the narrative deficiencies found in autism are consistent with the theory of mind explanation, they are not particularly supportive of the idea that deficits in understanding other persons are unique or central to autism.

Are there any characteristics of narrative language that *are* specific to autism? The little evidence that is available does not permit firm conclusions; however, we can suggest some directions that may be fruitful for further investigation.

It could be that autistic narrative deficits in understanding others are limited to difficulty understanding the cognitions of characters or listeners, as opposed to a broader difficulty understanding all internal states, including feelings, motives, and perceptions. This position has been argued by exponents of the metarepresentation model (Baron–Cohen *et al.* 1986; Baron–Cohen, this volume, Chapter 4). Studies of narrative language in LD and other developmentally disabled children have not focused on understanding cognitions *per se*, and thus we do not know whether this difficulty is present in those groups. On the other hand, there is not yet sufficient evidence to show that the deficiency shown by autistic persons is so narrowly circumscribed. Further research is needed to explore the boundaries between the understanding of cognitions and the understandings of other states, such as affective states, by autistic persons. It may be that these areas of understanding are not sharply divided, either in narrative performance or in development (Loveland 1991).

There are also aspects of autistic narratives that seem to require an explanation that goes beyond a failure of interpersonal understanding. There is reason to think that autistic persons do not fully share in the social and cultural context in which narration ordinarily takes place (see Bruner and Feldman, this volume, Chapter 13). As the examples of narrative language given in this chapter illustrate, autistic people tend to assign unusual and idiosyncratic meanings not only to words, but also to objects, persons, and events. In some cases, the very notion of a story, or of characters in a story, seems to be lost (Loveland *et al.* 1990*a*). These aspects of narrative language in autism reflect an apparent failure of acculturation, and perhaps also an incomplete appreciation of what might be called the 'human point of view'.

Loveland (1991) proposed an ecological approach to the development of autistic people using the notion of *social affordances* (Gibson 1986 [1979]; Reed 1988; McArthur and Baron 1983; Good *et al.* 1989). She suggested that autistic people are deficient in the ability to perceive the functional meaning (affordances) of those aspects of their environment that involve social and cultural information. Early failure of attunement to such meaningful aspects of the world as others' facial expressions may form a basis for later failure to share cultural attitudes and expectations. Thus, for example, the failure of some autistic people to grasp conventional aspects of the narrative, such as story form, the pragmatics of storytelling, or even the notion of 'story' may be an outgrowth of a more basic difficulty in grasping the meaning of non-linguistic human events that are normally a part of the infant's awareness of the social world. Looking at the relationship of narrativity to acculturation from a slightly different point of view, Bruner and Feldman (this volume, Chapter 13) argue not only that autistic narratives reflect a lack of acculturation, but that the failure of narrative modes of thinking in autistic people contributes to a more general failure to understand the world in conventionalized ways. It may therefore be that there is ordinarily a reciprocity between acculturation and narrative language, such that each normally supports the other in development; in the case of autism, this reciprocity is impaired, contributing to deficiencies in both areas.

The study of narratives in autism is very new. What little evidence we have suggests that autistic narratives are deficient on many levels, and that these deficiencies are potentially consistent with a number of different explanations. It is clear that not only the structure and content of narratives, but also their social and cultural context should receive more attention in future research.

Acknowledgement

This work was supported in part by grant number DC 00357–06 to the authors from NIDCD.

REFERENCES

Applebee, A. (1978). *The child's conception of story: ages two to seventeen.* The University of Chicago Press.

Baltaxe, C. (1977). Pragmatic deficits in the language of autistic adolescents. *Journal of Pediatric Psychology*, **2**, 176–80.

Baltaxe, C., and Simmons, J. (1977). Bedtime soliloquies and linguistic competence in autism. *Journal of Speech and Hearing Disorders*, **42**, 376–93.

Baron–Cohen, S., Leslie, A. M., and Frith, U. (1985). Does the autistic child have a 'theory of mind'? *Cognition*, **21**, 37–46.

Baron–Cohen, S., Leslie, A. M., and Frith, U. (1986). Mechanical, behavioral, and Intentional understanding of picture stories in autistic children. *British Journal of Developmental Psychology*, **4**, 113–25.

Bates, E. (1976). *Language and context: the acquisition of pragmatics.* Academic Press, New York.

Bryan, T. (1978). Social relationships and verbal interactions of learning-disabled children. *Journal of Learning Disabilities*, **11**, 107–15.

Bryan, T. S., and Pflaum, S. (1978). Social interactions of learning disabled children: a linguistic, social and cognitive analysis. *Learning Disabilities Quarterly*, **1**, 70–9.

Cohen, D. J., Paul, R., and Volkmar, F. R. (1987). Issues in the classification of pervasive developmental disorder and associated conditions. In *Handbook of autism and pervasive developmental disorders* (ed. D. J. Cohen and A. M. Donellan). Wiley, New York.

Donahue, M., Pearl, R., and Bryan, T. (1980). Learning-disabled children's conversational competence: responses to inadequate messages. *Applied Psycholinguistics*, **1**, 387–403.

Duchan, J. F. (1986). Learning to describe events. *Topics in Language Disorders*, **7**. 27–36.

Feagans, L. (1982) The development and importance of narrative for school adaptation. In *The language of children reared in poverty* (ed. Feagans, L., and Farran, D. C.). Academic Press, New York.

Feagans, L., and Short, E. J. (1984). Developmental differences in the comprehension and production of narratives by reading-disabled and normally achieving children. *Child Development*, **55**, 1727–36.

Feagans, L., and Short, E. J. (1986). Referential communication and reading performance in learning-disabled children over a 3-year period. *Developmental Psychology*, **22**, 177–83.

Fivush, R., and Slackman, E. A. (1986). The acquisition and development of scripts. In *Event knowledge: structure and function in development* (ed. K. Nelson). Erlbaum, Hillsdale, NJ.

Gerber, P. J., and Zinkgraf, S. A. (1982). A comparative study of social–perceptual ability in learning-disabled and nonhandicapped students. *Learning Disabilities Quarterly*, **5**, 374–48.

Gibson, J.J. (1986). *The ecological approach to visual perception.* Erlbaum, Hillsdale, NJ. (Original work published 1979.)

Good, J.M.M., Still, A.W., and Valenti, S.S. (1989). *Culture and the mediation of human affordances.* In J.M.M. Good, S.S. Valenti, and R. Van Acker (Chairs), *Social affordances and interaction.* Symposium conducted at the Fifth International Conference on Event Perception and Action, Oxford, OH.

Graybeal, C. (1981). Memory for stories in language-impaired children. *Applied Psycholinguistics*, **2**, 269-83.

Hansen, C.L. (1978). Story retelling used with average and learning-disabled readers as a measure of reading comprehension. *Learning Disabilities Quarterly*, **1**, 62-9.

Heath, S.B. (1986). Taking a cross-cultural look at narratives. *Topics in Language Disorders*, **7**. 84-94.

Hobson, R.P. (1989*a*). Beyond cognition: A theory of autism. In *Autism: nature, diagnosis and treatment* (ed. G. Dawson). Guilford, New York.

Hobson, R.P. (1989*b*). On sharing experiences. *Development and Psychopathology*, **1**, 197-203.

Kemper, S., and Edwards, L.L. (1986). Children's expression of causality and their construction of narratives. *Topics in Language Disorders*, **7**, 11-20.

Kontos, S., Mackley, H., and Baltos, J.G. (1986). Story knowledge in preschoolers: a comprehensive view. *Journal of Genetic Psychology*, **147**, 189-97.

Leslie, A.M. (1987). Pretense and representation: the origins of 'theory of mind'. *Psychological Review*, **94**, 412-26.

Liles, B.Z. (1985). Cohesion in the narratives of normal and language-disordered children. *Journal of Speech and Hearing Research*, **28**, 123-33.

Loveland, K.A. (1991). Social affordances and interaction II: autism and the affordances of the human environment. *Ecological Psychology*, **3**, 99-119.

Loveland, K.A., and Landry, S. (1986). Joint attention and communication in autism and language delay. *Journal of Autism and Developmental Disorders*, **16**, 335-49.

Loveland, K.A., and Tunali, B. (1991). Social scripts for conversational interactions in autism and Down syndrome. *Journal of Autism and Developmental Disorders*, **21**, 177-186.

Loveland, K.A., Tunali, B., Kelley, M.L., and McEvoy, R.E. (1989). Referential communication and response adequacy in autism and Down's syndrome. *Applied Psycholinguistics*, **10**, 301-13.

Loveland, K.A., McEvoy, R.E., Kelley, M.L., and Tunali, B. (1990*a*). Narrative story-telling in autism and Down's syndrome. *British Journal of Developmental Psychology*, **8**, 9-23.

Loveland, K.A., Fletcher, J.M., and Bailey, V. (1990*b*). Verbal and nonverbal communication of events in learning-disability subtypes. *Journal of Clinical and Experimental Neuropsychology*, **12**, 1-15.

Lucariello, J. (1990). Canonicality and consciousness in child narrative. In *Narrative thought and narrative language* (ed. B.K. Britton and A.D. Pellegrini). Erlbaum, Hillsdale, NJ.

McArthur, L.Z., and Baron, R.M. (1983). Toward an ecological theory of social perception. *Psychological Review*, **90**, 215-38.

McEvoy, R., Loveland, K., and Landry, S. (1988). The functions of immediate echolalia in autistic children: a developmental perspective. *Journal of Autism and Developmental Disorders*, **18**, 657-68.

Mundy, P., and Sigman, M. (1989). The theoretical implications of joint attention deficits in autism. *Development and Psychopathology*, **1**, 173–83.

Nelson, K. (1986). Event knowledge and cognitive development. In *Event knowledge: structure and function in development* (ed. K. Nelson). Erlbaum, Hillsdale, NJ.

Noel, M. M. (1980). Referential communication abilities of learning-disabled children. *Learning Disability Quarterly*, **3**, 70–5.

Oakhill, J. (1984). Inferential and memory skills in children's comprehension of stories. *British Journal of Developmental Psychology*, **54**, 31–9.

Ozols, E. J., and Rourke, B. P. (1985). Dimensions of social sensitivity in two types of learning-disabled children. In *Neuropsychology of learning disabilities: essentials of subtype analysis* (ed. B. P. Rourke). Guilford Press, New York.

Pellegrini, A. D., and Galda, L. (1990) The joint construction of stories by preschool children and an experimenter. In *Narrative thought and narrative language* (ed. B. K. Britton and A. D. Pellegrini). New Jersey: Erlbaum, Hillsdale, NJ.

Perner, J., Frith, U., Leslie, A. M., and Leekam, S. R. (1989). Exploration of the autistic child's theory of mind: knowledge, belief and communication. *Child Development*, **60**, 689–700.

Reed, E. S. (1988). The affordances of the animate environment: social science from the ecological point of view. In *What is an animal?* (ed. T. Ingold). Allen & Unwin, London.

Rommetveit, R. (1974). *On message structure*. Wiley, New York.

Roth, F. P. (1986). Oral narrative abilities of learning-disabled students. *Topics in Language Disorders*, **7**, 21–30.

Saloner, M. R., and Gettinger, M. (1985). Social inference skills in learning-disabled and nondisabled children. *Psychology in the Schools*, **22**, 201–7.

Searle, J. R. (1969). *Speech acts*. Cambridge University Press.

Spekman, N. J. (1981). Dyadic verbal communication abilities of learning-disabled and normally-achieving fourth- and fifth-grade boys. *Learning Disabilities Quarterly*, **4**, 139–51.

Stein, N. L., and Trabasso, T. (1981). What's in a story: critical issues in story comprehension. In *Advances in the psychology of instruction* (ed. R. Glaser). Erlbaum, Hillsdale, NJ.

Tager-Flusberg, H. (1989). An analysis of discourse ability and internal state lexicons in a longitudinal study of autistic children. Paper presented at the meeting of the Society for Research in Child Development, Kansas City, April.

Tager-Flusberg, H., and Quill, K. (1987). *Story-telling and narrative skills in verbal autistic children*. Paper presented at the Biennial Meeting of the Society for Research in Child Development, Baltimore, MD.

Vogel, S. A. (1974). Syntactic abilities in normal and dyslexic children. *Journal of Learning Disabilities*, **7**, 103–9.

Vogel, S. A. (1975). *Syntactic abilities in normal and dyslexic children*. University Park Press, Baltimore, MD.

Volkmar, F. R., and Cohen, D. J. (1985). The experience of infantile autism: a first-person account by Tony, W. *Journal of Autism and Developmental Disorders*, **15**, 47–54.

Weaver, P., and Dickinson, D. (1982). Scratching below the surface structure: exploring the usefulness of story grammars. *Discourse Processes*, **5**, 225–43.

Westby, C. E. (1985). Learning to talk — talking to learn: oral–literate language differences. In *Communication skills and classroom success: therapy methodologies for language-learning-disabled students* (ed. E. Simon). College-Hill Press, San Diego, CA.

Westby, C. E., Maggart, Z., and Van Dongen, R. (1984). *Language prerequisites for literacy*. Paper presented at the International Child Language Congress, Austin, TX., July.

13

Theories of mind and the problem of autism

JEROME BRUNER AND CAROL FELDMAN

This chapter is intended to fulfil three objectives. The first is to look critically at some current views about the 'theory of mind' upon which we base our judgements and appraisals of intentional states in others. This critical look focuses particularly upon how cultural and social factors may shape the ways in which human beings experience and represent the intentional states of others, and indeed of themselves — people's beliefs, desires, intentions, and the like. This part of the paper is principally the responsibility of the first author. The second part of the chapter, for which the second author is principally responsible, deals specifically with what it is that might conceivably produce that form of deficit that is seen in autistic children — their impaired ability to carry on ordinary social interactions, particularly in conversation. It will be plain later, however, that such a deficit cannot easily be attributed to this alleged inability to appreciate mental states in others or for that matter, in themselves. We shall take the position in the third and final section, for which both authors are jointly responsible, that at least part of the deficit derives from the inability or unwillingness of autistic children — high- and low-level alike — to be able to represent culturally canonical forms of human action and interaction by the vehicle of narrative encoding.

KNOWING ANOTHER MIND

What do we mean when we say that 'somebody shows signs of believing that a person or thing in their world is acting under the influence of an intentional state.'* Let me take some controversial examples first. I begin with some old experiments by Scaife and Bruner (1975) and by Butterworth and Castillo

---

* I am forswearing the ways of the 'intentional stance' proposed in Dan Dennett's book of that name. When people 'believe' that others are deliberately withholding information from them, they do not think of it 'as if' it were so. And while a sceptical philosopher might find it amusing to do so, I am not in the least impelled that way. People *have* notions about other minds, and sometimes risk their lives over them. I would like to repay them in kind by seriously acknowledging that fact.

(1976). A mother or care-taker *en face* and in eye-contact with a young child now brightens his or her face, says 'Oh look!' with the culturally (possibly biologically—Fernald and Kuhl 1987) canonical intonation, and turns outward some ninety degrees as if looking at a target along the line of regard. By the middle of the first year, the child partner in such a procedure turns her gaze in the same direction far in excess of any reasonable criterion of chance. If the child were merely imitating a bodily gesture, she might as reasonably have turned in the opposite direction, given the two were facing each other. Given our scientific scepticism, we cannot quite rule out a 'signposting' effect.

So we add another complication, or at least George Butterworth does—something about the 'intentionality' of the caregiver's looking, that it is directed *at* something. Now, on only half the trials, there is a target object placed out along the caregiver's line of regard. On the other trials, there is nothing there. When the child turns out to follow the experimenter's line of regard and finds no target there, we note that she turns back to re-examine the caregiver's eyes, face, orientation, after which checking she turns out again (a layman might say) to have another look. So what now? Shall we say that the young subjects might have thought they got their grown-up partner wrong and were coming back to check? Well, not quite. They might, as it were, be going back to check whether the sign was really pointing to Headington or somewhere.

So now I come to another experiment (these are all sets of experiments, by the way, and I am condensing but I hope not oversimplifying them). This set, more recent, is perhaps less well known. They are the products of Meltzoff's laboratory. Meltzoff (Meltzoff and Moore 1989; Meltzoff 1990; Meltzoff and Gopnik, this volume, Chapter 16) began by having the infant face two experimenters, one off to the right, the other to the left, all three with the same toy, which was changed on every trial. The children were fourteen months old. In the first experiment, one experimenter directly and immediately imitated certain predetermined target gestures characteristically performed by the child. The other experimenter, in effect, just hacked around with the toy. The child quickly glued on to the first experimenter: looked at him longer, smiled at him more, directed more acts toward him. But perhaps kids just prefer directed people to those just hacking around. So the next experiment: this time both experimenters were active with their toys, but the first one imitated the child's acts, while the second one imitated the acts of a previous subject exhibited on a concealed video screen. Both were responding contingently, as it were, but one by immediate imitation and the other with a non-matching response from the same repertoire. Neither response had any utilitarian consequence or extrinsic reinforcement value for the child.*

* There was a third experiment better to equate the timing of the two experimenters, which it duly did with no discernible effect on the main results.

The findings of the better-controlled second experiment, to paraphrase the old chestnut, were not surprising, but only baffling. The children cooed, oohed, smiled, made more eye-contact and seemed, as it were, to be considerably more appreciative toward the imitating experimenter than toward the equally active but non-imitative other one. I can even tell you that in a related set of experiments, immediately imitated children were more likely spontaneously to repeat an imitated gesture when brought back to the lab for further testing the next day. Now what shall we make of this? How did the children *know* they were being imitated? They can't see their *own* response, or at least not from the perspective from which they see the experimenter's matching response. So how could they have sensed some special match? Are they inferring or intuiting that the experimenter *intends* to imitate their own acts? And if so, how?

Well, you will immediately see a kinky connection between these studies and the well-known line of work that began with Gallup's (1970) study of mirror self-recognition in the chimpanzee. But in the present studies, the child is not only recognizing a match, as it were, between her own subjective intention and its bodily expression (a bodily expression previously unobserved and unobservable), but between that intention and its expression in another. And what kind of score does that performance get the child in the theory of mind standings?

It depends upon whose scoring form-sheet you want to use. I was struck at the European Conference on Developmental Psychology held in Scotland in 1990 that theory of mind researchers seemed to have got stuck on the criterion of false belief: you cannot be sure that a child has a proper theory of mind unless the child can appreciate the distinction between a true and a false belief in another. I want to call this a Graduation Day criterion: a child does not have a guaranteed notion of other minds until she can grasp the possibility of their holding false beliefs. To equate grasping other minds with getting a False Belief Diploma at Graduation Day is to oversimplify its form and function. To equate 'having' a theory of mind with grasping the epistemological distinction between true and false belief obscures the contribution of the three or four years of development that preceded it.

Let me turn back now to Meltzoff's interesting conclusions based on the studies I so briefly summarized above. His most general tacit conclusion is that the child's recognition of other minds and of their various intentional states comes gradually. The nature of its development, moreover, depends upon transactional processes between caregivers and children. These, as we shall see in the second section of this paper, are related to grasping narrative structure. The first of these is a *reciprocal attribution* process. Meltzoff (1990) notes that parents report in interviews after the experiment that they enjoy playing mutual imitation games with their infant because these accentuate 'for them that their infant is, like them, a sentient human being', an agent (p. 160). These imitative episodes, Meltzoff argues, also evoke similar attribution processes in the infants, who are led to make 'inferences about the sentience of adults'—or, if one takes the line of Robert Zajonc (1980), the attribution would occur by direct triggering and without the use of the mediated 'appraisals' whose involvement Lazarus (1981) claims.

There are good primatological evolutionary and even better human ontogenetic grounds for accepting this view of an initial, pre-programmed primitive readiness for attributing mental states, and particularly agentivity, to conspecifics (see Whiten and Byrne 1988; Bruner 1978; Trevarthen 1987; Humphrey 1986; Lewin 1989). Indeed, the absence of this pre-programmed readiness appears to be a crucial element in severe autism (Baron–Cohen *et al.* 1985; Leslie 1987; Mundy and Sigman 1989; Frith 1989). But we shall consider this in the following section. Meltzoff quite properly asks (1990, p. 160) in view of this, whether it even makes sense to tie the first recognition of mental states in others to the child's first use of mental-state words. Surely there are earlier indices than this. For my part, I find such a claim tantamount logically to asserting that the child's first visual discrimination of green and red dates from her first reliable use of those words. Does not the child need a mental interpretant, in Peirce's sense (1960), in order to acquire the verbal distinction?

This brings me to Meltzoff's second component process: a capacity to recognize a correspondence between one's own mental or intentional state and a mental state in another, simply on the basis of another's overt expressive behaviour—their facial expression or other gestural signs. The work Meltzoff cites in support of the operation of such a process (Ekman 1984; Ekman *et al.*, 1983; Meltzoff 1985; Meltzoff and Moore 1985, 1989) is very much in congruence with Zajonc's claim about the directness or non-inferentiality of the recognition of emotion in others. Ekman's work, you will recall, strongly suggests that there is a basic connection between subjective emotional states and their manifestation in facial expression. The Meltzoff studies indicate, moreover, that from very early on 'infants can relate the gestures they see on another's face to their own unseen facial behaviour' (1990, p. 161).

This set of findings can now be embodied in a four-step argument: (1) The infant recognizes a relation between her own emotional state and its overt expression, i.e., a primitive working belief that one signals one's inner states by one's overt actions. This belief is further confirmed, of course, by appropriate responses from others—a matter to which I will return later. (2) The infant recognizes a link between another's overt actions or expression and their corresponding subjective states, i.e., that from very early on there is a primitive belief that the acts of others also signal particular subjective states. There is also ample opportunity for the infant to confirm this hypothesis in interaction. (3) The infant can at least recognize an imitative correspondence between her own expression of a subjective affective state and a like expression in another, even when she cannot directly perceive her own expression *à la* Meltzoff and Moore's findings on facial imitation. And finally: (4) Under these circumstances, the child 'naturally' entertains the hypothesis that matching overt expressions in self and other have a special status by virtue of indicating matching subjective states — which accounts for the child's special reactions to being imitated by her caregiver.*

What appeals to me most about this general line of argument is not only that it presupposes little that is not demonstrable, but that it makes place from the very start for the operation of transactional, cultural factors in shaping the child's particular theory of mind. Moreover, it requires no special Graduation Days. As in a good progressive school, progress and challenge on joint attention, on mutual imitation, on peekaboo, all help get you there. And you will never get to propositionally mediated talk about true and false beliefs unless you climb this ladder of praxis.

Now let me say a word more about the power and utility of such a theoretical position in respect to the known effect of cultural and interpersonal influences on how we view mind — both in ourselves and others. First, about the primitive belief that one's overt expressions signal one's inner states and that these overt signs can, as it were, be correctly 'read' by others. Evidence from sensorially handicapped children is relevant here — both blind and deaf. In both instances, when the child is deprived of appropriate feedback by the failure of caregivers to respond appropriately, there is often a regression into tantrum-like behaviour, along with a severe failure in communicative development, including notable language delays (Wood *et al.* 1986; Urwin 1980; Keller 1928). Many have reported, indeed, that deaf–blind children often resembled autistic children, and were misdiagnosed as such.

But one finds comparable effects under less drastic conditions. Take the transformation of early demand-signalling in very young children as a case (see for example Pratt 1975; Bruner 1983*a*). Such signalling fairly early on takes the form of a full-blown, outstretched gesture of arm and hand accompanied by vocalized whining that increases in amplitude with delay of gratification. With appropriate adult response, the gesture and vocalization become ritualized and modulated, and at seven to nine months 'object pointing' is recruited into the pattern, to be followed in a few weeks or months by such accompanying semi-indexical holophrases as 'Apoo', 'Bada',

---

* Meltzoff presents his argument in three steps. I find it more satisfactory to add my first step to his three, rather than presupposing it, as he does.

or whatever. As these develop in the child, something else is developing in the mother–child dyad. The mother early places felicity conditions upon such primitive demand–requests: 'You can't have it if you yell like that!' The child, we know 'from many *in situ* studies of the development of children's social understanding (for example Dunn 1988) soon becomes very skilled in ingratiative techniques, in little deceptions, etc., etc. And as Dunn comments, they soon adopt various forms of alliance-making. When we observe such phenomena in higher primates, we proclaim that they have theories of mind (Whiten and Byrne 1988; Premack and Woodruff 1978). But we insist that propositional language is necessary before we can certify children as mental theorists. This may be convenient; but I think it obscures an important point.

Before turning to Part II, in which the topic of autism will be addressed directly, let me now trace out a few hypotheses that stem from what I have said — hypotheses that may be useful in considering the deficits found in autism.

If it is the case — and the evidence speaks very much in favour of it — that autistic children show very early deficits in such things as responding to adult facial expressions, to direction of adult gaze, to pointing and other attention-directing devices, then it follows that the child will not readily enter into those structured care-taker–child transactions that I mentioned earlier. In earlier work (Bruner 1983*a,b*) I referred to these as 'formats'. The characteristic shape of such formats is narrative in nature: there is a canonical steady state, followed by some precipitating event, followed by a restoration, followed by a coda in which the game is proclaimed to be over. The internal architecture of all such 'play' is laid out in terms of the intentional states of the participants as much as it is by the objects that are used in the play.

It is rather a chicken-and-egg matter to decide whether narrative structuring in this early period is driven by some innate push to narrativity, or whether it derives from early and possibly innate recognition of the so-called 'arguments of action' — that there are agents with intentions toward goals who use instruments in a means–end fashion, etc., etc. — or whether narrativizing is a universal cultural form (see Bruner 1990).* But it does not matter. For what this narrativization of social interaction does for the child is to permit him or her to build canonical representations of how the world of people-and-things works and should work. Anything that interferes with the mastery of such representations will not only interfere with the child's skill in dealing with social situations, but will more profoundly

---

* For an interesting discussion of the question of the cultural push toward narrative, see Loveland and Tunali, Chapter 12, this volume; Loveland *et al.* (1988, 1990); Loveland (1990, in press).

affect the capacity to organize and store information in a fashion relevant to such interaction. The effect would be to cut the growing child off from organized knowledge about the canonical cultural forms not only for interpreting knowledge from the cultural world, but for communicating such knowledge. And unless strenuous steps were taken to prevent it, through special care, the deficit that would develop would grow in severity as the child grew older. It should show itself in a failure to grasp the arguments of action which, at the macroscopic level, should reveal itself as a failure to use or grasp narrative.

SOME OBSERVATIONS ON AUTISM

The autistic child presents a singular puzzle: a human being whose conversational language and social skills generally lag way behind apparent intellectual ability. Everywhere else one looks in the human domain, even among those with low IQ, social contact and conversational fluency seem to develop so effortlessly that it looks as if they required no particular skill at all.

In recent years this unusual and puzzling pattern of the autistic child has attracted the interest of people in language and cognitive development, who have seen in autism an opportunity to discover a cognitive acquisition that in normal children happens so automatically we had failed to notice it. Whatever it is that we discover normal children know how to do but autistic children do not, it would have to explain the social strangeness, and especially the peculiar communicative patterns, that all observers agree are the cardinal feature of autism. The recent studies of autism have far more precise descriptions of the patterns of language-use (Paul and Cohen 1984; Tager-Flusberg 1981*a,b*, 1989*a,b*) and of cognition (Baron–Cohen *et al*. 1985) found in autism, and have given us a new theoretical account of the autistic child's difficulties. This had led to a 'theory of mind' theory of development.

The autistic child is said to lack a theory of mind, roughly by reason of one of two more fundamental deficits. The first is a primary affective disorder – that the attuned and intuitive understanding normal infants have of their mother's feelings is awry (Hobson, this volume, Chapter 10). The second is metacognition (Baron–Cohen *et al*. 1985) – that the normal child early on creates mental representations of mental states through a set of inferences covered under a theory of mind that is missing in autism. Lack of a theory of mind, or notions about intentionality, then, whether derived from an affective or a cognitive defect, is proposed to explain social strangeness, especially conversational strangeness.

Learning how to draw inferences about, or in some other way to know about others' (and especially one's interlocutor's) intentional states is plainly an extremely important acquisition of early, normal childhood, one over-looked in normal development before the deficit seen in autism made it salient (but see Bruner 1983; Dore 1974).

For some years, while thinking about the relation between language and thought in other contexts, I have been an interested amateur trying to understand through studies of my own or of my students precisely which aspects of language and conversational competence were affected in autism. Before reviewing those studies, let me report the conclusion they led me to. It is this: that in autism two important elements of productive language are missing: one is evident in dialogue, the other in speech generally. Both relate to narrative. In dialogue, autistic speakers seem unable to extend the interlocutor's previous comment. They seem not to know where it is 'going'. And in their speech generally, they seem not to know how to make a story. Both suggest a common failure in being able to encode the arguments of action into a structure: an agent pursuing a goal in some setting, etc.

In normal development, patterns of narration support or scaffold the kind of metacognition about intentions that lies at the heart of the 'theory of mind' theory. Perhaps in autism narration of this kind is interfered with. Perhaps autistic children are unable or unwilling to tell stories to themselves or others. If narration scaffolds metacognition, a lack of story-telling would have serious consequences for their developing theory of mind. But it would also have other consequences beyond those anticipated by the 'theory of mind' theory, because narration also scaffolds thinking in general, giving it a conventional encoding that makes it communicable. Thoughts, which normally appear not in the topic but in the comment of utterances, must have a reasonably conventional encoding if they are to be readily and successfully hooked up to shared topics in discourse. So a failure to narrate might also lead the autistic child to general difficulties in encoding and communicating about his experiences.

I am proposing, then, that knowledge of intentional states is encoded in the story-making of normal children from about two or three years of age, or even earlier in the (narratively) formatted 'play' that is so prominent among normal children and their caregivers (Bruner 1982). Thanks to Nelson's (1989) recent collection of one child's monologic talk, we now know much more about the rich role of story-telling in the life of the two- to three-year-old child. Emmy, a normal two-year-old, lies in her crib at night telling little stories to herself that make patterns out of the con-fusing events of the day. She poses puzzles to herself, and then resolves them by putting them into problem-solving narratives or temporal narrative frames (see Feldman 1989 and Bruner and Lucariello 1989, respectively). In effect, she uses narrative forms to create meanings for the events in her life. The musings frame events in linguistic forms that embody canonical patterns of the human culture in which she lives. For her, narratives are exercises in canonical encoding, and, specifically, in the making of non-idiosyncratic or culture-embodying meanings. These meanings therefore are connectable with one another and with those of others in a web of cultural associations.

I surmise that it is these early encodings that go awry in autistic children, because they have either no impulse or no gift to narrate, which leaves them unable to make new comments on a topic in discourse — a serious deficit. It would impede social life, and it would also have consequences for metacognition, since a certain form of topic–comment structure — namely, commenting on a prior comment taken as topic — is a fairly standard procedure used by normal children to bootstrap themselves into metacognition (Feldman 1991). We will return to this issue in the third part of this chapter. But now we must turn to the promised studies. Before we do, though, let me draw two quick distinctions needed to understand the language we shall be examining. The first is the difference in verbal fluency between high- and low-functioning autism that is already familiar in the literature. There is an enormous range in fluency in what counts as a conversation, and it needs to be taken into account. The second is less familiar in this literature, but it too can help to sort out some of the unexplained variability between various accounts of autistic language. It is the notion of *genre*.

Ordinary conversation, explaining a board game, describing a set of pictures that tell a story, retelling a story, and explaining a story someone else has told — all these are different genres with different language-patterns. To begin with, all but the first can be managed as monologues that dispense with the requirement of fitting my utterance to another's, and specifically its topic-and-comment structure into a framework negotiated with an interlocutor. On the other hand, monologues require more self-generated cohesion in fitting successive statements together, for the very reason that no interlocutor is there to help the speaker with the job. Indeed, since each one of these genres has its own special form of task-demand, each has also its own special forms of failure. One can fail at one while succeeding at the other, because of the different language skills they invoke. As we shall see, even moderately-functioning autistic children have more than one genre, and all but one of the genres mentioned above are within the repertoire of high-functioning autistic children.

Some years ago, one of my students, Naomi Driesen, made a visit to the Benhaven school, and mentioned that she had found the very low-functioning autistic adolescents there to be friendly and eager to 'talk' to her, even though their limited fluency made conversation all but impossible. Of what did their apparent sociableness consist? What conversational or discourse skills did they have? With Rhea Paul and Donald Cohen's collaboration, we set out to describe the 'conversational skills' of very low-functioning autistic adolescents who hardly spoke.

Briefly, the subjects were ten severely autistic young people aged 13–27, average age nineteen, who were diagnosed as children, and had an average Stanford–Binet IQ of 24. They were observed in 32 interactive episodes containing a total of 104 'utterances', which the children gave in response to adult staff persons. The adult utterances in these episodes were half interrogatives, 17 per cent declaratives, and 20 per cent imperatives. That the children plainly sought to interact is suggested by two findings. First, 91 per cent of the children's responses were socially accommodating, and indeed 84 per cent actually maintained the adult's topic. This was an astonishingly high percentage in the light of their limited language-production skills, and bespoke an effort to mobilize a very limited set of resources in the service of social contact. Second, children maintained topic most successfully in response to declarative utterances (89 per cent of adult declarative prompts were responded to in a topic-maintaining way), but somewhat less so in response to the other two (presumably more compelling) adult forms (74 per cent of responses to questions, 67 per cent of responses to commands); they were *choosing* to respond.

But of the topic-maintaining responses, it was notable that only 6 per cent went beyond simple maintenance to some form of comment-development. The rest were largely terminal appropriate answers (48 per cent) and physical compliance (25 per cent). We ended up convinced, on the basis of this finding, that our autistic subjects did have basic but rather undeveloped conversation skills, mostly at a simple topic-maintenance level. But we were left wondering about their inability to generate *new* comments. Was this lack the cardinal autistic conversation failure? Since in this case the lack of new comments might have been due to the subjects' generally low level of language-production, we had to turn to a higher-functioning group to learn more.

With Rhea Paul helping us, Bobbi Renderer then conducted lightly structured interviews with four high-functioning autistic adolescents resident at the Anderson school in Hyde Park, New York. She talked with them about their life, what they were doing, what they were going to do, and, especially, how they felt about it. They talked a good deal, in contrast to our previous low-level subjects, and they had some important additional language skills. For one, they showed a good deal of topic *initiation*. And, largely in response to the interviewer's queries, there also was a good deal of talk about intentional states: talk about what they wanted, hoped for, liked, and disliked, what they understood and found hard to learn, and even what they knew and wanted to know more about ('I want to learn about my face'). There was even some metacognitive talk — for example, about why the conversation went off track ('You said something wrong because you got the choice from me.').

However, even when they introduced their own topic ('Let's talk about X') they could seldom say anything beyond one rote comment. Presented with follow-ups like 'Why do you like it?' 'What do you do there?' 'When are you leaving?', they had difficulties. And when on rare occasions they did try to make an extended comment on a going topic, it often lay so far outside the canonical set of possible comments that it might as well have been a change of subject. An example is the sequence:

E: Are you leaving this school soon? I heard you were going away.

S: Yes, I am.

E: What's that all about?

S: I like Danny.

E: Yeah [pause]. Is Danny leaving, too?

S: Yeah.

The impression is one of general conversational willingness, but without much content to offer (for example, why he's leaving, where he's going, etc.) beyond an initial maintenance of topic ('Yes, I am'). And of a flawed effort to add *something* in the way of new comment that both sounds unconnected ('I like Danny' sounds as if the topic has changed to Danny) and is, in fact, strangely connected since liking Danny, or even going to the new school with Danny, are *not* what the move to the new school is 'all about'.

As for that special recursive form of comment-development that involves commenting on a prior comment which then gets taken as the new topic, it is virtually absent even in these high-level subjects. This could, of course, be a failure of recursive metacognition rather than of narrative encoding; and, if this is so, then both of these deficits may be responsible for the restricted creativity of autistic conversation.

Although all but one of the subjects manifestly enjoyed having the conversations, the interviewer felt she had failed. In spite of the appearance of so much talk, she nevertheless felt that she had been unable to make contact. And this may indeed be the crucial life problem for the high-functioning autistic person — that they make normal interlocutors so uncomfortable and unresponsive that the autistic children feel effectively cut off from participation in normal social life. Though they report experiencing many aspects of life as a struggle ('. . . emotional things . . . things I have to learn. It's not easy for me. I don't understand them. My way.'), their most plaintive desire is to be friends with normal people ('When can I come to your house?'). High-functioning autistic people's autobiographical reports are full of making and keeping friends (see for example the case 'Tony' reported by Volkmar and Cohen 1985). And I think that this problem, though it undoubtedly has other sources as well, is in large measure due to the particular form of conversational strangeness that is described above—the limited ability to say something both new and appropriate with respect to a topic under discussion. Their poor conversational skills convey to the normal listener that they simply do not have the essential human stuff of interpreted experience.

Tager-Flusberg (this volume, Chapter 7) must have something like this in mind when she proposes that autistic speakers do not have the idea that a conversation ought to entail an exchange of information. Loveland and Tunali (this volume, Chapter 12) lament the limited game-information offered by an autistic subject even while he is seated in front of a game that one knows he can play. Yet he can explain the game thoroughly when given suitable prompts. Finally, the description given above fits with Sigman's and my 'computational' surmise that we came to in trying to understand the unusual effort required by her very high-functioning autistic subjects in responding to questions about the affective states of characters in little videotaped stories. Although the answers given by Sigman's subjects were often correct, the rate, latency, and manner of approaching the questions were more like what one expects in a person trying to do a hard arithmetic computation in their head, than what one expects in a person reporting about how others might be feeling. This suggests that indeed they *do* know about affective states, but know about them in some unconventional (and non-intuitive) form that makes use of their knowledge in talk difficult (Yirmiya *et al.* 1992; see also Capps *et al.* 1992).

I proposed above that the lack of new comments in autistic conversation might reflect a deeper problem: that the autistic child has gone through life without giving his experience a requisite conventional narrative form that would make it available to be introduced into conversation. Emmy, as we have already noted, does this by telling herself little stories about what happened that day. We would expect that a lifelong failure to narrate would also leave the autistic child bereft of a normal complement of narrative skills—or perhaps they would more correctly be viewed as narrative *habits*. In the next study, we explored that possibility.

To see whether autistic children do in fact lack narrative skills, we abandoned conversation and chose a narrative task. Though successful conversation requires constant adjustments in topic and mode of discourse to the evolving communicative requirements of the interlocutor, it does not necessarily elicit the extended connected discourse of narrative, where cohesion has to be created by the speaker according to a model that serves as his or her organizational plan. These mental models can usefully be thought of as roughly corresponding to the narrative genres of literature found on offer in the speaker's own culture (Feldman 1991). Usually, there is a good deal of choice about which genre is selected, and the speaker partly exercises control over the manner in which his story is told by selection among a set of available alternatives (Bakhtin 1981). The distinctive organizational pattern of any genre is conveyed not so much by its lexicon or syntax, although these can play an important role, as by its tropes and, especially, its pragmatic markers. So, as we turn to the next study, we shall be especially interested in the pragmatic markers used by the autistic subjects in a story-telling task.

The pragmatic words that are of special interest in the present connection — those used to make a narrative text — are just beginning to be reported in the psychological literature. Among them are Emmy's problem-solving words: *but*, *so*, and *because*. Emmy also used temporal marking to give her stories sequence: *then*, *before*, *first*, *later*, *after*. A third kind of marking seems more dramatistic, conveying a sense of staging: *start*, *trouble*, *scene* (Feldman *et al.* 1994). There are more, and we'll meet them later.

In her Master's thesis, Heidi Scopinsky (1986) compared the story-telling ability of twelve normal subjects with fourteen autistic, moderately verbal subjects resident at the Anderson School. Ages ranged from 9–22 years, with a mean of seventeen years. Subjects were roughly matched for chronological age (normal controls being on average two years younger than the autistic children) and for gender, and both groups included the typical preponderance of boys, with three girls in each group. Subjects were presented four sets of six cartoon-like pictures that told a simple story. (For example, a boy plants some seeds, waters them, and picks them and gives them to his girlfriend.) In contrast with the more usual procedure, pictures were placed in line in narrative order in front of the child, one added at a time, and the child was asked to make up a story to go along with the pictures as they were added. Standard prompts were 'And now what happens'? and the like. The interviewer, in addition, prompted as freely as necessary to elicit responses from the autistic subjects. At the end of each of the stories, she gestured at the whole set of six pictures laid out before the child and asked what the whole story was, 'putting them all together'.

Autistic subjects, to start with, gave shorter and grammatically simpler responses. As one would expect for a simple narrative of action, neither group actually talked about intentional states. There were very few mentions of wants and thoughts occurring even in the control group, and no significant differences between groups. But actions can be described in a more or less intentional (that is, purposive) way (for example, *waiting* vs *sitting*), and there were many mentions of such intentional actions in both groups. As the 'theory of mind' theory would predict, autistic subjects attributed less intentionality to the actions in the pictures. The fourteen autistic subjects produced 28 responses that mentioned a character's intentional action (2 per child), to the twelve normal subjects' aggregate of 49 (about 4 per child).

There was a qualitative difference as well. Typical autistic intentional responses were: 'A boy planting seeds,' 'He made the flowers,' Put in water'. Normal subjects said: 'he's watering the seeds to make his garden grow,' and 'He's growing flowers to give the lady.' Normal responses in this task often link actions to the character's purpose in performing them, and they are much more richly marked.

Comparing the the two groups for frequency of word-use was complicated by the much larger number of words used by normal subjects. David Kalmar, in our lab, developed procedures in which raw word-frequencies generated by MacWhinney's (1989) CLAN program were first normalized as frequency per thousand. These relative frequencies for different subjects can then be compared across groups, using analysis of variance. This procedure helps us single out particular lexical markers used by our subjects to achieve narrative structuring.

When the two groups were compared for this relative frequency of use of words, 72 words were found to be significantly different at $p < .05$ or less. We shall concentrate on this set of 72 words. Since most of the words were used more often by the normal group, the few words that were used relatively more often by autistic subjects were of special interest. They were nearly all rather concrete and scene-related, and had a kind of flashy quality that made one think of 'attractive nuisances': *flowers, pool, swinging, swimming, water* and *put*. The word *put* is of special interest, and we will return to it in a moment. What is noteworthy for present purposes is that *none* of them are pragmatic markers.

In contrast, many of the pragmatic markers important to Emmy's narrative talk were used more often by the normal group: they included *just* and *really*, and the narrative markers, for problem-solving *because* and *so*, and for temporality, *next, starts, then*, and *while*. Indeed, except for *still* and *then*, this whole collection of pragmatic markers was virtually never used by the autistic group.

Of further interest for the narrative deficit hypothesis of autism was the greater usage of words for intentional action in the normal group—*brings, comes, goes, join, leave, look, offers, pushes*. The autistic subjects (as noted) used *put* more often, and they used it to convey a local purposiveness limited to the particular action that a character in a picture performs at a moment when he has something in hand. Hence, though intentional, *put* is a word that can be used to describe intentional action only in an impoverished way. Autistic subjects, then, seem particularly lacking in the causal, temporal, and intentional pragmatic markers needed for story-making.

Though these findings seemed to support the notion that there is a specific lack of narratorial skill in autism, they were at best suggestive, since this was indeed a low-functioning group. Indeed, both Tager-Flusberg (1981*b*) and Paul and Cohen (1984) had earlier noted that pragmatic aspects of language were reduced in autism. So we decided to do an identical linguistic analysis on Renderer's high-level autistic interview data discussed earlier, using as controls three unstructured but generally autobiographical interviews with non-autistic inner-city late adolescents. Of the twenty-five words differentiating the two groups, only two were used more often by the autistic subjects (*on* and *too*), but one of these, *too*, is a simple pragmatic marker, as in 'Some things are too hard for me' and 'I'm getting too old for school now.' Control subjects, on the other hand, more often used the pragmatic markers *but*, *just*, and *started*, that we encountered above. Consistently with our hypothesis, they also used more mental verbs (particularly *knew*). Just as in the story-telling data described above, these interview data are less pragmatically rich than those from the interviews with normal subjects, and indeed, generally less narrativized, though the differences between the two groups, when this high-functioning group is considered, are generally less extreme.

If the nature of the task affects the kind of talk produced by autistic subjects, then the form of talk used by our autistic subjects should differ across our tasks. Although task was confounded with subjects' level of functioning, we could nevertheless explore the particular nature of the non-narrative form each group gave to its particular task, and see whether on balance they had been affected by the task demands. This would tell us whether autistic subjects perhaps had a general ability to select a genre from the range of genres available to them, and something about what that range of genres was. We already knew from the low production of pragmatic devices that these genres, if any, would be non-narrative. The question remaining was what form they would take.

We ran the same analysis as before, this time comparing word-usage in the two autistic groups. What we found when we examined the word-usage distinctive of each group was that the two groups were using distinctively different genres. At the very least we could say that their verbal productions were, in fact, responsive to the nature of the task.

The interviews, in contrast to the story task, honoured the conventions of dialogue by exhibiting greater use of *I* and *you*, while the story-telling task seemed to evoke, in lieu of narration, a genre of *description*, for its more frequent words mainly concerned specific elements in the drawings used to evoke stories — *beach*, *castle*, *clouds*, *flowers*, *girl*, *plants*, *pool*, *rain*, *sand*, *swing*, and *water*. This, in effect, is descriptive rather than story-telling talk, the *content* of the presented story with the narration left out. Why had the autistic subjects used descriptive rather than story-telling talk? Was it due to a misunderstanding of the task, specifically a failure to understand the genre of talk it asked them for, or could it be that they lacked a specifically narrative skill either (a) in understanding the narrational element in the presented pictures or (b) in telling the story in a narrated manner?

The last study I want to report is still in its preliminary stages, but so far it suggests that high-level subjects at least can understand stories, but cannot tell them; they have some narrative resources, but these do not add up to a productive narrative genre. Marian Sigman, Lisa Capps, and I selected four folk-tales of deception to present to high-functioning autistic subjects, including 'The boy who cried wolf' and three other less familiar ones: 'The Navaho Blanket', wherein a Mohawk boy gets away with stealing a Navaho horse out of the camp in broad daylight by dressing up as a Navaho and pretending he is supposed to take the horse; 'The Wolf and the Donkey', wherein a donkey keeps a wolf from eating him by pretending he has a thorn in his foot, and by then kicking the wolf when he comes closer to examine the donkey's foot and running away; and 'The Sheep and the Goat', wherein two harmless animals fool three wolves into thinking they are wolf-killers by showing them a wolf's head that they have actually just found on the road. The picture story is called 'The Silver Pony': a farm boy sees a winged horse, and when he tells his father, who has come to the spot to see it, he is spanked because the horse is not there. (*All* our subjects said it was for lying.) The horse returns later, and the boy gets on his back and flies over his house to show his father (as one subject said 'try to show his folks there really is a horse like that').

The four very high-functioning autistic fifteen-year-old boys who served as subjects all had an average or greater than average WISC IQ. The stories were read to them, and the readings were interrupted to ask 'Now what do you think is going to happen?', and after they had responded, 'Why do you think that?' Finally, they were asked at the end what had happened, and then requested to tell the story back in their own words.

They were indeed high-level subjects, and all of them seemed to have had considerable exposure to stories. Two of them, for example, recognized the winged horse as a 'Pegasus'. Stories and story-types were familiar to them, and even the task of telling back. Indeed one child was both familiar and uncomfortable with the process of telling back: 'I don't usually like that . . . don't feel like telling any story in my own words.' And their performance surely suggests that these very high-functioning autistic subjects understand stories much of the time. Indeed, they even understand acts of deception and trickery in stories* and their likely effects ('the women thought that he knew the owner of the horse, so they didn't do anything'). If such matters are evidence for anyone's having a 'theory of mind', then

---

* But see Sodian and Frith (this volume, Chapter 8) for a contrary position.

these children have a theory of mind. Presumably, these children would fall into the minority of subjects with autism who can pass the false belief test (see Baron-Cohen, this volume, Chapter 4). They seem quite able to infer such mental states as deceptions, and even false attributions of deception, and the desire of a protagonist to set the record straight about his veracity. They get these matters from both prose and (even better) picture presentations. And they report them most vividly when answering the experimenter's interpretative questions while, and immediately after, the story is read.

But when they tell the story back 'in their own words', the deception they have just been explaining often either vanishes or is reduced to just another event. *They do not organize their retelling around it.* It is as if they cannot use in their own telling what they have previously recognized in the story, or as if they simply cannot make up a story for themselves — for every retelling is a making of some kind. Or, to put it another way, it is as if the narrative illocution of 'telling a story' were not in their productive repertoire. This is so despite the fact that they do produce some essential pragmatic markers that we have not seen in any of the other autistic data — for example, they use the basic narrative marker *then* often and correctly.

Their retellings preserve a good many events in the story, and even preserve their proper sequence; but they are strangely unweighted, as if without regard for the importance of events in motivating action in the plot. In this sense, their retellings *report* a sequence of actions or events seen in a story, rather than *composing* something that is a story itself. There seems to be no *act of story-telling* in their retellings. This is similar to the way the lower-functioning subjects in the Scopinsky study converted the story-telling task to one of describing what happened in the pictures. The present study strongly suggests that this was not due to her subjects' low level of functioning or to the way her particular *task* reduced the interpretative element of narrative. For as we see in the present study, the tendency to *describe* rather than to *narrate* is present even when the stories are full of implications of consequences that invite interpretation, and the subjects are very high-functioning.

On the other hand, as we noted above, they conduct an interesting conversation when asked about the stories. Would this not imply skill sufficient to manage everyday conversation in a fashion that would make ordinary social life possible? After all, it is probably not essential to be able to *tell* a story to enter into everyday talk (or is it?). But that is a question for another time. Or is the task format actually masking a second problem, an encoding problem, that would show up in everyday life: could their interesting answers to the interpretative questions depend on having the text and the questions right at hand — the questions to evoke an interpretative consideration of the *already* narrationally encoded events in the text?

One of the subjects was asked to chat a little after his brilliant performance on the interpretative questions. Nothing comes to mind. Nothing that happened in the last few weeks stands out. Then he says, 'Well, I do remember going to see movie X.' Interviewer prompts 'And that was . . . ?', and he tells when. The conversation limps along like this until she asks 'Did you like it?' Subject says 'Yeah. All right.' Terminal, punctate, two-to-eight-word responses appear in a transcript that had been filled, at least during the more conversationally scaffolded series of questions that preceded the retelling, with long paragraphs, complex grammar, and subtle insight. What was the difference between these two conversations?

I speculate that the story's narrative frame helped subjects to interpret events in the story they had been told immediately beforehand, aided of course by the interviewer's interpretative questions and the conversational setting. Loveland and Tunali (this volume, Chapter 12) similarly note of their subject who could not describe a board-game he knew how to play, that he did succeed with suitable prompts. Perhaps to be *high-functioning* is just exactly to understand enough about stories and questions and conversation to be able to make use of such support to bootstrap oneself into creating encoded narrative meanings that can be shared in conversation. At least for the duration of the task, they *do* have something to say, something that an interlocutor can understand, respond to, and even find interesting.

Even with such a prior scaffold these events can later only be told back as descriptions, but cannot be remade into a story; and without such scaffolding, conversations about the unnarrated events of ordinary life lack the richness of interpretation and even of report that appears in the scaffolded task context.

Let me sum up now. I have tried through a series of studies to discover specifically whether narratively-related language-production abilities might be undeveloped in autistic speakers, and which specific ones might be involved. It appears that general discourse skills, especially turn-taking and topic-maintenance, are not at issue, but that there is a particular problem in the development of new comments. I speculated that this has something to do with knowledge being held in a form that has not been encoded in a semiotic fashion that makes it accessible in the course of conversation. I suggested that this might be due to an early absence in the autistic child of the normal impulse to tell narratives to him- or herself or to others — an activity that ordinarily recreates raw experience into a symbolic form that is useful for entering real-life experience into discourse.

Then I suggested that early lack of an impulse to narrate might still be evident later, manifesting itself in a limited grasp of story-telling skills. This could be due both to a lack of early practice and also to a persistence of the original difficulty, whatever it is, that makes narration an unattractive or inaccessible activity in early autistic childhood. We saw evidence for this both in the picture-story narrations and in the interviews, for when compared with controls there was a real paucity of pragmatic marking. And by comparing these two autistic groups' performances we discovered a tendency to convert the story-telling task to a descriptive one.

Then we turned to a very high-functioning group to whom we read fables of trickery and deceit. There, despite a general competence in understanding narratives, there seemed to be a deficit both (1) in the ability to narrate a tale (as in a retelling) and (2) in the ability to carry on an interesting and socially appropriate conversation when unsupported by pre-narrated material to talk about provided in a context that specifically invited interpretation of it. They could not do the first in spite of an ability to use some narrative markers, and they could not do the second despite a demonstrated grasp of a 'theory of mind' when operating securely within the context of the story-interpretation task. This is all consistent with the speculation that one important deviation from normal experience in autism is the lack of generative skills of narrative organization, perhaps because they do not tell stories even to themselves.

Indeed, these same high-level subjects, when unsupported by narrative scaffolding, were unable to use the conversational skill we had seen them display when supported, finding themselves once again virtually mute. Without a narrative format for meaning-making, the autistic child cannot generate the meanings that make ordinary conversation, and social life, comfortable.

I want to close this section by raising a question. Is there a primary narrativizing impulse in the species that somehow goes awry early on in autism? We know from the work of Vivian Paley (1990), Judy Dunn (1988), Shirley Brice Heath (1983) and others how strong is the young child's impulse to tell stories. We also know that it gets a great deal of practice, and that it undergoes vast development with time, leading to real qualitative shifts in narrative patterns produced by children of different ages. Our own work (Feldman *et al*. 1994) shows big changes still going on between ten and sixteen-year-olds and even between sixteen and thirty-year-olds. And we know that narrative can frame the encoding or reworking even of 'raw' perceptual experience (Dasser *et al*. 1989). It scarcely needs an argument to claim that narrative as a semiotic form is useful for communication with others.

My conjecture, then, is that the communicative problem in autism is due to a weak or absent impulse to rework life experiences into narratives, beginning at a very young age, say two to three years, when most normal young children seem driven to encode their lives in story-talk. And that deficit continues into later life, manifesting itself even in high-functioning autistics as difficulty in telling a story and difficulty in encoding everyday experiences in story so as to give them a form that enters readily into everyday conversation. It is only somewhat mitigated by what I earlier called the 'computational' surmise — the possibility that the very high-functioning autistic individual may develop a laboured way of computing some of the intentional-state information that the narrative scaffolding available to normal people makes immediate and intuitive.

SOME CONCLUSIONS

We turn finally to the dual issue of what light theory of mind theories cast on the nature of autism and, equally interesting, what our knowledge of autism contributes to an understanding of how a theory of mind operates in human development.

With respect to the first, it seems difficult to make the case that it is exclusively a deficit in the child's capacity to 'know' others' mental states that produces or, even more strongly, 'constitutes' the autistic syndrome. Rather, it may well be that this deficit, the importance of which we are not questioning, is a *consequence* of a more pervasive (and perhaps ultimately even more maladaptive) deficit in general cognitive functioning. This more pervasive deficit, to which we turn next, has the developmental effect of progressively depriving the growing child of the support of those cultural 'prostheses' that make it possible for him or her to create the meanings with which to participate in ordinary social life.

How may we characterize this more general condition? It consists, we believe, of a sharply reduced ability to encode experience (or create meaning) in a narrative mode. The autistic infant is incapable (or highly resistant to) organizing interpersonal encounters into a canonical form that captures diachronic regularities in the way in which people's intentional states are situated in typical situations, are expressed in typical action sequences, and require reciprocal response from those who are interacting with them. Earlier in the chapter, we referred to these canonical sequences as 'formats', and noted the extent to which they are dependent upon a grasp by both infant and caregiver of intentional states in others. We use the word 'grasp' with a certain reluctance, for we do not mean to imply that either the infant or the caregiver need have 'accurate' or even conscious ways of knowing the other's mental state, but only, rather, *consistent* ways.

Here we must return for a moment to Meltzoff's (1990) account. Recall that he noted that both infant and caregiver are capable of recognizing not only reactions in the other that 'match' their own but that are contingent on their own (Ekman 1984; Meltzoff and Moore 1989). It would be theoretically extravagant (and quite unnecessary) to assume on the basis of this reciprocity that the infant therefore has a 'theory' of mind, any more than it is necessary to assume that the infant has a 'theory' of space by virtue of being able accurately to direct his reach at, say, five months, to an object at some distance removed. We do well, in this connection, to bear in mind

Quine's (1981) distinction between behaviour that *conforms* to a rule and behaviour that is guided by voluntarily *following* a rule, or the similar distinction made by Feldman and Toulmin (1976) with respect to whether Piagetian formal operational thinking 'follows' the logical rules of the propositional calculus or can simply be described by a psychologist as conforming to those rules. All that we need recognize is that, under ordinary conditions, a failure to conform to certain behavioural regularities may have striking consequences for later development — one of which may be serious difficulty in later being able to develop a conscious and voluntary form of rule-following behaviour in the domain affected by the deficit.

In the case of the autistic infant, there is ample evidence that he or she cannot or will not, from early on, enter into the interactive formats through which normal infants learn about canonical patterns of social behaviour in his or her immediate cultural milieu. We think this derives, as has already been noted, from the infant's failure to attend to or register extended sequences of interaction involving another human being. Formatted play sequences that are segmented in terms of such intentional states as surprise, restitution of a previously disrupted canonical state, etc., neither attract nor ordinarily delight the autistic infant. Part of this autistic deficit, to be sure, may possibly be an unwillingness or impaired ability to attribute intentional states to others or, in Meltzoff's sense, to recognize matching or contingent intentional states in others. But that cannot be the whole story, for there is more to be accounted for than that.

In some of the observations reported in the previous section, it was plain that some autistic children could *follow* narrative sequences, even those involving intentional states, when such narratives were well scaffolded by pictures and stories told by others. But as soon as it became necessary for them to put these narratives into their own words off their own cognitive bats, as it were, they would fail. What seems to be lacking, therefore, is some *productive* capacity to organize narratively a set of punctate events, even those that have a putatively easy narrative structure involving little attribution of intentionality (like the cartoon panels in the Scopinsky study). This leads us to believe that the deficit in question is not simply a failure to develop and use a theory of mind to interpret encounters with social events, but rather a more general semiotic deficit for organizing narratives generally.*

And that in turn leads us to the conclusion — and we realize it can only be a tentative one, given the scantiness of evidence — that the autistic child is rendered bereft of the canonical cultural expectations that are required for ordinary social interaction. For, without a narrative bent, the child does

---

* Whether this equates with what Frith (1989) calls a lack of drive for 'central coherence' remains to be explored.

not organize his or her own experience into the canonical forms, the 'folk psychology', by which people regulate their sociality (Bruner 1990).

This leads finally to the Sigman–Feldman 'computational surmise'— that high-functioning autistic children convert the personal world of intention-regulated social experience into an impersonal world of causally-driven events. Part of the 'strangeness' of the high-functioning autistic person, those characterized as suffering from Asperger's syndrome, is that their 'adequate' interpersonal functioning gives the impression of their doing arithmetic problems or figuring out cause–effect problems rather than operating intuitively with notions about how people ordinarily feel, think, believe in standard situations. Given sufficient time to brood about it (the kind of time ordinary sociality will not accommodate), they will come up with 'adequate' but bizarre interpretations of personal matters—as is so notably the case with the Asperger subjects whose autobiographies Happe (1991) has analysed. We think that this interesting and powerful compensatory manœuvre on the part of high-level autistic subjects (a blessing though it may be for them) may have led to a strangeness of talking and of manner that gives normal interlocutors a disturbing sense that they are interacting with someone who is, as it were, outside the culture.

Finally, given the picture we have tried to present, there seems to be no deep conflict between the two principal competing views of autism—the one that sees it principally as an affective disorder (Hobson 1991) and the one that sees it as a failure in the development of a theory of mind (Baron–Cohen *et al.* 1985). On our account, both views are consequences of a deeper deficit in the autistic person: a failure to be able to organize the narrative structures through which culture organizes its expectations about how people feel, think, and believe in typical extended situations in order to create conventional meanings communicable to other people who share the same culture.

Acknowledgement

This study was made possible by a grant to the authors at New York University by the Spencer Foundation.

REFERENCES

Bakhtin, M. (1981). *The dialogic imagination: four essays*. University of Texas Press, Austin.

Baron–Cohen, S., Leslie, A., and Frith, U. (1985). Does the autistic child have a theory of mind? *Cognition*, **21**, 37–46.

Bruner, J. (1978). Learning how to do things with words. In *Human growth and development* (ed. J.S. Bruner and A. Garton). Clarendon Press, Oxford.

Bruner, J. (1982). The organization of action and the nature of adult–infant transaction. In *The analysis of action: recent theoretical and empirical advances* (ed. M. von Cranach and R. Harré). Cambridge University Press.

Bruner, J. (1983*a*). *Child's talk: learning to use language*. Norton, New York.

Bruner, J. (1983*b*). Play, thought, and language. *Peabody Journal of Education*, **60**, 60–9.

Bruner, J. (1990). *Acts of meaning*. Harvard University Press, Cambridge, Mass.

Bruner, J., and Lucariello, J. (1989). Monologue as narrative recreation of the world. In *Narratives from the crib* (ed. K. Nelson). Harvard University Press, Cambridge, Mass.

Bruner, J., Feldman, C.F., and Kalmar, D. (1991). *Time, plight, and meaning*. A paper presented to the Piaget Society Twenty-first Annual Symposium, Philadelphia, PA.

Butterworth, G., and Castillo, M. (1976). Coordination of auditory and visual space in newborn human infants. *Perception*, **5**, 155–60.

Capps, L., Yirmiya, N., and Sigman, M. (1992). Understanding of simple and complex emotions in non-retarded children with autism. *Journal of Child Psychology and Psychiatry*, **33**, 1169–82.

Dasser, V., Ulbaek, I., and Premack, D. (1989). The perception of intention. *Science*, **243**, 365–7.

Dore, J. (1974). A pragmatic description of early language development. *Journal of Psycholinguistic Research*, **3**, 343–50.

Dunn, J. (1988). *The beginnings of social understanding*. Harvard University Press, Cambridge, Mass.

Ekman, P. (1984). Expression and the nature of emotion. In *Approaches to emotion* (ed. K. Scherer and P. Ekman). Erlbaum, Hillsdale, NJ.

Ekman, P., Levenson, R.W., and Friesen, W.V. (1983). Autonomic nervous system activity distinguishes among emotions. *Science*, **21**, 1208–10.

Feldman, C.F. (1989). Monologue as problem-solving narrative. In *Narratives from the crib* (ed. K. Nelson). Harvard University Press, Cambridge, Mass.

Feldman, C.F. (1991). I generi letterari come modelli mentali [Genres as mental models]. In *Rappresentazioni e narrazioni* (ed. M. Ammaniti and D. Stern). Laterza, Rome.

Feldman, C., and Toulmin, S. (1976). Logic and the theory of mind: *Nebraska Symposium on Motivation*, pp. 409–76. University of Nebraska Press, Lincoln.

Feldman, C., Bruner, J., Kalmar, D., and Renderer, B. (1994). Plot, plight, and dramatism: interpretation at three ages. In *The nature and ontogenesis of meaning* (ed. W.F. Overton and D.S. Palermo). Lawrence Erlbaum, Hillsdale, NJ.

Fernald, A., and Kuhl, P.K. (1987). Acoustic determinants of infant preference for motherese speech. *Infant Behavior and Development*, **10**, 279–93.

Frith, U. (1989). *Autism: explaining the enigma*. Blackwell, Oxford.

Gallup, G. (1970). Chimpanzees: self-recognition. *Science*, **167**, 86–7.

Happé, F.G.E. (1991). The autobiographical writings of three Asperger syndrome adults: problems of interpretation and implications for theory. In *Autism and Asperger syndrome* (ed. U. Frith). Cambridge University Press.

Heath, S.B. (1983). *Ways with words: language, life, and work in communities and classrooms*. Cambridge University Press.

Hobson, R.P. (1991). Against the theory of 'Theory of Mind.' *British Journal of Developmental Psychology*, **9**, 33–51.

Humphrey, N. (1986). *The inner eye*. Faber and Faber, Boston.

Keller, H.A. (1928). *The story of my life, by Helen Keller*. Houghton Mifflin, Boston.

Lazarus, R.S. (1981). The cognitivist's reply to Zajonc on emotion and cognition. *American Psychologist*, **26**, 222–3.

Leslie, A.M. (1987). Pretense and representation: the origins of 'theory of mind'. *Psychological Review*, **94**, 412–26.

Lewin, R. (1989). *Human evolution: an illustrated introduction*. Blackwell Scientific, Boston.

Loveland, K.A. (1990). Social affordances and interaction II: Autism and the affordances of the human environment. *Ecological Psychology*, **3**, 99–119.

Loveland, K.A. Autism, affordances, and the self. In *The perceived self: ecological and interpersonal sources of self-knowledge* (ed. U. Neisser). Cambridge University Press. (In press).

Loveland, K.A., Landry, S.H., Hughes, S.O., Hall, S.K., and McEvoy, R.E. (1988). Speech acts and the pragmatic deficits of autism. *Journal of Speech and Hearing Research*, **31**, 593–604.

Loveland, K.S., McEvoy, R.E., Tunali, B, and Kelley, M.L. (1990). Narrative story-telling in autism and Down's syndrome. *British Journal of Developmental Psychology*, **8**, 9–23.

MacWhinney, B. (1989). *CLAN programs of the Child Language Data Exchange System* [computer program]. Carnegie Mellon University, Department of Psychology, Pittsburgh, PA.

Meltzoff, A.N. (1985). The roots of social and cognitive development: models of man's original nature. In *Social perception in infants* (ed. T.M. Field and N.A. Fox), pp. 1–30. Ablex, Norwood, NJ.

Meltzoff, A.N. (1990). Foundations for developing a concept of self: the role of imitation in relating self to other and the value of social mirroring, social modeling, and self practice in infancy. In *The self in transition: infancy to childhood* (ed. D. Cicchetti and M. Beeghly). University of Chicago Press.

Meltzoff, A.N., and Moore, M.K. (1985). Cognitive foundations and social functions of imitation and intermodal representation in infancy. In *Neonate cognition: beyond the blooming, buzzing, confusion* (ed. J. Mehler and R. Fox). Erlbaum, Hillsdale, NJ.

Meltzoff, A.N., and Moore, M.K. (1989). Imitation in newborn infants: exploring the range of gestures imitated and the underlying mechanisms. *Developmental Psychology*, **25**, 954–62.

Mundy, P., and Sigman, M. (1989). The theoretical implications of joint-attention deficits in autism. *Development and Psychopathology*, **1**, 173–84.

Nelson, K. (ed.) (1989). *Narratives from the crib*. Harvard University Press, Cambridge, Mass.

Paley, V.G. (1990). *Bad guys don't have birthdays: fantasy play at four*. University of Chicago Press.

Paul, R., and Cohen, D.J. (1984). Responses to contingent queries in adults with autistic disorders and mental retardation. *Applied Psycholinguistics*, **5**, 349–57.

Peirce, C. S. (1960). *Collected papers of Charles Sanders Peirce*, Vol. 2. Harvard University Press, Cambridge, Mass.

Pratt, C. (1978). *The socialization of crying*. Unpublished doctoral dissertation, Oxford University.

Premack, D., and Woodruff, G. (1978). Does the chimpanzee have a theory of mind? *Brain and Behavioral Sciences*, **1**, 515-26.

Quine, W. V. (1981). *Theories and things*. Harvard University Press, Cambridge, Mass.

Scaife, M., and Bruner, J. S. (1975). The capacity for joint visual attention in the infant. *Nature*, **253**, 265-6.

Scopinsky, H. (1986). *A study of narrative patterns in autistic subjects*. Unpublished Master's thesis, New York University.

Tager-Flusberg, H. (1981a). Sentence comprehension in autistic children. *Applied Psycholinguistics*, **2**, 5-24.

Tager-Flusberg, H. (1981b). On the nature of linguistic functioning in early infantile autism. *Journal of Autism and Developmental Disorders*, **11**, 45-56.

Tager-Flusberg, H. (1989a). A psycholinguistic perspective on language development in the autistic child. In *Autism: new directions in diagnosis, nature, and treatment* (ed. G. Dawson). Guilford, New York.

Tager-Flusberg, H. (1989b). *An analysis of discourse ability and internal state lexicons in a longitudinal study of autistic children*. Paper presented at the SRCD, Kansas City.

Trevarthen, C. (1987). Sharing makes sense: intersubjectivity and the making of an infant's meaning. In *Language topics: essays in honour of Michael Halliday*, Vol. 1 (ed. R. Steele and T. Threadgold). Benjamins, Amsterdam.

Urwin, C. (1980). Unpublished doctoral dissertation, Cambridge University.

Volkmar, F. R., and Cohen, D. J. (1985). The experience of infantile autism: a first-person account by Tony W. *Journal of Autism and Developmental Disorders*, **15**, 47-54.

Whiten, A., and Byrne, R. W. (1988). Tactical deception in primates. *Behavioral and Brain Sciences*, **11**, 233-73.

Wood, D. J., Wood, H., Griffiths, A., and Howarth, I. (1986). *Teaching and talking with deaf children*. Wiley, New York.

Yirmiya, N., Sigman, M. D., Kasari, C., and Mundy, P. (1992). Empathy and cognition in high-functioning children with autism. *Child Development*, **663**, 150-60.

Zajonc, R. B. (1980). Feeling and thinking: preferences need no inferences. *American Psychologist*, **35**, 151-75.

14

The complexity of social behaviour in autism

CATHERINE LORD

Findings of autistic children's deficiency in understanding how other people think and feel have been impressive in their consistency across a variety of experimental tasks and investigators. Yet, in the end, their clinical and theoretical power will be determined by the extent to which they can explain the everyday social and cognitive deficits experienced by autistic individuals. The purpose of this chapter is to review the social experience of individuals with autism. Historical and diagnostic issues relating to social behaviour in autism are reviewed by Volkmar and Klin (Chapter 3, this volume). The focus of my chapter is on social experiences and contexts for social behaviours, including sociability, relationships, and behaviours that relate to successful social functioning. My aim overall is to describe and review the development of social behaviour in autism, from the perspective of descriptive and naturalistic studies, in order to clarify the nature of the social phenomena theories have to explain. I shall conclude that there is still a wide gulf between the cognitive deficit in theory of mind, and its purported relationship to social behaviour.

SOCIAL FUNCTIONING IN CHILDREN WITH AUTISM: A REVIEW

Sociability

Sociability has been defined as a person's interest and ease in being with other people (Cohen *et al.* 1987). Social motivation, social play and imitation, other prosocial behaviours, and antisocial behaviour are addressed in this section.

Social motivation. A common clinical experience is to observe almost no initial response when an autistic child is approached by a relatively unfamiliar person in a new or complicated setting. For example, autistic children are frequently described by their parents as not responding in usual ways to the approach of a friendly stranger while queuing or standing in line at a grocery store (Le Couteur *et al.* 1989). Parents of young autistic children often describe them as ignoring even familiar visitors, and failing to respond at all to friendly overtures by anyone except close family members. In fact, acting as if deaf in response to the speech of their parents is common in very young autistic children (Ornitz *et al.* 1978). In contrast, in familiar situations such as playing simple games or eating a meal with familiar people, autistic children are generally responsive to approaches by others (Lord 1984; Mundy *et al.* 1986). Observations of how much time autistic children spend watching other people have shown few significant differences from other populations (Dawson *et al.* 1990; Hermelin and O'Connor 1970); however, studies that have measured how far away from other children autistic youngsters place themselves during unstructured play suggest that there are real differences in behaviours, such as a lack of 'hovering' and of movement toward others by autistic children, that affect the likelihood of social encounters (Lord 1990). Differences in how autistic children act towards others in familiar, structured situations, and less structured, unfamiliar contexts illustrate the complexity of social behaviour. Age and developmental differences certainly seem likely to be one factor, but have not yet been documented. We do not know if the deficits in responsiveness in particular situations are due to lack of comprehension of what is expected, to lack of understanding of the goals of the other person, or to more basic factors such as level of arousal or goal-directedness, social or otherwise (Dawson and Galpert 1986; Walters *et al.* 1990).

As we have seen, the pattern of responses can be somewhat variable; but there is no doubt that autistic children, adolescents, and young adults in familiar but unstructured social contexts make far fewer approaches than other children (Attwood *et al.* 1988). For example, in observations carried out over eleven hours of free play during an integrated daycamp, two out of eleven autistic children of average non-verbal intelligence made only one or two approaches to other children (as compared to the equivalent of more than fifty for the non-handicapped and control children and at least ten for the other autistic children), and in two of these cases, the approaches were quite odd (for example, approaching another youngster to smell her hair) (Lord and Magill 1989).

Though autistic children and adolescents spend less time interacting in familiar, but unstructured environments, it is not the case that *all* autistic individuals are *uninterested* in *all* other people. Examples are readily available, particularly of high-functioning autistic children of school age or older, who can be unusually demanding of attention from adults, though these approaches may be unusual (Volkmar *et al.* 1989). Thus lack of motivation may be a factor in the social deficit of some autistic children; but it is not universal nor specific, and is not as good a discriminator between autism and other disorders as is the quality of approaches. Furthermore, one of the striking findings of the observational studies of non-retarded, autistic children and adolescents in unstructured settings (for instance, free play or free time during social groups or daycamp) was the paucity of directed activity of *any* sort beyond very simple physical manipulations (Lord 1984). One of the difficulties in judging social motivation in autistic individuals is that, when grouped together with other people with autism, they are required to deal with the handicaps of their playmates. On the other hand, when they are observed with familiar, supportive non-handicapped peers or family members, the other participants often structure the interaction to such an extent that *any* behaviour of the autistic person is sufficient to evoke a response, and the interaction manages to continue, even though it may be far from reciprocal.

Differences become even more marked when any sort of sustained and reciprocal interaction is observed (Lord 1990). Many years ago, Ferster (1961) proposed that autistic children had a specific inability to respond to 'secondary reinforcers' or to learn from rewards other than immediately gratifying experiences such as food or physical activity. While this view is no longer widely accepted, there is evidence that outside highly structured situations and specific kinds of interactions (for example, climbing on people, smelling hair), the sheer amount of time autistic children spend in interactions with parents or peers is much reduced compared to that spent by control groups (Konstantareas and Homatidis 1989; O'Neill and Lord 1982). Overall, consistency and appropriateness of social responsiveness change much more quickly with changes in the environment than do the quality or frequency of approaches (Lord 1984; Odom and Strain 1986).

Social play and imitation. Pre-school and school-age autistic children are less adept at elicited imitation (Curcio and Piserchia 1978; Stone and Caro-Martinez 1990) and show less frequent and less varied, spontaneous imitation than various control groups (Dawson and Adams 1984; Hertzig *et al.* 1898). In one study, 94 per cent of autistic children were described by their parents as having little or no spontaneous imitation, compared to 9 per cent of IQ-matched mentally handicapped children (Le Couteur *et al.* 1989). Parents of autistic children consistently describe them as lacking the elicited imitation skills typically shown by normally developing children at one year of age (Klin *et al.* 1992).

There is some evidence that in autism the ability to imitate correlates with social skills (Dawson and Adams 1984; DeMyer *et al.* 1972), but not with object permanence (Curcio and Piserchia 1978). Deficits in reciprocal social play, such as playing peek-a-boo and pat-a-cake, also seen in non-handicapped children under twelve months of age, are routinely described by almost all parents of autistic children (Klin *et al.* 1992; Van Berckelaer-Onnes 1983), though they have often not been found in experimental studies with older children in highly structured settings (Mundy *et al.* 1986). These differences may be developmental as much as methodological.

In another study, a turn-taking task involving completing a dot-to-dot picture discriminated mildly to moderately retarded autistic children from language-matched mentally-handicapped children, but did not discriminate high-functioning autistic children from non-handicapped controls. In contrast, the ability to use miniatures to play reciprocally with the examiner discriminated autistic students at all intellectual levels from non-autistic children and adolescents on an individual basis (Lord *et al.* 1989), suggesting that at least some social skills, such as simple turn-taking, may improve with development in autism (Baron–Cohen, this volume, Chapter 4), but others may not, or may improve more slowly (Harris 1991).

Deficits in imitation exist in autistic infants and toddlers long before traditionally conceived metarepresentational abilities can be tested (Klin *et al.* 1992; Rogers and Pennington 1991). Recently, these difficulties in imitation have been proposed as primary to autism, as beginning a path of disordered development that leads to deficits in both metarepresentation and affective relations (Rogers and Pennington 1991). Clinical reports of a lack of imitation of unintentional non-verbal behaviors, such as standing in the same posture as a parent, suggest that these problems are apparent below the level of metarepresentation (Le Couteur *et al.* 1989). However, whether or not autistic children show very early imitation is not known. A significant minority of parents report evidence of imitation during infancy in autistic children who are later quite unable to reproduce other's actions upon demand (Lord 1991). The question of the role imitation plays in autism is discussed at length by Meltzoff and Gopnik (this volume, Chapter 16).

Prosocial behaviour. Autistic children have traditionally been described as lacking almost all prosocial behaviours as youngsters, including giving, sharing, helping, offering comfort, offering affection, greeting others, and responding to humour (Dunn 1991 Ohta *et al.* 1987). Sometimes autistic children are considered by their parents to be affectionate, probably because affection is assumed to refer to any physical approach or desire for physical contact (Stone and Lemanek 1990). However, when questioned in detail in an investigator-based interview, 88 percent of parents of school-age youngsters described their autistic children as not showing affection spontaneously, compared to 7 per cent of parents of mentally handicapped children (Le Couteur *et al.* 1989). Similar results were found in comparisons of descriptions of affection on the Vineland Adaptive Behaviour Scale (Sparrow *et al.* 1984) for parents of autistic and mentally handicapped pre-school and early-school-age children (Klin *et al.* 1992; Volkmar and Klin, this volume, Chapter 3.).

In addition, as they get older, some autistic children become more consistent and willing to follow predictable social routines in highly routinized contexts, such as helping clean up, than other children (Lord *et al.* 1989). At the same time, they remain much more limited in other prosocial behaviours. In one study of greetings made by autistic children and adolescents to daycamp counsellors (Lord and Magill 1989), the autistic children either greeted the counsellor immediately upon seeing her, without waiting for her to look at them first, or did not greet her until she made a very clear overture to them; in contrast to the normal and language-delayed control groups, who were more likely to enter a room, look at the counsellor, wait for her to look at them, and then greet her. Autistic children in this study were also more likely to produce conventional greetings, such as saying 'Hi' or 'Hello', than the other children, who used a variety of statements and questions to acknowledge their counselors.

In a recent study of two-year-olds referred for possible autism, the absence of many prosocial behaviours at the age of two did not discriminate children who were judged to be autistic at the age of three from those who were not (Lord 1991). As is shown in Table 14.1, autistic children at the age of two did not differ significantly, according to their parents' descriptions at that time, in their likelihood of offering comfort or coming for comfort, or in spontaneous imitation, though they were significantly less likely to greet, share their enjoyment, and join in with others' expressions of pleasure than children who were not autistic. The lack of discrimination for these particular behaviours occurred because many of the non-autistic mentally handicapped children did not *consistently* show the relevant prosocial behaviours at the age of two; not because the autistic children did not show deficits. Even when analyses of individual data were not significant, group differences in mean scores often were reliable, because autistic children were more likely to show no, or almost no examples of the behaviour, compared to the non-autistic, mentally handicapped control children, who often showed some, but inconsistent, examples.

Thus, it is not that autistic children do *not* lack early prosocial behaviours, developing obvious abnormalities only when they fail to acquire metarepresentational abilities, but rather that evaluating the presence or absence of prosocial abilities in mentally handicapped children functioning at the infant or toddler level is a very complex task for both parents and professionals. Studying prosocial behaviours in contrived situations can be difficult, so much of the research available has relied on parental report. Yet the contradictory findings concerning affection reported above illustrate how careful one must be in working with parents to acquire information, rather than asking them to make judgements about these behaviours in their children.

Even in normally-developing toddlers, prosocial behaviours, though they certainly exist, are more the exception than the rule outside routines (Radke-Yarrow *et al.* 1983). Consistency in any behaviour — for example, waving goodbye or comforting a baby sister — is not necessarily found in normally developing toddlers. What may be different for the autistic

Table 14.1 Items discriminating autistic from non-autistic two-year-olds

Very reliable items (p < .01)
   Greeting
   Share own enjoyment and join in other's enjoyment
   Coordination of gaze and other behaviours in requesting
   Joint attention
   Use of another's hand as a tool
   Referential pointing
   Understanding gesture
   Hand and finger mannerisms

Reliable items (p < .05)
   Spontaneous imitation
   Coming for comfort
   Imaginative play
   Instrumental gesture
   Arms up to be lifted at twelve months
   Range of facial expression

Items without significant differences
   Affection
   Offering comfort
   Separation anxiety (current and at twelve months)
   Discrimination of parent
   Direct gaze (current and at twelve months)
   Social play
   Arms up to be lifted
   Smiling (current and at twelve months)
   Inappropriate facial expression
   Difficulties with change
   Compulsions and rituals
   Unusual attachments
   Abnormal idiosyncratic negative reactions

All analyses used Fisher Exact tests to compare scores from the Autism Diagnostic Interview — Revised Version (Lord *et al.* 1992) for 16 children diagnosed as autistic at the age of three with those of 13 children referred for autism at two years but not diagnosed as autistic at the age of three.

children is that social behaviours continue to be developed *as part* of routines, whereas for non-autistic children, spontaneous prosocial behaviours develop *out of* the routines. The spontaneous behaviours of non-handicapped children are truly responsive, at least some of the time, to the immediate needs of other people, but may remain less consistent in some ways than would more routinized responses to familiar situations, such as appear in the greetings study.

Sharing in infants has been described as primarily serving and arising out of the desire to continue an interaction (Breese and Caplan, in press). It is only later in development that it comes to represent a judicial decision to give up part of one's possessions. Three-year-olds with autism were described by their parents as much less likely to share than language-age-matched mentally handicapped children (Lord 1991; Lord *et al.* in preparation); however, no differentiation between sharing to continue an interaction or sharing to be fair was possible.

Prosocial behaviours are most evident in non-handicapped children in situations of threatened or heightened self-interest that are very much tied to affect (Dunn 1991). It is interesting that the prosocial behaviours that best discriminated autistic children at the age of two all involved communication of affect; greeting, seeking to share enjoyment of an event, and responding to others' indications of pleasure; more so than other prosocial behaviours that became significantly different only for the children at the age of three. The early lack of everyday prosocial behaviours in autistic children could be related to reduced or different affective responses or to a lack of interest or goal-directedness or to a lack of attention to social situations. Many prosocial behaviours seem likely to be learned through imitation, as is shown by the finding that the ways in which children offer comfort are similar to those they have seen at home (Zahn-Waxler and Radke-Yarrow 1982). Thus, deficits in imitation could have specific, as well as general, effects on the development of prosocial behaviours in young autistic children (Rogers and Pennington 1991).

Antisocial behaviour. Clinical descriptions of autistic children and adults abound with examples of their lack of most antisocial behaviours, particularly those that require deceit or some form of misrepresentation. One recent study found that autistic children had much more difficulty than other children learning how to play a game that required them to deceive their opponent than did other children (Baron–Cohen, this volume, Chapter 4; see also Sodian and Frith, this volume, Chapter 8). Many parents describe their older autistic children as being able to tease, but often this seems to mean that they are aware of a parent's attention when they do something they have been told not to do. Whether this means that they are actually aware of their parents' feelings or thoughts toward their actions, rather than simply anticipating a negative response, is not clear. This distinction is important because teasing and coping with teasing are good examples of how children learn the beginnings of social understanding in many families (Dunn 1991).

Individuals with autism at any age are rarely deliberately cruel in what they say, even to the extent possible given limited language. However, some autistic children experiment with hurting animals, for example, or make very blunt negative statements to or about people. These behaviours have generally been assumed not to have malicious intent, but rather to indicate a lack of understanding of how others, human or animal, feel. Whether this lack of empathy, even for animals, is mediated by lack of a theory of mind is not clear.

Aggression and tantrums are a problem for many young children with autism, and for a significant minority of adolescents and adults as well. In young children, aggressive behaviours are often directed toward care-givers, and can be unpredictable and difficult to interpret. Biting, pinching, and hitting in the pre-school autistic child may most easily be compared to the infant who bites his mother when overexcited or frustrated. Aggressive behaviours often decrease markedly with the beginnings of communication, and when goals and expectations are made developmentally appropriate (Schopler and Reichler 1971). Aggression in adolescents and adults, partic-ularly to people other than parents, tends to occur in autistic persons who are also severely mentally handicapped, though not always (Gillberg 1990). What provokes the aggression is not well understood, but it seldom seems to be purposeful in the sense of frightening or harming someone in order to accomplish a goal. Case studies of adults with Asperger's syndrome have suggested that violent behaviour (Mawson *et al.* 1985) and difficulties with the law can occur, linking these difficulties with the inability to understand the meaning of an act as others may see it (Wing 1981), though generaliza-tions from these case studies have also recently been called in question (Ghaziuddin *et al.* 1991).

Attachment and relationships

Attachment. Traditionally, one of the most notable characteristics of autism described in the popular literature has been the lack of attachment felt by autistic youngsters for their parents. Interview studies have consis-tently shown that most parents of autistic children feel that the attachments of their children to them are different from those of other children of the same age (Ohta *et al.* 1987). In one parent-report study (Le Couteur *et al.* 1989), parents of 73 per cent of autistic children, as against no parents of mentally handicapped children, described their children as failing to go through typical phases of separation or stranger anxiety. However, it is also interesting to note that, in two studies, most parents of pre-school children did not, in a questionnaire or in an interview, *spontaneously* express concern about their children's relationship to them, though detailed ques-tioning revealed that a minority were in fact distressed by what they saw as a lack of attachment (Lord 1991; Stone and Lemanek 1990).

Several recent laboratory studies have provided insight into these rather conflicting results. Across a number of different studies, autistic pre-school and early-school-age children have, on the whole, shown quite consistent responses to the departure of a parent in the strange-situation paradigm

(Rogers *et al*. 1991; Shapiro *et al*. 1987). Autistic children in these studies tended to direct more behaviour to their parent than to a stranger, and to show clear behaviour changes (such as a halt in crying) upon reunion (Rogers *et al*. 1991; Sigman and Mundy 1989). All these effects were smaller, however, for the autistic children than for variously matched control groups.

Most important, however, were qualitative differences in how the autistic children behaved when reunited with their parent. They were less likely to talk, or show or share an object, than other children (Sigman and Mundy 1989). In other contexts, autistic children show similar differences in how they interact with their parents, particularly in the ways in which they direct affect towards them (Dawson *et al*. 1990; Yirmiya *et al*. 1989). These differences in turn probably have real effects on how parents behave with their children, making an aberrant cycle. One of the difficulties in easily interpreting the laboratory studies of attachment is that the paradigm used was developed for younger children than it has been possible to study in the field of autism. Similarly, because of the age at which most autistic children are identified, parental accounts of attachment during the early years have generally been studied through retrospective accounts, so that we do not have detailed information of how attachments develop or fail to develop in very young autistic children. At the age of two, current separation anxiety and separation anxiety at twelve months of age (scored retrospectively) did not discriminate autistic from clinic-referred non-autistic children, because so many of the parents of the non-autistic children also felt that there was something not quite right about the ways in which their children reacted to separation (Lord 1991). On the other hand, these were the same children for whom failing to greet was a clear discriminator of autism.

Several interesting findings that may have some developmental implications have also emerged concerning the relationship between attachment and other factors in autism. In one study, while autistic pre-school children's attachment behaviours were not shown to be related to representational abilities, such as play or language (Sigman and Mundy 1989), they were shown to be related to chronological and mental age in a study of somewhat higher-functioning children (Rogers *et al*. 1991). Authors of both of these studies stressed the need to consider that 'routes' to acquiring attachment for young autistic children may differ from those of normally developing infants, but that they are probably built on more than representational abilities. As for the greetings discussed earlier, autistic children may develop attachment-like behaviours *within* routines, rather than 'out of' routines (Rogers and Pennington 1991). One implication of this possibility is that autistic children's behaviours would be much more limited across contexts and people (for example, perhaps excluding fathers, other relatives, family friends) than other children's responses.

Relationships with other family members and teachers. Very little is known about autistic children's relationships with their siblings. In one study of autistic children's spontaneous communicative attempts at home and at school, tremendous variability, from frequent to almost non-existent communicative interaction, between autistic children and their siblings was found across homes (O'Neill and Lord 1982). This variability did not seem to be related in any simple way to obvious factors such as the sex of children or their ages or age-differences. The most interesting outcome of that study, which was based on a very small sample, was that the autistic children who had normally developing siblings at home were more likely to produce spontaneous peer-directed language in their categorical classrooms at school, than were only children or children who lived in a group home for autistic youngsters. Obviously, these findings require more careful and systematic study, but suggest that social experience with siblings may play a role in increasing the likelihood of spontaneous peer-directed behaviours even in different, relatively unstructured settings.

Autistic children's relationships with their teachers are another aspect of social experience that seems likely to be important, but has received relatively little systematic attention. McHale *et al.* (1981) showed that autistic children directed more spontaneous behaviour both to classmates and to adults in teacher-structured small-group activities than in less structured or directed activities. Likewise, Clark and Rutter (1981) demonstrated more social behaviour and co-operation in autistic children in highly structured activities in which the adult was deliberately intrusive in a positive way, than in less structured situations. Studies of autistic children in therapeutic classrooms in which one of the teachers' primary goals is to establish a relationship with each autistic child have reported improvements in social and communicative behaviours and decreases in anxiety-related and self-stimulatory activity (Shapiro *et al.* 1987). Many parents report their school-age and older autistic youngsters having definite attachments to their teachers; the children ask for their teachers when they are gone, or show clear evidence of pleasure when seeing them at school, though not necessarily if they meet them in an unfamiliar context (Everard 1976).

Relationships with peers. Many clinicians feel that the social deficits of autism are most obvious when children are observed with peers. In a parent-interview study, even three- and four-year-old autistic children were discriminable from non-autistic mentally handicapped and language-impaired children in terms of their lower levels of interest in other children, responses to other children (such as at a park or in a grocery store), and consistent (over several weeks only) preferences for one other child above others (Lord *et al.*, manuscript in preparation). No autistic child (out of 31) was described by his parents as being interested in other children his own age, as against all but one of the non-autistic mentally handicapped three- and four-year-olds.

School-age autistic children in special classrooms for autistic children have been shown to have few interactions with each other without specific prompts (Guralnick 1976). When placed in classrooms with normally developing children with no specific supervision, autistic children have also generally shown very limited interaction. However, minimal interventions, such as giving an autistic child an opportunity to play with a normally developing child of similar age every day for 10 minutes for two weeks, have yielded dramatic improvements in the autistic children's responsiveness to the non-handicapped peers' approaches, and increases in the time over which individual interactions are sustained (Lord and Hopkins 1986). Results in integrated social groups comprising a wide age-range from school-age children to older adolescents and adults have been similar (Lord and Magill 1989).

Increased responsiveness was hypothesized to be related in part to the autistic children's becoming familiar with a particular peer and with the entire situation (such as the room and toys). However, just as much, changes seemed to be due to the peers, even as young as seven years old, adjusting their behaviour to the needs of the autistic children for repeated non-verbal approaches using objects (in contrast to many peers' initial attempts to interact by talking or by waiting for the autistic child to approach them). In fact, when peers are deliberately taught to engage autistic children in play using these methods, the level of responsiveness displayed by the autistic children is much higher, even with an unfamiliar playmate (McEvoy *et al.* 1988). However, generalization to new playmates who have not been taught extraordinary ways of behaving is better when autistic children and their playmates are allowed time to discover these strategies without particular training (but with the physical environment carefully modulated) (Lord 1984).

Unless very specific prompting procedures are employed (Odom *et al.* 1985), autistic children and adolescents make few initiations to peers in unstructured settings (Lord 1990). Even the significant minority of autistic children described by Wing and Attwood (1987) as 'active but odd', make many fewer, odd or ordinary, initiations to other children than various control groups (Volkmar *et al.* 1989). In the study of integrated daycamps, we found that overtures to peers made by autistic children were less likely to result in an interaction that was sustained longer than 30 seconds, and were more likely to receive no at all than approaches by other children (Lord and Magill 1989). Qualitative analyses suggest that, when autistic children do approach other children, they do so in most cases (though not all) in appropriate ways, but they are less likely to use a co-ordinated combination of non-verbal and verbal behaviours in doing so (Lord and Magill 1989).

When we move away from analyses of brief interactions in very sheltered situations, the differences in peer relationships between autistic children and other children become even greater. Thus, in one study, 75 per cent of autistic children and adolescents were described by their parents as never having had a friend of any sort (Le Couteur *et al.* 1989). In one observational study of social skills of relatively high-functioning autistic children, one of the clearest discriminators between autism and language-impairment was the autistic children's inability to describe what a friend was, though many of the children named people as friends, including their mothers, siblings, and all classmates (Lord *et al.* 1989).

On the other hand, it is important to remember that peer relationships, particularly those outside special environments (such as therapeutic social groups), can be difficult for other populations besides individuals with autism. Many moderately to severely mentally handicapped persons do not really have reciprocal friendships (Fombonne, in press; Le Couteur *et al.* 1989); and peer relationships have been shown to become a greater source of difficulty for severely language-impaired young adults than they were in childhood (Rutter *et al.* 1991). Given how different from the norm the social and educational experiences are of even high-functioning autistic children integrated into regular school programmes, care needs to be taken to separate out the role of experience, both on a molecular and on a molar level, from the role of basic social–cognitive deficits. Similar, the role of mental handicap and the behaviours that may arise from it, such as greater orientation to adults or staff than peers, must not be ignored.

Behaviours and skills associated with successful social interactions

Adaptive skills. Autistic children and adults have been found to have a unique pattern of scores on measures of adaptive skills compared to other standardized tests (Loveland and Kelley 1988; Venter *et al.* 1991), primarily because of extremely low scores in socialization and less impaired, but still relatively low scores in communication, given the overall level of mental handicap (see also Volkmar and Klin, this volume, Chapter 3). At school age, the self-help skills of autistic children are stronger than those of cognitively matched children; but this difference is not sustained for adolescents and adults (Jacobson and Ackerman 1990). Autistic adults surpass mental-age-matched controls in toileting skills and skills requiring gross motor activities, but are not as competent in any other respect.

Adaptive skills need to be considered as part of social skills because they reflect individuals' level of independent functioning and their access to social opportunities. For example, the ability to use public transportation or a telephone has implications for social contacts with peers or family members. The developmental trend of decreasing self-help skills in autism in comparison to other groups might be seen as part of two separate, though related deficits that become increasingly obvious in later childhood: a lack of interest or understanding in autistic children of 'growing up' and wanting to be like adults and older children (although this sense can develop eventually in older, higher-functioning, children and adolescents); and a lack of concern, or at least a lack of consistent concern, about social acceptability.

Play. Autistic children engage in less frequent functional play than matched controls (Lewis and Boucher 1988; Mundy *et al.* 1986), though it has been suggested that autism-specific deficits in play lie primarily in symbolic use of objects (Baron–Cohen 1987). Several investigators have suggested that the absence of early 'imaginative' play is one of the strongest indicators of autism in very young children (Baron–Cohen and Howlin, this volume, Chapter 21; Dahlgren and Gillberg 1989), although in our study of two-year-olds, deficits in imaginative play were not specific to autistic children until they had some language (Lord 1991; Lord *et al.*, manuscript in preparation), because of the high rate of delay in non-verbal non-autistic children.

Symbolic play has also been shown to be both less frequent and less complex for autistic children than other groups (Baron–Cohen 1987; Sherman *et al.* 1983). A lack of shared pretence has also been highlighted as being particularly indicative of autism (Leslie 1987), though opinions differ as to whether this is because of generally increased cognitive demands or specific deficits in metarepresentation (Harris 1991; Leslie 1987). Mundy, Sigman, and colleagues (Mundy, *et al.* 1987) found that the frequency of symbolic play in autistic children did not correlate with measures of joint attention, suggesting that the same social–cognitive deficit was not the source of both areas of difficulty. On the other hand, the frequency of autistic children's interactions with a non-handicapped peer in a dyad was found to be positively related to the amount of functional and/or symbolic play, which in turn was related to the amount of structure inherent in the materials (Dewey *et al.* 1988). This finding supports the importance of objects in buttressing 'scripts' that may underlie social interactions, even in older school-age children and adolescents with autism.

One of the most interesting results of our research concerning the impact of peer-based social interventions was an improvement in play (generally, in functional use of objects) in samples that consisted mostly of moderately retarded autistic youngsters of early school age, though the same phenomenon occurred even in high-functioning adolescents (Lord 1984; Lord 1990). Thus, simple activities with objects, observed initially in the autistic children only in structured interactions with non-handicapped age-mates, generalized to interactions with new non-handicapped children and with autistic classmates. Pretend-play sequences learned in role-play games generalized to free play with non-handicapped and autistic students. Dunn (1991) has emphasized the importance of play in the expression of intimacy and affection in non-handicapped youngsters. This is another example in which basic cognitive deficits may be one of the first components of a handicap that places the autistic child increasingly outside opportunities for social learning and experience in the form of reciprocal play and the expression of emotion and relationships within that context.

THE DEVELOPMENT OF SOCIAL BEHAVIOURS IN AUTISM

So far, the focus of this chapter has been on everyday contexts for social experience. However, another important perspective is the *acquisition* of social behaviours in autism, and the extent to which the theory of mind hypothesis or other theories emphasizing deficits in understanding persons can explain the pattern of development. There are many clinical accounts of changes with time in the social behaviours of autistic individuals, but few empirical investigations. The approach taken below is therefore to trace descriptions of specific social behaviours gleaned from various studies across developmental pathways. With a few notable exceptions (Mundy and Sigman 1989; Rogers *et al*. 1991; Snow *et al*. 1987), mildly to moderately retarded school-age to adolescent autistic youngsters constitute the bulk of samples in theoretically oriented studies. Relatively little is known about autistic children under age $3\frac{1}{2}$–4 years, autistic adolescents, adults, or profoundly retarded autistic individuals. The literature on high-functioning autistic children is rapidly increasing, but also remains smaller than that concerning the 'core' group of moderately to mildly handicapped school-age youngsters. Below, information from disparate studies across different age and IQ groupings is put together to describe some hypothetical pathways in social development.

Communicative use of gaze

As older pre-school-age children, autistic children have difficulty in following pointing (Loveland and Landry 1986; Mundy *et al*. 1986) and in using gaze to attract attention, but not in requesting (Dawson and Galpert 1986). Some studies have found no differences in the time autistic children spend looking at other people (Dawson and Adams 1984; Sigman and Mundy 1989). One possibility is that autistic children look at others for as much time as control children, but that their gazes are more brief (Hermelin and O'Connor 1970) and less well coordinated with both where others are looking (Lord and Magill 1989) and what the autistic child is doing (Mirenda *et al*. 1983).

At school age and adolescence, 100 per cent of relatively high-functioning autistic students were judged to have abnormal use of gaze, compared with 20 per cent of normal and mentally handicapped children in the Autism

Diagnostic Observation Schedule (Lord *et al*. 1989), a semi-structured series of tasks designed to allow a clinician to observe a variety of rather subtle social behaviours. Ninety per cent of the autistic children, compared with 20 per cent of the verbal-age-matched controls, were judged to have difficulty coordinating their use of gaze with vocalization and gesture. Similar results were found when parents were asked to describe how well their autistic or non-autistic, mentally handicapped children coordinated gaze and vocalization (Le Couteur *et al*. 1989). In contrast, school-age autistic children can generally follow pointing visually and understand another's visual perspective (Baron–Cohen 1989; Hobson 1984). In the study of greetings, older autistic children and adolescents had particular difficulty looking at their counsellor when she was looking at them and in coordinating their own looking and talking (Lord and Magill 1989).

Little direct information is available about use of gaze by autistic children as infants, but there are some clues. Kubicek (1980), in a study of three-month-old twins, one of whom was later dignosed as autistic, described the autistic infant as actively avoiding the gaze of his mother. Other studies asking parents about the development of infant gaze-patterns have not found parents of autistic children to describe them as unusual (Klin *et al*. 1992), though, in one case at least, this was because a substantial minority of parents of non-autistic, mentally handicapped children remembered their children's gaze as less socially directed than normal (Le Couteur *et al*. 1989). Studies of gaze in autistic children of less than three years of age that are based on parental reports have not found general differences reported in 'eye-contact' (Dahlgren and Gillberg 1989; Lord 1991); but when we asked parents of later-diagnosed autistic two-year-olds about their ability to coordinate eye-contact with vocalization or gesture *in order to request something*, 81 per cent of the autistic children had difficulty, compared with 11 per cent of the control group. Differences in joint attention, without any apparent instrumental goal, were also significant.

Socially directed vocalization, babbling, and response to speech

Early descriptions of autism reported that autistic children did not babble as infants (Rutter 1970), and that the vocalizations of autistic children were readily identifiable as odd, though parents of these children were able to interpret their intentions (Ricks and Wing 1976). On the Vineland Adaptive Behaviour Scale (Sparrow *et al*. 1984) autistic children were not described differently from IQ-matched controls in their ability to express vocally two or three emotions, or in their response to voice (Klin *et al*. 1992). However, older autistic children have been found to be unusual in preference for non-speech sounds (Klin 1991).

On retrospective accounts using the Autism Diagnostic Interview

(Le Couteur *et al.* 1989) the majority of verbal autistic children were described as lacking in the vocal expressiveness of their speech and abnormal in prosody, but were not described as different in the form or social directedness of their early babble. In contrast, not one parent of 16 two-year-olds later diagnosed as autistic described his or her child's babbling as normal in social directedness, though one-third of the parents of non-autistic children also felt their children's babble was unusual (Lord 1991). Parents' descriptions of the early babble of autistic three- and four-year-olds also differed from parents' descriptions of non-autistic mentally handicapped children's babble (Lord *et al.*, manuscript in preparation).

The phenomenon of loss of single words by autistic children during the second or third year of life also requires consideration. Careful study of this occurrence indicated that most autistic children who completely lose words had used them for only a few months before the loss, and had had fewer than ten words before the regression (Kurita 1985). No satisfactory account has yet been given for this phenomenon.

Gesture

The clearest discriminator of the two-year-olds later diagnosed as autistic from the non-autistic children was use of another person's hand as a tool (Lord 1991; Ornitz *et al.* 1978). Baron–Cohen (1989) has suggested that pointing for the purpose of sharing attention, rather than for requesting, is also an important marker for autism. In a school-age sample, he found that a significant proportion of autistic children did not point at all, but that, of those who pointed, none used pointing to communicate purely social intentions. Similarly, in our two-year-old sample, about half the autistic children had some instrumental pointing, and half had no pointing at all; but none showed any social pointing (Lord 1991). Even more interesting in some ways was that, of the non-autistic children, half had both social and instrumental pointing, and half had no pointing at all; none had only instrumental pointing. Parent-interview data (Le Couteur *et al.* 1989) indicated that no parents of autistic students and young adults felt that their child had normal use of pointing, while 80 per cent of parents of mentally handicapped controls described a 'normal' range of pointing. Only autistic children were described as using unusual parts of their body to point, that is, their elbows or noses, though this occurred in only about 10 per cent of the autistic group.

A failure to hold the arms up to be lifted has traditionally been considered one of the earliest indicators of autism (Klin *et al.* 1992). Parents of two-year-olds reported that 81 per cent of the children later diagnosed as autistic did not spontaneously hold their arms up to be lifted, in contrast to 36 per cent of the non-autistic children (Lord 1991). At two, there was no difference, though another retrospective study on somewhat lower-functioning children did find a difference for later pre-school years (Le Couteur *et al.* 1989). Understanding gesture was also felt to be very limited in the autistic, but not the non-autistic two-year-olds. However, in our later studies of older, high-functioning autistic children and adolescents, use of gesture did not characterize the autistic students, in part because the range of gesture in the normally developing and other control groups was so varied.

Facial expression

Facial expressions are both difficult to measure directly and to acquire information about from parents. Thus results have been inconsistent, probably to a great extent because of differences in measurement. Parents of autistic children have described them as failing to laugh or smile to positive comment, and to have smiled for the first time at later ages than mentally handicapped controls (Klin *et al.* 1992; Le Couteur *et al.* 1989). However, in our study of very young autistic children and non-autistic mentally handicapped children, we found no significant differences in descriptions of smiling in two-year-olds or retrospectively in one-year-olds, though more autistic children were described by their parents as never smiling (Lord 1991). In part, the failure to find differences, as for earlier studies on vocalization and gaze, was because many of the parents of the young non-autistic children felt their children did not smile as much as other children, a finding that has been supported in research with children with Down's syndrome (Cicchetti and Serafica 1981). 82 percent of parents of older autistic children, compared with 37 percent of parents of mentally handicapped children, described their children's facial expressions as limited in range, with 38 percent of the autistic group and 9 percent of the mentally handicapped group also describing frequent odd or inappropriate facial expressions (Le Couteur *et al.* 1989).

Observational studies have found more mixed and flat facial expressions and less positive and more negative affect expressed socially in pre-school autistic children than in various controls (Snow *et al.* 1987; Yirmiya *et al.* 1989), and less positive affect occurring as part of joint attention sequences (Dawson *et al.* 1990; Kasari *et al.* 1990). In our greeting studies, few of the autistic children and adolescents ever smiled, compared with the normally developing and psychiatric control children, who showed significant variability, but almost always smiled during at least one observation (Lord and Magill 1989). In the Autism Diagnostic Observation Schedule (Lord *et al.* 1989) 85 per cent of the autistic children and adolescents, versus 25 per cent of the mentally handicapped children, were judged to have a limited range of facial expression, with 30 per cent of the autistic children compared with 8 per cent of the mentally handicapped children showing odd or unusual facial expressions during the twenty-minute observation. Studies that have looked for differences in affect in particular situations, such as seeking, joint attention or making social overtures (Dawson *et al.* 1990; Hertzig *et al.* 1989), have been more able to find differences than those where expressions have been coded aside from any particular context.

CONCLUSIONS

Developmental information is critical to our understanding of the social experience of autism and the meaning of this experience for the identification of the central deficits in the disorder. Because of the typically relatively late age (four and five years) at which autism has been identified, this information has not been easily available; however, prospective studies currently in progress (see Baron–Cohen and Howlin, this volume, Chapter 21), detailed retrospective data from parents of younger children (Lord 1991), and careful use of fortuitously collected early information (Kubicek 1980) may begin to fill this gap. Another source of information that is critical to the interpretation of developmental findings concerns relationships between different areas of functioning (Ungerer 1989). Relationships between the development of joint attention and symbolic play (Hertzig *et al.* 1989; Ungerer 1989) and language comprehension and emotional recognition (Mundy and Sigman 1989) have been documented in pre-school and older children. However, to date, correlations have not been found, in either autistic or normally developing children, between age-related cognitive skills and affective sensitivity or prosocial behaviours (Shantz 1983; Rogers and Pennington 1991). Identification of such relationships, particularly across time, is critical to determining the extent to which a theory-of-mind hypothesis can be seen either as 'driving' other social behaviour, or as simply correlated to social development, or as not related at all. Adequate measures of social behaviour are crucial to this enterprise; checklists have not been sufficient.

Relationships over time, for example, among joint attention, social responsiveness, affective expression, and measures of understanding persons will be most important. It seems possible that the development of social behaviours in autism, such as reciprocal eye-gaze and vocalization, may not be linear, but may vary over time and among children within the first years of life. Yet further information is required, particularly to separate out whether behaviours begin to go awry when they become more socially rather than perceptually driven, or when they require representation or intention. Deficits in very basic behaviours, such as the use of gaze, prosody and vocal expressiveness, and gesture and coordination of non-verbal and verbal communication all abound in the clinical literature, but do not fit easily into explanations of autism as a highly specific cognitive deficit. Our knowledge of what the social deficits specific to autism really are in very young children and in older children, adolescents, and adults, is also limited by relatively little empirical research, excluding the notable exceptions described earlier.

Methodological issues, such as the care used in selecting control groups, the problems inherent in using multiple measures across a developmentally diverse sample, and the need for more systematic information about behaviour in natural settings can have a direct impact on results. It is not surprising that cognitive explanations have seemed so attractive, when what has been studied has for the most part been cognitive tasks that are set up in surroundings quite different from those of naturally occurring, affect-laden settings (Dunn 1991). Continued investigation of the extent to which early deficits remain in adolescence and adulthood is also of importance. Finally, some consideration needs to be given to the effects of possible deficits in understanding persons in the social environment experienced by autistic youngsters. If autistic children lack even the most basic infant behaviours that result in sustaining or promoting interactions from some point very early on, they will be exposed to fewer and different opportunities for social learning, perhaps even from the first year of life (Cohen 1980). Such a deficit may affect knowledge about social events — how they are organized, what others' goals might be, what are the rules for common types of interactions. It may limit opportunity for imitation. It may even affect very basic aspects of neurobiological development.

Mothers of older pre-school and school-age autistic children show less positive affect to them and are less interactive than parents of mentally handicapped children, probably to a great extent because of the lack of positive affect they receive (Dawson *et al*. 1990; Kasari *et al*. 1988; O'Neill and Lord 1982). Basic deficits in imitation (Dawson and Adams 1984) might have very early effects in the directedness of parents' behaviour to autistic children; basic deficits in joint attention may affect parents' likelihood of modulating their behaviour to their children's level of understanding (Dawson *et al*. 1990; Kasari *et al*. 1988; Kubicek 1980). If autistic children from very early on do not orient or respond, or are slower at responding to basic social overtures, they may receive less response-contingent feedback from others (Minshew *et al*. 1991). The overall result could be social isolation to some degree, even in the midst of a relationship with loving and responsive parents who, despite their willingness, have few behaviours to respond to. Social isolation, in terms of lack of response-contingent feedback, has been described in animal models as affecting arousal, perseveration, motivation, self-stimulation, and difficulties with complex tasks (Holson and Sackett 1984), all behaviours associated with autism.

Repeated failure or meaninglessness may also contribute to lack of motivation (Koegel and Mentis 1985). Lack of early friendships and peer relations removes the autistic child from opportunities to experience different kinds of intimacy and feedback, which have been seen as essential for the development of higher-order affective and social relationships (Breese and Caplan, in press). Failure to attend automatically to human language, perhaps a precursor to later deficits in understanding persons, could have serious ramifications for the early development of language comprehension, since comprehension of words in very young children often arises out of multiple sources of information, including parental facial expression, line of gaze, gesture, and context (Chapman 1978; Macnamara 1972). Thus, in some ways, absence of simple social behaviours can be seen as the beginning of a cycle of exclusion and absence of opportunity.

Theories that portray autism as arising from a deficit in theory of mind and other deficits in understanding persons have been the major impetus for research into the cognitive and social characteristics of autistic children in the last ten years. As theories are modified to take into account the results of the research they have generated, they become better able to address the realities of the full syndrome of autism across areas and levels of development. As more studies are driven by the need to sort out relationships between well-accepted clinical beliefs, new theories, and initial research, gaps in knowledge are being filled that allow fresh interpretation of old observations. The development of social behaviour in autistic children — and non-autistic children — is very complex. Theories must take this complexity into account, while continuing to seek parsimonious, testable explanations for the extraordinary spectrum of disorders associated with autism. It is from the interplay between theory and data, and between theory and clinical practice, that reliable and veridical answers to queries about the nature of autism and relationships between social and cognitive development seem most likely to arise.

REFERENCES

Attwood, A., Frith, J., and Hermelin, B. (1988). The understanding and use of interpersonal gestures by autistic and Down syndrome children. *Journal of Autism and Developmental Disorders*, **18**, 241–58.

Baron–Cohen, S. (1987). Autism and symbolic play. *British Journal of Developmental Psychology*, **5**, 139–48.

Baron–Cohen, S. (1989). Perceptual role-taking and protodeclarative pointing in autism. *British Journal of Developmental Psychology*, **7**, 113–27.

Breese, G.R., and Caplan, M. Inhibitory influences in development: the case of prosocial behavior. In *Precursors and causes in normal and abnormal development* (ed. D.F. Hay and A. Angold), (In press.)

Chapman, R.S. (1978). Comprehension strategies in children. In *Speech and language in the laboratory, school and clinic* (ed. J.F. Kavanagh and W. Strange). MIT Press, Cambridge, Mass.

Cicchetti, D.V., and Serafica, F.C. (1981). Interplay among behavioral systems: illustrations from the study of attachment, affiliation, and wariness in young children with Down syndrome. *Developmental Psychology*, **17**, 36–49.

Clark, P., and Rutter, M. (1981). Autistic children's responses to structure and to interpersonal demands. *Journal of Autism and Developmental Disorders*, **11**, 201–17.

Cohen, D.J. (1980). The pathology of the self in primary childhood autism and Gilles de la Tourette syndrome. *Psychiatric Clinics of North America*, **3** (3), 383–402.

Cohen, D.J., Paul, R., and Volkmar, F.P. (1987). Issues in the classification of pervasive developmental disorders and associated conditions. In *Handbook of autism and pervasive developmental disorders* (ed. D.J. Cohen and A.M. Donnellan). Wiley, New York.

Curcio, F., and Piserchia, E.A. (1978). Pantomimic representation in psychotic children. *Journal of Autism and Developmental Disorders*, **8**, 181–90.

Dahlgren, S.O., and Gillberg, C. (1989). Symptoms in the first two years of life: a preliminary population study of infantile autism. *European Archives of Psychiatric and Neurological Science*, **283**, 169–74.

Dawson, G., and Adams. A. (1984). Imitation and social responsiveness in autistic children. *Journal of Abnormal Child Psychology*, **12**, 209–26.

Dawson, G., and Galpert, L. (1986). A developmental model for facilitating the social behavior of autistic children. *Social behavior in autism* (ed. E. Schopler and G.B. Mesibov). Plenum, New York.

Dawson, G., Hill, D., Spencer, A., Galpert, L., and Watson, L. (1990). Affective exchanges between young autistic children and their mothers. *Journal of Abnormal Child Psychology*, **18**, 335–45.

DeMyer, M.K., Alpern, G.D., DeMyer, W.E., Churchill, D.W., Hingtgen, J.N., Bryson, C.Q., *et al.* (1972). Imitation in autistic, early schizophrenic, and non-psychotic abnormal children. *Journal of Autism and Childhood Schizophrenia*, **2**, 264–87.

Dewey, D., Lord, C., and Magill, J. (1988). Qualitative assessment of the effect of play materials in dyadic peer interactions of children with autism. *Canadian Journal of Psychology*, **42**, 242–60.

Dunn, J. (1991). Understanding others: evidence from naturalistic studies of children. In *Natural theories of mind* (ed. A. Whiten). Blackwell, Oxford.

Everard, M.P. (ed.) (1976). *An approach to teaching autistic children*. Pergamon Press, Oxford.

Ferster, C.B. (1961). Positive reinforcement and behavioral deficits of autistic children. *Child Development*, **32**, 437–47.

Ghaziuddin, M., Tsai, L., and Ghaziuddin, N. (1991). Brief report: violence in Asperger's Syndrome, a critique. *Journal of Autism and Developmental Disorders*, **21**, 349–54.

Gillberg, C. (1990). Outcome in autism and autistic-like conditions. *Journal of the American Academy of Child and Adolescent Psychiatry*, **30**, 375–82.

Guralnick, M.J. (1976). The value of integrating handicapped and nonhandicapped preschool children. *American Journal of Orthopsychiatry*, **45**, 236–45.

Hannan, T.E. (1987). A cross-sequential assessment of the occurrences of pointing in 3- to 12-month-old human infants. *Infant Behavior and Development*, **10**, 11–22.

Harris, P.L. (1991). The work of the imagination. In *Natural theories of mind: the evolution, development and simulation of everyday mindreading* (ed. A. Whiten). Blackwell, Oxford.

Hermelin, B., and O'Connor, N. (1970). *Psychological experiments with autistic children*. Pergamon Press, New York.

Hertzig, M.E., Snow, M.E., and Sherman, M. (1989). Affect and cognition in autism. *Journal of the American Academy of Child and Adolescent Psychiatry*, **28**, 195–9.

Hobson, R.P. (1984). Early childhood autism and the question of egocentrism. *Journal of Autism and Developmental Disorders*, **14**, 85–104.

Holson, R., and Sackett, G.P. (1984). Effects of isolation rearing on learning by mammals. *The Psychology of Learning and Motivation*, **18**, 199–254.

Jacobson, J.W., and Ackerman, L.J. (1990). Differences in adaptive functioning among people with autism or mental retardation. *Journal of Autism and Developmental Disorders*, **20**, 205–20.

Kasari, C., Sigman, M., Mundy, P., and Yirmiya, N. (1988). Care-giver interactions with autistic children. *Journal of Abnormal Child Psychology*, **16**, 45–56.

Kasari, C., Sigman, M., Mundy, P., and Yirmiya, N. (1990). Affective sharing in the context of joint attention interactions of normal, autistic, and mentally retarded children. *Journal of Autism and Developmental Disorders*, **20**, 87–100.

Klin, A. (1991). Young autistic children's listening preferences in regard to speech: a possible characterization of the symptom of social withdrawal. *Journal of Autism and Developmental Disorders*, **21**, 29–42.

Klin, A., Volkmar, F.R., and Sparrow, S. (1992). Autistic social dysfunction: some limitations of the theory of mind hypothesis. *Journal of Child Psychology and Psychiatry*, **33**, 861–76.

Koegel, R.L., and Mentis, M. (1985). Motivation in childhood autism: can they or won't they? *Journal of Child Psychology and Psychiatry*, **26**, 185–91.

Konstantareas, M.M., and Homatidis, S. (1989). Assessing child symptom severity and its stress in parents of autistic children. *Journal of Child Psychology and Psychiatry*, **30**, 459–70.

Kubicek, L.F. (1980). Organization in two mother–infant interactions involving, a normal infant and his fraternal twin who was later diagnosed as autistic. In *High-risk infants and children: adult and peer interactions* (ed. T.M. Field, S. Goldberg, D. Stern and A.M. Sostek). Academic Press, New York.

Kurita, H. (1985). Infantile autism with speech loss before the age of 30 months. *Journal of the American Academy of Child Psychiatry*, **24**, 191–6.

Le Couteur, A., Rutter, M., Lord, C., Rios, P., Robertson, S., Holdgrafer, M., and McLennan, J.D. (1989). Autism Diagnostic Interview: a semi-structured interview for parents and care-givers of autistic persons. *Journal of Autism and Developmental Disorders*, **19**, 363–87.

Leslie, A.M. (1987). Pretense and representation: the origins of 'theory of mind'. *Psychological Review*, **94**, 412–26.

Lewis, V., and Boucher, V. (1988). Spontaneous, instructed and elicited play in relatively able autistic children. *British Journal of Developmental Psychology*, **6**, 325–39.

Lord, C. (1984). The development of peer relations in children with autism. In *Advances in applied developmental psychology* (ed. F. J. Morrison, C. Lord and D. P. Keating). Academic Press, New York.

Lord, C. (1990). A cognitive–behavioral model for the treatment of social-communicative deficits in adolescents with autism. In *Behavior disorders of adolescence: research, intervention and policy in clinical and school setting* (ed. R. J. McMahon and R. DeV. Peters). Plenum, New York.

Lord, C. (1991). Follow-up of two-year-olds referred for possible autism. Paper presented at the Biennial meeting of the Society for Research in Child Development, Seattle.

Lord, C., and Hopkins, J. M. (1986). The social behaviour of autistic children with younger and same-age nonhandicapped peers. *Journal of Autism and Developmental Disorders*, **16**, 249–62.

Lord, C., and Magill, J. (1989). Methodological and theoretical issues in studying peer-directed behavior and autism. In *Autism: nature, diagnosis, and treatment* (ed. G. Dawson). Guilford Press, New York.

Lord, C., Rutter, M., Goode, S., Heemsberen, J., Jordan, H., and Mawhood, L. (1989). Autism Diagnostic Observation Schedule: standardized observation of communicative and social behavior. *Journal of Autism and Developmental Disorders*, **19**, 185–212.

Lord, C., Storoschuk, S., Parkerson, L., and Rutter, M. Diagnosing autism in preschool children. (Manuscript in preparation.)

Loveland, K. A., and Kelley, M. L. (1988). Development of adaptive behavior in adolescents and young adults with autism and Down syndrome. *American Journal of Mental Retardation*, **93**, 84–92.

Macnamara, J. (1972). Cognitive basis of language learning in infants. *Psychological Review*, **79**, 1–13.

McEvoy, M. A., Nordquist, V. M., Twardosz, S., Heckaman, K. A., Wehby, J. H., and Denny, R. K. (1988). Promoting autistic children's peer interactions in an integrated early childhood setting using affection activities. *Journal of Applied Behavior Analysis*, **21**, 193–200.

McHale, S. M., Olley, J. G., and Marcus, L. M. (1981). Variations across settings in autistic children's play. Paper presented at the Biennial meeting of the Society for Research in Child Development, Boston.

Mawson, D., Grounds, A., and Tantam, D. (1985) Violence and Asperger's Syndrome. *British Journal of Psychiatry*, **147**, 566–9.

Minshew, N., Pettegrew, J., Rattan, A., Phillips, B. A., and Panchalingam, K. (1991). Correlations between brain metabolism and cognitive function in autism. Paper presented at the annual meetings of the American Academy of Child and Adolescent Psychiatry, San Francisco, 1991.

Mirenda, P. L., Donellan, A. M., and Yoder, D. E. (1983). Gaze behavior: a new look at an old problem. *Journal of Autism and Developmental Disorders*, **13**, 397–410.

Mundy, P., Siaman, M., Ungerer, J., and Sherman, T. (1986). Defining the social deficits of autism: the contribution of nonverbal communication measures. *Journal of Child Psychology and Psychiatry*, **27**, 657–69.

Mundy, P., Sigman, M., Ungerer, J., and Sherman, T. (1987). Nonverbal communication and play correlates of language development in autistic children. *Journal of Autism and Developmental Disorders*, **17**, 349–64.

Odom, S. L., and Strain, P. S. (1986). Comparison of peer-initiation and teacher antecedent interventions for promoting reciprocal social interaction of autistic preschoolers. *Journal of Applied Behavior Analysis*, **19**, 59–71.

Ohta, M., Nagai, Y., Hara, H., and Sasaki, M. (1987). Parental perception of behavioral symptoms in Japanese autistic children. *Journal of Autism and Developmental Disorders*, **17**, 549–64.

O'Neill, P. J., and Lord, C. (1982). Functional and semantic characteristics of child-directed speech of autistic children. In *Proceedings from the International Meetings for the National Society for Autistic Children* (ed. D. Park). National Society for Autistic Children, Washington, DC.

Ornitz, E. M., Guthrie, D., and Farley, A. J. (1978). The early symptoms of childhood autism. In *Cognitive defects in the development of mental illness* (ed. Kittay Scientific Foundation). Brunner–Mazel, New York.

Radke-Yarrow, M., Zahn-Waxler, C., and Chapman, M. (1983). Children's prosocial dispositions and behavior. In *Handbook of child psychology: Vol 4. Socialization, personality, and social development* (ed. P. H. Mussen). Wiley, New York.

Ricks, D. M., and Wing, L. (1976). Language, communication and use of symbols. In *Early childhood autism* (ed. L. Wing). Pergamon Press, Oxford.

Rogers, S. J., and Pennington, B. F. (1991). A theoretical approach to the deficits in infantile autism. *Development and Psychopathology*, **3**, 137–62.

Rogers, S. J., Ozonoff, M. A., and Maslin-Cole, C. (1991). A comparative study of attachment behavior in young children with autism or other psychiatric disorders. *Journal of the American Academy of Child and Adolescent Psychiatry*, **30**, 483–8.

Rutter, M. (1970). Autistic children: infancy to adulthood. *Seminars in Psychiatry*, **2**, 435–50.

Rutter, M., Mawhood, L., and Goode, S. (1991). Adult follow-up of boys with autism or with a specific developmental disorder of receptive language. Paper presented at the Biennial meeting of the Society for Research in Child Development, Seattle.

Schopler, E., and Reichler, R. J. (1971). Developmental therapy by parents with their own autistic child. In *Infantile antism: concepts, characteristics and treatment* (ed. M. Rutter). Churchill, London.

Shantz, C. V. (1983). Social cognition. In *Handbook of child psychology: cognitive development*, Vol. 3 (ed. P. H. Mussen). Wiley, New York.

Shapiro, T., Sherman, M., Calamari, G., and Koch, D. (1987). Attachment in autism and other developmental disorders. *Journal of the American Academy of Child Psychology*, **26**, 480–4.

Sherman, M., Shapiro, J., and Glassman, M. (1983). Play and language in developmentally disordered preschoolers: a new approach to classification. *Journal of the American Academy of Child Psychiatry*, **22**, 511–24.

Sigman, M., and Mundy, P. (1989). Social attachments in autistic children. *Journal of the American Academy of Child Psychiatry*, **28**, 74–81.

Snow, M., Hertzig, M., and Shapiro, T. (1987). Expression of emotion in young autistic children. *Journal of the American Academy of Child Psychiatry*, **26**, 836–8.

Sparrow, S., Balla, D., and Cicchetti, D. (1984). *Vineland Adaptive Behavior Scales.* American Guidance Service, Circle Pines, Maine.

Stone, W. F., and Caro-Martinez, L. M. (1990). Naturalistic observations of spontaneous communication in autistic children. *Journal of Autism and Developmental Disorders*, **20**, 437–54.

Stone, W. L., and Lemanek, K. L. (1990). Parental report of social behaviors in autistic preschoolers. *Journal of Autism and Developmental Disorders*, **20**, 513–22.

Ungerer, J. A. (1989). The early development of autistic children. In *Autism: nature, diagnosis and treatment* (ed. G. Dawson). Guilford Press, New York.

Venter, A., Lord, C., and Schopler, E. (1991). A follow-up study of high-functioning autistic children. *Journal of Child Psychology and Psychiatry*, **32**.

Walters, A. S., Barrett, R. P., and Feinstein, C. (1990). Social relatedness and autism: current research, issues, directions. *Research in Developmental Disabilities*, **11**, 303–26.

Wing, L. (1981). Asperger's syndrome: a clinical account. *Psychological Medicine*, **11**, 115–29.

Wing, L., and Atwood, A. (1987). Syndromes of autism and atypical development. In *Handbook of autism and pervasive developmental disorders* (ed. D. J. Cohen and A. M. Donnellan). Wiley, New York.

Yirmiya, N., Kasari, C., Sigman, M., and Mundy, P. (1989). Facial expressions of affect in autistic, mentally retarded and normal children. *Journal of Child Psychology and Psychiatry*, **30**, 725–36.

Zahn-Waxler, C., and Radke-Yarrow, M. (1982). The development of altruism: alternative research strategies. In *The development of prosocial behavior* (ed. N. Eisenberg-Berg). Academic Press, New York.

15

The development of individuals with autism: implications for the theory of mind hypothesis

AMI KLIN AND FRED VOLKMAR

In his original description of the syndrome of early infantile autism, Kanner (1943) emphasized the centrality of a very basic, and pervasive, disturbance in social development. Over the next two decades research on the condition was impeded by various controversies, for example, those centring around continuities with other conditions and the role of experiential factors in pathogenesis. By the 1970s it was generally agreed that the disorder was indeed distinctive in a host of ways, and probably reflected the operation of some biological insult to the developing central nervous system. Although the nature of the underlying psychological deficit remained (and remains) obscure, there was general agreement that a pervasive impairment of symbolic functioning was involved (Ricks and Wing 1975; Prior and Bradshaw 1979). Early studies failed, however, to consider the social context in which the acquisition of symbolic capacity was firmly grounded (see for example Bates *et al.* 1979), and its implications for autism (Klin 1989).

With a few notable exceptions (Cohen 1980), studies of autistic social dysfunction were to come only in the later 1980s. A few experiments exploring autistic children's ability to take the perspective of others (for example Hobson 1984) revealed that while they showed no deficits in Piagetian visual perspective-taking tasks, there was marked impairment in more social or conceptual perspective-taking tasks (Baron–Cohen *et al.* 1985). It seemed that the former could be solved using solely visual–spatial skills, whereas in order to succeed in the latter the child would inevitably have to take the role of another person, exploring the other person's conceptual alternatives and acting accordingly. Although role-taking skills and their importance for social and communicative development had been originally described and studied by investigators working within a pragmatist or social-interactive framework (for example Flavell *et al.* 1968; Light 1979; Mead 1934), these important findings in autism research were conceptualized within a cognitive framework (Leslie 1987). The various studies that followed (see Baron–Cohen, this volume, Chapter 4, for a review) substantiated and expanded the notion that the verbal autistic child's ability to conceive of other people's subjectivity, or to have a theory of mind, was markedly impaired.

In the past years the theory of mind hypothesis (Baron–Cohen 1990*a*) has provided a new conceptualization of autistic social dysfunction. Grounded on the emerging literature on children's metacognitive capacities, this theoretical stance follows and supersedes the seminal work of Hermelin, O'Connor, and Frith (Frith 1989), who, utilizing carefully designed experimental studies of school-aged autistic children, delineated a specific profile of perceptual and cognitive skills and disabilities in autism. The theory of mind hypothesis suggests that the social disability in autism reflects a very specific, innate, and primarily cognitive incapacity to impute mental states to others and to the self (Leslie 1987; Baron–Cohen 1990*a*). It has grown out of experimental findings in which, for example, autistic children were shown to be unable to attribute a false belief to others (Baron–Cohen *et al.* 1985), or to understand picture-stories which required a grasp of mental states with content (Baron–Cohen *et al.* 1986). Autistic children were said to lack a 'theory of mind' (Premack and Woodruff 1978), i.e. an implicit capacity which involves 'the person postulating the existence of mental states and then using these to explain and predict another person's behavior' (Baron–Cohen 1988, p. 392).

In Leslie's (1987) view, this capacity for a theory of mind implies a need to represent other people's representations and, accordingly, the implicit 'theory of mind' requires the usage of 'second-order representations' or 'metarepresentations'. Metarepresentational skills are thought to presuppose a cognitive capacity (Leslie 1987), which presumably is evident, if in its most rudimentary form, by late in the first year of life (Baron–Cohen 1990*b*). The lack of this capacity has been hypothesized to account for autistic children's disabilities in joint attention, declarative pointing, symbolic play, pragmatics in speech, empathy and other aspects of social development and functioning (Baron–Cohen, this volume, Chapter 4). In its absence, autistic individuals are indeed autistic in the strictest sense of the term—they are in their own world. A single-deficit hypothesis, this cognitive impairment has been hypothesized to reflect a circumscribed brain disorder affecting the specific ability to conceive of other people's mental states (Baron–Cohen 1990*a*), or its cognitive prerequisities (Leslie 1987). In sum, 'theory of mind' provides a competence model of intersubjectivity, in that the ability to impute mental states with content is presupposed for intersubjective occurrences to take place. Devoid of this capacity, autistic individuals are faced with an inscrutable puzzle, namely how to understand and predict other people's social behaviour without the ability to infer that they have beliefs and desires, i.e. without an implicit theory that other people have minds.

The theory of mind hypothesis of autistic social dysfunction has made an important contribution to the field by refocusing our attention on an issue which is at the heart of the autistic disorder: the striking incapacity to conceive of other people's minds. While its general applicability to persons with autism (or at least higher-functioning persons with autism) as a clinical phenomenon seems reasonably clear, the power of this hypothesis as an *explanation* of autistic social dysfunction has been more controversial. In other words, the issue of whether deficits in theory of mind are responsible for the fundamental social disability, or only interesting correlates of a more general problem, remains unresolved. Accordingly, predictions derived from the hypothesis have particular importance in evaluating its power as a general theory of autism. In this chapter we briefly outline some aspects of the theory of mind hypothesis which either appear to limit its utility as a general *explanation* of social dysfunction in autism, or seem to be incongruent with research findings. These areas include the expression of social dysfunction in very young autistic children, the discrete (i.e. the hypothesized, essentially metacognitive) nature of social deficits in autism, and whether theory of mind deficits are best viewed as 'primary' or 'secondary' aspects of syndrome expression.

THE ONSET OF AUTISM AND THEORY OF MIND CAPACITIES

In normative development, there are a host of perceptual, affective, and neuroregulatory mechanisms which predispose and ensure the young infants' engagement in social interaction from very early on in their lives (Tronick 1980; Cohn and Tronick 1987). The onset of shared emotional states is behind the magic of the three-month-old little human being. As Knobloch and Pasamanick (1975) put it: 'Nothing stirs the heart of a misanthrope more than a 16-week-old baby whose eyes crinkle and lips curl spontaneously in an appealing smile as he initiates social play. So absorbed is the infant with people at these tender ages that it might be difficult to divert his attention to test objects during a developmental evaluation' (p. 182).

Against the background of normative development, the social deficits of autistic infants are striking, appearing to be present from birth, or at least from the first few months of life (Caparulo and Cohen 1977; Clancy and McBride 1969; Prior and Gajzago 1974; Rutter 1974; Volkmar and Cohen 1988; Volkmar *et al.* 1985). In stark contrast to what happens with normally developing infants, the human face appears to hold little interest or have little salience for the autistic child (Volkmar 1987). Similarly, young autistic children appear to lack a differential preference for speech sounds (Klin 1991, 1992), a behaviour documented in the first month, if not the first days of life (Mills and Melhuish 1974). Typical forms of early non-verbal interchange are deviant, so that usually very early emerging forms of 'intersubjectivity' (Stern 1985; Trevarthen 1979) are absent. Affected children may not seek physical comfort from parents, and may be difficult to hold (Ornitz *et al.* 1977). Similarly, while some autistic children exhibit differential social responsiveness to familiar adults (Shapiro *et al.* 1987; Sigman and Ungerer 1984), the quality of such behaviors is highly deviant (Mundy and Sigman 1989). These early social deficits are striking not only in contrast to normative human development, but in contrast to that of other primates as well (Plooij 1978): for example, orphaned infant monkeys develop strong and persistent affectional responses even towards inanimate surrogate mothers (see for instance Harlow and Zimmermann 1959).

In normally developing human infants, the motivation and capacities for social interaction are generally extremely robust from the first weeks of life, even in the face of severe maternal deprivation (Bowlby 1969; Rutter 1981). In contrast, although it is now clear that a few children appear to develop autism after some months, or even a few years of normal development, most do not (Volkmar and Cohen 1988). The theory of mind hypothesis rests on notions of interpersonal subjectivity that typically are not observed until after the first year of life (see for example Bates *et al.* 1979). If, for example, the age-range of nine to twelve months is taken (rather liberally) as the earliest possible point at which metarepresentational capacities develop, it remains unclear why individuals functioning below this level would exhibit deficits in basic social behaviours. Similarly, the theory of mind hypothesis would seem to imply that no differences in those social skills which ordinarily emerge before nine months would be observed in autistic and MA-matched mentally retarded children. At least one study, using carefully matched groups and employing an additional MA control (Klin *et al.* 1992) has, however, revealed a number of such differences.

Several possible solutions to this problem are immediately apparent: (1) perhaps theory of mind capacities are indeed present from *very* early in life; or (2) theory of mind capacities are assumed to emerge on the basis of one or more maturational processes; or (3) perhaps these capacities are themselves dependent on the development of even earlier cognitive processes. Sufficient data are not yet available to support any of these notions. If, however, such processes are hypothesized to appear in the first weeks or months of life, or if they are presumed to appear at some point in development on, essentially, a biological (Leslie 1987) basis, it is unclear to what extent the theory-of-mind hypothesis truly differs from Kanner's original view, although it may add to it by focusing research attention on more discrete processes.

In sum, intersubjectivity, so much impaired in autism, is already present in very young normally developing infants. The theory of mind hypothesis has successfully modelled the cognitive prerequisites for intersubjective experiences. Its greatest challenge is to provide a psychogenetic (Piaget 1972) account of the theory of mind, i.e. to describe its precursors, so that early infancy intersubjectivity, and indeed, many of the early social deficits exhibited by autistic children, will not lie outside its scope. The issue of possible precursors of theory of mind capacities clearly represents an important topic for future research.

SOCIAL DEVIANCE IN AUTISM: GENERALIZED OR DISCRETE?

The theory of mind hypothesis provides a highly specific model (and prediction) of the dysfunction underlying the social disabilities in autism, i.e., the underlying dysfunction is hypothesized to be *discrete*. Accordingly, only certain social skills (those presupposing theory of mind capacities) should be impaired secondary to the more specific cognitive impairment. An alternative view, consistent with Kanner's original (1943) report, suggests that the social disability is *general* in nature, i.e., that it affects basically all fundamentally social processes. This is a complex issue for several reasons: (1) far from being a homogeneous group, autistic persons vary widely in the severity and range of their social disabilities (Volkmar and Cohen 1988; Wing 1988); (2) some social skills do emerge in autistic children over the course of development (Lotter 1978; Rutter *et al.* 1967) as a result, at least in part, of the intensive social-skills training adopted in most schools, as well as of dedicated, round-the-clock, parental intervention; and (3) clearly social, communicative, and cognitive skills are closely interrelated.

In characterizing the underlying disability as either discrete or general, three approaches might help minimize the various confounding variables. Firstly, we may focus on the early social development of autistic children. From this standpoint, Kanner's (1943) initial description views the social abnormalities in autism as a generalized phenomenon, not only present from the beginning of life, but also affecting very early and basic social contacts, such as babies' anticipatory postures upon being picked up, and their body-moulding to caregivers. In contrast, the theory of mind hypothesis apparently predicts that some social behaviours may not be impaired in autism, namely those behaviours which do not presuppose the need to attribute beliefs, intentions, or desires to others, i.e. the use of a 'theory of mind' (Baron–Cohen 1988). The social dysfunction is more circumscribed or discrete as a result of the rather specific cognitive impairment hypothesized to account for it. Accordingly, one should expect that autistic children should fail to exhibit social behaviours mediated by the metarepresentational capacity, whereas social behaviours with no such demands should be observed in their social functioning.

Secondly, naturally occurring, rather than experimentally obtained, social behaviours may be considered. Although various clinical features of autistic persons (for example pragmatic deficits, dearth of pretend-play, lack of embarrassment and empathy, etc.) have been accounted for in terms of the metarepresentational disability (Baron–Cohen 1988), the evidence in support of the theory of mind hypothesis has been, to date, primarily experimental (see Baron–Cohen 1990*a* for a review). In order to ascertain the ecological validity (Bronfenbrenner 1979) of the hypothesis, the proposed specificity of the underlying mechanism should be evident in real-life social situations.

Thirdly, one may examine normative, rather than deviant, social behaviour. Until recently, the characterization of the autistic social dysfunction was approached by focusing on specific types of *deviance* peculiar to autism, rather than on impaired social functioning *per se* (Rutter and Schopler 1987). However, with the advent of an operational approach using *norm-referenced* measures, it was demonstrated that social deficits could significantly discriminate autistic from other developmentally disabled individuals, both as a group (Volkmar *et al.* 1987) and individually (Volkmar *et al.* 1990).

In a study designed to investigate very early, naturally occurring and normative social functioning in young autistic children (Klin *et al.* 1992), various behaviours were found to discriminate the autistic subjects from their MA-matched controls. These behaviours, normatively occurring between the ages of less than two months to seven months, included failures in anticipating being picked up by caregiver, showing affection toward familiar people, showing interest in children or peers other than siblings, reaching for a familiar person, playing very simple interaction games with others, showing interest in the activities of others, and imitating simple adult movements (for example clapping hands or waving goodbye, in response to a model).

These findings appear to suggest that, rather than affecting only social behaviours mediated by theory of mind capacities, the underlying disability in autism has a more pervasive impact on these children's social functioning. Such early social behaviours cannot be described as 'theory-like' (cf. Wellman 1990), because they do not aid the child in explaining or predicting other people's behaviour, nor do they have to be inferred from the social situation. Rather, they refer to very early social reactions shown by infants in the presence of their caregivers. They also seem to refer to an impairment in social responsivity or in early imitative reactions rather than in a cognitive capacity necessary for reciprocal social interaction.

THE NATURE OF SOCIAL DEFICITS IN AUTISM: PRIMARY OR SECONDARY?

A third set of issues has to do with the fundamentally cognitive nature of the theory of mind hypothesis (Hobson 1989). In this respect the theory-of-mind hypothesis is diametrically opposed to Kanner's original view of what was fundamentally disordered in autism. Clearly there is considerable potential for becoming over-concerned with the issue of what aspects of the disorder (social–affective, cognitive, communicative, attentional–motivational) are 'primary'; as a practical matter all developmental skills emerge with some relationship to each other and with varying degrees of relative importance attached to experiential vs 'maturational' factors. Many basic processes (for example attention, perception, cognition) are involved even in what otherwise appear to be very early-emerging social activities. Similarly many different processes are subsumed under overarching terms like 'social development', and there has been a tendency to equate certain aspects of social development (for instance affective development) with social development as a whole. In some sense, this latter process would be equivalent to equating performance on one selected kind of cognitive skills, such as visual–spatial orientation, with the entirety of cognitive development.

The theory of mind hypothesis fundamentally emphasizes the cognitive aspects of interpersonal transaction. However, it is clear that the social world differs in a host of ways from the non-social environment, i.e. given that relationships with people, rather than things, are involved, basic social needs, issues of affective expression and understanding, cultural and personal context, and the fact that other people (rather than other things) are being understood, are all factors in social functioning throughout the life-span. Although historically cognitive factors in autism have been assumed to be primary because some social skills develop over time, it is just as reasonable to argue that since some (and sometimes quite sophisticated) cognitive skills develop, it is the social factors which more parsimoniously account for the clinical manifestations of autism. This latter view would be more consistent with the observation that even the very highest-functioning autistic adults, who are able to achieve relatively high levels of social–cognitive understanding, still exhibit such marked problems in relating to others. Also, given the very considerable body of research which suggests the early importance of fundamental social factors in the development of many basic processes, it will be important to understand how this hypothesis can account for the severity and early onset of the autistic child's social disturbance.

Similarly, one might expect, according to the hypothesis, that intellectually more able children would be to more likely to manifest the social deficit, i.e. in comparison to more cognitively-impaired children. For example, since various sensorimotor skills have to be achieved before the final advent of protocommunicative behaviours (which appear with the onset of Piaget's Stage V of sensorimotor abilities: Bates *et al.* 1979), and given that the sensorimotor development of mentally retarded children usually lags behind that of normal children, one reaches the paradoxical conclusion that expressions of deviant social behaviour should have an earlier onset for non-retarded than for retarded autistic children, a prediction which is not supported by available evidence (Volkmar *et al.* 1985).

Another issue to be addressed involves the natural course of autistic social dysfunction. Following the theory of mind hypothesis, one might expect that attraction to social stimuli should decrease over time, given that in normative development social transactions become increasingly more dependent upon the capacity to impute mental states to others and the self. Such a prediction would contradict the commonly observed tendency of autistic aloofness to decrease, not increase, with time (Volkmar *et al.* 1989; Wing 1988).

In sum, the failure of verbal autistic individuals to exhibit basic capacities for representing the feelings, wishes, and thoughts of others seems reasonably clear; what is less clear is whether the theory of mind hypothesis can sufficiently account for social deficits in their entirety, and through the tremendous range (in age, developmental level, and severity) of the syndrome expression.

ISSUES FOR FUTURE RESEARCH

The theory of mind hypothesis has made a significant contribution to the field of autism research, as it has refocused the attention of researchers and clinicians on the nature of the social dysfunction observed in autism. An appraisal of this hypothesis has suggested several potential lines of research that may further clarify the nature of this dysfunction.

As has been noted above, several predictions derived from the hypothesis appear to contradict the available evidence; the contradictions include the nature of the very early onset and pervasiveness of early autistic social dysfunction, and the fundamental nature (cognitive vs social) of the observed deficits. In this respect, it remains unclear whether deficits in theory of mind capacities are best conceptualized as cause or as correlate of autistic social dysfunction. Several factors also appear potentially to limit the applicability of the hypothesis to the entire range of the autistic syndrome: for instance, how does the theory of mind hypothesis explain the profound social deficits exhibited by the sizeable group of severely retarded and/or mute autistic persons, or the social difficulties experienced by high-functioning, relatively social-cognitively sophisticated, autistic adults who may exhibit at least rudimentary theory of mind capacities.

The general ecological validity (Bronfenbrenner 1979) these approaches have for autistic people remains to be established. More generally, and in the context of the theory of mind framework, it is the case that, outside highly artificial experimental settings, we usually operate on not two, but various representations of our conversational partners (see Mayes, Cohen, and Klin, this volume, Chapter 20).

An important contribution of the theory of mind framework is the proposition of a competence model of social skills. By abstracting the cognitive processes from other factors which together establish the context of a social situation, such as facial and body gestures and tone of voice, it imposes the rigour necessary for the establishment of testable hypotheses. However, we should not overlook the importance of these other factors. Wittgenstein (1968) once described the acquisition of language as a process in which the child learns to play language-games. Similarly, we may describe the acquisition of social skills as a process in which the child learns to play social games. Whether metacognitive capacities are products of the games, or presupposes them, remains an interesting issue to be explored.

Clearly, as has been noted, there is considerable potential for becoming over-involved in the issue of what aspects of the disorder (such as social–affective or cognitive) are 'primary'; as a practical matter, all developmental skills emerge with some relationship to each other. At this stage in our understanding of the autistic disorder, the various elements of social functioning should neither be prematurely discarded nor sidestepped in our research efforts. For example, even if social factors are not, ultimately, proved to be 'primary' in the pathogenesis of autism, the explication of autistic social dysfunction has considerable importance for the understanding of the pathogenesis, natural course, and wide range of expression of the social disturbance in autism, and, indeed, for the understanding of the role of social factors in normative child development.

Similarly, the subgroups of autistic individuals studied have tended, to date, to be rather higher-functioning; as we have noted, it remains unclear how, or whether, the hypothesis can account for the social deficits of the very large group of autistic individuals who are severely mentally handicapped, and remain essentially mute throughout their lives. This issue assumes particular significance given that at least one study (Prior *et al.* 1990) has suggested that deficits in theory of mind capacities might be more parsimoniously viewed as a function of overall developmental level rather than categorical diagnosis. Methodologically, it has also been the case that the large majority of experimental studies have matched autistic children with controls based on level of mental functioning. Given that the implied objective of these studies has been to characterize the specific underlying mechanism of autistic social dysfunction, it would appear that the metric of social developmental level should be employed instead.

Another area to be explored relates to those high-functioning autistic individuals who, as noted, remain unable to engage socially and emotionally with other people despite their apparent intellectual understanding of social situations. For these individuals 'the world of feelings and the language of emotions may be impossible or very hard to decipher. They lack the ability to empathically understand social situations and intuitively act upon their formal language' (Cohen 1991, p. 35). In Bemporad's (1979) report on one of Kanner's original patients who was now thirty-one and of normal intelligence, and had graduated from college, the major deficit pervading this person's life was a glaring lack of empathic relatedness, which went hand-in-hand with obvious intellectual knowledge of people. Bemporad noted that the stories accompanying Figure Drawings told by this man showed an awareness of how people should feel, without any evidence that he shared these feelings (p. 189). In response to a Thematic Apperception Test card, which showed a young woman's head against a man's shoulder, he commented: 'This is a husband and wife embracing each other — typical of one's so-called human need for companionship' (p. 189). Bemporad concludes that his subject, even as an adult, appeared to understand intellectually how another person might feel without seeming to be able to automatically sense in himself another's inner state (p. 195). Autistic individuals showing similar problems were seen and followed up for more than twenty years at the Child Study Center (Cohen 1980; Volkmar and Cohen 1985). Unfortunately, many of those individuals may recognize this impairment and become depressed (Cohen 1991). The absence of true intersubjective experiences in the presence of the prerequisite intellectual skills poses another challenge to the theory of mind conceptualization.

The possible variability in metacognitive capacities and the suggestion of a maturational delay in some cases (Baron–Cohen 1989) do go a long way in accounting for the observed phenomena. And yet the autistic population seems rather too varied for us to explain the heterogeneity of the phenotype solely by adopting a discrete cognitive incapacity as the only source of variance. Besides the factors already mentioned, such as IQ distribution, other aspects of social development might be implicated. For example, by considering the various early social adaptive mechanisms and their biological underpinnings, as well as interactions with the environment, one might be at a vantage-point to understand better the wide range of expression of the social dysfunction. It is also possible, and quite probable, that various different factors work synergistically to bring about the autistic phenotype. To paraphrase Bellac's (1958) proposal in the context of schizophrenia, autism might be a syndrome of multiple causation that works through a final common pathway. Better and early assessments, and longitudinal follow-up, might help create profiles of development, given certain developmental starts.

Finally, an important area for future research relates to the neurobiological processes that underlie the observed psychological deficit(s). The theory of mind hypothesis has successfully described a discrete cognitive deficit which is hypothesized to be a core deficit in autism. The importance of singling out discrete core deficits in autism has been repeatedly underscored (by, for example, Rapin 1987). Clearly, a core deficit might not, necessarily, be directly connected to aetiology (Rutter and Garmezy 1983), but it certainly provides better directives for biological research of the syndrome. Since there are a variety of processes involved in early social development, we should not overlook the possibility that various biological mechanisms might be involved in bringing about the autistic social dysfunction. Given the early plasticity of the central nervous system and the marked alterations in its structure occurring over the first years of life (see for example White and Held 1967), it is possible that CNS alterations in autism are reflected in changes in aspects of the fine structure or in the regulating chemicals of the brain, rather than in specific, readily localized, neuroanatomical sites or in specific neurotransmitter systems, which might be hypothesized as correlates of cognitive capacities. For example, the amygdala, a highly interconnected cluster of neurons lying deep in the medial temporal lobes, has been hypothesized to impart emotional tone to analysis of sensory data (Nauta and Feirtag 1986) and integrate affect and memory via a few memory-regulating neuropeptides (Leckman 1991). Abnormalities in this system could result in severe disruptions in socialization. In Brothers' (1989) thought-provoking review of the biological aspects of empathy, various other suggestions for research are explored. As he put it, 'during the evolution of the primate CNS, organization of neural activity has been shaped by the need for rapid and accurate evaluation of the motivations of others' (p. 10).

SUMMARY

There are many unanswered questions in our understanding of autistic social dysfunction. The theory of mind hypothesis has refocused our attention on the right issues, and further refined our concepts and methodologies. But, clearly, much remains ahead. And yet, for those of us who have participated in parent conferences after diagnostic evaluations, and followed the dramas of families who have courageously battled the disorder for years, it is a worthy enterprise.

REFERENCES

Baron-Cohen, S. (1988). Social and pragmatic deficits in autism: cognitive or affective? *Journal of Autism and Developmental Disorders*, **3**, 379–402.

Baron-Cohen, S. (1989). The autistic child's theory of mind: a case of specific developmental delay. *Journal of Child Psychology and Psychiatry*, **30**, 285–97.

Baron-Cohen, S. (1990*a*). Autism: a specific cognitive disorder of 'mind-blindness'. *International Journal of Psychiatry*, **2**, 81–90.

Baron-Cohen, S. (1990*b*). Precursors to a theory of mind: understanding attention in others. In *Natural theories of mind* (ed. A. Whiten). Basil Blackwell, Oxford.

Baron-Cohen, S., Leslie A.M., and Frith, U. (1985). Does the autistic child have a 'theory of mind'? *Cognition*, **21**, 37–46.

Baron-Cohen, S., Leslie, A.M., and Frith, U. (1986). Mechanical, behavioural and Intentional understanding of picture stories in autistic children. *British Journal of Developmental Psychology*, **4**, 113–25.

Bates, E., Benigni, L., Camaioni, L., Bretherton, I., and Volterra, V. (1979). *The emergence of symbols: cognition and communication in infancy*. Academic Press, New York.

Bellak, L. (1958). *Schizophrenia: a review of the syndrome*. Logos Press, New York.

Bemporad, J.R. (1979). Adult recollections of a formerly autistic child. *Journal of Autism and Developmental Disorders*, **9**, 179–97.

Bowlby, J. (1969). *Attachment and loss, Vol.I: Attachment*. Basic Books, New York.

Bronfenbrenner, U. (1979). *The ecology of human development*. Harvard University Press, Cambridge, Mass.

Brothers, L. (1989). A biological perspective on empathy. *American Journal of Psychiatry*, **146**, 10–19.

Caparulo, B., and Cohen, D.J. (1977). Cognitive structures, language and emerging social competence in autistic and aphasic children. *Journal of the American Academy of Child Psychiatry*, **16**, 620–45.

Clancy, H., and McBride, G. (1969). The autistic process and its treatment. *Journal of Child Psychology and Psychiatry*, **10**, 233–44.

Cohen, D.J. (1980). The pathology of the self in primary childhood autism and Gilles de la Tourette syndrome. *Psychiatric Clinics of North America*, **3**, 383–402.

Cohen, D.J. (1992). Finding meaning in one's self and others: Clinical studies of children with autism and Tourette's syndrome. In *Contemporary constructions for the child* (ed. F. Kessell and A. Sameroff). Erlbaum, Hillsdale, NJ.

Cohn, J.F., and Tronick, E.Z. (1987). The sequence of dyadic states at 3, 6, and 9 months. *Developmental Psychology*, **23**, 68–77.

Flavell, J., Botkin, P.T., Fry, C.L., Wright, J.W., and Jarvis, P.E. (1968). *The development of role-taking and communication skills in children*. Wiley, New York.

Frith, U. (1989). *Autism: explaining the enigma*. Basil Blackwell, Oxford.

Harlow, H.F., and Zimmermann, R.R. (1959). Affectional responses in the infant monkey. *Science*, **130**, 421–32.

Hobson, R.P. (1984). Early childhood autism and the question of egocentrism. *Journal of Autism and Developmental Disorders*, **14**, 85–104.

Hobson, R. P. (1989). Beyond cognition: a theory of autism. In *Autism: nature, diagnosis, and treatment* (ed. G. Dawson). Guilford Press, New York.

Kanner, L. (1943). Autistic disturbances of affective contact. *Nervous Child*, 2, 227–50.

Klin, A. (1989). Understanding early infantile autism: an application of G. H. Mead's theory of the emergence of mind. *L.S.E. Quarterly*, 3, 336–56.

Klin, A. (1991). Young autistic children's listening preferences in regard to speech: a possible characterization of the symptom of social withdrawal. *Journal of Autism and Developmental Disorders*, 21, 29–42.

Klin, A. (1992). Listening preferences in regard to speech in four children with developmental disabilities. *Journal of Child Psychology and Psychiatry*, 33, 763–69.

Klin A., Volkmar, F. R., and Sparrow, S. S. (1992). Autistic social dysfunction: some limitations of the theory of mind hypothesis. *Journal of Child Psychology and Psychiatry*, 33, 861–76.

Knobloch, H., and Pasamanick, B. (1975). Some etiologic and prognostic factors in early infantile autism and psychosis. *Pediatrics*, 55, 182–91.

Leckman, J. (1991). Genes and developmental neurobiology. In *Child and adolescent psychiatry* (ed. M. Lewis). Williams and Wilkins, Baltimore, Maryland.

Leslie, A. M. (1987). Pretense and representation: the origins of 'Theory of Mind'. *Psychological Review*, 94, 412–26.

Light, P. (1979). *The development of social sensitivity: of social aspects of role-taking in young children*. Cambridge University Press.

Lotter, V. (1978). Follow-up studies. In *Autism: a reappraisal of concepts and treatment* (ed. M. Rutter and E. Schopler). Plenum, New York.

Mead, G. H. (1934). *Mind, self and society*. University of Chicago Press.

Mills, M., and Melhuish, E. (1974). Recognition of mother's voice in early infancy. *Nature*, 252, 123–4.

Mundy, P., and Sigman, M. (1989). Specifying the nature of the social impairment in autism. In *Autism: nature, diagnosis, and treatment* (ed. G. Dawson). Guilford Press, New York.

Nauta, U. J. H., and Feirtag, M. (1986). *Fundamental neuroanatomy*. Freeman, New York.

Ortnitz, E. M., Guthrie, D., and Farley, A. H. (1977). Early development of autistic children. *Journal of Autism and Childhood Schizophrenia*, 7, 207–29.

Piaget, J. (1972). *The principles of genetic epistemology*. Routledge & Kegan Paul, London.

Plooij, F. X. (1978). Some basic traits of language in wild chimpanzees? In *Action, gesture, and symbol* (ed. A. Lock). Academic Press, New York.

Premack, D., and Woodruff, G. (1978). Does the chimpanzee have a 'theory of mind'? *Behavioral and Brain Sciences*, 4, 515–26.

Prior, M. R., and Bradshaw, J. L. (1979). Hemisphere functioning in autistic children. *Cortex*, 15, 73–81.

Prior, M. R., and Gajzago, C. (1974). Recognition of early signs of autism. *The Medical Journal of Australia*, August 3.

Prior, M., Dahlstrom, B., and Squires, T. L. (1990). Autistic children's knowledge of thinking and feeling states and other people. *Journal of Child Psychology and Psychiatry*, 31, 587–601.

Rapin, E. (1987). Searching for the cause of autism: a neurologic perspective. In

Handbook of of autism and pervasive developmental disorders (ed. D. J. Cohen and A. M. Donnellan). Wiley, New York.

Ricks, D. N., and Wing, L. (1975). Language, communication and symbols in normal and autistic children. *Journal of Autism and Childhood Schizophrenia*, **5**, 191–222.

Rutter, M. (1974). The development of infantile autism. *Psychological Medicine*, **4**, 147–63.

Rutter, M. (1981). *Maternal deprivation reassessed* (2nd edn). Penguin, New York.

Rutter M., and Garmezy, N. (1983). Developmental psychopathology. Chapter in *Handbook of child psychology, Vol. 4* (ed. E. M. Hetherington). Wiley, New York.

Rutter, M., and Schopler, E. (1987). Autism and pervasive developmental disorders: cancepts and diagnostic issues. *Journal of Autism and Developmental Disorders*, **17**, 159–86.

Rutter, M., Greenfeld, D., and Lockyer, L. (1967). A five- to fifteen-year follow-up of infantile psychosis: II. Social and behavioural outcome. *British Journal of Psychiatry*, **113**, 1183–9.

Shapiro, T., Sherman, M., Calamari, G., and Koch, D. (1987). Attachment in autism and other developmental disorders. *Journal of the American Academy of Child and Adolescent Psychiatry*, **26**, 480–4.

Sigman, M., and lingerer, J. (1984). Attachment behaviors in autistic children. *Journal of Autism and Developmental Disorders*, **14**, 231–44.

Stern, D. (1987). *The interpersonal world of the human infant*. Basic Books, New York.

Trevarthen, C. (1979). Communication and cooperation in early infancy: a description of primary intersubjectivity. In *Before speech: the beginning of interpersonal communication* (ed. M. Bullowa). Cambridge University Press, New York.

Tronick, E. (1980). The primacy of social skills in infancy. In *Exceptional infant, Vol. 4: Psychosocial risks in infant–environment transactions* (ed. D. B. Sawin, R. C. Hawkins, L. Olzenski Waker and J. H. Penticuff). Brunner–Mazel, New York.

Volkmar, F. R. (1987). Social development. In *Handbook of autism and pervasive developmental disorders* (ed. D. J. Cohen and A. Donnellan). Wiley, New York.

Volkmar, F. R., and Cohen, D. J. (1985). A first-person account of the experience of infantile autism by Tony W. *Journal of Autism and Developmental Disorders*, **15**, 47–54.

Volkmar, F. R., and Cohen, D. J. (1988). Diagnosis of pervasive developmental disorders. In *Advances in Clinical Child Psychology, Vol. 11* (ed. B. B. Lahey and A. E. Kazdin). Plenum, New York.

Volkmar, F. R., Stier, D. M., and Cohen, D. J. (1985). Age of recognition of pervasive developmental disorder. *American Journal of Psychiatry*, **142**, 1450–2.

Volkmar, F. R., Sparrow, S. A., Goudreau, D., Cicchetti, D. V., Paul, R., and Cohen, D. J. (1987). Social deficits in autism: an operational approach using the Vineland Adaptive Behavior Scales. *Journal of the American Academy of Child Psychiatry*, **26**, 156–61.

Volkmar, F. R., Bregman, J., Cohen, D. J., Hooks, M., and Stevenson, J. (1989). An examination of social typologies in autism. *Journal of the American Academy of Child and Adolescent Psychiatry*, **28**, 82–6.

Volkmar, F.R., Carter, A., Sparrow, S., and Cicchetti, D.V. (1992). Towards an operational definition of autism. *Journal of the American Academy of Child and Adolescent Psychiatry*. (In press.)

Wellman, H.M. (1990). *The child's theory of mind*. MIT Press–Bradford, Cambridge, Mass.

White, B.L., and Held, R. (1967). Experiences in early human development. II. Plasticity of sensorimotor development in the human infant. In *Exceptional infant*, Vol. 1 (ed. J. Hellmuth). Brunner–Mazel, New York.

Wing, L. (1988). The continuum of autistic characteristics. In *Diagnosis and assessment in autism* (ed. E. Schopler and G. Mesibov). Plenum, New York.

Part IV: Wider perspectives

16

The role of imitation in understanding persons and developing a theory of mind

ANDREW MELTZOFF AND ALISON GOPNIK

So soul into the soul may flow, though it to body first repair — John Donne

Normal adults share a network of ideas about human psychology that are often described as 'common-sense' psychology. Although we directly observe other people's behaviour, we think of them as having internal mental states that are analogous to our own. We think that human beings want, think, and feel, and that these states lead to their actions. Our ideas about these mental states play a crucial role in our interactions with others and in the regulation of our own behaviour.

Deepening our understanding of mind is a lifelong enterprise (Bruner, 1990); but recent research has shown that by the age of five years, children operate with many of the key elements of a common-sense psychology. By five years old, children seem to know that people have internal mental states such as beliefs, desires, intentions, and emotions. Moreover, they understand that a person's beliefs about the world are not just recordings of objects and events stamped upon the mind, but are active interpretations or construals of them from a given perspective. This allows five-year-olds to realize that people can have mental states that are different from their own, and that people act according to their mental representations of the world, rather than according to the way the world actually is.

Such a model explains a lot of otherwise baffling human behaviour; it allows children to predict and comprehend many events within the interpersonal sphere. This model of the way people work has been referred to as a 'representational model of mind' (Forguson and Gopnik 1988). Although there is some debate about details of timing, there is a consensus that such a model develops somewhere between three and six, and supplants an earlier 'non-representational' understanding of mind (see Gopnik 1990; Flavell 1988; Perner 1991; Wellman 1990; Whiten 1991).

Understanding the way other people's minds work, and knowing how those minds are similar to or different from your own mind, is crucial if you want to interact with people. A particularly dramatic example of this is the suggestion that the pervasive social–communicative impairments of people with autism are rooted in an inability to develop this kind of psychological understanding (Baron–Cohen *et al.* 1985). Autism has been likened to a kind of 'mindblindness' (Baron–Cohen 1990), in that autistic children seem unable to conceptualize another person as an entity with interpretative mental states.

The principal goal of this chapter is to inquire about the earliest developmental history of the *normal* child's understanding of the mind. How does common-sense psychology ever get off the ground? One way of putting this might be to say that we are interested in the earliest precursors of the child's 'theory of mind'. What sorts of things in infancy set the normal child on a developmental trajectory for eventually thinking of people as having interpretative minds — the level of psychologizing that seems so natural for five-year-olds, and so out of reach for most children with autism?

If we want to find the origins of common-sense psychology a good place to look might be in infant interactions with and understanding of persons. We will argue that the bedrock on which a commonsense psychology is constructed is the apprehension that others are similar to the self. Infants are launched on their career of interpersonal relations with the primary perceptual judgement: **'Here is something like me.'** One of the aims of the chapter is to explore the basis and cascading developmental effects of this sort of judgement.

It is sometimes held that normal infants are innately endowed with a special attentiveness to the human facial pattern. This may be so, but we will argue that this is not the only, or even the most critical basis for the 'like me' judgement. Such pattern detectors might direct visual attention, but in themselves they do not provide a link between the self and the other. The infant might see the adult as a particularly interesting entity; but because infants cannot see their own facial features, why should they think of the adult as relating to themselves? Similarly, others have seen the roots of intersubjectivity in the early temporal co-ordination of infant and adult behaviour, the 'conversational dances' that infants and care-takers perform. But again there seems no clear reason why these behaviours, by themselves, should lead the infants to think of other people as similar to themselves in deep ways. Infants, for example, also engage in temporally contingent interplay with objects.

We propose that infants' primordial 'like me' experiences are based on their understanding of bodily movement patterns and postures. Infants monitor their own body movements by the internal sense of proprioception, and can detect cross-modal equivalents between those movements-as-felt and the movements they see performed by others. Indeed, we will suggest that one reason normal infants preferentially attend to other people is the perceptual judgement that those entities are 'like me'. Without such a judgement, other humans might have interesting visual or temporal characteristics, but they would not have the unique place they do in our world.* It is this fundamental relatedness between self and other that we wish to explore in this chapter.

Until comparatively recently, there was no reason to suppose that young infants could apprehend cross-modal equivalences between body-movements-as-felt in the self and body-movements-as-seen in others. Indeed, classical theories of infant development explicitly denied this capacity to young infants, portraying the infant as 'solipsistic', 'radically egocentric', and so on. Among the recent experiments that served to change this view are those showing that normal infants are more proficient *imitators* than was previously thought. As we shall see, these findings suggest that infants can, at some basic level, process the correspondence between self and other (Meltzoff 1985).

The news is not just that infants imitate, for that has been known for some time (Baldwin 1906; Piaget 1962), but that they can imitate facial movements at an early age. Why is early facial imitation so important for developmental theory, and particularly for accounts of the ontogenesis of common-sense psychology? One reason is that it informs us about the 'starting state' of social cognition in normal infants. A second reason derives from the unique nature of facial imitation itself. Facial movements are special because infants cannot make a direct visual comparison between their own faces and those of adults. We will argue that early imitation is relevant to developing theories of mind because it provides the first, primordial instance of infants' making a connection between the visible world of others and the infants' own internal states, the way they 'feel' themselves to be.

Early imitation also provides a mechanism for infants' learning about other people and distinguishing them from things. In order for a common-sense psychology to get off the ground infants must make a basic cut between people and things, and respond to them differently. What is a person for a young infant, for a newborn? How would a newborn recognize one when he or she sees one? For the youngest infants, persons may not be defined solely in terms of salient facial features like the presence/absence of eyes. We suggest that infants at first rely on more functional rules (Meltzoff and Moore 1992). We suggest that for the youngest infants, persons are: 'entities that can be imitated and also who imitate me', entities that pass the 'like me' test. Such a rule would be effective in sorting the world into people versus things, and could be operative in the opening weeks of life—because the data show that infants imitate at birth.

---

* This reverses the standard developmental relation; but it is as easy, perhaps easier, to see how a primordial 'like me' apprehension might determine the direction of perceptual preferences than how raw looking preferences in and of themselves would ever lead to a means of making the 'like me' connection of interpersonal relatedness. We return to this issue in the conclusions of this chapter; also see Meltzoff and Moore 1993.

Moreover, as increasingly complex imitative interactions take place, this basic knowledge may be extended. In particular, at least by nine months of age or so, infants will not only imitate pure body movements but will also duplicate specific object-manipulations, and will do so after extended delays. Such deferred imitations provide an important source of information about objects in the world and the shared relation to those objects that people can hold. As we will show, imitation is not only an indicator of early common-sense psychology, but may itself be a mechanism for developing and elaborating this framework.

Imitation in infancy also runs in the reverse direction: parents mimic their infants as well as infants imitating parents. Why should this be so enjoyable to both parties? Trevarthen (1979), Bruner (1975, 1983), Stern (1985), and others have shown that infants seem to take pleasure in the temporal aspects of early interactions; the interactions can be likened to gestural dialogues, because of their turn-taking nature and overall rhythm. Without denying these temporal characteristics, we want to highlight a different aspect of the gestural dialogues. In particular, we will focus on a subset of interactive games that are imitative in nature. Mutual-imitation games may be an especially meaningful avenue of early communication because both partners can recognize the common acts — the self–other equivalences that exist when the body movements of one person match the other. We will suggest that over and above turn-taking and temporal factors, infants take special pleasure in mutual-imitation episodes because the adult's acts become more 'like me' in their form. Mutual-imitation games ratify the identity between adult and child.

BODY AND SOUL

The kind of 'like me' equivalences that we have discussed so far all involve equivalences between the child's *body* and the body of others. In contrast, the aspect of common-sense psychology that has attracted so much recent attention is the development of the understanding that people have *mental* states of a certain character. Is it helpful to think of infants' understanding of bodily movements as the bedrock for 'like me' judgements, and this in turn as being connected up to the ascription of 'like me' human minds? Quite apart from the infant data, there are philosophical reasons for thinking that some understanding of a 'like me' equivalence, indeed one centred on body equivalences, is wrapped up in our ascription of mind. Although 'philosophy of the body' has always been a neglected area of inquiry, several philosophers have suggested that such abstract mentalistic notions as reference may have their origins in the perception and understanding of bodies (for example Evans 1982). From this viewpoint it makes sense that infants are engaged in mapping out 'like me' equivalences in the bodily realm as the first step toward understanding persons.

Two aspects of the psychology of early imitation are particularly relevant here. First, the child maps externally perceived behaviour on to a set of *internal* bodily impressions. Second, the mapping is not only to internal states alone, but also to motor intentions and plans. We suggest that both internal proprioceptive sensations and motor intentions may be interesting half-way stations between behaviour on the one hand, and mental states on the other.

In common-sense psychology, one classical characteristic of mental states that distinguishes them from physical states is their spatial location. Mental states are located inside the skin (or the head or the body), while physical objects, including the bodies of others, are located outside it. In Wellman and Estes' (1986) work, this 'inside/outside' distinction is one of the first children use in differentiating the mental and the physical. Similarly, the paradigmatic example of behaviour is the body movements of others. The work on early imitation shows that even newborn infants recognize some equivalences between externally perceived behaviour — that is, perceived body movements — and literally internal proprioceptive states. Moreover, such proprioceptive sensations, in addition to being spatially located 'inside', would seem to have much of the character of mental states. In particular, they are not publicly observable, and are private experiences. Indeed, on many philosophical accounts, pains and other internal sensations, which are phenomenologically similar to proprioceptive sensations, are *the* quintessential mental states *par excellence*.

Moreover, in order to imitate, infants must not only recognize the similarities between externally perceived bodily movements on the one hand and internal proprioceptive sensations on the other, they also must map those externally perceived movements on to intentions of a sort. The child must not only know that this visually perceived movement maps on to that motor plan, but also know how to go about producing the motor plan in question; and in the case of deferred imitation the child must produce this motor plan in the absence of any visual guidance from the model.

These motor plans, like the internal proprioceptive sensations themselves, are an interesting midpoint between the physical and the mental. It seems difficult to draw a hard and fast line between such simple motor plans and, say, 'simple desires', which themselves are viewed in the theory of mind literature as providing legitimate instances of very early and primitive mentalism (Wellman 1990; Astington and Gopnik 1991). The new findings on imitation strongly imply that motor plans and intentions are mapped on to the behaviour of others from the start. It is as if children, in the case of simple desires, immediately recognize that the other person's behaviour implies desires similar to their own. This would be grounds for attributing a simple common-sense psychology capacity to the child. In the same way, in seeking the most primitive building-blocks of common-sense psychology, we see it as relevant that the young infant apprehends a similarity between a particular pattern of externally perceived behaviour, a particular internal proprioceptive sensation, and the motor plan that is necessary to produce both the sensation and the behaviour.*

Infants are, apparently, never strict behaviourists: one fundamental assumption of mentalism — that external, visible behaviours are mapped on to phenomenologically mental states — is apparently given innately. Clearly infants have much to learn about the nature of mind, but apparently they need not learn that it, or something like it, exists, and perhaps not even that it is shared by themselves and others. Ironically, given the great Platonic philosophical tradition of devaluing bodies in favour of minds, it may, quite literally, be our knowledge of the body that leads us to knowledge of the mind. From a developmental viewpoint, knowing that we inhabit similar bodies to others, and assuming that they share our internal bodily states, might be an important precursor to assuming that they share more abstract mental states as well. A person is, after all, both a body and a mind, and for very young infants these two aspects of personhood may not be divorced. (See Hobson, Chapter 10, this volume, for a similar view).

THE ORIGINS OF INFANT IMITATION AND THE NOTION OF A SUPRAMODAL BODY SCHEME: RECENT DATA AND THEORY

The last ten to fifteen years have seen the establishment of a new area of infant research, that of early infant imitation. Classical developmental theories had considered the imitation of facial actions to be a milestone in social–cognitive development that was first passed at about one year of age (Piaget 1962). Although other types of imitation, notably hand movements and vocal imitation, were said to occur earlier, facial imitation was classically viewed as a late achievement because infants cannot see their own faces. If they are young enough they will never have seen their own face in a mirror. How can infants possibly match a gesture they see with an action of their own that they cannot see? How can infants come to bridge the gap between visible and invisible experiences? Because this question is so baffling for developmental theory, researchers for many years were content with the analysis that facial imitation first became possible at about one year.

Meltzoff and Moore (1977) challenged the consensus that facial imitation was late to emerge by reporting that twelve- to twenty-one-day-old infants

---

* This capacity to map one's internal sensations on to the behaviour of others might form the aboriginal basis for a simulation device of the sort that has been proposed by Harris (1989, 1991), Johnson (1988), Gordon (1986), and Goldman (1987).

imitated tongue-protrusion, mouth-opening, and lip-protrusion. Beyond the raw fact that young infants imitate, there are several subtle points raised in this study and the ones that followed that are relevant to theories about the origins and early development of common-sense psychology.

First, the facial gestures used were picked to help assess the specificity of the imitative effects and distinguish it from a general arousal response. If infants were simply being aroused by the sight of a human face (but could not imitate) then they might make more oral movements when they saw a human face than when they saw no face at all. This would not support the inference of imitation; but the increased oral movements might be confused with imitation if the correct control conditions were not employed. In Meltzoff and Moore's work true imitation was demonstrated, because infants responded differentially to two types of lip-movements (mouth-opening vs lip-protrusion) and two types of protrusion actions (lip-protrusion vs tongue-protrusion). In other words, the results showed that when the body part was controlled, when lips were used to perform two subtly different movements, infants responded differentially. Likewise, when the same general movement pattern was demonstrated, a 'protrusion in space', but with two different body parts (lip- vs tongue-protrusion), they also responded differentially. The response was not global or a general reaction to the mere presence of a human being or a human face, because the same face was present in all these conditions, yet the infants responded differentially.

Another issue concerns the psychological basis of the imitation. It is critical to determine if young infants are restricted to some sort of reflexive mimicry, a kind of Gibsonian 'resonance' in which perception of human acts somehow 'directly' lead to their motor production with no intervening mediation. To test this notion experimentally a pacifier was put in infants' mouths as they watched the display, so that they could only observe the adult demonstration, but not duplicate the gestures. After the infant observed the display, the experimenter assumed a passive-face pose, and only then removed the pacifier. Infants were then given 2.5 minutes to respond, during which the adult maintained this passive face regardless of the infant's response. The pacifier was effective in disrupting imitation while the adult was demonstrating. Infants' sucking reflexes took precedence over any tendency to imitate. In Gibsonian terms, it was as if the second tuning-fork was bound and forbidden to resonate while the first tuning-fork was sounding. In such a situation there would, of course, be no transfer of the tone from one fork to the other. However, the infants imitated the displays. The finding suggests that imitation, even this very early imitation, could be mediated by memory of the absent display (Meltzoff 1990*a*; Meltzoff and Moore 1977, Study 2; Meltzoff and Moore 1989, 1992, 1994).

There are also other data showing that the early imitation is not well characterized as a simple reflex. In particular, the imitative response was not simply triggered, or fired off by the sight of the adult display. The data showed that the infants did not produce exact matches early in the response period. The first responses of the infants were often with the correct body part, but were only an approximation of the adult's act. Infants would move their tongues, but not produce full tongue-protrusions. Infants then appeared to home in on the detailed match, gradually correcting their responses over successive efforts to correspond more exactly to the details of the display. The adult was sitting with a passive face all this time; thus the infant was comparing his or her motor performance against some sort of internal model or representation of what had been seen. For these reasons and others it seems more accurate to think of early imitation as intentional matching to the target provided by the other, rather than as a rigidly-organized purely reflexive response (Meltzoff *et al.* 1991).

Learning theorists could argue that all this is unnecessary. The subjects were twelve to twenty-one days old. Perhaps they had been trained to imitate during the first weeks of life. Infants could be conditioned to poke out their tongues to a ringing sound, or to an adult tongue-protrusion. Perhaps the conditioning of a few oral gestures is part of the natural interaction between mother and baby. To resolve the point, Meltzoff and Moore (1983) tested 40 newborns in a hospital setting. The average age of the sample was 32 hours. The youngest infant was only 42 *minutes* old. The results showed that the newborns imitated both the gestures shown to them, mouth-opening and tongue-protrusion. We can infer that a primitive capacity to imitate is part of the normal child's innate endowment.

These findings of early infant imitation were originally considered surprising, and sparked lively debate in the literature. Surprising though they may be, they have now been replicated and extended in well over 20 different studies. Early imitation is a cross-cultural phenomenon: positive results have been reported in the US (Abravanel and Sigafoos 1984; Field *et al.* 1982); Canada (Legerstee 1991); France (Fontaine 1984); Switzerland (Vinter 1986); Sweden (Heimann and Schaller 1985; Heimann *et al.* 1989); Israel (Kaitz *et al.* 1988); and rural Nepal (Reissland 1988). In short, the basic phenomenon reported by Meltzoff and Moore has now been documented by independent investigators, in different settings, using a variety of different procedures. At a phenomenological level, the finding of early facial imitation seems secure. Attention has now shifted from debates about the existence of early matching to a search for the mechanisms underlying this behaviour and its role in development (Meltzoff and Moore 1992, 1994).

There are several psychological mechanisms that might underlie this behaviour. The hypothesis suggested by Meltzoff and Moore is that imitation is based on infants' capacity to register equivalences between the body transformations they see and the body transformations they only feel themselves make. On this account there is a primitive *supramodal body scheme* that allows the infant to unify acts-as-seen and acts-as-felt into a common framework. Meltzoff and Moore have argued that early imitation fits in with a larger network of perceptual and social-cognitive abilities that is also tapped by studies showing infant matching of facial movements and speech sounds (Kuhl and Meltzoff 1982, 1984) and other intermodal phenomena (Bower 1977, 1982, 1989; Meltzoff and Borton 1979). We suggest that the supramodal body scheme revealed by early imitation provides the foundation for the development of the notion of persons and self-other equivalences in infants, as elaborated later in this chapter (for further analysis see Meltzoff 1990*b*; Meltzoff and Moore 1992, 1994).

USING OTHERS AS A SOURCE OF INFORMATION ABOUT ACTIONS ON OBJECTS: DEFERRED IMITATION AND MEMORY

The foregoing research with neonates concerns imitation of basic body movements. Such imitative behaviours reveal a capacity to map internal states on to externally perceived behaviour, a kind of aboriginal mentalism. The states that are so mapped, however, could not be construed as referential in any way. There is no sense in which either the bodily movements that are imitated, or the proprioceptive sensations and motor plans, involve anything outside the child or the other person.

Later in development, however, we can see signs of what might be called 'proto-referential' imitation: imitation begins to be used as a mechanism for learning about how objects work. Children treat adults as a source of information about objects—they look to adults for guidance when they are uncertain how a particular novel object works, in a manner somewhat analogous to more traditional cases of social referencing (Campos and Stenberg 1981; Klinnert *et al.* 1983). Adult pedagogy often takes the form of showing the child that the object can be used in a peculiar new way. Certainly before language can be used with the child, much of the explicit teaching about the world by parents is done via showing the child what to do, and trying to elicit a decent reproduction of the activity. Of course, the adult's goal is not just to get the child to 'mindlessly' perform the act on-line, merely mimicking the act when the adult is performing it and failing to access this new information at a later time, after a significant delay. The parents' goal is to bequeath something to the child, to have the child incorporate it into his or her repertoire, in a sense to truly make it his or her own. In the experimental literature, the ontogenesis of these phenomena is deeply related to the problem of 'deferred imitation'.

Children who do not treat adults as a source of information about the world, who do not learn from observing the acts of others (perhaps because they cannot map between the self and the other), would be at a developmental disadvantage. At what age do normally-developing children begin to profit from deferred imitation? The classical view derives from Piaget, who thought that deferred imitation emerged contemporaneously with pretend-play, high-level object permanence, and productive language, at about eighteen to twenty-four months of age. We shall return to this potential connection between deferred imitation and pretend-play, partly motivated by Leslie's (1987, 1988, 1991) thesis that pretend-play is related to children's theory of mind, and partly because the recent data provide some new insights about the relation between play and imitation. To set the stage for this discussion we first provide a brief overview of some new studies on deferred imitation in normal children.

Meltzoff conducted a series of studies on deferred imitation in infants ranging from nine to twenty-four months old. One of the studies with fourteen-month-olds has three interesting features: (a) it tested imitation after an exceedingly long delay, one week; (b) infants were required to remember not just one demonstration, but to keep in mind multiple models—six different displays; (c) at least one of the acts was completely novel to the children. In particular, one object was a small wooden box with a translucent orange plastic panel for a top surface. The novel act demonstrated was for the experimenter to bend forward and touch the panel with the top of his forehead.

In this study, six different actions, each involving a different object, were shown to the infants (Meltzoff 1988*a*). Infants in the imitation group were shown all six actions on the first day of testing. They were then sent home for the one-week delay. Upon returning to the laboratory, the infants were given the objects one at a time to play with, and their behaviour was video-taped to determine how many of the target actions they reproduced. Two types of control groups were used. The control infants followed the same procedure as infants in the imitation condition, except that they did not see the target actions modelled on day 1, and so they had no memory of what to do with the toys. Like the infants in the imitation group, these control infants also visited the lab after a one-week delay. For the 'baseline' control group, the adult did not show the children the test toys on day 1, and simply talked pleasantly to the mother and child. This group assessed the spontaneous likelihood of the infants producing the target acts when they returned to the lab for the second visit. For the 'adult-manipulation' control group, the adult actively played with each of the objects during the first visit, but did not demonstrate the target acts themselves. This controlled for the possibility that infants might be induced into producing the target behaviour if they saw the adult approach and play with each object, even if the exact target action was not modelled.

The results provided clear evidence for deferred imitation. Of the 12 children in the imitation group, 11 duplicated three or more target behaviours on day 2, whereas only 3 of the 24 control subjects did so ($p < 0.0001$).

What is most striking is the aptitude these young infants exhibited for duplicating the novel act of using the forehead. Fully 67 per cent of the infants in the imitation condition produced this behaviour, as against none in the control conditions ($p < 0.0001$). Similar results have been reported showing deferred imitation in nine-month-old infants (Meltzoff 1988*b*), and these basic effects of imitation after a delay have been replicated and extended by Bauer and Mandler using a variety of tasks in infants between one and two years of age (Mandler 1990).

In the research discussed so far, an adult served as the model. In such cases the infants are directly mimicking with their own bodies acts that were seen in 3-D space with a minimum of differences between the adult's actions and the imitative act. It is also of interest whether infants can perform deferred imitation when there is 'distancing' (Werner and Kaplan 1963) between the self and the display to be copied. Television presents a miniature, two-dimensional depiction of actions in three-dimensional space. Meltzoff (1988*c*) found that fourteen-month-olds could also perform deferred imitation (24-hour delay) of particular object manipulations they had seen on TV, even when they had only seen the novel object on television and were not exposed to the real, 3-D variant until twenty-four hours later. These results suggest that for toddlers imitation is not highly stimulus-bound, and can be accomplished even in the face of some distancing and generalization.

More speculatively, the argument can be offered that these results also begin to address the developmental roots of children's capacity to use 'models' of reality to guide their action in space (DeLoache 1987, 1989; Perner 1991). The imitation-from-TV test would seem to be related to, but be a developmentally lower-order task than, DeLoache's intriguing studies on the use of scale model analogies by children. In the case of TV displays, the child needs to learn something in one problem space, a miniaturized depiction of reality by the TV, and project it on to its own actions in 3-D space with no direct comparison between the two. (The children first saw the act done by an adult on TV and then after a 24-hour delay they were given the real object for the first time. During the test, the TV model was absent. So children had to apply what they had learned from seeing the 'other' act in miniaturized, 2-D format to their own behaviour with a 3-D toy in a new situation. (For further discussion about what is involved here, see Meltzoff 1990*a*).

Older children and even adults learn more easily when the model is perceived to be more 'like me'. Hanna and Meltzoff (1993) conducted studies of peer imitation, in which infants were given the opportunity to watch and learn from other similar-aged playmates. In these studies some infants were trained to become 'infant experts' at particular tasks. Other infants, 'infant novices', observed these experts. In one experiment, the novice fourteen-month-old infants watched the expert fourteen-month -olds manipulate objects. A five-minute delay period was interposed, and then the novices were presented with the test objects. The results showed that of the infants who watched the experts, 80 per cent produced three or more of the five targets modelled, as opposed to only 1 of 20 control infants ($p < .0001$). The striking level of success in these peer-modelling studies raises the (somewhat counterintuitive) possibility that in some cases infants may actually learn better from observing their peers than from the pedagogical forays of parents. Perhaps toddlers perceive peers as more 'like me', and therefore the incorporation of the other's action as a basis for self-action is facilitated.

One wonders whether deferred imitation might be highly context-dependent at this age. Perhaps toddlers later re-enact actions only if they are in the same environment as they were when they first saw the demonstration. Imitation would be highly situation-bound. To test this we extended the peer-imitation paradigm (Hanna and Meltzoff 1993). The novices saw the expert perform actions in the laboratory. After a two-day delay, an adult experimenter went into the child's home and laid out the test objects. The results again showed strong evidence of deferred imitation. Infants who had previously watched the peer produced significantly more of the target acts than did controls — this despite the displacement in time (a two-day delay), space (the home context differed from the lab), and associated cues (the adult experimenter who tested the child at home was different from the experimenter used in the lab). This type of flexibility in observing others and then applying this knowledge in new settings is characteristic of normal infants. It seems quite likely that children with autism would be more context-bound and would be less likely to generalize to novel situations anything they managed to pick up from another's modelling.

In these cases of deferred imitation, children not only map perceived movements on to their own internal proprioceptive sensations and motor plans, they also do so with reference to objects. Not only do they seem to think 'this person is like me', but also to think that their responses to this object ought to be like the other person's. This suggests the beginnings of a shared attitude toward objects, in a way that is similar to the social referencing (Campos and Stenberg 1981; Klinnert *et al.* 1983) and joint-attention behaviours (Butterworth 1991; Butterworth and Jarrett 1991) that also appear at about this age in normally developing children. This synchrony in development may not be fortuitous.

MUTUAL IMITATION GAMES: A TEST OF 'LIKE ME' RECOGNITION IN INFANTS

Thus far we have shown that infants, from birth on, respond to the behaviour of others by producing similar behaviour of their own, and we have suggested that this indicates a mapping between the behaviour of the other and the infant's own behaviour and internal states. If this is true the process should, as it were, run both ways. That is, infants should not only imitate adults, but should also recognize when the adult is imitating them. It is, after all, equally true in this case that the infant's behaviour and the adult's are equivalent.

A series of experiments were conducted in which an adult purposely imitated the child, with the goal of determining if the child could recognize that his or her own behaviour was being adopted by the adult (Meltzoff 1990*b*). We wanted to know if fourteen-month-old infants could recognize such self-other correspondence, and if so, the psychological basis for this recognition.*

There were three converging experiments. The first investigated whether or not infants showed any interest in seeing that their own behaviour was adopted by another person. Two adults sat across a table from the child. All three participants were provided with replicas of the same toys. Everything the child did with his toy was directly mimicked by one of the adults, who had been assigned as the imitator. If the child slid the toy on the table, the imitating adult slid his toy on the table in the same manner. It was as if the adult were tethered to the child, a puppet under the child's control. The second adult was not so tethered. This adult sat passively, holding the toy loosely on the table top.

We thought that if children could recognize that their actions were being matched, they would prefer to look at the imitating adult and also smile at him more. We also thought that children would investigate this relationship between the self and the other by experimenting with it. For example, children might modulate their acts by performing sudden and unexpected movements to check if the imitating adult was still conforming to their actions. This is a way of 'catching the adult out', a way of experimenting with the relationship between self and world.

The results showed that infants had an overwhelming preference for the imitating adult over the non-imitating adult. Infants looked significantly longer at the imitating adult, there were more smiles directed toward the imitating adult, and infants directed more 'test' behaviour at the imitating adult. Of course, this study alone does not establish that infants can recognize the self–other equivalence engendered when another human acts just 'like me'. Infants may simply be attracted to any adult who actively manipulates a toy, without invoking any detection of like-me equivalence.

* In a loose sense, we set up an experiment in which we could study infants' reactions to what Searle (1983) calls 'world to mind' relationships. The world (the imitating adult) could be modified and manipulated in accordance with the child's whim. Is the infant interested in this? What criteria does the child use to determine that the events in the world correspond to the child's own action: temporal contingency information, or the structure of the action?

In a follow-up study, the general procedure was similar to that of the first study, but the control experimenter did not remain passive. Instead, this adult actively manipulated the toys. Furthermore, we wanted the adult not only to be active, but to do 'baby-like' things with the toys, so that no preference for the imitating adult could be based solely on a differentiation of adult versus infantile actions. We accomplished this by putting two TV monitors behind the infants, one monitor displaying the current infant and the other displaying the video record of the immediately preceding infant. The job of each adult was to mimic one of the infants on the TV monitors. Both adults performed in perfectly infantile ways, but only one matched the perceiving infant. Could the infants recognize which adult was a reflection of themselves, and which was acting like another baby? The results again showed that infants looked longer at the person who acted just like them, smiled more often at that person, and directed more testing behaviour toward him.

These effects cannot be explained as simple reactions to activity, for both adults were active. Nor can they be explained by saying that the infants recognized a generic class of baby-like actions, for both experimenters were copying the acts of babies. It would seem that the subjects are recognizing the *relatedness* of the actions of the self and the actions of the imitating other.

What is the basis for recognizing this sort of interpersonal relatedness? Two kinds of information are available, temporal contingency information and structural equivalences. On the first alternative, the child need only detect that whenever he does X the adult does Y. The child need not detect that X and Y are in fact equivalent, only that they are temporally linked. The second alternative is that the child can do more than recognize the temporal contingency between self and other. In particular, the child may be able to recognize that the actions of the self and other have the same form — that the adult is behaving 'just like me', not 'just when I act'.

To distinguish these alternatives, a third study was conducted in which the purely temporal aspects of the contingency were controlled by having both experimenters act at the same time. This was achieved by having three predetermined pairs of target actions. Both experimenters sat passively until the infant performed one of the target actions on this list. If and only if the infant exhibited one of these target actions, both experimenters began to act in unison. The imitating adult performed the infant's act, and the control adult performed the other behaviour that was paired with it from the predetermined target list. What differentiates the two experimenters is not the purely temporal relations with the acting subject, but the structure of their actions *vis-à-vis* the subject.

The results showed that the infants looked, smiled, and directed more testing behaviour at the adult who imitated them. Thus even with temporal contingency information controlled, infants can recognize the structural equivalence between self and other. In a very real sense, infants can recognize the reflection of themselves in an 'other'.

Normal children's games with their parents are often reciprocal in nature. The infant bangs a table top, the parent bangs in return, and so on. Theorists have emphasized the *temporal patterning* of these exchanges, the conversation-like turn-taking they embody (Bruner 1975, 1983; Stern 1985; Trevarthen and Marwick 1986). Without minimizing the importance of timing, our experiments highlight the importance of the *commonality in the structure of the bodily movements*. The new data show that when temporal contingency information is equated, young children still can detect which of two adults is conforming to the child's own behaviour. Moreover, these data demonstrate that when normally-developing children are given a choice, they preferentially attend to the adult who is matching them, and also smile more at this adult. The children respond socially, with increased looking and smiling, to an adult who is acting in the same way/manner/form that the infant is. Even before spoken language, normal infants seem to notice and appreciate this 'meaningful contact' with an other.

It is possible that children with autism have, among other deficits, an impairment in the capacity for recognizing the cross-modal isomorphisms between their own body movements and the movements of others; this would be compatible with Rogers and Pennington's (1991) theory of autism. If so, such children might find such interactions less predictable and enjoyable than normally developing children. This would be unfortunate, because mutual-imitation games are a unique and important constituent of early interpersonal growth. Adults are both selective and interpretative in the behaviour they reflect back to the child. They provide interpretative imitations to their infants: reflections that capture aspects of the infant's activity, but then go on beyond it to read in intentions and goals to that behaviour. The infant may wave an object, but the parent interprets this as waving in order to shake, and therefore waves intensely enough to shake the toy and produce a sound. This, in turn, leads the infant beyond his or her initial starting-point. Likewise, selected actions, especially those that are potentially meaningful in the culture, will be reflected back more often than others, as part of a larger process that Bruner (1975, 1983) has called 'parental scaffolding'. Children who had a disturbance in the ability to recognize interpersonal sharing at the level of motor imitation would not profit from such scaffolding in the same way as normally developing children. Thus impairments in motor imitation and/or the supramodal body scheme that underlies it could have extended developmental consequences.

IMITATION AND DEVELOPMENTAL PSYCHOPATHOLOGY— DOWN'S SYNDROME AND AUTISM

Unlike autistic children, children with Down's syndrome seem quite social; they smile at people and seem to enjoy interactions with them. Imitation has rarely been tested experimentally in this population (but see Dunst 1990), and the capacity for deferred imitation in particular has not been assessed.* Rast and Meltzoff (1993) adapted Meltzoff's deferred-imitation paradigm so that it could be used with young children Down's syndrome. A total of 48 children between the ages of twenty and forty-four months old were tested. A five-minute delay was used between the modelling period and the test of imitation. All the children were also given object-permanence tests to evaluate relations between the emergence of deferred imitation and high-level object-permanence skills.

As expected, the children were delayed in their understanding of object permanence. On average these children passed the A-not-B task (passed at about one year of age in normally developing children), and failed more complex tasks. Despite this retardation on object permanence, there was strong evidence for deferred imitation within the sample. We also divided the sample into 'young' (20–24 months) and 'old' (25–43 months) children. Deferred imitation was strongly evidenced in both age-groups. It is highly relevant for Piagetian theory that the young group showed strong evidence of deferred imitation, because not one of the young children passed high-level object-permanence tasks typical of 'stage 6' functioning (serial invisible displacements).

This pattern of results is quite baffling for classical theory, which postulates that deferred imitation (re-enacting a now invisible action from memory) and high-level object permanence (determining the location of a now-invisible object) emerge contemporaneously and are developmentally interdependent (Piaget 1952, 1962). On classical theory, one is led to ask why deferred imitation should be spared in Down's syndrome children and object permanence retarded. At a more general level, such a pattern of results would support the idea of 'deviance' in Down's syndrome children, inasmuch as two achievements that are synchronized in normally-developing children are broken apart in this syndrome. On this view, Down's syndrome children do not progress through the normal stages in a slowed-down manner, but

---

* There are many studies of object permanence, play, categorization, memory, and other infant skills in Down's syndrome children, but fewer experimental studies of imitation. In the studies that have assessed immediate imitation, controls of the type discussed in the foregoing sections have not been used, which means that it is difficult to distinguish true imitation from simpler types of social learning (see Meltzoff 1988*d* for a detailed discussion of the necessary controls for isolating true imitation versus social facilitation, stimulus enhancement, and so on).

rather show selective retardation in some areas (object permanence) and not others (deferred imitation).

Looked at from the viewpoint we have developed here and elsewhere (Meltzoff 1990a), however, the Down's syndrome pattern does *not* show 'deviance' from the normal pattern. We have presented evidence that the classical theory had profoundly underestimated imitative capacities which were once thought of as late to emerge, are actually building-blocks for development, and occur far earlier than has been assumed. In particular, we found that deferred imitation did not first arise in the 18–24-month age-group, but could be readily elicited in 9–14-month-old children. What this means is that the Down's syndrome results match the pattern found in normally-developing children quite closely: infants can perform deferred imitation well before solving 'serial invisible displacement' tasks on object permanence, and this appears to be true both in the normal and in this atypical population.* This underscores the necessity for interdisciplinary collaborations between those working with normal and atypical populations (Cicchetti 1989, 1990; Rutter and Garmezy 1983). If we are misinformed about the 'normal pattern' of psychological growth, we may mistake delay for deviance, and obscure underlying developmental patterns.

This immediately raises the question of autism, which does seem to be a case of developmental deviance. In relation to matched controls (often Down's syndrome children), autistic children show an impairment in social relations and communicative functioning. Autism seems to be a syndrome in which there are specific deficits, and not merely general retardation, although there is debate about the specificity of the impairments, as well as their origins and development (Baron–Cohen 1988, 1989, 1990, 1991a; Dawson and Lewy 1989a,b; Frith 1989; Hobson 1989, 1990a,b,c, 1991; Leslie 1987, 1988, 1991; Rogers and Pennington 1991; Mundy and Sigman 1989; Sigman 1989; other chapters in this volume). As frank 'outsiders' to the field, we tread with caution; none the less, there does seem to be something that can be added to the current debate by taking seriously the lessons from normally-developing and Down's syndrome children that have here been discussed.

In particular, we have presented data and thoughts as to the foundational role that imitation and cross-modal coordination play in the normal development of social and cognitive abilities. In a nutshell, we have proposed that the first act of common-sense psychology is the perception: 'here is something like me.' A disturbance to this primordial sense of kinship should

---

* Indeed the nine- to fourteen-month-old normally developing children who solved our deferred imitation tasks would be predicted to be at about the A-not-B stage of object permanence, just as was found in the Down's syndrome population. For a discussion of differences between object permanence and deferred imitation, see Meltzoff 1990a.

have cascading consequences for social–communicative development. Might autistic children have such a deficit?

This question is particularly relevant because the new data on normally-developing infants show that the perception of 'like me' relatedness to other human beings has a biological basis. Normal children are innately endowed with the capacity to imitate others. This provides a social bridge between the newborn and caregivers. It is as if humans are provided with an innate mechanism for social learning. This Janus-like quality of imitation, its biological basis coupled with its social implications for linking self with other, make it a key capacity to explore in autism.

There are no studies of neonatal imitation in autistic children. In part this is because of the recency of the discovery that neonates imitate; but it is also because children who will later be diagnosed as autistic are not born with genetic markers (at least not ones yet discovered) identifying themselves. One can study neonatal imitation in Down's syndrome children because they are genetically identifiable; but it would take broad-based screening and later follow-up testing to discover the actual neonatal status of later-identified autistic children — not an altogether uninteresting project (cf. Gillberg *et al.* 1990; Rogers and DiLalla 1990).

Is there any empirical support for the notion that children with autism might indeed have imitative deficits? There is accumulating evidence for an imitation impairment in autistic children (Rogers and Pennington 1991). A review of seven empirical studies done between 1972 and 1989 makes the point.

DeMyer *et al.* (1972) assessed imitation in autistic and mental-age-matched retarded children. The study compared imitation of pure body movements with actions on objects. Autistic children performed more poorly than the controls on both types of imitation tasks, but were particularly impaired on the imitation of simple body movements. Curcio (1978) tested autistic children using the Piagetian-based Uzgiris and Hunt (1975) scales. The study is of interest because of the sharp divergence between the children's performance on object-permanence and imitation tasks. Although 83 per cent of the subjects solved object-permanence tasks of the type passed by normally-developing 18–24-month-olds (serial invisible displacements), 5 of the 12 children did not imitate at all, and the majority could not imitate simple facial gestures, tasks we have shown to be within the capacity of normal newborns. Dawson and Adams (1984) also reported extreme deficits in motor imitation in autistic children who showed high levels of object-permanence understanding. Sigman and Ungerer (1984) tested imitation, general sensorimotor intelligence, and play in autistic, mentally retarded, and normal children. In relation to the mental-age-matched controls they found poor performance on imitation and play in the autistic children. The authors argued that these were specific impairments inasmuch as the results showed no differences in general levels of sensorimotor intelligence between the groups. Moreover, the normal and mental-age-matched retarded children did not significantly differ from each other in imitation.

Jones and Prior (1985) compared the imitation of simple body movements in autistic and chronological- and mental-age-matched normal children, and, in line with the foregoing studies, the autistic children showed significantly poorer imitative performance. Ohta (1987) reported a study of Japanese autistic children. In relation to control groups the autistic children showed an impairment in body-movement imitation, and when there were imitative attempts, children often performed odd, partial versions of the adult's display, as if they did not register human actions within the same body-scheme framework as normally-developing children. Hertzig *et al.* (1989) found imitative impairments, especially for imitating affect-related actions of people, in autistic children in relation to appropriately matched normal and non-autistic mentally retarded children.*

In contrast to these studies of imitative deficits in the gestural realm is the classic finding of inappropriate verbal echoing in children with autism (Rutter 1983), which occurs as part of a more general pattern of deviant language. There are many reasons why vocal imitation may differ from gestural imitation, including the obvious point that language is a highly canalized system that recruits specialized neurophysiological mechanisms.

At a more psychological level, it is intriguing to consider the possibility that the self–other mapping in the verbal sphere is quite different from that in gestural acts. For vocalizations, the actor can hear both the model and the self-productions. The behaviour of the self and the other are both picked up through the same modality, and are directly comparable. In the case of certain body gestures — for example facial acts, or even bringing objects to the head and so on — the child cannot make a direct comparison between self and other, because self and other are perceived through different modalities. The subject can see the model, but cannot *see* his or her own face, neck, back, etc. The imitation of these acts involves cross-modal mapping, and implicates a body scheme to coordinate the intercorporeal correspondences. Moreover, gestural imitation entails a kind of primitive perspective-taking that is quite unlike vocal imitation. Even in the case of non-facial body acts, such as manipulating an object, there is a kind of *perspective-taking in action*, because you see your own acts literally from a different perspective than you see the bodily act of the other. It is relevant to theories of mind that normal infants perform this motor-level perspective-taking with facility. The cross-modal nature of gestural imitation and the

---

* Some literature suggests that elementary imitations can be elicited in some children, but that there is a special, intensified difficulty in imitating more 'symbolic' gestural acts (Bartak *et al.* 1975; Curcio and Piserchia 1978; Hammes and Langdell 1981; Riguet *et al.* 1981).

primitive perspective-taking it entails may contribute to the imitative impairment in children with autism.

Over and above the imitative deficit *per se*, we are intrigued by the strikingly deviant pattern of mental organization that emerges if we compare the foregoing studies with autistic children to the results we found with Down's syndrome and normally-developing children. Our results with Down's syndrome and normal children reveal essentially the opposite mental make-up from that reported from children with autism: in the former populations, imitation (both simple body movements and deferred object-related tasks) is readily accomplished far in advance of any signs of 'stage 6' object permanence (serial invisible displacements); yet several of the studies with autistic children show the reverse pattern: object permanence in advance of imitation. This is generally compatible with the view that autistic children have deficits in social cognition.

Imitation after a significant delay — deferred imitation — has not been investigated in autistic children. It would be useful to compare autistic, Down's syndrome, and normally developing children on both object-permanence tests and Meltzoff's (1988*a,b*) new deferred-imitation paradigm. Children with autism show robust memory for *objects* that are no longer in view (object permanence); what of their capacity to imitate human *actions* that are no longer in view (deferred imitation)? Moreover, even if some children with autism can imitate in some tasks, it seems likely that extending the tests recently developed with normally-developing infants to this population would yield more refined information than is currently available.

For example: (a) children with autism may be limited to duplicating familiar routines and not perform deferred imitation of novel behaviours as exhibited by normally-developing infants Meltzoff 1988*a*; (b) they may be further handicapped by not generalizing actions they observe in one context to their own behaviour in a new context (cf. the Hanna and Meltzoff, 1993, laboratory-home generalization study); and (c) they may be disrupted by 'symbolic distancing' and not be able to use a representation of a human form as a model for their own actions (cf. the Meltzoff, 1988*c*, TV imitation in normal fourteen-month-olds). We expect that within the autistic population there will be individual differences in imitative functioning and, for the population as a whole, success will vary according to the precise nature of the task. Three imitation tasks that draw on differentiable psychological mechanisms and therefore may vary in children with autism are the following: pure body-movements, object-manipulations, symbolic gestures. All three could be assessed in immediate versus deferred tests and with or without changes in context. Systematic tests along these lines would deepen our grasp of the nature and extent of the imitation deficit that exists in people with autism.

CONCLUSIONS AND EXTENSIONS

Mutual imitation as a 'tutorial' in early common-sense psychology

Imitative games between parent and child have been reported in widely differing cultures. Do they serve any psychological function over and above the shared enjoyment that is experienced? As one parent expressed it to us: 'After playing these games I feel so happy — like I've been able to reach my baby and communicate with her.' Is there anything to this intuition about child-rearing?

We suggest that mutual-imitation games provide children with a kind of 'primer in common-sense psychology', a private tutorial in person-related versus thing-related interaction. Physical causality in the ordinary world of middle-sized objects has both spatial and temporal characteristics; there is physical contact between the cause and effect. In the imitation game the child 'causes' the adult to move in a particular way, but there is no physical contact between child and adult. Why does the child perceive his own actions as the cause of the adult's movements if there is no physical contact? It is because of the way that the parent arranges the game. The causal nature of the interaction is heightened not only because of the Hume-like temporal contingencies, but because of the cross-modal structural information of the parent's *imitation*. The child may interpret this 'action at a distance' — the cause–effect perception that is devoid of physical contact — as something like 'psychological control', or even communication. This ascription of communication might be especially motivated when the agent is the self and the recipient is another like-me human who can move just as I move. Just as hitting objects and watching them bump provides opportunities for exercising and enriching the child's naïve physics, the imitation game provides opportunities for the exercise and development of the child's naïve psychology.

A child who lacked the aboriginal capacity for perceiving self–other equivalences in such games might enjoy them less. Certainly, the game of predicting the adults' actions (which can be known, within limits, because they are copies or transformations of the child's own acts) would yield fewer successes. Rather than a tutorial in sharing and in predicting human behaviour, such interactions could easily become overwhelming. Children with autism may have a disturbance in the core mechanism for detecting the commonality in body movements between self and other. To the extent that mutual-imitation exchanges are tutorials in common-sense psychology, their absence or diminution might lead to deficits in social understanding and communicative functioning.

Imitation, empathy, and emotions

Emotions are mental states that have intrigued philosophers of mind, and their place within the theory of mind literature has been considered especially by Harris (1989; Harris and Gross 1988), Wellman (1990), Perner (1991), and Hobson (this volume, Chapter 10). There are no definitive answers, certainly no simple answers, to the questions of how and when children become able to 'give meaning to' the emotional expression of another and feel empathy with him or her. One is not surprised that models range from learning theories to innate pattern-decoders. The aim of this section is not to review these alternatives, but to highlight the special role that imitation might play. The particular idea we wish to discuss is that infants' imitation of facial movements is the substrate for early empathic reactions. A connection between imitation and empathy was early championed by Lipps (1906); the new empirical findings allow us to elaborate interesting developmental implications.

It has long been thought that there may be deep connections between body and mind in the case of human emotions (Darwin 1965 [1872]; James 1844; Tomkins 1962). The nature and strength of these connections has recently been analysed in an interesting series of converging studies. The importance of this new empirical work is that it goes beyond the ordinary claim that causation runs from subjective state (underlying emotional feeling) to behavioural expression. That has been known at least since Darwin's (1872) insightful claim about the innateness and cultural universality of certain basic emotional expressions. The importance of this new work is the empirical demonstration that causality runs in the opposite direction as well: the adoption of certain facial poses actually causes the corresponding mental states and physiological reactions.

For example, Ekman *et al.* (1983) discovered that if people produce certain facial muscular actions — certain muscle contractions around the eyes, brows, and mouth — this results in the corresponding emotion-specific physiological changes that naturally go with those facial patterns. Zajonc *et al.* (1989) measured self-reported emotional states directly after the subject was asked to produce a speech sound that brought his or her face into accord with a smiling position (saying the vowel 'ee') versus a different face. The results showed that adopting a facial pose influenced the underlying felt emotional state.

Thus, in the case of emotions, the body configuration does not just indicate or express or specify (depending on one's theory) how one feels, but can actually influence it. In this sense, emotions are quite different mental states from 'beliefs'. One can hold the belief that 'the chocolate is in the blue cupboard' regardless of one's facial expression or bodily configuration. For beliefs, the body does not mould the mind, at least not to the extent that a certain facial configuration can 'reach inside the mind' and alter the mental state. But this is just what happens in the case of emotions. The face bone is connected to the mind bone.

There is as yet no definitive research indicating that the bodily configuration influences emotional state in infants, but there is no compelling reason to dismiss this idea. Infants are known to produce basic emotional expressions in appropriate contexts—they smile when stroked, show fear faces at monstrous toys, produce disgust faces to bitter liquid, and so on. This suggests that there are connections between certain basic emotional states and facial expressions. This much does not have to be learned, and there is no reason to think that the bi-directionality of this connection is solely a product of learning.*

The question is, of course, how any of this would help to account for infant empathic reactions—feeling sadness when they see another being sad or feeling fear when they see another's fear. The view typically held is that this *emotional contagion* is somehow direct and unmediated. It has been said that there is an innate decoder for the meaning of basic emotional expressions, or an innate sympathy for conspecifics, or maybe a Gibsonian innately-based 'direct pick-up' of the distal variable, the emotional state. Nativism abounds.

Darwin (1965, p. 358) taught us that some sort of emotional–empathic reactions occur surprisingly early in development, and may have an innate basis, with the following observation of his infant son:

> When a few days over six months old, his nurse pretended to cry, and I saw that his face instantly assumed a melancholy expression, with the corners of the mouth strongly depressed; now this child could rarely have seen any other child crying, and I should doubt whether at so early an age he could have reasoned on the subject. Therefore it seems to me that an innate feeling must have told him that the pretended crying of his nurse expressed grief; and this through the instinct of sympathy excited grief in him.

One reason the unmediated empathy view is put forward is that young infants were classically thought to be incapable of facial imitation. If they cannot imitate the expressions they see, then they have no way of connecting the facial-expressions-as-seen in another with the corresponding mental state. The other's facial expression is 'out there' in space and publicly available to be seen. But the other's emotional state is 'inside', invisible. The only way they could make this emotional state their own was if there was 'direct perception' of the invisible mental state (the Gibsonian solution), or some sort of direct stamping of emotion into the heart of the baby via 'innate sympathy'.

---

* We do not wish to imply that the precise tuning of young infants' emotional-state categories is identical to that of adults or that there is no sharpening or learning involved in the development of emotions.

The discovery that young infants can imitate facial movements affords us another interpretation of Darwin's report. Perhaps infants do not begin by directly experiencing the emotional state of the other, but at first merely imitate the other's facial movements. A conservative view would be that they could perform such motor imitation even without (before) recognizing that it was an emotional expression *per se* that was being copied. For example, infants might even imitate components of the facial expression — just the lip position or the brow position, which is well within their capabilities. Motor imitation is not dependent on their knowing that the facial configuration carries emotional information and specifies an underlying emotional state in the other. Having imitated, having conformed their faces to the emotional expression, would then influence the child's own emotional state (as was the case with adults).

Thus imitation of the visible behaviour could be the avenue by which the invisible emotional state is transmitted. In other words, imitation of behaviour provides the bridge that allows the internal mental state of another to 'cross over' to and become one's own experienced mental state. Such a mechanism was untenable as an account of Darwin's observation of empathy in early infancy, if one adopted the traditional view that facial imitation was a late achievement. We now see that facial imitation is present in newborns. We are not postulating that imitation is the sole mechanism for empathy, especially in adults, but the findings of infant imitation make it plausible that it is one primitive mechanism for the interpersonal transfer of affect between parent and child.*

Children with autism show a relative lack of empathy — little indication that another's sadness touches them, that another's joy makes them feel happy. Children with autism also have an impairment in imitation. These two characteristics, lack of empathy and a deficit in behavioural imitation, may be causally related.

Imitation, autism, and two kinds of nativism

The empirical and theoretical issues here addressed have implications for recent proposals about the innate basis for common-sense psychology and a 'theory of mind'. The existence of a profound deficit in autistic children's theory of mind (see for example Baron–Cohen, Chapter 4, this volume) and the new data on infancy discussed here, provide evidence that *some* aspects of common-sense psychology are innately determined. One-hour-old infants map behaviour on to internal phenomenological states; you certainly cannot

* Something like the proposed process could occur in adults, as Lipps (1906) foresaw, because adults clearly can imitate facial expressions; the new information is that facial imitation is innate, and therefore could underlie empathic reactions right from the earliest phases of infancy.

get much more direct evidence of innate capacities than that. However, the more delicate problem is the question as to which particular aspects of common-sense psychology/theory of mind are innate, and the form that that innate knowledge may take.

For Leslie (1987, 1988, 1991; Chapter 5, this volume) the innate aspect of theory of mind involves the maturation of various metarepresentational abilities, particularly the representational 'decoupling' found in symbolic play of certain kinds. If taken at face value, this would predict that simple imitative abilities, which are non-metarepresentational in character, would be relatively unaffected in autism. In contrast, the aspect of common-sense psychology/theory of mind that we suggest is innate is not its referential or representational character, but the very idea of mentalism itself. At a very primitive level, normal children seem innately to map behaviours on to internal states; this is a starting-point for the later elaboration of common-sense psychology in the normal case. Children who lack this primitive sense of mentalism may well develop along different paths than normals because their construal of interpersonal encounters will be so very different.

A distinction can and should be drawn between two different forms of the nativist position — 'modularity nativism' and 'starting-state nativism' (Astington and Gopnik 1991; Gopnik and Wellman, in press). On the modularity view, well represented by Leslie's (1991) postulate of a 'theory of mind module', there are innate constraints on the form that a theory of mind may take. Certain kinds of cognitive architectures are innately determined, and represent indefeasible and unchanging constraints on the form of a final theory of mind. On the starting-state view, children are innately equipped with certain kinds of information about the nature of persons. In particular, we have here adduced data to show that they innately apprehend other human beings as 'like me' in fundamental ways. On the starting-state view this information itself may be modified or revised as the child learns more about the world and the people in it.

Modularity nativism and starting-state nativism lead to rather different views of the nature of the 'theory of mind' deficit discovered in people with autism. On the first view, the deficit represents a failure of the growth of a particular piece of cognitive architecture, in Leslie's account of the growth of a 'decoupler'. One might think of children with autism as psychological thalidomide victims, in whom a particular mental organ (to use Chomsky's phrase) fails to mature. On the second view, the absence of the initial starting-state means that the evidence available to children elaborating their understanding of mind is seriously limited in a way that it is not for normally developing children. Children with autism, on this view, might be seen as more analogous to astronomers who try to develop theories of the stars without telescopes. There appear to be innate mechanisms that allow newborn infants to accumulate particularly relevant kinds of evidence about mental life. In particular, these innate mechanisms allow them to map at least some of their internal states on to the behaviour of others. These mechanisms provide an important beginning point for constructing notions of mind and persons, even if they do not specify the final state of those notions (Gopnik and Meltzoff, in press).

Our view is closer in this respect to views proposed to account for autism by Rogers and Pennington (1991), Hobson (1990*b*, 1991), and Baron–Cohen (1991*b*). For example, Hobson's idea that infants are innately equipped with the ability to see others as persons rather than as objects, and that this capacity is damaged in autism, is one that fits well with what we are proposing here. We suggest, however, that the evidence for that ability, and the mechanism by which it takes place, may be rather different from that proposed by Hobson for autistic children, and by others such as Trevarthen and Stern for the normally-developing child. These accounts rely heavily on evidence of an early 'affective attunement' between infants and others. This attunement is seen as evidenced in the temporal synchrony of early infant and adult behaviours—the coordinated 'conversational dance' or 'proto-conversation' typical of very early mother–infant interaction. Although such behaviours may indeed be the result of an early concept of the person, there is, as we have argued, nothing in the fact of temporal synchrony itself that seems to require such a concept.

Early imitative interactions, on the other hand, require intersubjective attunement in a deeper sense, because they literally involve a mapping of the behaviour of the other and the child's own internal state. In the final analysis, there may be an interconnected web of different proclivities— imitation and 'like me' apprehension being key (or so we would argue), but in conjunction with temporally synchronized action, preferential attention to faces and voices, etc.—that together constitute a 'starting state' in which infants recognize that they themselves, and the others around them, are all persons together.

There is another respect in which imitation informs the debate about innateness and children's understanding of mind. As we have repeatedly emphasized, imitation is not only a sign of certain common-sense psychological capacities, it is also a mechanism by which the understanding of mind might be developed. We have already suggested, for example, how imitation might play a role in differentiating persons and objects, in distinguishing physical and psychological causality, and in establishing empathy. The early existence of this powerful technique for learning about other people may help infants to elaborate a common-sense psychology that goes far beyond their innate endowment.

Acknowledgments

We thank P. K. Kuhl for comments on an earlier draft, and M. K. Moore for long-term collaboration and insightful ideas on both the empirical and theoretical issues raised in this chapter. We thank J. S. Bruner for initially sparking our collaboration and for pointing out new paths to explore, and S. Baron–Cohen for his intellectual generosity. We gratefully acknowledge the financial support of the following agencies: NIH (HD-22514), Washington Association of Retarded Citizens, University of Washington Graduate School Research Fund, Bloedel Hearing Research Center, Joseph P. Kennedy Foundation, and James McKeen Cattell Fund.

REFERENCES

Abravanel, E., and Sigafoos, A. D. (1984). Exploring the presence of imitation during early infancy. *Child Development*, **55**, 381–92.

Astington, J. W., and Gopnik, A. (1991). Developing understanding of desire and intention. In *Natural theories of mind* (ed. A. Whiten). Basil Blackwell, Oxford.

Baldwin, J. M. (1906). *Mental development in the child and the race* (3rd edn). Augustus M. Kelley, New York.

Baron–Cohen, S. (1988). Social and pragmatic deficits in autism: cognitive or affective? *Journal of Autism and Developmental Disorders*, **18**, 379–402.

Baron–Cohen, S. (1989). The autistic child's theory of mind: a case of specific developmental delay. *Journal of Child Psychology and Psychiatry*, **30**, 285–97.

Baron–Cohen, S. (1990). Autism: a specific cognitive disorder of 'mind-blindness'. *International Review of Psychiatry*, **2**, 81–90.

Baron–Cohen, S. (1991a). The theory of mind deficit in autism: how specific is it? *British Journal of Developmental Psychology*, **9**, 301–14.

Baron–Cohen, S. (1991b). Precursors to a theory of mind: understanding attention in others. In *Natural theories of mind* (ed. A. Whiten). Basil Blackwell, Oxford.

Baron–Cohen, S., Leslie, A. M., and Frith, U. (1985). Does the autistic child have a 'theory of mind'? *Cognition*, **21**, 37–46.

Bartak, L., Rutter, M., and Cox, A. (1975). A comparative study of infantile autism and specific developmental receptive language disorder. *British Journal of Psychiatry*, **126**, 127–45.

Bower, T. G. R. (1977). *The perceptual world of the child*. Harvard University Press, Cambridge.

Bower, T. G. R. (1982). *Development in infancy* (2nd edn). Freeman, San Francisco.

Bower, T. G. R. (1989). *The rational infant: learning in infancy*. Freeman, New York.

Bruner, J. S. (1975). From communication to language — a psychological perspective. *Cognition*, **3**, 255–87.

Bruner, J. S. (1983). *Child's talk: learning to use language*. Norton, New York.

Bruner, J. S. (1990). *Acts of meaning*. Harvard University Press, Cambridge, Mass.

Butterworth, G. (1991). The ontogeny and phylogeny of joint visual attention. In

Natural theories of mind (ed. A. Whiten). Basil Blackwell, Oxford.

Butterworth, G., and Jarrett, N. (1991). What minds have in common is space: spatial mechanisms serving joint visual attention in infancy. *British Journal of Developmental Psychology*, **9**, 55–72.

Campos, J. J., and Stenberg, C. R. (1981). Perception, appraisal and emotion: the onset of social referencing. In *Infant social cognition* (ed. M. E. Lamb and L. R. Sherrod). Erlbaum, Hillsdale, NJ.

Cicchetti, D. (1989). *The emergence of a discipline: Rochester symposium on developmental psychopathology*. Erlbaum, Hillsdale, NJ.

Cicchetti, D. (1990). A historical perspective on the discipline of developmental psychopathology. In *Risk and protective factors in the development of psychopathology* (ed. J. Rolf, A. Masten, D. Cicchetti, K. Nuechterlein, and S. Weintaub). Cambridge University Press, New York.

Curcio, F. (1978). Sensorimotor functioning and communication in mute autistic children. *Journal of Autism and Childhood Schizophrenia*, **8**, 281–92.

Curcio, F., and Piserchia, E. A. (1978). Pantomimic representation in psychotic children. *Journal of Autism and Childhood Schizophrenia*, **8**, 181–9.

Darwin, C. 1965 [1872]. *The expression of the emotions in man and animals*. University of Chicago Press.

Dawson, G., and Adams, A. (1984). Imitation and social responsiveness in autistic children. *Journal of Abnormal Child Psychology*, **12**, 209–26.

Dawson, G., and Lewy, A. (1989*a*). Arousal, attention, and the socioemotional impairments of individuals with autism. In *Autism: nature, diagnosis, and treatment* (ed. G. Dawson). Guilford, New York.

Dawson, G., and Lewy, A. (1989*b*). Reciprocal subcortical–cortical influences in autism. In *Autism: nature, diagnosis, and treatment* (ed. G. Dawson). Guilford, New York.

DeLoache, J. S. (1987). Rapid change in the symbolic functioning of very young children. *Science*, **238**, 1556–7.

DeLoache, J. S. (1989). Young children's understanding of the correspondence between a scale model and a large-scale space. *Cognitive Development*, **4**, 121–39.

DeMyer, M. K., Alpern, G. D., Barton, S., DeMyer, W. E., Churchill, D. W., Hingtgen, J. N., *et al.* (1972). Imitation in autistic, early schizophrenic, and non-psychotic subnormal children. *Journal of Autism and Childhood Schizophrenia*, **2**, 264–87.

Dunst, C. J. (1990). Sensorimotor development of infants with Down syndrome. In *Children with Down syndrome: A developmental perspective* (ed. D. Cicchetti and M. Beeghly). Cambridge University Press.

Ekman, P., Levenson, R. W., and Friesen, W. V. (1983). Autonomic nervous system activity distinguishes among emotions. *Science*, **221**, 1208–10.

Evans, G. (1982). *The varieties of reference*. Clarendon Press, Oxford.

Field, T. M., Woodson, R., Greenberg, R., and Cohen, D. (1982). Discrimination and imitation of facial expressions by neonates. *Science*, **218**, 179–81.

Flavell, J. H. (1988). The development of children's knowledge about the mind: from cognitive connections to mental representations. In *Developing theories of mind* (ed. J. W. Astington, P. L. Harris, and D. R. Olson). Cambridge University Press.

Fontaine, R. (1984). Imitative skills between birth and six months. *Infant Behavior and Development*, **7**, 323–33.

Forguson, L., and Gopnik, A. (1988). The ontogeny of common sense. In *Developing theories of mind* (ed. J. W. Astington, P. L. Harris, and D. R. Olson). Cambridge University Press.

Frith, U. (1989). *Autism: explaining the enigma*. Basil Blackwell, Oxford.

Gillberg, C., Ehlers, S., Schaumann, H., Jakobsson, G., Dahlgren, S. O., Lindblom, R. *et al.* (1990). Autism under age three years: a clinical study of 28 cases referred for autistic symptoms in infancy. *Journal of Child Psychology and Psychiatry*, **31**, 921–34.

Goldman, A. I. (1989). Interpretation psychologized. *Mind and Language*, **4**, 161–85.

Gopnik, A. (1990). Developing the idea of intentionality: children's theories of mind. *Canadian Journal of Philosophy*, **20**, 89–114.

Gopnik, A., and Meltzoff, A. N. (In press). Minds, bodies, and persons: young children's understanding of the self and others as reflected in imitation and 'theory of mind' research. In *Self-awareness in animals and humans: Developmental perspectives* (ed. S. Parker, M. Boccia, and R. Mitchell). Cambridge University Press.

Gopnik, A., and Wellman, H. M. The 'theory' theory. In *Domain-specificity in cultural cognition* (ed. L. Hirshfield and S. Gelman). Cambridge University Press, New York. (In press.)

Gordon, R. M. (1986). Folk psychology as simulation. *Mind and Language*, **1**, 158–171.

Hammes, J. G. W., and Langdell, T. (1981). Precursors of symbol formation and childhood autism. *Journal of Autism and Developmental Disorders*, **11**, 331–45.

Hanna, E., and Meltzoff, A. N. (1993). Peer imitation by toddlers in laboratory, home, and day-care contexts: Implications for social learning and memory. *Developmental Psychology*, **29**, 701–10.

Harris, P. L. (1989). *Children and emotion: the development of psychological understanding*. Basil Blackwell, Oxford.

Harris, P. L. (1991). The work of imagination. In *Natural theories of mind* (ed. A. Whiten). Basil Blackwell, Oxford.

Harris, P. L., and Gross, D. (1988). Children's understanding of real and apparent emotion. In *Developing theories of mind* (ed. J. W. Astington, P. L. Harris, and D. R. Olson). Cambridge University Press.

Heimann, M., and Schaller, J. (1985). Imitative reactions among 14–21-day-old infants. *Infant Mental Health Journal*, **6**, 31–9.

Heimann, M., Nelson, K. E., and Schaller, J. (1989). Neonatal imitation of tongue protrusion and mouth opening: methodological aspects and evidence of early individual differences. *Scandinavian Journal of Psychology*, **30**, 90–101.

Hertzig, M. E., Snow, M. E., and Sherman, M. (1989). Affect and cognition in autism. *Journal of the American Academy of Child and Adolescent Psychiatry*, **28**, 195–9.

Hobson, R. P. (1989). Beyond cognition: a theory of autism. In *Autism: nature, diagnosis, and treatment* (ed. G. Dawson), Guilford, New York.

Hobson, R. P. (1990*a*). On acquiring knowledge about people and the capacity to pretend: response to Leslie (1987). *Psychological Review*, **97**, 114–21.

Hobson, R. P. (1990*b*). On the origins of self and the case of autism. *Development and Psychopathology*, **2**, 163–81.

Hobson, R. P. (1990c). On psychoanalytic approaches to autism. *American Journal of Orthopsychiatry*, **60**, 324–36.

Hobson, R. P. (1991). What is autism? *Psychiatric Clinics of North America*, **14**, 1–17.

James, W. (1844). What is an emotion? *Mind*, **9**, 188–205.

Johnson, C. N. (1988). Theory of mind and the structure of conscious experience. In *Developing theories of mind* (ed. J. W. Astington, P. L. Harris, and D. R. Olson). Cambridge University Press.

Jones, V., and Prior, M. (1985). Motor imitation abilities and neurological signs in autistic children. *Journal of Autism and Developmental Disorders*, **13**, 37–46.

Kaitz, M., Meschulach-Sarfaty, O., Auerbach, J., and Eidelman, A. (1988). A re-examination of newborn's ability to imitate facial expressions. *Developmental Psychology*, **24**, 3–7.

Klinnert, M. D., Campos, J. J., Sorce, J. F., Emde, R. N., and Svejda, M. (1983). Emotions as behavior regulators: social referencing in infancy. In *Emotion: theory, research, and experience, Vol. 2* (ed. R. Plutchik and H. Kellerman), Vol. 2. Academic Press, New York.

Kuhl, P. K., and Meltzoff, A. N. (1982). The bimodal perception of speech in infancy. *Science*, **218**, 1138–41.

Kuhl, P. K., and Meltzoff, A. N. (1984). The intermodal representation of speech in infants. *Infant Behavior and Development*, **7**, 361–81.

Legerstee, M. (1991). The role of person and object in eliciting early imitation. *Journal of Experimental Child Psychology*, **51**, 423–33.

Leslie, A. M. (1987). Pretense and representation in infancy: the origins of 'theory of mind'. *Psychological Review*, **94**, 412–26.

Leslie, A. M. (1988). Some implications of pretense for mechanisms underlying the child's theory of mind. In *Developing theories of mind* (ed. J. W. Astington, P. L. Harris, and D. R. Olson). Cambridge University Press, New York.

Leslie, A. M. (1991). The theory of mind impairment in autism: evidence for a modular mechanism of development? In *Natural theories of mind* (ed. A. Whiten). Basil Blackwell, Oxford.

Lipps, T. (1906). Das Wissen von fremden Ichen. *Psychologische Untersuchungen*, **1**, 694–722.

Mandler, J. M. (1990). Recall of events by preverbal children. In *The development and neural bases of higher cognitive functions* (ed. A. Diamond). *Annals of the New York Academy of Sciences*, **608**, 485–516.

Meltzoff, A. N. (1985). The roots of social and cognitive development: models of man's original nature. In *Social perception in infants* (ed. T. M. Field and N. A. Fox). Ablex, Norwood, NJ.

Meltzoff, A. N. (1988a). Infant imitation after a 1-week delay: long-term memory for novel acts and multiple stimuli. *Developmental Psychology*, **24**, 470–6.

Meltzoff, A. N. (1988b). Infant imitation and memory: nine-month-olds in immediate and deferred tests. *Child Development*, **59**, 217–25.

Meltzoff, A. N. (1988c). Imitation of televised models by infants. *Child Development*, **59**, 1221–9.

Meltzoff, A. N. (1988d). The human infant as *Homo imitans*. In *Social learning: psychological and biological perspectives* (ed. T. R. Zentall and J. B. G. Galef). Erlbaum, Hillsdale.

Meltzoff, A. N. (1990a). Towards a developmental cognitive science: the implications of cross-modal matching and imitation for the development of representation and memory in infancy. In *The development and neural bases of higher cognitive functions* (ed. A. Diamond). *Annals of the New York Academy of Sciences*, **608**, 1–31.

Meltzoff, A.N. (1990*b*). Foundations for developing a concept of self: The role of imitation in relating self to other and the value of social mirroring, social modeling, and self practice in infancy. In *The self in transition: infancy to childhood* (ed. D. Cicchetti and M. Beeghly). University of Chicago Press.

Meltzoff, A.N., and Borton, R.W. (1979). Intermodal matching by human neonates. *Nature*, **282**, 403–4.

Meltzoff, A.N., and Moore, M.K. (1977). Imitation of facial and manual gestures by human neonates. *Science*, **198**, 75–8.

Meltzoff, A.N., and Moore, M.K. (1983). Newborn infants imitate adult facial gestures. *Child Development*, **54**, 702–9.

Meltzoff, A.N., and Moore, M.K. (1989). Imitation in newborn infants: Exploring the range of gestures imitated and the underlying mechanisms. *Developmental Psychology*, **25**, 954–62.

Meltzoff, A.N., Kuhl, P.K., and Moore, M.K. (1991). Perception, representation, and the control of action in newborns and young infants: toward a new synthesis. In *Newborn attention: biological constraints and the influence of experience* (ed. M.J.S. Weiss and P.R. Zelazo). Ablex, Norwood, NJ.

Meltzoff, A.N., and Moore, M.K. (1992). Early imitation within a functional framework: the importance of person identity, movement, and development. *Infant Behavior and Development*, **15**, 479–505.

Meltzoff, A.N., and Moore, M.K. (1993). Why faces are special to infants — On connecting the attraction of faces and infants' ability for imitation and cross-modal processing. In *Developmental neurocognition: Speech and face processing in the first year of life* (ed. B. de Boysson-Bardies, S. de Schonen, P. Jusczyk, P. Mac-Neilage, and J. Morton). Kluwer Academic Publishers, Dordrecht, Netherlands.

Meltzoff, A.N., and Moore, M.K. (1994). Imitation, memory, and the representation of persons. *Infant Behavior and Development*, **17**, 83–99.

Mundy, P., and Sigman, M. (1989). Second thoughts on the nature of autism. Development and Psychopathology, **1**, 213–17.

Ohta, M. (1987). Cognitive disorders of infantile autism: a study employing the WISC, spatial relationship conceptualization, and gesture imitations. *Journal of Autism and Developmental Disorders*, **17**, 45–62.

Perner, J. (1991). *Understanding the representational mind*. MIT Press, Cambridge, Mass.

Piaget, J. (1952). *The origins of intelligence in children*. International Universities Press, New York.

Piaget, J. (1962). *Play, dreams and imitation in childhood*. Norton, New York.

Rast, M., and Meltzoff, A.N. (1993). Imitation from memory in young children with Down's syndrome. Presented at the biennial meeting of the Society for Research in Child Development. New Orleans, LA.

Reissland, N. (1988). Neonatal imitation in the first hour of life: observations in rural Nepal. *Developmental Psychology*, **24**, 464–9.

Riguet, C.B., Taylor, N.D., Benaroya, S., and Klein, L.S. (1981). Symbolic play in autistic, Down's, and normal children of equivalent mental age. *Journal of Autism and Developmental Disorders*, **11**, 439–48.

Rogers, S. J., and DiLalla, D. L. (1990). Age of symptom onset in young children with pervasive developmental disorders. *Journal of the American Academy of Child and Adolescent Psychiatry*, **29**, 863–72.

Rogers, S. J., and Pennington, B. F. (1991). A theoretical approach to the deficits in infantile autism. *Development and Psychopathology*, **3**, 137–62.

Rutter, M. (1983). Cognitive deficits in the pathogenesis of autism. *Journal of Child Psychology and Psychiatry*, **24**, 513–31.

Rutter, M., and Garmezy, N. (1983). Developmental psychopathology. In *Handbook of child psychology* (ed. P. Mussen). Wiley, New York.

Searle, J. R. (1983). *Intentionality*. Cambridge University Press.

Sigman, M. (1989). The application of developmental knowledge to a clinical problem: the study of childhood autism. In *The emergence of a discipline: Rochester symposium on developmental psychopatholgy* (ed. D. Cicchetti). Erlbaum, Hillsdale, NJ.

Sigman, M., and Ungerer, J. A. (1984). Cognitive and language skills in autistic, mentally retarded, and normal children. *Developmental Psychology*, **20**, 293–302.

Stern, D. N. (1985). *The interpersonal world of the infant*. Basic Books, New York.

Tomkins, S. S. (1962). *Affect, imagery, and consciousness*. Springer, New York.

Trevarthen, C. (1979). Communication and cooperation in early infancy: a description of primary intersubjectivity. In *Before speech* (ed. M. Bullowa). Cambridge University Press.

Trevarthen, C., and Marwick, H. (1986). Signs of motivation for speech in infants, and the nature of a mother's support for development of language. In *Precursors of early speech* (ed. B. Lindblom and R. Zetterström). Stockton Press, New York.

Uzgiris, I. C., and Hunt, J. M. (1975). *Assessment in infancy: ordinal scales of psychological development*. University of Illinois Press, Urbana.

Vinter, A. (1986). The role of movement in eliciting early imitations. *Child Development*, **57**, 66–71.

Wellman, H. M. (1990). *The child's theory of mind*. MIT Press, Cambridge, Mass.

Wellman, H. M., and Estes, D. (1986). Early understanding of mental entities: a reexamination of childhood realism. *Child Development*, **57**, 910–23.

Werner, H., and Kaplan, B. (1963). *Symbol formation: an organismic developmental approach to language and the expression of thought*. Wiley, New York.

Whiten, A. (1991). *Natural theories of mind*. Basil Blackwell, Oxford.

Zajonc, R. B., Murphy, S. T., and Inglehart, M. (1989). Feeling and facial efference: implications of the vascular theory of emotion. Psychological Review, **96**, 395–416.

17

Evolving a theory of mind: the nature of non-verbal mentalism in other primates

ANDREW WHITEN

INTRODUCTION: MENTALISM VERSUS BEHAVIOURISM IN WILD BOYS AND WILD APES

By the nature of this volume, I shall assume that the majority of readers are likely to be students of human developmental psychology, perhaps unfamiliar with the social life of non-human primates. They may experience an initial reaction of puzzlement about the very idea that such creatures might utilize anything like a theory of mind (ToM), and accordingly deserve the attention of those who study autistic or normal young children's first steps to mentalism. After all, it can be argued forcefully that children cannot be said to have a true ToM until passing a test of false-belief attribution, which is generally not until they are over four years of age (Perner 1988), following several years of exposure to talk about 'belief' and all the other elements of an elaborate and publicly recognized folk psychology. What chance then has the ape, with no communally discussable system of folk psychology, and only rarely exceeding the cognitive achievements of a three-year-old child (Premack 1988)?

I begin, therefore, by inviting the reader to engage in a thought experiment. The scenario is something of a pastiche of *Lord of the Flies* and *The Wild Boy of Aveyron*. This time, the boys deposited on the uninhabited island are prelinguistic, or at any rate their vocabulary has not yet incorporated any words remotely resembling mental terms. Ripe fruit falls into their laps, and several of the wild boys survive to middle childhood. It is a relatively uncontentious speculation that they would be motivated to interact with each other and form some sort of society. Just what sort of language (if any) they would jointly invent, and what sort of ToM (if any) each would construct, is less obvious. But imagine that these boys reach middle childhood without any linguistic terminology for states of mind. Would that mean that each of them had nothing we would recognize as a child's ToM?

To the contrary, I suggest that it is not implausible that such children would, for example, from time to time intentionally create false *beliefs* in their companions, or help them obtain things they were thought to *want*:

in short, each child would develop something of a natural psychology or *personal* theory of mind. Perhaps the evolving language would eventually develop terms to describe certain elements of it, and so a *public*, linguistically explicit folk psychology would follow, in turn shaping the personal theories held — to this extent probably mimicking aspects of the elaboration of folk psychologies which have occured in human evolution and history.

The main point is simply that before language played its part the children could plausibly become fairly competent natural psychologists, presumably with consequences for their social expertise in both cooperation and competition. Indeed, there seems no reason why in the prelinguistic phase they should necessarily fail a 'standard false-belief task' if it were presented non-linguistically (see Bennett 1976, p. 110; Whiten and Perner 1991, p. 11). I suggest it is a mentalism of this character which we need to entertain in the case of monkeys and apes, who are increasingly recognized as sophisticated social tacticians (de Waal 1982; Byrne and Whiten 1988*a,b*).

Being thus forced to consider the nature of non-verbal mindreading has several implications for the study of theory of mind in humans, particularly in the case of child development and the problems of autism.

First, at a methodological level, it has been necessary (and continues to be so) to generate new experimental techniques to test for ToM without recourse to language (details will be given below). This is important both in ordinary children, where it is important to be sure that the experiment is not just testing the child's understanding of the meaning of a linguistic term like 'thinking', and in work with autistic children, where speech and language are so often delayed or qualitatively different from normal (Wing 1976; Schreibman 1988). Studies of autism and ToM have nevertheless focused almost exclusively on children with quite high verbal mental ages (four and above). The majority of autistic children, who lack language, remain unstudied, so methods for examining non-verbal mentalism should continue to be an important shared concern of primate and child psychologists. As Frith (1989, p. 155) acknowledges, it was Premack's and Woodruff's (1978) 'ground-breaking' experiments asking 'does the chimpanzee have a theory of mind?' which laid the foundation-stone for the current research industry in children's ToM.

A separate methodological link is that natural non-verbal social interactions in animals may suggest optimal contexts for tests in children: Chandler *et al.* (1989), for example, used this rationale in their focus on *deceptive behaviour*, widespread in animals, as a natural context for the expression of children's false-belief attribution. This methodological approach yielded results suggesting a much more precocious competence than had been previously claimed. Whether the conclusion of Chandler *et al.* turns out to be true or not (see for example Sodian 1991, and Sodian and Frith, this volume, Chapter 8), the controversy generated cannot help but define better the true nature of childhood mentalism, underlining the value of imaginatively varied perspectives at this early stage of ToM research (Whiten 1991).

In addition to these methodological links, and perhaps even more important in the long run, the notion of a non-verbal ToM encourages us to be more thoughtful in specifying the nature of 'mentalism' and/or the varieties of it we might wish to distinguish.

One important issue here is how correctly to characterize the distinction between a five-year-old's mindreading and that implied by 'mutual adjustments . . . to one another's mental states' (Trevarthen 1977), as some researchers describe *infant*–adult social interaction (Trevarthen 1980; Stern 1985; and see Bruner and Feldman, this volume, Chapter 13). It is common to describe the infant's achievement as signifying 'implicit' mindreading, by contrast with the older child's 'explicit' (= verbalizable?) knowledge of mind (Bretherton *et al.* 1981). Should we then expect our *non-verbal* wild boys' understanding of mind to remain only at the infantile level? Or is there some other, non-linguistic sense in which they would be expected to exhibit progressively more sophisticated mindreading? With respect to wild boys, we may never know: but we can ask the question of non-verbal, non-human primates who *are* mature, and using whatever mindreading abilities they possess at full stretch in social tactics critical to their survival and reproductive success in the face of natural selection.

At a more fundamental level yet, the behaviour of both non-human primates and human infants poses tough questions about the sense in which they are ever mentalists as opposed to behaviourists. This is a question we shall examine below, particularly from a functional perspective. Looking at the individual as an information-processor, we need to consider how mentalism can facilitate efficient analysis of others' actions: and looking at the individual as a social interactant, we must consider how mentalism can make them more successful. The study of non-human primates may here provide rather different insights to those derived from research on human infants, because of differences both in the nature of the beasts and of their social contexts (Whiten and Perner 1991).

The infants are of course inherently immature, with all the motoric and cognitive weaknesses that entails, whereas much of our interest in potential non-human primate mindreaders focuses on them at the peak of their adult powers. In this latter case it is often in the cut and thrust of severe social competition that we most expect mindreading to be manifested, giving an individual the edge in socially outmanoeuvring others (Whiten and Byrne 1988*a*): human infant 'intersubjectivity', by contrast, is typically analysed in the cosy, protective context of parent–infant preverbal games and 'conversations' (Trevarthen 1977; Reddy 1991). Of course, to see mindreading as functioning to compete in the one case (primates), and cooperate in the other (human infants), would be an oversimplification: cooperation is a functionally important part of social life in monkeys and apes (Harcourt 1988), and conversely in human parent–offspring conflict there is evidence of exploitative tactics by infants (Trivers 1974), while naturalistic studies emphasize deceit and evasion as early manifestations of mentalism in two-year-olds' everyday life (Dunn 1988, 1991). Nevertheless, because of the major differences which do exist, non-human primate societies provoke us to think in a fresh way about the origins, nature, and functions of mentalism as it differs from behaviourism. The primate perspective thus complements that based on studies of human development. It may also give us glimpses of earlier stages in the *evolution* of that human mindreading which nowadays can go so much beyond anything of which an ape is capable, and which is so 'natural' to most of us that understanding humans who are different in this respect forces us to contemplate the architecture of possibly quite alien mentalities.

MINDREADING IN NON-HUMAN PRIMATES: THE STORY SO FAR

In this section I shall provide an overview of the main primatological findings to date, concentrating on evidence for the reading of particular states of mind. First, however, it is important to understand why some primatologists have thought such possibilities worthy of pursuit.

Machiavellian intelligence and the utility of mindreading

The answer has to do with the nature of society amongst the anthropoid primates—monkeys and apes.* Recent research has revealed that these primates exhibit a degree of social expertise which is at the least unusual, and possibly unique, amongst animals: and this means that each individual is confronted by a social environment which is particularly complex on a number of dimensions (Kummer 1982), which include, of course, the expert social skills of its competitors (Whiten and Byrne 1988a). Humphrey (1976) has speculated that this recursive situation represents a spiralling selection pressure for that extra bit of social intelligence which will allow one monkey to outwit the others,† resulting in what Richard Byrne and I (1988a) have

---

* A note for non-biologists: primates are traditionally divided into the *prosimians* ('before monkeys', which include the earliest fossil primates and modern forms such as lemurs and bushbabies), and the *anthropoids* (the monkeys and apes, which evolved later); more recently, a slightly different *strepsirhine/haplorhine* division has been used (Passingham 1982). The modal social pattern in prosimians is basically solitary, and, on the present balance of evidence, it seems that only anthropoids show unusual social intelligence (Whiten and Byrne 1988b, 1989).

† The hypothesis that intelligence in primates (including humans) evolved principally through the selection pressures of the social, rather than the physical, environment has a history preceding Humphrey's influential paper. For a summary see Whiten and Byrne (1988b).

dubbed 'Machiavellian intelligence'. This, as the reader may anticipate, is where mindreading might enter the picture. Let us first, however, make more concrete the Machiavellian nature of primate societies which is being referred to here. I shall select just two illustrative examples — the use of alliances, and deception. For more extensive reviews on this topic and detailed accounts for particular species, the reader may start by consulting de Waal (1982), Cheney *et al.* (1986), Byrne and Whiten (1988*a,b*), Whiten and Byrne (1989), and Cheney and Seyfarth (1990*a*).

Alliance formation

As in many social animals, an individual's closest competitors for limited resources (principally for the best foods and the best reproductive partners) will be other members of the individual's social group. However, in primate groups contests between pairs of competitors are complicated in a particular way which seems to be relatively rare amongst other sorts of animal: they are complicated by the intervention of a third member (or more) in the group, who allies him- or herself with one of the protagonists, and can thus be the determining factor in the outcome (Harcourt 1987, 1988). Such alliances tend to be stable over long periods, but important changes occur with new opportunities or constraints such as those produced by the maturation, decline, death, or migration of various group members. Consequently, an individual's course of action will often need to be informed not merely by assessments of the potential actions of opponents, but by assessments of the likely actions of several other companions, including potential allies of both self and opponent: in short, the social world which needs to be grappled with is transformed from one made up of a number of discrete dyadic possibilities (between self and each other member) to one which is much more of a dynamic network, *defined by the (interacting) attitudes of various other group members to each other*, as well as to oneself (Whiten and Byrne 1988*a*).

A blow-by-blow account of the events leading to a change in the alpha (top) ranking male in a group of chimpanzees has therefore been aptly titled 'chimpanzee politics' (de Waal 1982), for it is a subtle chronicle of multiple, shifting allegiances. It is in an interwoven, volatile social world like this that Humphrey (1980) suggested that any competitor would be at an advantage if it could become a mindreader, or in Humphrey's words a 'natural psychologist', getting one step ahead of the game by estimating the intentions and attitudes of the other participants, which are so critical to almost every power-bid it makes.

Tactical deception

Deception is widespread in nature, but apparently not so the *tactical deception* observed by Byrne and Whiten (1985) and defined as 'acts from the normal repertoire of the agent, deployed such that another individual is likely to misinterpret what the acts signify, to the advantage of the agent' (Byrne and Whiten 1988a). Our surveys have, however, revealed a significant number and variety of cases for many different species of monkey and ape, and we have classified these cases into a number of functional categories (Whiten and Byrne 1986, 1988c; Byrne and Whiten 1990). Consider the following two examples of the category 'creating an image':

If Puist is unable to get a hold of her opponent during a fight, we may see her walk slowly up to her and then attack unexpectedly. She may also invite her opponent to reconciliation in the customary way. She holds out her hand and when the other hesitantly puts her hand in Puist's, she suddenly grabs hold of her. This has been seen repeatedly and creates the impression of a deliberate attempt to feign good intentions in order to square accounts. Whether we regard it as deceit or not, the result is that Puist is unpredictable. Low-ranking apes hesitate when she approaches: they mistrust her (de Waal 1982 (chimpanzees): case no. 239 in Byrne and Whiten 1990).

In agonistic situations an individual (aggressor) may approach another (victim) in an apparently distracted way (e.g. not looking at him/her), but when it comes close enough it launches an attack. In doing so, the aggressor may either avoid being attacked by a third party or prevent the victim's escape, or both. The context for this kind of interaction is triadic. For example, two individuals may fight with each other, and a third-party (intervener) may come and intervene on the behalf of one of them (beneficiary) against the other (target). When the intervener is dominant over both but subordinate to the target's allies it may use this behaviour to attack the target (F. Colmenares (baboons): case no. 85 in Byrne and Whiten 1990).

In both cases, we as observers naturally question the real intention of the 'untrustworthy' players in these interactions. It follows that the other animals in these groups might avoid some attacks if they could read the mind — in this case the true intention — of those companions who act in this way.

Of course, in addition to favouring sensitivity *to* others' intentions, the existence of deception suggests another advantage to the mindreader: that of intentional deception, the deliberate *creation* in others of the state of mind of false belief. But note that tactical deception as we have defined it does not necessarily *require* any such mindreading at all — whether of beliefs, intentions, or whatever. My aim so far has just been to indicate some of the special features of anthropoid society which have led researchers seriously to entertain the proposition that the evolution of some form of mindreading might be selected for.

Having spent some time observing both non-human primate societies and autistic children in social (school) groups, I am struck by how much the children (despite their vast overall cognitive and linguistic superiority) fall behind the primates I study in those aspects of social manoeuvring which I have picked out above to illustrate potential contexts for the utility of mindreading. Autistic childrens' lack of manifestation of cooperative, triadic, and deceptive interactions are well documented (Volkmar *et al.* 1987; Frith 1989).

Having outlined such potential hotbeds for mindreading in primates, however, we still need to examine the evidence for its existence.

Evidence for mindreading by chimpanzees

The two species of chimpanzee represent our closest living relatives in the animal kingdom. Current estimates put the date of evolutionary divergence between our ancestors and those of these apes at only 4–7 million years ago. This may be contrasted with the split between our line (i.e. the ape lineage) and that of the Old World monkeys, which occurred as long ago as 20–30 million years (Sibly and Ahlquist 1984; Hasegawa *et al.* 1989). The ancestors of the other Great Apes (the orang-utan and the gorilla) and the Lesser Apes (gibbons), split off in between. The closeness of our relationship with chimpanzees is only one reason for focusing on them here, however; they are simply the only apes for which significant evidence for attribution of mental states exists.

Reading others' motivational states

Premack and Woodruff (1978) originally tested for a theory of mind by showing the chimpanzee Sarah brief videotaped sequences in which a human actor appeared to be faced with one of a number of different problems. The problems included trying to obtain out-of-reach fruit, trying to undo a padlock, and trying to light a fire. Each problem took several different forms: thus the fruit, for example, was sometimes outside the cage, and sometimes suspended above the person's head. After viewing each video segment, Sarah had to choose one photograph from a small selection, only one of which showed the correct solution to the problem shown in the videotape. Climbing on a chair was a correct solution when the video had shown fruit suspended overhead, whereas reaching out with a stick would be incorrect (that would be a solution to one of the other problems).

Sarah usually succeeded in making a correct choice, which is remarkable at the very least because this appeared to be her spontaneous decision: when young children were tested they equally often chose a photograph which had some physical feature matching one in the videotape (for example, a yellow flower in the case of a video with yellow bananas). But did Sarah's behaviour necessarily imply mindreading? Premack and Woodruff's interpretation was that it did. Premack and Dasser (1991) have recently summarized the rationale succinctly:

a problem is not an entity that is physically instantiated by the videotape. The videotape shows no more than a sequence of events, such as, for example, the actor jumping up and down with bananas overhead. This is not a problem, nor does it become one unless it is interpreted in a particular way. Once the jumping individual is perceived as having the *intention* of getting the bananas, as *wanting* the bananas, as *trying* to get them, the sequence becomes a problem, and it then becomes sensible to choose solutions. But not until the sequence is interpreted in this general manner does the consistent choice of solutions make any sense.

Not everybody has agreed with this controversial conclusion (see peer commentary in Premack and Woodruff 1978; Bennett 1988; Dennett 1988). Amongst the behaviourist alternatives to Sarah's purported mentalism, Premack and Woodruff themselves considered the possibility that Sarah succeeded by knowledge of behavioural sequencing—she knew 'what would tend to happen next'—and they proposed a thought experiment to test this. This has now been turned into reality, although, since the chimpanzee laboratory has closed, young children were the subjects (Premack and Dasser 1991).

The experiment mimicked the earlier one except that subjects were familiarized with a videotape in which the actor interrupted his efforts at jumping for the fruit to hitch up his trousers. With the tape then stopped just before the trouser-hitching, subjects were presented with three photographs and had to 'choose what comes next in the movie'. The choices were 'next'— the actor hitching up his trousers; 'relevant'—stepping up on to a chair; and 'irrelevant'—reaching out with a stick.

Whether the subjects were three, four or five years of age, nearly all of them (19) chose the 'relevant' alternative (just one three-year-old chose 'irrelevant': 'next' was chosen by one three-year-old, two four-year-olds, and one five-year-old). Children are not chimpanzees, of course, and Premack and Dasser admit, as they must, that this tells us nothing final about the chimpanzee's theory of mind: they note only the 'rule of thumb' (which we must remember is no more than that) that three-year-old child and chimpanzee have been found to perform similarly on a range of non-verbal cognitive tests. Although replication of these experiments is badly needed, the working conclusion at present is thus that the chimpanzee attributes at least *motivational states*, such as we label *want* or *intend*. Note, however, that the experiments do not discriminate between these latter two alternatives, although there are theoretical and empirical reasons for considering the attribution of intention to be a more sophisticated achievement than the attribution of wants (Astington and Gopnik 1991; Whiten 1991).

For nearly a decade Premack and Woodruff's pioneering investigation has stood by itself. Moreover, there was no attempt to examine any evidence for mindreading in the context which might make sense of its existence and nature—social groups operating in the wild. Theoretical analyses were elaborated on this topic (Humphrey 1981, 1983, 1986; Dennett 1983; Krebs and Dawkins 1984), but there was next to nothing empirical. This was a situation which Richard Byrne and I attempted to remedy in our observations and surveys on deception, referred to above. Although our central aim has been simply a *functional* analysis of primate tactical deception in the context of natural social interactions, we have where possible examined in some detail the incidental evidence for mindreading (Whiten and Byrne 1988*c,d*; Byrne and Whiten 1991).

Only a subset of the records we have collected are relevant, and even there our conclusion has been a fairly cautious one: that much of the circumstantial evidence is suggestive of various forms of mindreading, but so far little of it is compelling. Much of our writing has focused on what are the critical points for attention in future observations! Having said this, it has turned out to be chimpanzees who have provided the most impressive evidence.

That bearing most strongly on the reading of others' motivational states comes from records in which one chimpanzee appears to recognize that another's actions were deceptive. One type of reaction indicating this is counterdeception, which has been observed a number of times. The most graphic example remains that described by Plooij, which started with one chimpanzee's coming to a feeding station, and as soon as another appeared, acting as though there was no food available, when actually there was. The second chimpanzee appeared to have been fooled and departed the scene, leaving the first individual to start helping itself to the bananas. However, the second chimpanzee had not really left: it hid in the undergrowth peering at the first chimp, emerging to snatch the food only when the unsuspecting animal started to feed (record quoted in full in Whiten and Byrne 1988*c*; Byrne and Whiten 1990, case no. 253).

Plooij also described a different action, which emerged most clearly when he himself repeated a tactic he had seen the chimpanzees use. He pretended to stare intently at a (non-existent) object of interest in the distance, in order to distract a young female chimp who had been bothersomely over-attentive. She was taken in by this, and went off in the intended direction. However, she later returned, 'walked over to me, hit me over the head with her hand and ignored me for the rest of the day . . . she must have realised that I deceived her' (case no. 222, Byrne and Whiten 1990).

Taken together, the observational and experimental evidence is in accord with Plooij's interpretation in this particular case — some chimpanzees at least seem capable of discriminating between the apparent and real *intentions* of others that occur in cases of potential deception. What is important here is the *convergence* of evidence. First, the tactical deception records come from several different observers and locations, and they involve different categories of diagnostic behaviour (in the above examples, counter-deception and 'righteous indignation' — Byrne and Whiten 1991). Secondly, these observations of spontaneous social interaction are consistent with the conclusions of the experimental work of Premack and Woodruff (1978), Premack (1988), and Premack and Dasser (1991), discussed above.

Indeed we now come full circle, for Povinelli and Boysen (cited in Povinelli 1991 and Gallup, in press) have now performed what appears to be an experimental elicitation of 'righteous indignation'. In one condition of the experiment, a person approached the chimpanzee's cage with a cup of highly prized juice, but 'accidentally' spilled it; whereas in the alternative condition, a person 'deliberately' poured the juice on the floor. When then given the choice of selecting one of these people to get the next cup of juice, the chimp consistently chose the one who had only 'accidentally' spilled the juice, suggesting an ability to discriminate intentional from non-intentional action in others.

Reading what others see

There is again convergence of laboratory experiments and ethological observations when we turn to the ability to take account of what others can see. In the simplest experiment, Woodruff and Premack (as summarized in Premack 1988) first trained each of four juveniles to lead a person across a large compound to help them unlock a food container. The person was then blindfolded, and accordingly rendered resistant to being led. One of the chimps removed the blindfold at the first opportunity, and further tests showed that to elicit this response the blindfold had to be over the eyes (rather than over the nose and mouth, for example). This individual appeared to understand the role of unobscured eyes in seeing, although the other three failed altogether to remove the blindfold, and tried to drag the unfortunate helper across the compound!

Testing the ability to take another's visual perspective required a more complex training programme (Premack 1988). In its final stage, the subject could select one of two (human) helpers to signal which of two containers was baited with food. The chimp was not itself able to see in which container the food was hidden, but it watched while the baiting was done, and at the same time it could see the two helpers, one of whom *could* see which container was baited, and one of whom could not because of a screen placed in the way. Faced with this choice, two of four chimpanzees correctly and consistently chose the seeing/knowing helper from the start, an achievement little devalued by the performance of the other two subjects, one of whom chose correctly but did not always follow the helper's advice, and the other of whom chose randomly. Povinelli *et al.* (1990) have since performed a related series of experiments, ignorance in one of the potential helpers being imposed in this case not by a screen but by leaving the room. In addition a transfer test was incorporated, in which the ignorance was produced by the helper covering his head with a bag, and here three of four chimpanzees correctly sought the advice of the 'knower' rather than the 'guesser' (although no mention is made of a control condition which would have been desirable — wearing the bag in a way which does not impede vision).

Convergent evidence from naturalistic observations comes particularly from cases of tactical deception involving visual concealment. Much of primate tactical deception is concerned with the monitoring and manipulation of others' visual attention (Whiten and Byrne 1988*b*), and chimpanzees have been described as very skilled at taking into account rather precisely the visual perspective of another: de Waal (1985) describes a number of instances in which a male who is courting a female by exhibiting a penile erection nevertheless selectively hides his sexual state from a competing male. Thus:

precisely at the point when Dandy was exhibiting his sexual urge in this way, Luit, one of the older males, unexpectedly came round the corner. Dandy immediately dropped his hands over his penis concealing it from view (p. 49); . . . often a low-ranking male will sit with his upper arm resting on his knee and his hand loosely hanging down so that a female in front of him can see his erect penis, but apes on the side cannot see it. This inconspicuous form of concealment occurs together with quick glances at dominant males. Needless to say, the subordinate always uses the hand on the body side which is turned towards dominants (p. 113); . . . Luit looked round to watch Nikkie's progress and then he looked back at his own penis, which was gradually losing its erection. Only when his penis was no longer visible did Luit turn around and walk towards Nikkie (p. 49).

Taken together, these observations and the experimental results suggest that chimpanzees may take into account a number of things about the nature of 'seeing': some seem to know that the eyes must be unobscured; some can accurately compute the visual perspective of another individual; and some know that what another individual sees has further, particular implications (discussed further below).

The 'seeing' involved here, however, can be argued not to be a mental state — not something truly requiring a theory of mind, or indeed any mind-reading at all: it seems to fail on the philosopher's touchstone of logical or referential opacity (Whiten and Perner 1991), and it can be exhaustively described at the 'exterior' level of eye and head movements coupled with the geometry of the physical and social environment.

When we begin to talk about taking account of others' visual *attention*, however, it can be argued that a more significant step in mentalism has been taken. Baron–Cohen (1991; see also this volume, Chapter 4, and Gómez, Sarriá, and Tamarit, this volume, Chapter 18) has suggested that it is in representing others' attention that one might see early stages in autistic children's difficulties with mentalism. In the case of chimpanzees, I have argued that when one individual has noticed a choice piece of food that others have not, and inhibits its otherwise tell-tale gaze until the competitors depart and leave it to snatch the food (Whiten and Byrne 1988*d*) the selective deployment of this strategy only with respect to such 'secret' food items suggests the attribution of attentional states to others (Whiten 1991). The subject is mindreading in the sense that it is making judgements, not directly about observed behaviour, but rather about what its competitors *might or might not notice*.

Reading what others know

What is the status of the 'further implications' of what another can see which I earlier suggested chimpanzees were capable of grasping? Are they just behavioural predictions (for example, that if a foe *sees* one's erect penis he will *attack*)? Or are they mental—'to see is to know'?) Does the chimpanzee hiding his penis somehow mentally represent *ignorance* in his foe? And does the subject in the food-hiding experiments described above do likewise with respect to the knowledge or ignorance of the helpers? Both Premack (1988) and Povinelli *et al.* (1990) seem to suggest so, for the title of the Povinelli *et al.* paper is 'Inferences about guessing and knowing by chimpanzees'; and Premack says that his two successful subjects 'evidently understand that knowing depends on seeing'. I do not think that the inference is that straightforward (and indeed nor does Povinelli 1991); in fact, just what is going on in these episodes goes to the heart of the difficult question of just how mentalism really differs from behaviourism, an issue we shall examine more closely below. For the moment we shall just record the experimentalists' own positive conclusions, and continue with our over-view of the empirical findings.

Reading others' beliefs

To date there has only been one single-shot experiment which could be interpreted as testing whether a chimpanzee will take account of another's false beliefs. Premack and colleagues (Premack 1988) arranged that one of Sarah's favourite trainers should irregularly restock a cabinet mounted in front of Sarah's cage. The cabinet had two halves, one consistently used for goodies which were shared with the ape at tea-time, and the other stocked with faeces, rubber snakes, and other items clearly repellent to Sarah and the trainer alike. Sarah was in control of opening the door of the cabinet by remote control. One day, before the trainer's visit, Sarah watched as a 'villain' entered, forced open the cabinet, and switched the locations of the nice and nasty contents.

Now when the trainer entered, falsely believing the contents were in their usual positions, she was at risk of putting her hand into the disgusting mess. Did Sarah take account of this in her remote control of the door? Certainly not in any clear way: the latency to push the button was as usual, as was her 'general demeanour' (Premack 1988, p. 178). She did do one unusual thing, pressing the button again once the door was open; but this is difficult to interpret. As Premack (p. 178) concludes, 'negative outcomes seldom lend themselves to diagnosis. Sarah could have failed for any of dozens of reasons'. So far there is no experimental evidence for what developmentalists would call 'false-belief attribution' in primates; but it is equally clear that the work has scarcely begun.

The same conclusion applies to evidence from tactical deception. Richard Byrne and I are often misunderstood as claiming that in tactical deception primates intentionally deceive each other—i.e. they *intend to create false beliefs* in others. In fact our now large corpus of records still offers no convincing evidence for this, in chimpanzees or any other primates. This, however, should not be an insurmountable difficulty of observational evidence. What is required is evidence that, once one deceptive tactic has failed, another alternative (or more) is tried, the one common factor between them being their potential to create a specific false belief (Chandler *et al.* 1989). Until such behaviour is recorded, tactical deception does not in itself imply an intent to create false beliefs.

Monkey mindreading versus ape mindreading: a metarepresentational Rubicon?

When we turn to monkeys, with whom we shared an ancestor much longer ago, we find less evidence for mindreading at the levels suggested above for chimpanzees.

Cheney and Seyfarth (1990*a*,*b*; 1991) have conducted experimental tests for the attribution of *knowledge*, using macaque monkeys. In one series of experiments, a mother macaque watched as food was placed in a bin in a test arena, and then her offspring was released into the arena. In the 'knowledgeable' condition the offspring sat with the mother while the food was put in the bin; in the 'ignorant' condition, however, only the mother could see the food being made available, the infant being visually isolated and separated from the mother by a steel partition. The key question was whether mothers would 'inform' their infants, when the latter were ignorant, by giving more of their characteristic 'coo'-like food-calls. In fact there was no significant difference in calling rate, and ignorant infants accordingly took longer to acquire the food. In a related test involving ignorance versus knowledge of the presence of a predator, mothers evidenced no tendency preferentially to warn an ignorant infant.

In a second series of experiments, a mother was separated from her infant by one of a number of different types of screen, the infant being left in an arena with another female monkey. When the screen was clear glass, the apparent presence of the mother caused the other female to show less aggression to the infant than in a condition where an opaque screen hid the mother. In the critical tests, the screen was a one-way mirror, so arranged that the mother could not see into the arena, although the other female and the infant could see her. All animals were exposed to the properties of one-way mirrors beforehand. Cheney and Seyfarth reasoned that if the female was influenced only by the presence of the mother, their behaviour in the mirror condition should have been similar to that with clear glass. By contrast if they were influenced by what the mother could *see* and *know*, their behaviour in the one-way mirror condition would become more like that in the opaque condition.

The complexity of this experiment — in particular the likelihood that the monkeys would not appreciate the implications of a one-way mirror in the first place — counsels against optimism about the results. But, in the event, several measures of behaviour did come out in favour of the mindreading hypothesis. For example the basic measure of aggressive behaviour by females was greater under both opaque and mirror conditions, compared to glass. In Cheney and Seyfarth's own words, 'some of our results argue against a "theory of mind" in monkeys, while others support it and still others are inconclusive' (1991, p. 193).

This is not the place for a detailed critique of these experiments (see Cheney and Seyfarth, in press; Whiten, in press). The authors themselves note that even the apparently positive evidence for mindreading in the second series of experiments may have been caused through the female and infant reacting to subtle differences in the behaviour of the mother in the glass and one-mirror conditions, for in the latter she could not see the other animals. And, as in the case of Premack's negative results, it is easy to think of several excuses for the monkeys' failure to evidence mindreading in these tests, even if they are truly mindreaders: for example, in experiment one, how did a mother know the infant could *not* see what *she* could?; and in experiment two, did they understand the nature of one-way mirrors? Nevertheless, these first experiments on monkey mindreading are an excellent start. One thing which particularly deserves emphasis is that the authors' experience as fieldworkers (Cheney and Seyfarth 1990*a,b*) shines through in the way in which the first series of experiments creates situations which correspond plausibly to those in which mindreading, if it exists, could be used to advantage in the animals' natural world.

Results which converge with Cheney and Seyfarth's have been obtained in an experiment in which Povinelli *et al.* (1992*a*) substituted macaques for chimpanzees in the knowledge/ignorance test described earlier. Unlike chimpanzees, the monkeys did not learn to discriminate the knower and guesser, even after 600 trials, although one monkey was shown to be able to do so readily if the knower wore a distinctive glove. The failure of the monkeys leads Povinelli to conclude that the chimpanzees' performance cannot be explained by any simple learning mechanism. That the test is really diagnostic of knowledge attribution is further supported by parallel tests with three-year-old children, who performed at the level of the monkeys, and with four-year-olds, who performed at or above the level of the chimpanzees (Povinelli and deBlois 1991).

Turning back to the natural social world of the monkeys and the implications of tactical deception, we find no records which suggest the reading by monkeys of others' *intentions*, evidenced in the case of chimpanzees by their facility in counterdeception. On the other hand, there is some evidence for visual perspective-taking in records of monkeys' partial hiding of body parts, corresponding to the penis-hiding of chimpanzees referred to earlier (Whiten and Byrne 1988c; Byrne and Whiten 1990).

A (necessarily very tentative) interpretation of the evidence available to date is thus that in monkeys we see only shallow mindreading, involving an ability to take another's visual perspective, whereas chimpanzees go beyond this to read the intentions and perhaps the knowledge of others. It could thus be argued that the chimpanzees are mentalists in a more real sense. Whiten and Byrne (1991) have pointed out that this presents the picture of an elaboration of mindreading functions in the course of our phylogenetic history which correponds to the progression we see in human development, from early visual perspective-taking (see for example Flavell *et al.* 1981) to the later attribution of the mental states of intent (for example Astington and Gopnik 1991), and knowledge/ignorance (for example Leslie and Frith 1988).

Indeed, there may be much deeper links between the evolutionary and developmental perspectives. Whiten and Byrne (1991) recall Leslie's (1987) argument that *pretend-play* rests upon the same facility in meta-representation which must underlie a theory of mind, evidence for which includes deficits in both pretence and theory of mind in autistic children (Leslie 1987, 1991; see also Harris 1991 and this volume, Chapter 11 for related analyses of the role of imaginative and pretending abilities in mentalism). Whiten and Byrne show that, as one would predict from their superiority in mindreading, chimpanzees provide by far the most startling evidence for pretend-play amongst non-human primates. The conjecture is thus that there may exist within the order of primates a 'mental Rubicon' — not the familiar one, with *Homo* on one side and the rest on the other, but one with *Homo* and at least *Pan* on the same side, marked by a facility for mental metarepresentation (see also Gallup 1982, 1983, 1985 and Povinelli 1991 for different, although related, analyses of this pattern; and see Gómez, Sarriá, and Tamarit, this volume, Chapter 18 for a converse interpretation of what the evidence has to say about where the Rubicon lies).

Whiten and Byrne suggest that what underlies pretence and mindreading may explain differences in other competencies as well, including imitation, in which again there is evidence for chimpanzee superiority (Whiten 1989; Visalberghi and Fragaszy 1990).*

To imitate in the visual mode involves B copying an action pattern of A's which was originally organised from A's point of view. It is necessarily a different pattern from B's point of view, yet it has then to be re-represented in its original organisational form so as to be performed from A's point of view. The expression 're-represented' seems unavoidable and it is used advisedly: it translates as second-order representation . . . To put the idea more graphically, we might say that B has to get the program for the behaviour out of A's head: in other words, to engage in a form of mindreading (Whiten and Ham 1992).

It might thus be instructive to pursue further work on autistic children's difficulties with imitation (Curcio 1978; Jones and Prior 1985) in conjunction with ToM research (see also Meltzoff and Gopnik, this volume, Chapter 16, and Bruner and Feldman, this volume, Chapter 13).

A further contrast which seems to extend this account of the essential monkey–ape difference has emerged in recent work on role-reversal. Both macaque monkeys (Povinelli *et al* 1992*c*) and chimpanzees (Povinelli *et al* 1992*b*) were trained to co-operate with human partners, either indicating to them which of several food containers truly contained food, or playing the opposite role of using such information. When in each case these roles were reversed, three of the four chimpanzees, but none of the monkeys, showed immediate comprehension of their new social role. Again, then, the apes but not the monkeys appear capable of representing the world from the mental perspective of another individual.

THE NATURE OF NON-VERBAL MENTALISM

In the foregoing I have attempted to lay before the reader many of the important empirical findings available. It should now be possible more seriously to consider just what sort of process non-verbal mindreading could be.

Mentalism versus behaviourism

I noted earlier that both Premack (1988) and Povinelli *et al.* (1990) discussed their data as evidence simultaneously for the attribution of *seeing* (visual perspective-taking) and *knowing*. This may strike the reader as contrasting markedly with the situation in developmental psychology, where it is assumed that attribution of seeing and attribution of knowing are very dif-

---

* Where other apes might fall with respect to these distinctions is not discussed here, for the relevant evidence is even more patchy.

ferent and separable achievements: in autistic children, for example, the two seem to be dissociated, with the first usually mastered when the second is not (Hobson 1984; Leslie and Frith 1988; Baron–Cohen 1989). Also, it can be argued that *knowing* is a mental state, whereas *seeing* is not, so that attribution of *knowing* to others requires metarepresentational ability, ascription of *seeing* only primary representation (see for example Baron–Cohen 1991). The latter can be tested verbally in a fairly straightforward fashion with children, by the experimenter asking them questions along the lines of 'what can I see now?' as he or she looks in turn at different objects (Baron–Cohen 1989).

At first pass, it may therefore seem just unfortunate that with non-verbal subjects it has appeared impossible to assess separately the ability to take account of what others see, and the ability to attribute knowledge/ignorance to others (i.e. the only way to test that **A** knows that **B** *saw* **X** is where **A** modifies its behaviour to allow for **B**'s acting on the *knowledge* so gained). But perhaps there is something more profound to learn here: that for the practical mindreader, the only function achieved in recognizing that **B** saw **X** *is* to modify one's behaviour to allow for **B**'s acting on the knowledge gained. This seems to suggest that one should not expect a non-verbal mindreading creature to have any distinctions between *seeing* and *knowing* in its mindreading apparatus (see also discussion in Povinelli 1991).

The question remains, however, of how such *seeing/knowing* relates to the concepts of mentalism or of behaviourism. It can be argued that in the Premack/Povinelli experiments, for example, the subject need not be a mentalist, but instead merely make *observations of behaviour* by the helpers, which allow it to *predict further behaviour*, using behavioural rules. Thus, a helper whose behaviour is describable in terms of a particular gaze configuration relative to screen and food-bins (corresponding to seeing where food is hidden) can be predicted to *generate further behaviour* of a particular sort, namely that which tends to indicate accurately which bin contains food (for you or me this would correspond to 'they know where the food is').

The question is, just how is this 'behaviourist analysis' really different to what goes on in a child who takes account of knowledge in others? When we talk of mindreading or ToM even in adults we are not meaning telepathy. We do not look directly into people's minds, but rather we can watch only external observables — principally others' behaviour and various features and states of the physical environment — and we label certain complex patterns of these with mental-state terms (cf. Wittgenstein 1958). As Bennett noted in 1976, all these data are in principle observable by animals too. In the end, we might say that mentalism has to be just a particular sort of behaviourism! The question is, just what sort? Note that success on a classic ToM test, the Sally-Anne test (Baron-Cohen *et al.* 1985), can be interpreted quite reasonably as indicating attribution of *belief*: but alternatively, the child's success could be described as requiring recognition of the behavioural–physical pattern corresponding to seeing an object hidden in location **X**, and knowing that this predicts certain further actions, such as searching in location **X** when there is a need to recover the hidden object.

Gómez's (1991) formulation in the case of a gorilla's mindreading corresponds closely to this: 'she seemed to understand that in subjects perceiving is causally related to acting. And here is where the mind appears, since the coordination between perception and action is carried out by the mind' (p. 201). Gómez suggested that this type of practical understanding constitutes an 'implicit' theory of mind, perhaps characterizing apes and human infants, and in later human development being superseded by a theory of mind in which mind 'eventually becomes explicit (or can be made explicit)' (p. 205).

Gómez's analysis is subtle and sophisiticated (see also this volume, Chapter 18), but I suspect that an implicit/explicit dichotomy will not finally be up to the job of discriminating all the varieties of what we might wish to call mindreading.* There is a danger that 'explicit' will be taken just to mean 'can be talked about (using the terminology of folk psychology)'. Conversely, where 'implicit' mindreading is defined as 'adaptation to mental phenomena through the perception or representation of the external manifestations of the mind' (ibid., p. 206), there is a danger that this may be argued to be true of all social interactions amongst animals: i.e. to interact behaviourally is automatically to interact with external manifestations of mind; the mother (whether human or rat) who allows the striving infant to suckle is discerning the external manifestation of the infant's *want*, for example. Would the mindreading of wild boys or other linguistically impoverished children be expected to achieve implicit or explicit levels – and what would this distinction mean for them?

In what follows I shall explore some middle ground, elaborating on the start made by Gómez (1991) and Whiten (1991), and regarding mentalism as a form of behaviour analysis that is sophisticated in particular ways (see also Dennett 1987; Heyes 1987; Gómez 1990; Whiten and Byrne 1990; Beer 1991 for recent, varying perspectives on the underlying issues here). Although I hope what follows is a novel contribution, particularly given common assumptions in the 'new ToM' literature of recent years, it has much older precursors both in the philosophical theory of mind (Ryle 1949; Wittgenstein 1958) and behaviouristic psychology (Tolman 1932).

Mental states as intervening variables

My starting-point is the way in which some behavioural scientists have dealt with the ascription to animals of internal states like hunger and thirst

---

* See Reddy (1991) for an important complementary critique of the implicit/explicit distinction in child development.

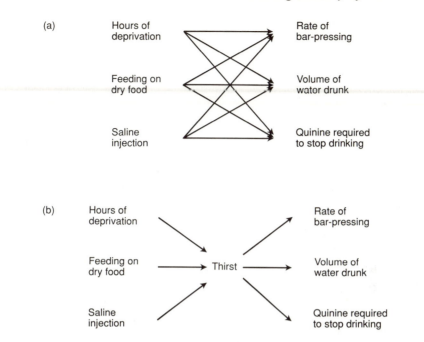

Fig. 17.1. (a) Relationships between three independent and three dependent variables (from Hinde 1970 after Miller 1959); (b) relationships between three independent variables, one intervening variable, and three dependent variables (from Hinde 1970 after Miller 1959).

(Tolman 1932; and see the discussion in Hinde 1970, pp. 194–202). Figure 17.1a shows, on the left, a number of stimulus conditions, which can lead to the observable outcomes on the right. Nine different, specific links are required to define how each stimulus can generate each response. However, ascription to the animal of an intervening variable, (an unobservable variable posited to lie between the S's and the R's) is justified when it allows us to predict the S–R pattern more economically, as exemplified by the need to show only six links rather than the original nine (Fig. 17.1b; Miller 1959). This particular intervening vairable could be called 'thirst'; but note that it would do its simplifying job for us regardless of what (verbal) label is finally stuck on it: it would be useful as a distinct representational code, or decision-point, even in a non-verbal brain.

The same logic applies to the ascription of mental states as applies to such physiological states. When thought of as intervening variables, neither is directly observable, nor, for that matter, do they need to be 'seen as' being somewhere inside the animal, let alone necessarily in its head. They are just intervening variables, the values of which are computable by an onlooker on the basis of stimulus observations (on the left of Fig. 1), thence permitting

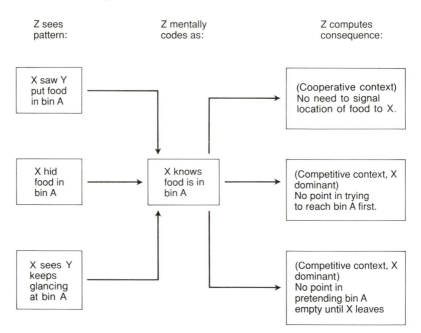

Fig. 17.2. Coding another's mental state as an intervening variable. Here **Z** reads **X**'s mental state of knowledge.

prediction of checkable dependent variables (on the right). If the behavioural scientist finds it useful to state-read or mind-read in this fashion, so might an animal mindreader, who has access to the same observations of contingencies between the observables on either side of the black box which is the mind-read individual.

Figure 17.2 shows a hypothetical example which relates the sort of chimpanzee mindreading capacities cited earlier to types of natural social interaction. Various perceptual arrays shown on the left can all be coded (categorized, classified, represented) by the mindreader's brain in the same way, amounting to the decision that **X** now *knows* a certain thing. This unitary categorization of **X**'s state can then be used with an improvement in economy like that occurring between Figs. 17.1a and 17.1b to make a number of different predictions, some of which are shown on the right of Fig. 17.2. I suggest this is *one* sense in which it would be valid to label a non-verbal animal (whether human or not) as a mentalist rather than behaviourist. But two further claims should be emphasized for this model.

First, it is plausible from a functional point of view. As Dawkins (1976) showed so lucidly, hierarchically organized information-processing is a very general principle of economy in biological control systems. In the present model, it is suggested that the often rapid, finely tuned intelligent responses required of a Machiavellian primate could be facilitated if such perceptions as those on the left of Fig. 17.2 are coded as a value of a mental-state variable (in this case, the equivalence class is some state of *knowledge*); for in the situations on the right this is precisely the quantity which is relevant for the decisions which need to be made.*

Second, in the same way that it makes sense for the behavioural scientist studying an animal's behaviour to describe unobservable states as intervening variables when that is the most economical way of accounting for the phenomena under study (as it is for the animal mindreader **Z**, observing the behaviour of another animal **X**, described in the previous paragraph), so *mindreading itself* represents the behavioural scientist's best description of what an animal **Z** is doing when it accounts most economically for the animal's ability to make *a novel* prediction like one of those about **X** on the right of Fig. 17.2, on the basis of observations of one of the circumstances on the left. This, of course, is quite a tall order for the scientist in practice (see Bennett 1991).

How does this model relate to the implicit/explicit distinction? Clearly it does not need to be explicit in the sense of being verbal – it is meant to be applicable to non-verbal mindreaders. However, I see no reason why it should not apply to the way in which features of mentality are represented prior to the acquisition of mental-state language by children, mental-state language later coming to be used by children to label the mental-state categories – the intervening variables – which they may already be using to guide social interactions. If this were true, it underlines the fact that mental states in this model should be thought of as 'explicit' in so far as they would be specifically coded for by unique brain states: indeed, the formulation in Fig. 17.2 is designed to avoid the charge that mentality is here merely 'implicit' in the sense that in classifying another's behaviour an animal must (merely) automatically and simultaneously be classifying mental states.

Mentalism as the sort of categorization process suggested here does not necessarily warrant the description 'theory of mind', although intervening variables are indeed *hypothetical*, one of the key characteristics Wellman (1991) requires of a mental entity. Just when the expression 'theory of mind' is warranted is rightly a contentious issue (Butterworth 1990, 1991; Harris 1991; Whiten and Perner 1991; Whiten 1991): I suggest that it is best limited to the ability to *integrate* judgements relating to more than one mental state

---

* This logic should apply to all social creatures to an extent; but the degree of encephalization which the species can afford (see for example Dunbar 1992; Foley and Lee 1992) may place limits on the depth of hierarchical processing feasible, and thus on the existence of mindreading. In the case of human infants, there may be additional or alternative functional reasons for the existence at this stage of mindreading of the character proposed here: for example, preparation for communication by speech, or for the development of a representational ToM of the form discussed later.

(see for example Wellman 1990; 1991), whether these derive from the processes suggested above, those discussed below, or others. In science a *theory* is usually multi-componential, by contrast with, for example, a more unitary *hypothesis*.

Mentalism as yet more sophisticated behaviourism: insight into mental representation

In Fig. 17.2, **Z** mentally codes three different percepts as exemplars of a single equivalence class, which correponds to what **X** knows. **Z** is thus mentally representing what is in effect a representation of **X**'s (**X**'s knowledge): thus we have a case of *metarepresentation*. However, two senses of 'metarepresentation' are used in the literature:

1. *A representation of a representation* (where a representation may, for example, be a perception or a belief). This is the sense I just employed, and is also used by Leslie (1987), for example.

2. *A representation of a representation as a representation*. This has been interpreted as 'a particular concept of representation which someone grasps' (Leslie 1991, p. 77), or more briefly as *understanding* the idea of representation, as is implied in the ability to attribute false beliefs (Perner 1988, 1991; Leekam and Perner 1992). This would seem to be *another* sense in which a particularly advanced grade of mentalism could be labeled as 'explicit', independently of verbalization about either mentality or representation.

It seems possible that 'grasping the concept of representation' constitutes an *insight*: the (sudden?) realization that (along the lines of the leap in economy illustrated in Figs. 17.1b and 17.2) an equivalence class something like '**X** *misrepresents*' can act as a powerful explanatory intervening variable, linking observable behaviour and circumstances to the predictable actions of others acting on, for example, false beliefs ('in-sight' is particularly apt here for its connotation of 'seeing into' the mind!). Using this idea in a similar way in the context of language-learning, McShane (1980) proposed that an acceleration in the acquisition of object-names at a certain stage resulted from the child's achieving an insight into a general rule like 'all things have names', which subsumes the previously isolated instances of label-learning.*

---

* In McShane's own words: 'The appeal to the concept of insight is quite deliberate. Insight is a psychological process of common experience (though much ignored theoretically, with a few exceptions, notably Kohler 1925). The basic achievement of an insight is a relatively sudden realisation of some previous unseen structural relationship . . . that the words are names for objects'. With respect to an implicit-to-explicit progression, the following logic is also relevant (substitute *mentalizing* for *naming*): 'On the antecedent side the appeal to insight is motivated by the fact that the ritual nature of many naming games will be such that the child's use of names will have the characteristics and context appropriate to naming even if the activity is not conceptualised by the child as one of naming . . . On the consequent side the appeal to insight is motivated by the sudden increase in naming that is frequently observed toward the middle of the second year . . . the child acts as if he or she had discovered that things have names' (McShane 1980).

Insight into the ways in which aspects of psychological causality permit efficient analysis of the actions of others could be the basis of non-verbal or pre-verbal discovery of many mental states as intervening variables in the sense discussed in the previous section above, but I emphasize it here because insight into the phenomenon of (mental) *representation itself* must be recognized as a qualitative leap in the evolution of a theory of mind.

Ever since Kohler's (1925) pioneering studies, chimpanzees have been credited with some special capacity for insight (although see Epstein *et al.* 1984). Do they ever achieve insight into the representational nature of mind? If we juxtapose the current controversy over just when children usually achieve this in the 2–4-year-old period (for example Chandler *et al.* 1989; Sodian *et al.* 1991; Dunn 1988, 1991; Freeman *et al.* 1991) with Premack's rough equivalence rule for chimp and $3\frac{1}{2}$-year-old child (Premack 1988), it is as yet far from apparent what we should expect the answer to be.

Reporting on intervening variables

To recap: in the above discussion I have questioned the traditional implicit/explicit dichotomy. In the literature, 'explicit' usually seems to reduce to 'can be talked about'. 'Implicit' mindreading (as attributed by some to infants, for example) is not usually defined, but the idea seems to be that the child is somehow acting only 'as if' recognizing mental states. I have suggested instead that even in a non-verbal mentalist, there can be explicit representation of mental states. First, in the case of an individual holding *metarepresentations* which act as intervening variables, an *explicit* representation in that individual's brain is implied. Second, where we move up one level of sophistication to *representation of representation as representation*, with intervening variables such as 'X misrepresents . . .', the representational nature of mind itself finds *explicit* representation.

Given the third and conventional usage of 'explicit' to refer to talking about mental states, however, a finding from the series of experiments on discrimination of knowing versus guessing by Povinelli and colleagues becomes of particular interest. Recall that the three-year-old children performed at the level of monkeys, and four-year-olds at or beyond the level of the chimpanzees. The important additional finding is that the four-year-olds who succeeded in distinguishing the knowledgeable from the ignorant partner could give verbal accounts of *why* they succeeded which incorporated the linkage between seeing and knowing. Children who failed tended not to be able to give such accounts. To be provocative, one might suggest the implication of this linkage of successful mindreading with verbalization about the underlying representation is that, if the chimpanzees could talk, they should likewise be expected to be able to *report* on the contents of their representations such as are schematized in Fig. 17.2: in other words, that a dependence of knowledge on seeing underlies the success they show in tasks failed both by monkeys and by younger children who express no verbalized understanding. However, such a conclusion would ignore other findings that children can succeed on some ToM tests at a non-verbal level which they are much more likely to fail verbally (Freeman *et al.* 1991).

CONCLUSION: VARIETIES OF NON-VERBAL AND
PRE-VERBAL MENTALISM

It will be apparent from the foregoing that the value of non-human primate research in this area for students of autistic and normal children does not lie in the traditional advantages of superior experimental control or the exploitation of simple animal models. Quite the contrary: research in anthropoid mindreading is proving extremely taxing, by comparison with the ease of working with normal, biddable children at least! The developmental psychology of ToM is already better documented than the comparative psychology of ToM.

Important links between these two research areas at present lie elsewhere. As was noted at the outset, the majority of autistic children who are non-verbal remain to be studied in any detail and, given the language difficulties of others, the efforts of primatologists in devising methodologies appropriate to non-verbal creatures reviewed above may be found to be adaptable for these human cases also.

No less important is cross-fertilization at the conceptual level. On the one hand, we have noted that the 'metarepresentational conjecture' (Leslie 1991) in the developmental work of Leslie, Baron–Cohen, and Frith has been extended into theories which may also make sense of phylogenetic differences amongst non-human primates in a cluster of cognitive capacities including mindreading, pretend-play, and imitation.

On the other hand, the relevance of the previous few pages for autistic research rests on our knowledge that these children suffer an enormous variety of cognitive and linguistic delays and deficits, such that the full range of mental processes whereby they interpret the behaviour of others is quite likely to be greater than the diversity of performance already apparent in verbal ToM tests (Baron–Cohen, this volume, Chapter 4). In this chapter and Whiten (1991) I have therefore attempted to indicate some of the rather different ways in which non-verbal mindreading might operate, and why it might do so in the first place. The non-human primate and child research together provoke contemplation of very many grades of mindreading, ranging from processes which scarcely merit that title (low levels of sophistication in behavioural analysis) through higher levels, to mature human folk psychology. In some autistic children and some or all non-human primates, mindreading may even take forms which are different from any to be found along the 'normal' human developmental route. At this stage of our research, whether on monkeys, apes, autistic or non-autistic children, I suggest that it will be productive to take advantage of all the means available to appreciate the nature of minds and mindreading that are possibly different from our own.

REFERENCES

Astington, J. W., and Gopnik, A. (1991). Developing understanding of desire and intention. In *Natural theories of mind* (ed. A. Whiten). Basil Blackwell, Oxford.

Baron–Cohen, S. (1989). Perceptual role-taking and protodeclarative pointing in autism. *British Journal of Developmental Psychology*, 7, 113–27.

Baron–Cohen, S. (1991). Precursors to a theory of mind: understanding attention in others. In *Natural theories of mind* (ed. A. Whiten). Basil Blackwell, Oxford.

Baron–Cohen, S., Leslie, A. M., and Frith, U. (1985). Does the autistic child have a 'theory of mind'? *Cognition*, 21, 37–46.

Beer, C. G. (1991). From folk psychology to cognitive ethology. In *Cognitive psychology* (ed. C. A. Ristau). Erlbaum, Hillsdale NJ.

Bennett, J. (1976). *Linguistic behaviour*. Cambridge University Press. Reissued in 1989: Hackett, Indianapolis.

Bennett, J. (1988). Thoughtful brutes. *American Philsophical Association Proceedings and Addresses*, 62, 197–210.

Bretherton, I., and Beeghly, M. (1982). Talking about internal states: the acquisition of an explicit theory of mind. *Developmental Psychology*, 18, 906–21.

Bretherton, I., McNew, S., and Beeghly-Smith, M. (1981). Early person knowledge as expressed in gestural and verbal communication: when do infants acquire a 'theory of mind'? In *Infant social cognition* (ed. M. E. Lamb and L. R. Sherrod) Erlbaum, Hillsdale, NJ.

Butterworth, G. (1990). Book review of *Developing theories of mind* (ed. Astington *et al.*). *Perception*, 19, 135–8.

Butterworth, G. (1991). The ontogeny and phylogeny of joint visual attention. In *Natural theories of mind* (ed. A. Whiten), Basil Blackwell, Oxford.

Byrne, R. W., and Whiten, A. (1985). Tactical deception of familiar individuals in baboons. *Animal Behaviour*, 33, 669–73.

Byrne, R. W., and Whiten, A. (1986). *Machiavellian intelligence: social expertise and the evolution of intellect in monkeys, apes, and humans*. Oxford University Press.

Byrne, R. W., and Whiten, A. (1988b). Towards the next generation in data quality: a new survey of primate tactical deception. *The Behavioral and Brain Sciences*, 11, 267–73.

Byrne, R. W., and Whiten, A. (1990). Tactical deception in primates: the 1990 database. *Primate Report*, 27, 1–101.

Byrne, R. W., and Whiten, A. (1991). Computation and mindreading in primate tactical deception. In *Natural theories of mind* (ed. A. Whiten). Basil Blackwell, Oxford.

Chandler, M., Fritz, A. S., and Hala, S. (1989). Small scale deceit: deception as a marker of 2-, 3- and 4-year-olds' early theories of mind. *Child Development*, **60**, 1263–77.

Cheney, D. L., and Seyfarth, R. M. (1990*a*). *How monkeys see the world*. University of Chicago Press.

Cheney, D. L., and Seyfarth, R. M. (1990*b*). Attending to behaviour versus attending to knowledge: examining monkeys' attribution of mental states. *Animal Behaviour*, **40**, 742–53.

Cheney, D. L., and Seyfarth, R. M. (1991). Reading minds or reading behaviour? Tests for a theory of mind in monkeys. In *Natural theories of mind* (ed. A. Whiten). Basil Blackwell, Oxford.

Cheney, D. L., and Seyfarth, R. M. Multiple book review of *How monkeys see the world*. *The Behavioral and Brain Sciences*, **15**, 135–82.

Cheney D. L., Seyfarth, R. M., and Smuts, B. B. (1986). Social relationships and social cognition in nonhuman primates. *Science*, **234**, 1361–6.

Curcio, F. (1978). Sensorimotor functioning and communication in mute autistic children. *Journal of Autism and Developmental Disorders*, **8**, 281–92.

Dawkins, R. (1976). Hierarchical organisation: a candidate principle for ethology. In *Growing points in ethology* (ed. P. P.G. Bateson and R. A. Hinde). Cambridge University Press.

Dennett, D. C. (1983). Intentional systems in cognitive ethology: the 'Panglossian paradigm' defended. *The Behavioral and Brain Sciences*, **6**, 343–90.

Dennett, D. C. (1987). *The intentional stance*. Bradford Books, MIT Press, Cambridge, Mass.

Dennett, D. C. (1988). The intentional stance in theory and practice. In *Machiavellian intelligence: social expertise and the evolution of intellect in monkeys, apes, and humans* (ed. R. W. Byrne and A. Whiten). Oxford University Press.

de Waal, F. (1982). *Chimpanzee politics*. Jonathan Cape, London.

de Waal, F. (1986). Deception in the natural communication of chimpanzees. In *Deception: perspectives on human and non-human deceit* (ed. R. W. Mitchell and N. S. Thompson). State University of New York State, Albany.

Dunbar, R. I. M. (1991). Neocortex size as a constraint on group size in primates. *Journal of Human Evolution*, **20**, 469–93.

Dunn, J. (1988). *The beginnings of social understanding*. Basil Blackwell, Oxford.

Dunn, J. (1991). Understanding others: evidence from naturalistic studies of children. In *Natural theories of mind* (ed. A. Whiten). Basil Blackwell, Oxford.

Epstein, R., Kirshit, C. E., Lanze, R. P., and Rubin L. C. (1984). 'Insight' in the pigeon; antecdents and determinants of an intelligence performance. *Nature*, **308**, 61–2.

Flavell, J. H., Everett, B. A., Croft, K., Flavell, E. R. (1981). Young children's knowledge about visual perception: Further evidence for the Level 1–Level 2 distinction. *Developmental Psychology*, **17**, 99–103.

Foley, R. A., and Lee, P. C. (1992). Ecology and energetics of encephalisation in hominid evolution. In *Foraging strategies and natural diet of monkeys, apes, and humans* (ed. A. Whiten and E. Widdowson). Oxford University Press.

Frith, U. (1989). *Autism: explaining the enigma*. Basil Blackwell, Oxford.

Gallup, G. G. (1982). Self-awareness and the emergence of mind in primates. *American Journal of Primatology*, **2**, 237–48.

Gallup, G. G. (1983). Toward a comparative psychology of mind. In *Animal cognition and behaviour* (ed. R. E. Mellgren). North Holland, New York.

Gallup, G. G. (1985). Do minds exist in species other than our own? *Neuroscience and Biobehavioral Reviews*, **9**, 631–41.

Gallup, G. G. Toward a comparative psychology of self-awareness; species limitations and cognitive consequences. In *The self: an interdisciplinary approach* (ed. A. Goethals and J. Strauss). Springer-Verlag, New York. (In press.)

Gómez, J. C. (1990). Primate tactical deception and sensorimotor social intelligence. *Behavioral and Brain Sciences*, **13**, 414–15.

Gómez, J. C. (1991). Visual behaviour as a window for reading the mind of others in primates. In *Natural theories of mind* (ed. A. Whiten). Basil Blackwell, Oxford.

Harcourt, A. H. (1987). Cooperation as a cooperative strategy in primates and birds. In *Animal societies, theories and facts* (ed. Y. Ito and J. L. Brown). Japan Scientific Societes Press, Tokyo.

Harcourt, A. H. (1988). Alliances in contests and social intelligence. In *Machiavellian intelligence: social expertise and the evolution of intellect in monkeys, apes, and humans* (ed. R. W. Byrne and A. Whiten). Oxford University Press.

Harris, P. L. (1991). The work of the imagination. In *Natural theories of mind* (ed. A. Whiten). Basil Blackwell, Oxford.

Hasegawa, M, Kishino, H., and Yano, T. (1989). Estimation of branching dates among primates by molecular clocks of nuclear DNA which slowed down in Hominoidea. *Journal of Human Evolution*, **18**, 461–76.

Heyes, C. M. (1987). Contrasting approaches to the legitimation of intentional language within comparative psychology. *Behaviorism*, **15**, 41–50.

Hinde, R. A. (1970) *Animal behaviour: a synthesis of ethology and comparative psychology* (2nd edn.). McGraw-Hill, London.

Hobson, R. P. (1984). Early childhood autism and the question of egocentrism. *Journal of Autism and Developmental Disorders*, **14**, 85–104.

Humphrey, N. (1976). The social function of intellect. In *Growing points in ethology* (ed. P. P. G. Bateson and R. A. Hinde). Cambridge University Press.

Humphrey, N. K. (1980). Nature's psychologists. In *Consciousness and the physical world* (ed. B. Josephson and V. Ramachandran). Pergamon, Oxford.

Humphrey, N. K. (1983). *Consciousness regained*. Oxford University Press.

Humphrey, N. K. (1986). *The inner eye*. Faber and Faber, London.

Jones, V., and Prior, M. (1985). Motor imitation abilities and neurological signs in autistic children. *Journal of Autism and Developmental Disabilities*, **15**, 37–46.

Kohler, W. (1927). *The mentality of apes*. Routledge and Kegan Paul, London.

Krebs, J. R., and Dawkins, R. (1984). Animal signals: mind reading, and manipulation. In *Behavioural ecology: an evolutionary approach* (ed. J. R. Krebs and N. B. Davies). Blackwell, Oxford.

Kummer, H. (1982). Social knowledge in free-ranging primates. In *Animal mind-human mind* (ed. D. R. Griffin). Springer-Verlag, New York.

Leslie, A. M. (1987). Pretense and representation in infancy: the origins of 'theory of mind'. *Psychological Review*, **94**, 84–106.

Leslie, A. M. (1991). The theory of mind impairment in autism: evidence for a modular mechanism of development? In *Natural theories of mind* (ed. A. Whiten). Basil Blackwell, Oxford.

Leslie, A. M., and Frith, U. (1988). Autistic children's understanding of seeing, knowing and believing. *British Journal of Developmental Psychology*, **6**, 315–24.

McShane, J. (1980). *Learning to talk*. Cambridge University Press.

Menzel, E. W., and Johnson, M. K. (1976). Communication and cognitive organisation in humans and other animals. *Annals of the New York Academy of Sciences*, **280**, 131–42.

Passingham, R. E. (1982). *The human primate*. Freeman, New York.

Perner, J. (1988). Developing semantics for theories of mind: from propositional attitudes to mental representation. In *Developing theories of mind* (ed. J. W. Astington, P. L. Harris, and D. R. Olson). Cambridge University Press.

Povinelli, D. J. (1991). *Social intelligence in monkeys and apes*. Ph. D. thesis, Yale University.

Povinelli, D. J., and deBlois, S. (1991). Young children's (*Homo sapiens*) understanding of knowledge formation in themselves and others. Unpublished manuscript, Yale University.

Povinelli, D. J., Nelson, K. E., and Boysen, S. T. (1990). Inferences about guessing and knowing by chimpanzees (*Pan troglodytes*). *Journal of Comparative Psychology*, **104**, 203–10.

Povinelli, D. J., Parks, K. A., and Novak, M. A. (1992a). Do rhesus monkeys (*Macaca mulatta*) attribute knowledge and ignorance to others? *Journal of Comparative Psychology*, **105**, 318–25.

Povinelli, D. J., Nelson, K. E., and Boysen, S. T. (1992b). Comprehension of role reversal in chimpanzees: evidence of empathy? *Animal Behaviour*, **43**, 633–40.

Povinelli, D. J., Parks, K. A., and Novak, M. A. (1992c). Role reversal by rhesus monkeys; but no evidence of empathy. *Animal Behaviour*, **44**, 269–81.

Premack, D. (1988). 'Does the chimpanzee have a theory of mind?' revisited. In *Machiavellian intelligence: social expertise and the evolution of intellect in monkeys, apes, and humans* (ed. R. W. Byrne and A. Whiten). Oxford University Press.

Premack, D., and Dasser, V. (1991). Perceptual origins and conceptual evidence for theory of mind in apes and children. In *Natural theories of mind* (ed. A. Whiten). Basil Blackwell, Oxford.

Premack, D., and Woodruff, G. (1978). Does the chimpanzee have a theory of mind? *The Behavioral and Brain Sciences*, **1**, 515–26.

Reddy, V. (1991). Playing with other's expectations: teasing and mucking about in the first year. In *Natural theories of mind* (ed. A. Whiten). Basil Blackwell, Oxford.

Ryle, G. (1949). *The concept of mind*. Hutchinson, London.

Schreibman, L. (1988). *Autism*. Sage, London.

Sibley, C., and Ahlquist, J. (1984). The phylogeny of the hominoid primates, as indicated by DNA hybridisation. *Journal of Molecular Evolution*, **20**, 2–15.

Sodian, B. (1991). The development of deception in young children. *British Journal of Developmental Psychology*, **9**, 173–88.

Sodian, B., Harris, P. L., Taylor, C., and Perner, J. (1991). Early deception and the child's theory of mind: false trails and genuine markers. *Child Development*, **62**, 468–83.

Stern, D. N. (1985). *The interpersonal world of the infant*. Basic Books, New York.

Tolman, E. C. (1932). *Purposive behaviour in animals and men*. Century, New York.

Trevarthen, C. (1977). Descriptive analyses of infant communicative behaviour. In *Studies in mother-infant interaction* (ed. H. R. Schaffer). Academic Press, London.

Trevarthen, C. (1980). The foundations of intersubjectivity: development of interpersonal and cooperative understanding in infants. In *The social foundation of language and thought: essays in honor of J. S. Bruner* (ed. D. Olson). Norton, New York.

Trivers, R. L. (1974). Parent–offspring conflict. *American Zoologist*, **14**, 249–64.

Visalberghi, E., and Fragaszy, D. (1990). Do monkeys ape? In *Comparative developmental psychology of language and intelligence in primates* (ed. S. Parker and K. Gibson). Cambridge University Press.

Volkmar, F. R., Sparrow, S. A., Goudreau, D., Cicchaetti, D. V., Paul, R., and Cohen D. J. (1987). Social deficits in autism: an operational approach using the Vineland Adaptive Behaviour Scales. *Journal of the American Academy of Child Psychiatry*, **26**, 156–61.

Wellman, H. M. (1990). *Children's theories of mind*. Bradford/MIT press, Cambridge, Mass.

Wellman, H. M. (1991). From desires to beliefs: acquisition of a theory of mind. In *Natural theories of mind* (ed. A. Whiten). Basil Blackwell, Oxford.

Whiten, A. (1989). Transmission mechanisms in primate cultural evolution. *Trends in Ecology and Evolution*, **4**, 61–2.

Whiten, A. (1991). The emergence of mindreading: steps towards an interdisciplinary enterprise. In *Natural theories of mind: evolution, development and simulation of everyday mindreading* (ed. A. Whiten). Basil Blackwell, Oxford.

Whiten, A. (1992). Mindreading, pretence and imitation in monkeys and apes. *The Behavioral and Brain Sciences*, **15**, 170–1.

Whiten, A., and Byrne, R. W. (1986). The St Andrews catalogue of tactical deception in primates. *St Andrews Psychological Reports*, no. 10.

Whiten, A., and Byrne, R. W. (1988*a*). Taking (Machiavellian) intelligence apart. In *Machiavellian intelligence: social expertise and the evolution of intellect in monkeys, apes, and humans* (ed. R. W. Byrne and A. Whiten). Oxford University Press.

Whiten, A., and Byrne, R. W. (1988*b*). The Machiavellian intelligence hypotheses. In *Machiavellian intelligence: social expertise and the evolution of intellect in monkeys, apes, and humans* (ed. R. W. Byrne and A. Whiten). Oxford University Press.

Whiten, A., and Byrne, R. W. (1988*c*). Tactical deception in primates. *Behavioral and Brain Sciences*, **11**, 233–73.

Whiten, A., and Byrne, R W. (1988*d*). The manipulation of attention in primate tactical deception. In *Machiavellian intelligence: social expertise and the evolution of intellect in monkeys, apes, and humans* (ed. R. W. Byrne and A. Whiten). Oxford University Press.

Whiten, A., and Byrne, R. W. (1989). Machiavellian monkeys: cognitive evolution and the social world of primates. In *Cognition and social worlds* (ed. A. R. H. Gellatly, D. R. Roger, and J. A. Sloboda). Oxford University Press.

Whiten, A., and Byrne, R. W. (1990). Primates' appreciation of physical and psychological causality. *Behavioral and Brain Sciences*, **13**, 415–19.

Whiten, A., and Byrne, R. W. (1991). The emergence of metarepresentation in human ontogeny and primate phylogeny. In *Natural theories of mind* (ed. A. Whiten). Basil Blackwell, Oxford.

Whiten, A., and Ham, R. (1992). On the nature and evolution of imitation in the animal kingdom: reappraisal of a century of research. In *Advances in the study of behaviour*, 21 (ed. P. J B. Slater, J. S. Rosenblatt, C. Beer, and M. Milinski). Academic Press, New York.

Whiten, A., and Perner, J. (1991). Fundamental issues in the multidisciplinary study of mindreading. In *Natural theories of mind* (ed. A. Whiten). Basil Blackwell, Oxford.

Wing, L. (1976). *Early childhood autism: clinical, educational and social aspects.* Pergamon Press, Oxford.

Wittgenstein, L. (1958). *Philosophical investigations.* Blackwell, Oxford.

18

The comparative study of early communication and theories of mind: ontogeny, phylogeny, and pathology

JUAN CARLOS GÓMEZ, ENCARNACIÓN SARRIÁ, AND JAVIER TAMARIT

INTRODUCTION

This paper is about the origins and precursors of theory-of-mind abilities in infancy. It has been suggested that early communicative behaviours appearing in one-year-olds or even younger infants could be *precursors* to a theory of mind. There are at least three sources to this idea. The first is related to research on normal infants. In an early paper, published prior to the current outburst of research in theory of mind, Bretherton *et al.* (1981) suggested that normal infants develop 'explicit, verbally-expressible theories of mind' from, in their own words, 'a theory of interfacible minds, which is implicit in infants' first attempts at intentional communication at the end of the first year'. Secondly, a number of people conducting research in autism (among them, Mundy, Sigman, and Kasari 1990, this volume, Chapter 9; Mundy and Sigman 1989a; Baron–Cohen 1989a, 1991, this volume, Chapter 4) have proposed that since deficits in preverbal communication seem to be as characteristic of autistic children as theory of mind deficits, it seems likely that some relation might exist between both abilities.

Finally, recent studies of communication and cognition in non-human primates have posed the question of whether they need to represent mental states in their forcefully non-linguistic communicative interactions (see Whiten, this volume, Chapter 17). Some authors have suggested that the abilities shown by non-human primates involve an *implicit* theory of mind that could be considered as an evolutionary precursor to explicit ones (Gómez 1991; Whiten, this volume, Chapter 17).

All in all, the hypothesis of the relation between preverbal communication skills and theory of mind abilities remains 'an interesting but still open question' (Leslie 1987). In this paper we want to explore this 'open question' from a comparative perspective. This approach expands the traditional comparison of autistic children and normal children, resuming the line of inquiry that was at the origin of theory of mind research – the study of non-human primates. Indeed the first to ask if anybody had a theory of mind were Premack and Woodruff, and they asked this question of a chimpanzee. The debate raised by their paper (see for example Dennett's (1978) commentary to Premack and Woodruff's paper) inspired the first experiments with normal children (Wimmer and Perner 1983). This, in turn, led other authors to ask the theory of mind question with respect to autistic persons (Baron–Cohen *et al.* 1985). Ever since, the study of the theory of mind has become a well-established area of inquiry in developmental psychology and psychopathology.

In this paper we will adopt a triple comparative perspective, in which data from the three populations relevant for theory of mind research will be considered. We will specifically propose that the study of non-human primates can be useful to help us to understand autism. Let us first summarize what we know about each population.

THE COMPARATIVE APPROACH

Normal infants

At around four years of age normal children pass the acid test of theory of mind: they are able to understand that the *false belief* of other persons will determine their subsequent behaviour in situations such as the Sally-Ann task (Wimmer and Perner 1983; Baron–Cohen *et al.* 1985) or the Smarties task (Perner *et al.* 1987, 1989). Before this, three-year-olds seem to be able to understand limited aspects of desires and beliefs and their relation to emotions (see Wellman, this volume, Chapter 2). On the other hand, two-year-olds (who at best are credited with a more limited theory of mind, one concerning only people's desires [Wellman, this volume, Chapter 2]) show the ability to pretend in symbolic-play behaviours. On theoretical grounds, Leslie (1987) has argued that the ability to pretend is based upon the same cognitive capacity as a theory of mind: both would involve the cognitive ability to 'decouple' and form metarepresentations. Thus pretend-play would be an early manifestation of the line of development that leads to four-year-olds' theories of mind.

But prior to pretend-play, at around 8–12 months, normal infants develop *communicative actions* or *gestures* to regulate interactions with others (usually adults) in relation to external objects (Schaffer 1984). This is usually referred to as 'prelinguistic intentional communication'. This occurs after a first period (under eight months) during which infants are only able to interact in face-to-face situations, i.e., those which do not involve external referents (for example, a peek-a-boo game). We are going to focus upon the period of 'triangular' interactions that begins at 8–12 months. The communicative actions developed by children in this period present a number of characteristics, including the following:

1. Their topography is not designed to be mechanically effective upon the world, but is *open to completion* by other people (Gómez 1990*a*; Gómez, 1992; Rivière and Coll 1987): for example an infant pointing to a desired object is not actually trying to take it; he or she seems to understand that his or her action is effective only to the extent that it provokes a response from other people.

2. Communicative actions are also characterized by including *joint-attention behaviours*. *Eye-contact* with an adult is typically introduced before, during, or after the performance of a gesture. For example, an infant alternately looks at the person and at the object he is pointing to (Seibert and Hogan 1982; Bretherton and Bates 1979).

3. There are at least two domains of gestural communication in normal infants: gestures may be used either to make requests or to show objects or events to others. *Requests* are also called *protoimperatives* (Bates 1976), and they consist of asking a person to carry out a particular action for the infant's benefit (for example, having an object given). *Showing gestures* are also called *protodeclaratives* (Bates 1976),* and they consist of calling the attention of a person to a particular object or event in the environment without requiring the person to do anything else but to attend to the object.

4. The development of preverbal communication is accompanied by developments in other aspects of cognition, notably developments in *sensorimotor intelligence*. However, strong correlations with other aspects of early cognitive development (object permanence, spatial skills, tool-use), although occasionally reported (especially in the domains of causality and imitation), are not consistently found (Bates and Snyder 1987). Intentional communication seems to be a *relatively* independent domain of development (Sarriá and Rivière 1991).

When considering together the characteristics of prelinguistic intentional communication, some authors have emphasized that they seem to involve some understanding of the mental processes of other people. Intentional communication, according to some definitions, must involve some understanding of the other person's understanding (Grice 1956). This has led some people to consider the possibility that these communicative behaviours are the earliest manifestation of a theory of mind (Bretherton *et al.* 1981; Wellman, this volume, Chapter 2).

* Showing gestures have received a variety of names: among them: *expressive*, *indicative*, or *joint-attention* gestures. Note that in this volume, Mundy, Sigman, and Kasari use the label 'joint-attention behaviours' roughly to refer to what we are going to call 'protodeclaratives'.

Thus, the possibility arises that theory of mind development in normal children would involve a developmental sequence leading from infant intentional communication at the end of the first year, to pretence in the second year, then to desire/belief-understanding in the third year, and finally to a mature form of theory of mind that would include the understanding of belief and false belief at the age of four.

Autistic children

When Baron–Cohen *et al.* (1985) asked the question 'Does the autistic child have a theory of mind?', they found a negative answer. Most autistic subjects were unable to pass the false-belief task. Since autistic children also present a severe impairment in their symbolic play abilities (Baron–Cohen 1988), this constitutes crucial evidence for the model connecting the theory of mind and pretence via metarepresentation (Leslie 1987). Autistic children would have an impairment in their metarepresentational capacities, and this would explain both their symbolic-play deficits and their inability to understand false belief and, in general, other people's minds (Leslie, this volume, Chapter 5; Baron–Cohen 1988).

But the most relevant point for our purposes in this paper is that autistic children also present deficits in the prelinguistic communicative behaviours that in normal children appear at around one year of age (Curcio 1978; Hammes 1982; see also Baron–Cohen, this volume, Chapter 4; and Mundy, Sigman, and Kasari, this volume, Chapter 9). This pattern of impairment supports the idea suggested by the study of normal infants that prelinguistic communication may have some connection to theory of mind.

Let us try to characterize what seems to be the typical impairment pattern in the preverbal communication of autistic children, although the complexity of the autistic profiles and growing evidence in this area will make this effort only a tentative one.

1. Autistic children tend to develop *contact gestures*, i.e., gestures that involve establishing contact with persons, such as the well-known behaviours of leading a person by the wrist or the hand towards a desired place or object (Ricks and Wing 1976). Reaching and giving are described in autistics, but *pointing*, the emblematic gesture in normal infant communication, is very rare in them (Baron–Cohen 1989*b*).

2. Concerning *joint-attention behaviours*, the traditional view was that in autistic children there was a lack, even an avoidance, of eye-contact (Frith 1989*a*). However, research in recent years shows that the deficit is not in eye-contact *per se* (Sarriá and Rivière 1986; Hermelin and O'Connor 1970), but in the coordination of eye-contact to gestures and actions, i.e., in *joint attention* proper. Some autistic children, though, seem to be capable of at

least some amount of joint-attention behaviours (Mundy, Sigman, and Kasari, 1990, this volume, Chapter 9).

3. Autistic communication is at best *requestive*. Autistic children produce only 'protoimperatives', very rarely 'protodeclaratives' (Curcio 1978; Hammes 1982; Wetherby and Prutting 1984; Rivière *et al*. 1988; Baron–Cohen 1989b; Mundy *et al*. 1986). Even the autistic children who are able to point or exhibit some joint-attention behaviours seem to do so only to request objects or actions (Baron–Cohen 1989b).

4. Autistic children seem to be proficient in other aspects of cognitive development not related to social cognition. For example, they do not seem to be delayed in tool-use or object permanence, although sometimes a possible deficit in the domains of sensorimotor causality understanding and imitation has been reported (Curcio 1978; Rivière *et al*. 1988).

Thus the study of normal development suggests, on theoretical grounds, that prelinguistic communication could be connected to theory of mind. The impairment pattern in autism (deviant prelinguistic communication, impaired symbolic play, and lack or severe impairment of a theory of mind) provides empirical support for that claim, but also suggests that, since only some aspects of prelinguistic communication seem to be damaged, perhaps only these are connected to later developments in theory-of-mind abilities. Specifically, it has been claimed that only *protodeclarative gestures* should be considered as precursors to theory of mind abilities (Mundy and Sigman 1989a; Mundy *et al*., this volume, Chapter 9; Baron–Cohen 1989a, b, 1991, this volume, Chapter 4; Leslie and Happé 1989; Frith 1989b).

Thus the comparison between normal and abnormal human development has produced an interesting and specific hypothesis concerning the early ontogeny of theory-of-mind abilities. Let us now turn to the phylogenetic comparison.

Non-human primates

Despite Premack's and Woodruff's (1978) optimistic initial interpretation of their results, so far nobody has been able to demonstrate theory of mind abilities in non-human primates that are comparable to those shown by four-year-old human children (Premack 1988). Recent evidence by Povinelli *et al*. (1990) points to the possibility of chimpanzees' having some understanding of 'knowing'. We have no experimental evidence, however, demonstrating that non-human primates can pass the 'acid test' of theory of mind, namely, understanding 'false beliefs'.

Furthermore, non-human primates are known not to develop symbolic play spontaneously. There are only occasional reports of possible instances of symbolic play, and most of them come from apes that were subjected to 'linguistic' training. Linguistic apes constitute a special case. There is evidence that symbolic training may significantly affect the cognitive processes of anthropoids (Premack and Premack 1983; Premack 1984). In this paper, however, we are going to concentrate upon 'untrained' apes and leave to another occasion (see Tamarit *et al.* in press) the by no means trivial question of 'linguistic' apes. Considering the available evidence, our conclusion must be that symbolic play is not a typical achievement of anthropoid infants.

What about the non-linguistic communication of anthropoid infants? How can it be compared to that of human infants? Unfortunately, most studies on non-human primate communication are carried out from theoretical assumptions radically different from those adopted in human infant studies (Gómez 1992).* This is why to address the question of anthropoid communication we are going to rely upon our own studies with captive *hand-reared* apes, i.e., anthropoid infants who, because of the unavailability of conspecifics, are reared in zoo nursery environments by human adults. Although these developmental conditions are not identical to those enjoyed by a human infant, they reproduce many aspects of typical adult-infant interactions, such as bottle-feeding, play with human adults, opportunity to manipulate and interact around typical human objects, such as toys, and so on. This makes them an ideal population in which to study the 'prelinguistic' communicative abilities of non-human primates. So what follows is mainly based upon our own studies (Gómez 1990*a*,*b*, 1992).

Gestural communication in hand-reared gorillas

Figure 18.1 shows the communicative profile of a hand-reared gorilla at about twenty-four months of age, as assessed using a set of interactive categories originally designed to describe interactions in human infant–adult dyads. The categories were developed by Hubley and Trevarthen (1979) to describe what they call 'secondary intersubjectivity' in human infant–adult pairs.

The results show that these categories work well to describe interactions between gorillas and human adults. This gorilla offered and gave objects; followed instructions about objects; accepted or rejected assistance in object manipulation; and so on. And these patterns were coordinated with interpersonal behaviours (those on the right of the figure), notably *look at face*.

---

* An exception is a field study by Plooij (1978), in which he showed that concepts borrowed from studies of human infants could be useful to analyse mother–infant interactions in wild chimpanzees. His use of terms like 'protoimperative' and 'protodeclarative' is, however, questionable.

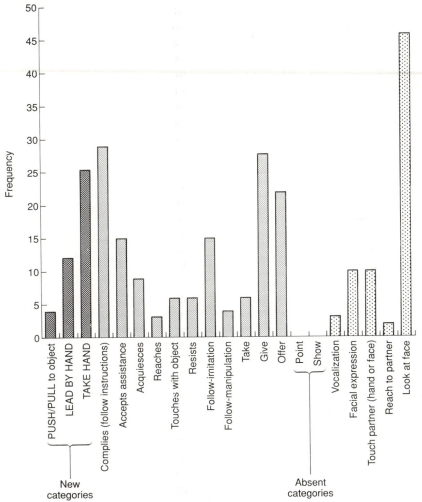

Fig. 18.1. Communicative behaviours of a female baby gorilla aged two years described using Hubley and Trevarthen's (1979) categories for human infant–adult interaction. Note that three new categories were added to account for important gorilla patterns, and that two of the original categories (pointing to and showing objects) did not appear in the gorilla (from Gómez 1990*b*).

By and large, this subject could be said to be able to engage in communicative interactions much as human infants do (Gómez 1990*a*,*b*; 1992).

But the communicative behaviour of this gorilla presented a number of peculiarities when compared to that of normal human infants.

1. Two categories, *point to object* and *show object*, do not appear in the gorilla's repertoire. Our gorilla never used the typical pointing gesture that

has been described as species-specific to humans, nor would she take objects to show them to humans.

2. Three new categories had to be created to account for important behaviour patterns of the gorilla apparently not present in Hubley and Trevarthen's subjects. These were: 'leading the human by hand or wrist to a place or object'; '*taking the hand* of the human to an object', and '*pushing/pulling* the human to a place', i.e., they were categories for *contact gestures*. They constituted one of the most frequent methods of communication in this gorilla (Gómez 1992).

3. These contact gestures as well as the other prominent gestures (object giving and offering) were used almost exclusively to *make requests*. The interactions analysed contained no instance of protodeclarative gesturing. Although other observations on the same subject provided examples close to some form of protodeclarative function (such as leading a person to a window and making him stay while she watched through it: Gómez 1992), they were far from being clear-cut examples identifiable with the gestures usually described in human infants as protodeclaratives.

Concerning other cognitive developments concurrent with gestural communication, gorillas have been reported to present a developmental profile in the Uzgiris and Hunt Scales of sensorimotor development very similar to that of humans (Redshaw 1978). Interestingly, the only possible differences seem to appear in the imitation and, perhaps, the causality scales, where gorillas seem to do less well than normal human infants.

These data refer to a single gorilla, but preliminary observations on four other hand-reared gorillas and a number of hand-reared chimpanzees essentially confirm this pattern (Gómez 1992). Assuming that the results with this gorilla accurately reflect the communicative abilities of hand-reared apes with human adults, it seems that, although gorillas and chimpanzees are able to engage in communicative interactions with humans, their spontaneous gestures present a number of remarkable peculiarities, namely, the tendency to use contact gestures, the absence of pointing, and the prevalence of protoimperative or requestive functions, with an almost total absence of protodeclarative gesturing.

When we compare this profile of prelinguistic communication with the profiles of normal and autistic children, the most remarkable result is the striking similarity between the patterns of hand-reared apes and of autistic children. The lack of protodeclarative functions and the tendency to use contact gestures are common salient features of their communicative strategies. If we add to this (1) the absence or severe impairment of symbolic play in both populations, (2) the failure in the case of autistic children to develop a theory of mind, and (3) the lack of evidence in the case of apes about their having a well-developed theory of mind, then we seem to be confronted with strong evidence in favour of the view that the three sets of abilities have something in common. In the case of apes, they would lack the three of them (if they are shown not to have theory of mind) because that 'thing' in common would not be part of their evolutionary endowment. In the case of autistic children, the 'thing' is assumed not to be operative because of severe damage. For different reasons apes and autistic children would lack an essential component for the development of theory of mind and related abilities. This would explain why the pattern of deficiencies shown by autistic children is similar to the pattern of divergencies shown by hand-reared apes.

Although the most obvious candidate to be the 'thing' underlying protodeclaratives, pretence, and theory of mind is metarepresentation or the decoupling mechanism, as proposed by Leslie (1987), there is little agreement about the exact nature of the connection between the three abilities, beyond the fact that there seems to be some connection (see Mundy and Sigman 1989*a,b*; Baron–Cohen 1989*a*; Harris 1989; Hobson 1989; Leslie and Happé 1989; Rogers and Pennington 1991). In the remainder of this paper we will try to introduce a little more controversy into this already complex debate.

THE 'PROTODECLARATIVE AS THEORY OF MIND PRECURSOR' HYPOTHESIS

One version of this hypothesis states that it is protodeclarative and not protoimperative or requestive abilities which are precursors to a theory of mind. The argument goes that there is a qualitative difference between protodeclaratives and requests. Protodeclaratives would be a more complex form of communication, one that would require meta-representational abilities.

Requests would be instrumental behaviours whose aim is to elicit certain behaviours in others. To make requests an organism would just need to be able to understand that its own actions may be instrumental in provoking certain reactions in the others. In protoimperatives 'the infant's goal is limited to obtaining some state of affairs in the *physical* world' (Baron–Cohen 1989*b*). It has even been suggested that requests rely upon a physical causal understanding of others. According to this view (illustrated in Fig. 18.2A with the image of a 'headless' human), there would be no essential difference between using a stick or using a person to obtain an out-of-reach object. Furthermore, this hypothesis would explain the *contact gestures* used by autistic children and apes as a reflection of the physical–causal understanding underlying requests.

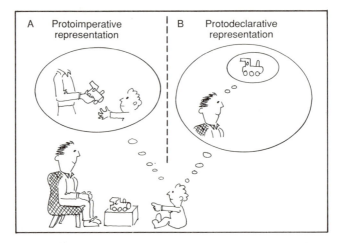

Fig. 18.2. Illustration of the hypothesis that requests or protoimperatives require simple representations of the behaviour of other people, whereas protodeclaratives involve an understanding of other persons' mental experiences.

In contrast, *protodeclaratives* would be based upon a more complex cognitive understanding. The goal of an infant who produces a protodeclarative gesture would be to 'share experiences of events' with others, to 'share interest in objects and events in the world' (Baron–Cohen 1989*b*, 1991). This, obviously, would imply an ability to hold a representation of the other person's mental states — at the very least, the ability to represent other people as perceiving and being interested in something (Fig. 18.2B). Briefly, whereas requests would involve simple representations of contingent physical events, protodeclaratives would require a more complex form of representation — second-order representations or representations of other people's mental experiences.

This would explain the absence of protodeclaratives in autistic children (a population with impaired metarepresentational capacities) and in apes (a population that presumably never evolved such metarepresentational capacity). And this would also support the notion that the protodeclarative is the earliest manifestation of a theory of mind. According to this view, then, the connection underlying protodeclaratives, pretence, and theory of mind would be the ability to form metarepresentations (Leslie 1987; Leslie and Happé 1989).

In what follows we are going to argue that the picture above painted of requests is too simple, and that painted of protodeclaratives is too complex, and we will explore the alternative hypothesis that protodeclaratives and protoimperatives do not differ in the level of representation they require.

THE NATURE OF REQUESTS IN GORILLAS

Let us begin by having a closer look at the nature of protoimperatives, and specifically those conveyed through contact gestures.* In a study of contact-gesture development in gorillas we concluded that requests can be executed at at least three different levels (Gómez 1990a, 1992). The three levels are schematized in Fig. 18.3.

1. *Person as OBJECT request*. Requests can consist literally of applying to another a physical–causal act, such as mechanically displacing someone when one wants him or her to be in a different place. This is what happens with the ape in Fig. 18.3A2 in her attempt to get the latch of a door opened. She wants to move the human towards the door, and she pushes him as she would do with a chair or a box (Fig. 18.3A1). The gorilla in this case would be relying upon only a physical–causal understanding of others: 'if you want someone to be in a particular place, just push them in the appropriate direction with the appropriate degree of effort'. This remarkable procedure, in which persons are treated as objects, is rare. We observed only a few pure instances of it in a gorilla under one year of age (Gómez 1992).

2. *Person as AGENT request*. A second possibility is that one understands at least the *agentive* side of persons, i.e., that they are able to move and act by themselves. In this case, the subject would act upon the other just to trigger off the other's actions. We could recognize this level of requesting by at least two signs: (a) The subject would not apply an act capable of producing the required movement mechanically (she would, for example, not apply the appropriate force (Gómez 1990a); and (b) there would be an 'expectant pause', leaving time for the other to act (Sarriá 1989). Note that in this case no conception of the psychological processes of the other would be necessary or implied. The other seems to be conceived of as 'a person [who] exerts power or produces an effect', to use the *Concise Oxford Dictionary* definition of 'agent'. The procedure of Fig. 18.3B illustrates a request in which the person is treated as an 'agent'. These forms of making requests appeared in our gorilla after twelve months of age

---

* Contact gestures such as hand-leading and hand-taking are not totally absent in the development of normal infants. These gestures seem to appear at least for a brief period of time during the infant's first year. They are also used later in combinations with pointing and other gestures to convey complex requests. Piaget (1937) described them as typical of Stage IV of cognitive development. His interpretation was thus that they represented a transitional period in the development of externalized causality, one in which infants still considered an agent's action as dependent upon their own. This is probably why they are sometimes seen as a 'primitive' form of request, one relying upon physical causality rather than on social understanding. A study of contact gestures in normal infants remains to be carried out; perhaps they are more frequent and important in the infant's communication system than is usually assumed.

Fig. 18.3. Three different ways of treating a person when making a request, as exemplified by the behaviour of a gorilla. In procedure A, the person is treated like an *object*. In procedure B, the person is treated as an *agent*, i.e., as someone who will act by him- or herself. In procedure C, the gorilla takes into account not only the human's agency but also his or her attention, treating him or her as a *subject* (from Gómez, forthcoming).

(Gómez 1992).* Probably there are different degrees and ways of understanding people as agents: from conceiving of them as a sort of mechanical toy that is released by means of a gesture, and once activated performs fixed action patterns, to being closer to a behaviouristic account of them, or having some sort of goal-directed conception of their movements (cf. Baron–Cohen, this volume, Chapter 4).

---

* In Gómez (1990*a*) the category 'person as agent' is not considered. Only the distinction between treating the human as an object and as a subject is included.

3. *Person as SUBJECT request.* The third level in making requests appeared when the gorilla gave signs not only that she expected human adults to move by themselves, but also that she understood something of the reasons why they would do so. As is illustrated in Fig. 18.3C, at around one and a half years old our gorilla began to look at the human's eyes while she performed her contact gestures, i.e., she began to include the eyes of the human among the things that were important to control visually when she wanted the human to move to the door. Our interpretation is that this behaviour reveals some understanding that to have the attention of the person is as important as the gesture itself for the latter to become effective (Gómez 1990a, 1991, 1992). Requests that involve such joint-attention behaviours imply *some* ability to recognize the cognitive or information-processing side of persons. Our suggestion is that these are the true requestive behaviours.

Requests of this kind, involving the combination of gestural activity and joint-attention behaviour, are also the same as those described in human infants as 'protoimperatives' (Bates 1976). They cannot be considered to be based upon a mere physical–causal understanding of the other (this would explain only type-I 'requests'), nor can they be reduced to a simple understanding of the contingency between one's own behaviour and other people's reactions (this would correspond to some forms of type-II requests). The joint-attention (attention-checking and attention-contact) patterns shown by gorillas and human infants when they perform a request of the third type demonstrate that they also understand something about the causal link that makes possible this contingency. Since this causal link is made of mental processes, our conclusion must be that true requests can involve the sort of implicit understanding of other people's minds that led Bretherton *et al.* (1981) to claim that prelinguistic communication required an 'implicit theory of mind'.

Thus, a closer look at requests reveals that they are complex communicative behaviours that may involve a sophisticated understanding of people. We want to emphasize that this understanding can refer not only to other people's attentional behaviours (as we have been stressing thus far) but also to their moods and emotions (see Bruner *et al.* 1983 for a study demonstrating the subtlety and complexity of requestive behaviours in human infants and children).

If this analysis is correct, a consequence would be that requests or protoimperatives would seem as good candidates to be theory-of-mind precursors as protodeclaratives. Both involve those joint-attention behaviours that seem to imply *some kind* of understanding of other people's mental processes. Does this imply that we should extend the metarepresentational explanation of protodeclaratives to requests? This is an interesting possibility, and we think that it makes better sense than attributing metarepresentation only to protodeclaratives. In this paper, however, we are going to explore a different alternative.*

One of us has proposed elsewhere that the understanding of attention shown in preverbal requests is based upon the *first-order* representation of the external manifestations of other people's attentional processes (Gómez 1991). Attention is a mental state that is directly reflected in overt, externally perceivable behaviours, so that an organism that had available only first-order representational devices would be able to represent attentional states. It is in this sense that infants and gorillas devoid of metarepresentational abilities could still be able to 'understand' attention. The overt signs of joint-attention (for example, eye-contact and gaze) would be captured by infants as first-order representations of causal links connecting their gestures with adults' behaviour (Gómez 1991).

The main point of this non-metarepresentational explanation of requests is that some mental processes like attention may have a direct overt expression (for example, gaze), and this opens the possibility of perceiving and having a first-order representation of them (cf. Hobson 1989). This is not the case with other mental processes, such as believing, that need to be *inferred* from the behaviour of people. According to this interpretation, an organism that engages in a true request seeks to establish *attention contact* with the addressee (which is usually achieved through eye-contact), and exhibits some kind of *attention-monitoring* to check if the addressee attends both to its gesture and to the target (which can be achieved by observing the other's gaze). This the infant can do using first-order representations of the other's attentional manifestations, notably of the other's gaze. Indeed Butterworth (1991) has proposed a geometric mechanism to explain the development of the ability to follow the gaze of others in young infants.

Thus we have seen that requests are not such simple behaviours; like protodeclaratives, they involve a sophisticated understanding of people. We have proposed, however, that this sophisticated understanding is not based upon metarepresentations of their mental processes, but upon a first-order understanding of the external manifestations of attention (and, very probably, some concomitants of attention, such as emotional expressions). Our next task will be to apply a similar analysis to protodeclaratives, trying to explain them as sophisticated behaviours, but behaviours that *only* require first-order representations of people.

---

* Note that the requests and protodeclaratives we are referring to are those shown in early infancy. No doubt there must come a time when communicative actions, including requests, are performed with the help of metarepresentation. But then they would no longer be *pro-to*imperatives and *proto*declaratives, but real declaratives and imperatives.

Fig. 18.4. Two alternative interpretations of the protodeclarative: (a) *intellectual* interpretation with metarepresentation: the child seeks to provoke a mental experience about that particular object in the adult; (b) *expressive* interpretation with first-order representations: the child seeks to provoke attentional and emotional responses about the object.

PROTODECLARATIVES AS FIRST-ORDER REPRESENTATIONAL BEHAVIOURS

The best way to present our view of protodeclaratives is to say that, for us, the goal of an infant that shows an object to a person is not to provoke a mental experience in him or her (Fig. 18.4A), but an *emotional and attentional* reaction (Fig. 18.4B). The infant is not interested in the intellectual side of people's experiences when they attend to an object, but in the expressive manifestations (both emotional and attentional) concomitant to those experiences. The infant seeks to provoke in the person the external signs of having an experience, and all she or he is able to represent are those overt signs that are manifested in the attentional and emotional behaviour of the adult, notably in his face: looking at, smiling, saying things . . . A part of the reaction of a person to an object or event is directly reflected in facial expressions and other overt behaviours. Our contention is that it is the overt correlates of experience young infants are interested in, because these are the only ones they are able to represent and, therefore, understand. In the hypothetical example of Fig. 18.4B, the child is not interested in what the adult will think or mentally experience about the object she is pointing at, but in the facial and vocal expressions he will exhibit *about* that toy.

We wish to make it clear that our interpretation of protodeclaratives is not that infants are just interested in the contingency between their pointing and, say, the adult's smiling. They are interested in the adult's smiling *at* the pointed-out object. And they can also be interested in any other expressive reaction oriented to the target they are pointing to—expressions of loathing, surprise, verbal commentaries, etc. Infants probably perceive and represent the adult's set of attentional and emotional reactions in a *Gestalt* way, i.e., integrated in a global pattern that may have emergent properties such as the orientation or *aboutness* of the adult's expressions towards an object. For instance, the whole reaction of the adult in Fig. 18.4b may be perceived and represented as '*being interested in*'.

Baron–Cohen (1989*a*), pursuing a metarepresentational hypothesis, proposes that what an infant might need to do in order to perform a protodeclarative is to *represent another's person representation* of an object as being tagged with a positive or negative valence (i.e., interesting or uninteresting)' (ibid., p. 198). From our point of view, this could be reformulated as 'to represent another's person *attentional reaction about* an object as being tagged with a positive or negative valence (i.e., interesting or uninteresting)'. The value of the valence is probably provided by the emotional reactions concomitant to the attentional orientation. These reactions could add further shades to the adult's orientation towards the object (not only 'interested/uninterested' but also 'amused', 'frightened', 'angry', etc.). The point is that all these representations are of directly perceivable events—the overt expressions of the adult, not yet his covert mental correlates—and they therefore require only first-order representational abilities.

REQUESTS AND PROTODECLARATIVES DIFFERENTIATED

Thus we have proposed an interpretation of protodeclarative and requestive gestures that makes them depend upon the same, first-order level of representational abilities. This interpretation has the advantage of explaining two behaviours that apppear at around the same time, and that present many superficial similarities (Tamarit *et al.*, in press) while only resorting to a single level of representation. But it has a serious disadvantage. Since autistic children and hand-reared apes develop requestive behaviours, why do they not develop protodeclaratives too, if these are said to involve the same level of representation? The communicative profile of autistic children and apes suggests that even if requests and protodeclaratives do not differ in the representational level they require, they *must* differ in some very important respect. What then is the difference between protodeclaratives and protoimperatives? Our duty, having disclaimed the metarepresentational hypothesis, is to propose an alternative explanation. This we try to do now.

There is a difference between requests and protodeclaratives that is not due to the level of representation required by each kind of gesturing, but to the role attention and its emotional correlates play in the communicative act. In *requests*, attention and emotion (i.e., their overt manifestations) are always a means, a link in the causal chain that will or will not lead to a desired behavioural effect (for example, getting an object). This is true both of the infant who emits the gesture whilst looking and perhaps smiling or whining at the adult, and of the adult whose gaze and emotional reaction give information to the infant about her progress in receiving the request. For example, imagine that an infant reaches to an old Chinese vase on a shelf while he emits his typical requestive whining vocalization and looks back at his mother. The infant is using joint attention accompanied by emotional expressions to make a request. When he sees that his mother looks at him and looks at the vase, this informs him that he has been successful in engaging his mother in the request (she is attending to his gesture and to his object of attention). But when he sees the horrified expression of his mother upon understanding that it is the old Chinese vase that her son wants, he probably understands that something is going wrong with his request. In the whole sequence, the overt signs of attention and emotion about an object are subordinated to the goal of getting that object through the mother's help.

In *protodeclarative gestures*, however, the attentional and emotional reactions seem to be themselves the goal of the communicative act. The infant's aim is to get the other's attention upon an object and to explore the reactions (emotional, vocal, behavioural) of the other towards it, perhaps even engaging in an emotional exchange around the object. For example, consider the case of an infant who is specially interested in a fountain in the park where he is taken for a daily walk by his mother. He looks at the fountain, and after a few seconds he points to it and looks back at his mother smiling and laughing; the mother looks at the fountain, smiles and tells him 'Yes, it is your favourite fountain!'. The infant looks back at the fountain and back again at his mother. Then he directs his attention to something else. The attentional and emotional expressions of the infant in this hypothetical example can be described as a means to engage the mother's attention in the fountain, as in the case of requests. However, the attentional and expressive reactions of the mother now seem to be the very *goal* of the infant's efforts. He seems to be interested in directing the attention of his mother to the fountain and having some emotional expressions from her about the object. Attentional and emotional manifestations in others are no longer an index of the relative success or failure of the children's efforts to engage the adults in something, but the very aim of these efforts.

Whereas in requests the infant uses his or her power to influence another's attention to get things done, in protodeclaratives infants seem to explore, sometimes even to play with, this power. Thus, what in requests are always a means — the infant's first-order representations of attention and emotion in others — in protodeclaratives becomes the very goal of communication.

In the discussion of the differences between requestive and protodeclarative communication we are confronted with an important limitation: the absence of a detailed study of protodeclaratives. We think that different kinds of communicative behaviours coexist under this label (or, for that matter, under the label used by Mundy *et al.* of 'joint-attention behaviours'; see this volume, Chapter 9). Perhaps part of the discrepancy in the interpretation of protodeclaratives is due to the fact that different authors might be using this label to refer to different kinds of behaviours. What the infant has in his or her head when producing them may be, however, different in each case. It is probably not the same to point routinely to the nice fountain in the park every day as it is to point to a familiar object that now appears in a new state (for example, a broken toy), or to a totally new object that appears unexpectedly. Probably we should instead speak of 'the protodeclarative *family* of gestures'.

Mundy *et al.* (this volume, Chapter 9) also address the issue of the difference between requestive and protodeclarative behaviours ('joint-attention behaviours' in their terminology). Their finding that positive affect expression is a common component of protodeclaratives in normal infants has led them to propose that 'joint-attention behaviours [protodeclaratives] and requesting behaviours may be distinguished on the basis of affect. Joint attention not only involves the coordination of attention *vis-à-vis* some object or event, but also involves the conveyance of affect' (Mundy *et al.*, this volume, Chapter 9). This hypothesis is, of course, related to the one we propose here. In fact, both can be considered complementary, in that Mundy *et al.* stress the infant's acts when producing a protodeclarative, whereas we mainly focus upon what the infant seeks to provoke in the other. However, we would like to emphasize that the production and perception of emotional expressions are not exclusive to protodeclaratives: they also seem to be important components of normal requests. In fact, the data of Mundy *et al.* (this volume, Chapter 9) indicate that showing positive affect, although most characteristic of protodeclaratives, is also common in the requests of normal children (but, interestingly, not in those of autistic children).

In summary, we suggest that gestures belonging to the protodeclarative family (at least those appearing at around twelve months) have in common their focus upon the external manifestations of attention and emotion as goals in themselves, whereas gestures belonging to the requestive family use those manifestations as means to regulate a variety of behaviours in other people (object-giving, play, being taken into the arms, etc.). Is this difference important enough to explain why some organisms can have requests but not protodeclaratives? We address this question in the next section.

REQUESTS, BUT NOT PROTODECLARATIVES

Different kinds of requests?

Let us consider first the case of autistic children. They seem to develop requests, but not protodeclarative gestures. But are their requestive behaviours totally similar to those of normal infants? Thus far we have been assuming that the requests of autistic children are identical to those of normal children. We have, however, no detailed study that warrants this assumption. All we know is that at least some autistic children are able to use gestures accompanied by eye-contact in apparently requestive functions (Mundy *et al.*, this volume, Chapter 9; Canal 1991).

What kind of differences could we expect to find between normal and autistic requests? From the point of view that we have been defending here, we should look for differences in the use of attentional and emotional cues. Is the use of joint attention and attention contact in autistic requests consistent and similar to that in the requests of normals? Are they able to use emotional cues (both productively and receptively) in their requests? These questions have not yet been fully answered, but some facts, such as the finding of Mundy, Sigman, and Kasari (this volume, Chapter 9) that autistic children use fewer expressions of positive affect in their requests, or Canal's (1991) report that autistic children do not use eye-contact as a cue to initiate requests, point to the relevance of the question.*

Concerning the requests of Great Apes, we are faced with a similar problem. Gómez's (1990*a*) single-case study discussed above demonstrates that anthropoid requesting is very different to tool-use, and cannot be attributed to a mere understanding of others in physical–causal or contingency-detection terms. The gorilla exhibited some degree of understanding of the role of attention in others (she considered persons as 'subjects'). However, is this understanding equivalent to that of normal infants? Is it comparable to that of autistic children? Perhaps the taxonomy of requests discussed above could be extended to include different subclasses of 'treating others as subjects'. Again we need more detailed studies of requests in the three populations to answer these questions.

If autistic and anthropoid requests were found not to be equivalent to

---

* Phillips *et al.* (1992) have recently compared the requestive strategies of normal and autistic children, with special reference to the ability to use attention-regulation behaviours in requesting.

those of normal children, then the problem of 'requests, but not pro-todeclaratives' would have to be at least partially reconsidered. But until we have the appropriate evidence, let us continue our discussion on the basis of what we currently know. And what we know is that normal infants, autistic children, and anthropoid apes are able to use *some* attentional cues to perform requestive functions, but only the first group use them to effect protodeclaratives.

A motivational difference?

Our above description of the difference between the two kinds of com-municative acts could be interpreted as leading to the conclusion that the difference between protodeclarative and protoimperative gesturing is *motivational* rather than *representational* in nature. To carry out pro-todeclaratives one should be interested in the other's manifestations of atten-tion (eye-ontact and gaze-alternation to objects) and interest (emotional reactions and affective expressions) *in themselves*, and not as intermediaries to influence the behaviour of others. If one were not intrinsically interested in these aspects of a subject's behaviour (whatever the cause of this inability), one would never produce protodeclaratives. What does one need to be intrin-sically interested in the attention and emotion of others?

First of all, one needs to be able to perceive and to represent them. Note that in the metarepresentational interpretation of protodeclaratives autistic children could not be interested in using pointing to provoke *mental* experiences in others, because they are held to be *unable* to represent those experiences. Mental states cannot perforce be the goal of their gesturing, because they do not exist for them (cf. Baron–Cohen's (1990) notion of 'mind-blindness'). It is much like not being interested in music because one is deaf, or not being interested in painting because one is blind.

The non-metarepresentational view we have proposed, however, states that to produce protodeclaratives all one needs to represent is the 'external manifestations of emotion and attention'. This means that the kind of deficit we should look for in autistic children would be one affecting the first-order representation of expressive behaviours. Since blindness and deafness are not part of the autistic syndrome, autistic children must be able to perceive and represent such overt events as facial and vocal expressions. But do they see and hear exactly the same things as normal infants?

Gestalt psychologists demonstrated long ago that our perception of simple physical configurations occurs in a complex way: we directly perceive organized wholes out of aggregated elements. We perceive a square, and not four lines arranged at right angles. When watching a film we perceive movement, and not a sequence of frames. There are emergent properties in our perception of things. It is, then, no wonder that we can speak of our perception of persons in *Gestalt* terms: i.e., we do not perceive unstructured aggregates of units (eyes, brows, nose, mouth, vocal sounds, etc.), but organized wholes with emergent properties. What are the emergent properties one perceives (and represents) in emotional and attentional expressions?

Let us illustrate this issue with a primatological example. An inexperienced person visiting a primate centre is likely to wonder why the caged monkeys around him seem to express surprise (opening their mouths and raising their brows) when he or she looks at them. The monkeys, however, are not surprised, they are *threatening* the person who stares at them, because for them sustained gaze is an expression of threat, and they respond to it with another threat—staring plus open mouth plus brows raised—that bears a striking resemblance to the human expression of surprise. A naïve human is no doubt able to discriminate between this facial expression and other species-specific expressions of the monkey without, however, being able to represent their expressive 'meanings'. A facial expression does not only consist of a physical configuration; it also conveys an 'expressive message', a 'meaning' that is previous to any metarepresentational capacity, and does not require an assessment of the intentions or mental states behind it.

Do autistic children normally perceive and represent the emotional and attentional 'meanings' beyond the physical appearance of other people's expressions? This is an empirical question that we cannot yet answer on the basis of the available evidence. But some possible anomalies have already been pointed out in their appreciation of facial and vocal expressions of emotion (Hobson 1986*a,b*, this volume, Chapter 10; Klin 1991).

The reason why normal infants find expressive behaviours interesting in themselves (thereby producing protodeclaratives) might lie in some of the emergent properties they perceive in them. A deficit in the perception or representation of those emergent properties could render them less adequate to become objects of intrinsic interest. It is interesting to note that such a deficit would presumably appear before nine months of age (cf. Klin and Volkmar, this volume, Chapter 15; Mundy *et al.*, this volume, Chapter 9; Hobson, this volume, Chapter 10). Normal infants are said to show from very early on a special attunement to and preference for human expressions. This means that for infants, human expressions are conspicuous stimuli that engage their interest. The expressions of other people seem to offer specific *affordances*, i.e. 'information for action, rather than . . . signs of another's phenomenal experience', to infants as young as three or five months (Walker-Andrews 1988).

Thus the consideration of protodeclaratives as first-order representational behaviours allows us to connect them to the rich body of studies on adult–infant interactions at less than nine months of age. Of special interest as possible precursors of protodeclaratives seem to be face-to-face interactions whose goal is described as the interchange of attentional and emotional expressions: for example looking at each other's eyes, smiling and cooing in turns (see Schaffer 1984, for an extensive review of the field). Do autistic children engage in face-to-face interactions of this kind? And if so, do they interact as normal infants do? Do they perceive the 'affordances' of expressive behaviours? Answering these questions will probably be part of future inquiry about the autistic children's *sensitivity* to other people's expressions.

The expression of emotion and attention in anthropoids

Let us now turn again to our third population – non-human primates. Our interpretation of protodeclaratives has led us to suggest that their absence in autistic children can be explored in relation to some putative anomaly in their *sensitivity* to emotional and attentional expressions. Would a similar 'difference in sensitivity' also explain the absence of protodeclaratives in anthropoid apes?

One of the features of gorilla and chimpanzee societies is the complexity of their social lives: they are able to engage in sophisticated polyadic interactions, such as coalition-formation, reconciliation behaviours, or even deception (de Waal 1982, 1989; Whiten, this volume, Chapter 17), and they seem to be highly sensitive to the emotional manifestations of others in such situations. Although the appropriate empirical evidence remains to be collected, there are reasons to predict that apes would show nothing akin to a 'deficit' in their sensitivity to conspecifics' expressions of emotion and attention.

But in the case of hand-reared apes it could be argued that they would quite naturally lack 'sensitivity' to *human* expressions, and, therefore, fail to develop protodeclaratives addressed to humans. Do normally-reared apes show protodeclarative behaviours among themselves? They certainly do not seem to develop pointing or showing gestures, although perhaps we should not totally rule out the possibility that some alternative versions of protodeclarative behaviours could be found among them. However, available evidence points to the absence of protodeclaratives as a species typical behaviour in anthropoids, whereas requests and complex social strategies involving elaborate emotional displays are well documented in their everyday interactions (de Waal 1982, 1989; Whiten, this volume, Chapter 17). They are highly 'sensitive' to expressive behaviours, yet they do not develop what we usually identify as protodeclaratives. How can we reconcile this with our suggestion about why autistic children fail to develop protodeclarative behaviours?

Remember that the kind of interest manifest in protodeclarative behaviours, as we describe them, is an *intrinsic* interest in expressive behaviours, i.e., interest in the expressions themselves, not in what they announce or are concomitant with. Consider, for example, a typical episode of rough-and-tumble play among gorillas in which two animals try to catch and bite each other while producing play-faces and vocalizations (Gómez 1986). Are the gorillas interested in the play-faces and vocalizations as such, or rather as regulators of the physical activities the game consists of? Could gorillas play just to interchange play-faces? It is one thing to find a play-face interesting and amusing in itself, and different altogether to find the play-bout it announces and regulates of interest. Some preliminary evidence (González del Yerro, unpublished manuscript) suggests that infant anthropoids tend to engage in 'rough-and-tumble' games whose main components are body contact and movement, whereas human infants seem to devote more time to interactions consisting mainly of attentional and expressive interchanges. In anthropoids attention-contact and vocal and facial expressions seem to play the role of regulators of interactions, whereas in human infants they may be the very content of interactions. This possible difference in face-to-face interactions is similar to that described for requests and protodeclaratives. It is again a 'motivational' difference, affecting the role of attentional and emotional expressions in interactions. In this respect, it is interesting to note that autistic children have been described as willingly engaging in 'rough-and-tumble' games such as tickling (Ricks and Wing 1976).

What can we conclude from this? First, an interesting area of research concerns face-to-face interactions in anthropoids and autistic children, especially examining the role of attentional and emotional expressions in the interaction. Second, the kind of 'sensitivity' to expressions needed to produce protodeclaratives is an *intrinsic one*, i.e., one allowing the infant to be interested in expressions in themselves. To answer the question of autistic children's hypothetical deficit in 'sensitivity' to expressions, we have to be prepared to distinguish between at least two possible kinds of deficits: one affecting the general sensitivity to expressions, and another affecting only the intrinsic sensitivity. An anomaly in the general sensitivity would probably favour the appearance of deficits not only in protodeclaratives but also in requestive-like behaviours, whereas an anomaly in 'intrinsic sensitivity' would essentially affect protodeclarative-like behaviours.

To summarize, when we say that the difference between protodeclarative and requestive gestures could be motivational rather than representational, we do not imply that autistic persons (or apes) are unwilling to choose some goals that they are able to conceive of for their actions, but rather that they are unable to *see* certain things as possible goals. Our aim in this section has not been to defend a specific explanation of the autistic deficit in protodeclarative behaviours, but rather to show that our first-order-representation interpretation of protodeclaratives points to some promising directions of research that at least partially coincide with those proposed by other authors (Mundy *et al.* 1990, this volume, Chapter 9; this volume, Chapter 10; Trevarthen 1989).

PROTODECLARATIVES AND THEORY OF MIND

In the preceding section we have defended the idea that a first-order-representation interpretation of protodeclaratives still confers on them some unique properties that can explain why they are not equivalent to requestive gestures, and have pointed to some possible antecedents for them. The question we want to address in the final section is the one that opened this paper — are protodeclaratives precursors to a theory of mind? Does our first-order interpretation of them still warrant any relation between both abilities? The metarepresentional view tries to explain protodeclaratives as a consequence of the existence of an ability called 'metarepresentation' in normal human infants (Leslie and Happé 1989; Baron–Cohen 1989*a*). Protodeclaratives, then, are considered to be precursors to a theory of mind in the sense that they are one of the earliest manifestations of the metarepresentational ability underlying theory of mind.

There are two possible ways of being a *precursor* according to the *Oxford English Dictionary*. The first one is indicating, announcing the coming of something else, and the example provided in the dictionary is John the Baptist in relation to Christ. Indeed, this seems to be the sense in which the metarepresentational theory states that protodeclaratives are precursors to the theory of mind (Fig. 18.5A). They somehow announce the advent of metarepresentation's beloved son — theory of mind. What both abilities have in common is that both are the product of the metarepresentational father, and perhaps that the first fulfils some of the functions of the second in a less elaborate way.

But there is a second way of being a 'precursor'. This is linked to chemistry, where a substance is said to be a precursor of another when the former can be transformed into the latter. Chemical transformations can give rise to substances with new emergent properties. This is the sense in which we think the question of protodeclaratives as precursors to a theory of mind should be addressed. Protodeclaratives reflect a first-order representational ability from which a metarepresentational capacity will be developed later. Rather than being a *consequence* of metarepresentation, protodeclaratives might be one of the *causes* that make possible a metarepresentational theory of mind (see Fig. 18.5B). The cognitive skills involved in protodeclarative behaviours, not being themselves metarepresentational, could however be a necessary component of the developmental process leading to a theory of mind.

Obviously the most urgent task of a hypothesis like this is precisely to explain how this step — the passage from first-order representation to

Two ways of being a precursor

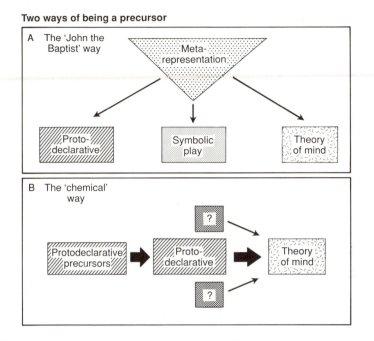

Fig. 18.5. Two ways of being a precursor. A. The *John The Baptist* way: the protodeclarative announces the advent of more complex metarepresentational capacities, being itself a consequence of a common developmental father: metarepresentation. B. The *chemical* way: the protodeclarative develops from its own precursors and then, in combination with other unspecified factors, contributes to growth of a metarepresentational theory of mind.

metarepresentation — can be carried out. In this paper, however, we are not going to solve this riddle. We will just point out something that must be taken into account in this task. The chemical version of precursivity requires other substances and factors acting on the precursor for it to be transformed (the unknowns in Fig. 18.5B). The ability to produce protodeclaratives could be just one more factor in the development of a theory of mind, perhaps one whose main effect is to highlight the evidence — the external manifestations of mental processes — from which theories of mind can be formulated.

We are aware that this view leaves open many questions and problems: for example, what is the relation of protodeclarative gesturing to symbolic play? Could protoimperatives fulfil, at least partially, the functions of protodeclaratives in the development of a theory of mind? Is it still necessary to postulate an independent 'metarepresentational ability' to account for the passage from the data — protodeclaratives — to the theory — the mind? What are the most likely candidates for the 'unknowns' of Fig. 18.5B —

symbolization, imitation, perhaps metarepresentation? A more detailed consideration of these problems is carried out in Tamarit *et al.* (in press). Mundy, Sigman, and Kasari (this volume, Chapter 9) and Rogers and Pennington (1991) also provide approaches to the issue that seem to fit well with a 'chemical' version of precursivity.

The interpretation of protodeclaratives as first-order representational behaviours does not imply, therefore, that they are not connected to theory of mind, but rather that we should look for a different kind of connection, one in which prelinguistic communicative behaviours are regarded as components, causes of cognitively more complex abilities, rather than as early reflections of them.

CONCLUSION

The aim of this paper has been to show the usefulness of a comparative perspective (including normal children, autistic children, and anthropoid apes) in addressing the issue of the relation between theory of mind abilities and early communicative development. Although many pieces of this comparative puzzle are still missing, there are some interesting results that suggest its resolution is worth pursuing.

We have shown that protodeclarative and requestive gestures can be interpreted as requiring the *same* level of representational abilities, and we have suggested that this level is a first-order one. In trying to explain protodeclaratives as behaviours based upon an intrinsic interest in attentional and emotional expressions, we have been led to the consideration of some possible precursors of protodeclaratives themselves.

Autistic children could suffer an impairment in an ability ontogenetically prior to a metarepresentational theory of mind — one related to the perception and representation of the external signs of mental activities. The study of anthropoid apes, whose 'prelinguistic' communication apparently shares important characteristics with that of autistic children, suggests that extremely sophisticated social interactions can be developed from essentially protoimperative abilities, and perhaps without a metarepresentational theory of mind. On the other hand, the study of apes suggests that to develop protodeclaratives an organism may need something that is best described as an 'intrinsic sensitivity' to attentional and emotional expressions.

Finally, we have tried to show that the first-order interpretation of protodeclaratives (and of infant prelinguistic communication in general) does not rule out the possibility that they are precursors to theory of mind abilities, but rather suggests a stronger form of precursivity, in which they would become causes, rather than consequences, of those abilities.

All in all, the aim of this paper has been not to propose specific solutions to problems, but to point to new interpretations of facts and some possible lines of future research that can help to answer the open question of theory of mind and its precursors. Since our task in this paper has been mainly one of 'pointing', the reader must take our attempts just as 'proto-explanations' of the issues addressed.

Acknowledgements

This paper has partly benefited from a grant of the CIDE to the authors. A. Brioso, E. León, and A. González del Yerro participated in discussions related to the topic of this paper. A first version of it was presented at the IV European Conference of Developmental Psychology in Stirling, September 1990. Our thanks to S. Baron–Cohen for fruitful discussions and commentaries. We are grateful to the director and staff of Madrid Zoo for their collaboration in the work with anthropoid apes.

REFERENCES

Baron–Cohen, S. (1988). Social and pragmatic deficits in autism. *Journal of Autism and Developmental Disorders*, **18**, 379–402.

Baron–Cohen, S. (1989a). Joint-attention deficits in autism: towards a cognitive analysis. *Development and Psychopathology*, **1**, 185–9.

Baron–Cohen, S. (1989b). Perceptual role taking and protodeclarative pointing in autism. *British Journal of Developmental Psychology*, **7**, 113–27.

Baron–Cohen, S (1990) Autism: a specific cognitive disorder of 'mind-blindness'. *International Review of Psychiatry*, **2**, 79–88.

Baron–Cohen, S. (1991). Precursors to a theory of mind: understanding attention in others. In *Natural Theories of Mind* (ed. A. Whiten). Blackwell, Oxford.

Baron–Cohen, S., Leslie, A.M., and Frith, U. (1985). Does the autistic child have a 'theory of mind'? *Cognition*, **21**, 37–46.

Bates, E. (1976). *Language and context: the acquisition of pragmatics*. Academic Press, New York and London.

Bates, E., and Snyder, L. (1987). The cognitive hypothesis in language development. In *Infant performance and experience: new findings with the ordinal scales*. (ed. I.C. Uzgiris and J.M. Hunt). University of Illinois Press, Urbana.

Bretherton, I., and Bates, E. (1979) The emergence of intentional communication. In *Social interaction and communication during infancy* (ed. I.C. Uzgiris). Jossey-Bass, San Francisco.

Bretherton, I., McNew, S., and Beeghly-Smith, M. (1981). Early person knowledge as expressed in gestural and verbal communication: when do infants acquire a 'theory of mind'? In *Infant social cognition* (ed. M.E. Lamb and M.R. Sherrod). Erlbaum, Hillsdale, NJ.

Bruner, J., Roy, C., and Ratner, N. (1982). The beginnings of request. In *Children's language*, Vol. 3 (ed. K.E. Nelson). Erlbaum, Hillsdale, NJ.

Butterworth, G. (1991). The ontogeny and phylogeny of joint visual attention. In *Natural theories of mind* (ed. A. Whiten). Blackwell, Oxford.

Canal, R. (1991). *Problemas de la comunicación prelingüística en niños autistas*. Unpublished Ph.D. dissertation, Universidad Autónoma de Madrid.

Curcio, F. (1978). Sensorimotor functioning and communication in mute autistic children. *Journal of Autism and Childhood Schizophrenia*, **8**, 281–92.

Dennett, D. (1978). Beliefs about beliefs. *Behavioral and Brain Sciences*, **1**, 568–70.

de Waal, F. (1982). *Chimpanzee politics*. Jonathan Cape, London.

de Waal, F. (1989). *Peacemaking among primates*. Harvard University Press, Cambridge, Mass.

Frith, U. (1989*a*). *Autism: explaining the enigma*. Blackwell, Oxford.

Frith, U. (1989*b*). Autism and 'theory of mind'. In *Diagnosis and treatment of autism* (ed. C. Gillberg). Plenum, New York.

Gómez, J.C. (1986). *Algunos aspectos del juego social de lucha y persecución en un grupo de gorilas cautivos*. Unpublished Master's thesis, Universidad Autónoma de Madrid.

Gómez, J.C. (1990*a*). The emergence of intentional communication as a problem-solving strategy in the gorilla. In *'Language' and intelligence in monkeys and apes: comparative developmental perspectives* (ed. S.T. Parker and K.R. Gibson). Cambridge University Press.

Gómez, J.C. (1990*b*). The development of secondary intersubjectivity in infant gorillas. Paper presented at the 13th Annual Meeting of the American Society of Primatologists.

Gómez, J.C. (1991). Visual behavior as a window for reading the minds of others in primates. In *Natural theories of mind* (ed. A. Whiten). Blackwell, Oxford.

Gómez, J.C. (1992). *El desarrollo de la comunicación intencional en el gorila*. Ph.D. dissertation, Universidad Autónoma de Madrid.

Grice, H.P. (1957). Meaning. *Philosophical Review*, **66**, 377–88.

Hammes, J. (1982). *Taal noch teken: Een studie over de symboolvorming bij zwakzinnige autistische kinderen*. Swets & Zeitlinger, Lisse.

Harris, P. (1989). The autistic child's impaired conception of mental states. *Development and Psychopathology*, **1**, 191–5.

Hermelin, B. and O'Connor, N. (1970). *Psychological experiments with autistic children*. Pergamon Press, Oxford.

Hobson, P. (1986*a*). The autistic child's appraisal of expressions of emotion. *Journal of Child Psychology and Psychiatry*, **27**, 321–42.

Hobson, P. (1986*b*). The autistic child's appraisal of expressions of emotion: a further study. *Journal of Child Psychology and Psychiatry*, **27**, 671–80.

Hobson, P. (1989). On sharing experiences. *Development and Psychopathology*, **1**, 197–203.

Hubley, P. and Trevarthen, C. (1979). Sharing a task in infancy. In *Social interaction and communication during infancy* (ed. I.C. Uzgiris). Jossey Bass, San Francisco.

Klin, A. (1991). Young autistic children's listening preferences in regard to speech: a possible characterization of the symptom of social withdrawal. *Journal of Autism and Developmental Disorders*, **21**, 29–42.

Leslie, A. (1987). Pretense and representation: the origins of 'theory of mind'. *Psychological Review*, **94**, 412–26.

Leslie, A.M. and Happé, F. (1989). Autism and ostensive communication: the relevance of metarepresentation. *Development and Psychopathology*, 1, 205-12.

Mundy, P. and Sigman, M. (1989*a*). The theoretical implications of joint attention deficits in autism. *Development and Psychopathology*, 1, 173-183.

Mundy, P. and Sigman, M. (1989*b*). Second thoughts on the nature of autism. *Development and Psychopathology* 1, 213-17.

Mundy, P., Sigman, M., Ungerer, J., and Sherman, T. (1986). Defining the social deficits of autism: the contribution of non-verbal communication measures. *Journal of Child Psychology and Psychiatry*, 27, 657-69.

Mundy, P., Sigman, M., and Kasari, C. (1990). A longitudinal study of joint attention and language development in autistic children. *Journal of Autism and Developmental Disorders*, 20, 115-28.

Perner, J., Leekam, S.R., and Wimmer, H. (1987). Three-year-olds' difficulty with false belief: the case for a conceptual deficit. *British Journal of Developmental Psychology*, 5, 125-37.

Phillips, W., Laà, V., Gómez, J.C., Baron-Cohen, S., and Rivière, A. (1992). Treating people as objects, agents or subjects: what can we learn from the Muni task about development and autism? Unpublished MS, Institute of Psychiatry, and paper presented at the European Developmental Psychology Conference, Seville.

Piaget, J. (1987). *La construction du réel chez l'enfant*. Delachaux et Niestlé, Neuchâtel.

Plooij, F.X. (1978). Some basic traits of language in wild chimpanzees? In *Action, gesture and symbol: the emergence of language* (ed. A. Lock). Academic Press, London.

Povinelli, D., Nelson, K.E., and Boysen, S.T. (1990). Inferences about guessing and knowing by chimpanzees (*Pan troglodytes*). *Journal of Comparative Psychology*, 104, 203-10.

Premack, D. (1984). Upgrading a mind. In *Talking minds: the study of language in the cognitive sciences* (ed. T.G. Bever, J.M. Carroll, and L.E. Miller). MIT Press, Cambridge, Mass.

Premack, D. (1988). 'Does the chimpanzee have a theory of mind?' revisited. In *Machiavellian intelligence: social expertise and the evolution of intellect in monkeys, apes, and humans* (ed. R.W. Byrne and A. Whiten). Oxford University Press.

Premack, D., and Premack. A.J. (1983). *The mind of an ape*. Norton, New York.

Premack, D., and Woodruff, G. (1978). Does the chimpanzee have a theory of mind? *Behavioral and Brain Sciences*, 1, 515-26.

Redshaw, M. (1978). Cognitive development in human and gorilla infants. *Journal of Human Evolution*, 7, 133-41.

Ricks, D.M., and Wing, L. (1976). Language communication and the use of symbols. In *Early childhood autism: clinical, educational and social aspects* (2nd edn, ed. L. Wing). Pergamon Press, New York.

Rivière, A., and Coll, C. (1987). Individuation et interaction dans le sensorimoteur: notes sur la construction génétique du sujet et de l'objet social. In *Comportement, cognition, conscience* (ed. M. Siguán). Presses Universitaires de France, Paris.

Rivière, A., Belinchón, M., Pfeiffer, A., Sarriá., *et al.* (1988). *Evaluación y alteraciones de las Funciones psicológicas en autismo infantil*. CIDE, Madrid.

Rogers, S.J., and Pennington, B.F. (1991). A theoretical approach to the deficits in infantile autism. *Development and Psychopathology*, **3**, 137–62.

Sarriá, E. (1989). *La intención comunicativa preverbal: observación y aspectos explicativos*. Unpublished Ph.D. dissertation, UNED.

Sarriá, E., and Rivière, A. (1991). Desarrollo cognitivo y comunicación intencional preverbal: un estudio longitudinal multivariado. *Estudios de Psicología*, **46**, 35–52.

Schaffer, R. H. (1984). *The child's entry into a social world*. Academic Press, London.

Seibert, J. M., and Hogan, A. E. (1982). A model for assessing social and object skills and planning intervention. In *Infant communication: development, assessment and intervention* (ed. D. P. McClowry, A. M. Guilford, and S. O. Richardson). Grune and Stratton, New York.

Tamarit, J., Gómez, J. C., Sarriá, E., Brioso, A., León, E., and González del Yerro, A. *La génesis de la comunicación y del símbolo: estudio comparado con poblaciones normales, autistas y primates no humanos*. CIDE, Madrid. (In press.)

Trevarthen, C. (1989). Les relations entre autisme et développement socioculturel normal: arguments en faveur d'un trouble primaire de la régulation du développement cognitif par les émotions. In *Autisme et troubles du développement global de l'Enfant* (ed. G. Lelord, J. P. Muh, M. Petit, and D. Sauvage). Expansion Scientifique Française, Paris.

Walker-Andrews, A. (1988). Infant's perception of the affordances of expressive behaviors. In *Advances in infancy research*, Vol. 5 (ed. C. Rovee-Collier and L. P. Lipsitt). Ablex, Norwood, NJ.

Wetherby, A. M., and Prutting, C. A. (1984). Profiles of communicative and cognitive–social abilities in autistic children. *Journal of Speech and Hearing Research*, **27**, 364–77.

Wimmer, H., and Perner, J. (1983). Beliefs about beliefs: representation and constraining function of wrong beliefs in young children's understanding of deception. *Cognition*, **13**, 103–28.

19

Autism and theory of mind: some philosophical perspectives

JERRY SAMET

INTRODUCTION

My goal in this paper is to explore the bearing of some central themes in the philosophy of mind and the philosophy of science on the theory of mind research programme in psychology, and on the recent attempts to understand the cognitive deficit in autism as rooted in theory of mind. This is a potentially large area, and there is much of relevance that will not be covered here. In order to give a feel for this larger picture, the treatment of individual topics will be programmatic; my hope is more to stimulate interest and discussion than to defend a single point of view.

Since this particular sort of interdisciplinary approach may be novel to some, I begin with a word of caution. One might suppose that since theory of mind research focuses on the highly abstract concepts of mind, representation, belief, theory, etc., it would be a simple matter of the division of intellectual labour to call on philosophy to say exactly what these concepts come to, and have the developmental psychologists worry about the course of their acquisition. The relation between linguistics and developmental psycholinguistics comes to mind as a model: the linguists specify the grammar, and the developmental psychologists work out how it is mastered. The word of caution is that this model will not work here. Despite 2000 years of self-promotion, philosophy is *not* the queen of the sciences, and there are very, very, few questions that it can answer in an authoritative and final way. 'What is a theory?', 'What is a mind?', and 'What is a mental representation?' are not among those questions. All of these questions are still very much up for grabs in philosophy, and perhaps even more so now than a generation ago.

Despite this, I am convinced that there is much to be gained by considering the developmental issues against the broader philosophical landscape. Although philosophers have not settled these questions, they have thought hard and productively about them for 2000 years, and the fruits of this work provide a ready starting-point for developing interesting empirical hypotheses. What's more, the influence can work in the other direction as well: what we learn about the empirical development of theory of mind will influence and constrain our philosophical conception. So perhaps progress will proceed not by division of labour but by working hand-in-hand.

THE PROBLEM OF OTHER MINDS

How do we know other people have minds? This is the sort of question that gives philosophy a bad reputation. But the reason for asking it is usually not to call our knowledge into question, but to try to understand what its basis must be. So the most common parsing of the question is this: given that we do know that other people have minds, what can consideration of this particular secure piece of knowledge teach us about our concept of knowledge, and of mind, in general?*

There are of course philosophers who really *do* doubt the existence of other minds, but this sort of scepticism is rare. The received view—due to Descartes—is that there is a duality in our knowledge of mind. Knowledge of our *own* minds is direct. There is no inference involved; we simply know *that* we think and *what* we think by direct inspection. What holds for thought holds for sensation, perception, affect, and the rest of the mental realm as well. But in the case of *other* minds, our knowledge rests on analogical inference and more—in fact, Descartes thought it was a quite complex inference, involving God and the absence of evil deceivers. The idea that our knowledge of other minds is mediated by knowledge of the existence of God and of God's properties is no longer part of the philosophical landscape, but the notion that *some* inference (perhaps unconscious) is involved remains part of the received view, despite objections from very different directions.

The problem of other minds made a surprise return to the philosophical scene about forty years ago, and the issue involved was the nature of the analogical inference. At first blush, it looks as if our belief in other minds depends on an inference from our own case, based on the principle 'Like effects, like causes'. We know that our actions (the effects) are the consequence of a mental life in us (the causes). When we observe the behaviour of others, we project a similar unobserved cause—i.e., a mental life—going on inside them as well. The well-known problem is that we would seem here to be making an inductive inference on the basis of a single observed instance, on the fact that *our* action is mind-driven, and projecting it on to the rest of humanity. If this were in fact the implicit support for the belief in other minds, we would expect that the belief would be more tentative than it actually is. So there must be something wrong with this picture.

As a matter of historical fact, this attack on the received view—viz, that

---

* This approach to the problem of other minds and to epistemology in general has recently been challenged in an interesting way in Kaplan (1991).

it can't easily account for our certainty that others have minds — was thought to challenge mentalism and support behaviourism. The behaviourists diagnosed the problem as founded in the mentalist conception of mind as private and internal. According to the behaviourist, once it is recognized that having a mind is simply a matter of evincing certain perceptually observable patterns of behavioural response, the mystery of our knowledge of other minds is cleared up. We *can* be confident in our judgement that others have minds, because we can observe the evidence of the pieces of these patterns. Suffice it to say that this behaviourist argument did not win the day. Most philosophers accept mentalism, but reject the idea that we make an inductive inference from our own case. One prominent view is that the belief in other minds is based on a pattern of inference to the best explanation. We will return to this view in the next section.

To see this bit of philosophical landscape in its proper light, one must understand that (rightly or wrongly) philosophers have usually *not* seen the concern with our belief in other minds as a matter of psychology. 'Psychologizing' has functioned as a term of derision in epistemological circles; the aims of psychology and philosophy have been taken to be independent. Philosophers have been interested in what *justifies* our belief that others have minds, in its rational support. They have left it to psychologists to figure out the empirical causes. Although this way of conceptualizing the relation between philosophy and psychology has dominated the philosophical landscape, it too has come under attack. Many have abandoned the purist conception of epistemology, and moved closer to a 'naturalized' epistemology that is more responsive to the facts of empirical psychology.*

But whether one accepts a sharp epistemology–psychology distinction or not, the issues that arise in the controversy about other minds raise some suggestive possibilities about autism and theory of mind. A leading hypothesis — the one that brings together the contributors to this volume — is that the autistic child does not have a 'normal' conception of people as having minds. Once we think of acquisition of this conception as possibly inferential, we might consider more subtle hypotheses. Perhaps the autistic child has *some* primitive conception of other minds, but needs to construct it by painstaking and tentative inductive inference of a sort, while the normal child arrives at a theory of mind by a much easier route. Perhaps autistic children form the beginnings of a theory of other minds, but are unable to carry it far enough to serve as a useful basis for understanding others.

---

* An important call for change is Quine (1969); some recent work in this area is collected in Kornblith (1985).

THE DISTINCTIVENESS OF SELF-KNOWLEDGE

Just as there is a philosophical problem about other minds, there is a closely related and long-standing philosophical concern with self-knowledge. The received view in the history of philosophy, again associated with Descartes, is that self-knowledge has a special status: we know our own minds better than we can know anything else. This doctrine crosses ideological boundaries between rationalists and empiricists; it is often treated as a datum that every theory must accommodate. In this century this notion of relative superiority has sometimes been transformed into the absolutist doctrine of the incorrigibility of the mental — the idea is that we can't be wrong in our sincere judgements about our own mental states. In fact, in the debate over physicalism and dualism, many dualist philosophers took this incorrigibility as *the* mark of the mental, and argued that the mind could not be physical, because every statement we make about the physical brain *is* corrigible, but some statements about the mind (our own particular judgements about our own mental states) are *in*corrigible.

Why would anyone think that our judgements about our own mental states must be authoritative in just this way? Leaving theory aside, it is hard to think of a serious challenge to such claims as 'That hurts' or 'It looks blue to me', or 'I believe I'm in Seattle'. But there is a theoretical story that is supposed to support this incorrigibility claim. The idea comes out in the contrast between introspection and perception. In the case of perception, our knowledge of the world seems to be mediated by internal representations. This allows the possibility of misrepresentation, and our perceptual judgements might therefore be mistaken. Our access to our own mental states, on the other hand, cannot be via further representations of those states. If it were, we would have a regress of representations. Therefore, our access to our own mental states in introspection must be unmediated, in some sense 'direct', and therefore error-proof. This is how the philosophical story was told, anyway.

But this story has been challenged, and the ammunition for the challenge has come from psychology. Some have argued that the Freudian revolution shows us that introspection is not only fallible, but chronically so. Others have drawn on cognitive psychology, and cited presentation effects, expectation effects, and so on, as showing that introspection might be no more foolproof than is perception. We perceive objects; but we can also misperceive them. So too, we introspect our mental states; but can also misintrospect them. The theory behind this alternative position to the 'direct access' view — let's call it the 'mediated access' view — is, very roughly, that introspection, like perception, provides us with something like what James referred to as a 'blooming and buzzing flow' of conscious experience, and that we make sense of this flow by bringing mental concepts to bear. This account leaves open the question of what 'bring to bear' really means, and of how the threatened regress is to be avoided. The important point, for now, is that on this view, one's *own* mind, and the mental states and processes that we take to constitute it, is not an object of direct experience.

Common sense seems to agree with the philosophical tradition in assuming an unbridgeable gap between self-knowledge and knowledge of others. The mediated-access approach I have just sketched can also accept this intuition: it only *narrows* that gap. It does *not* say that we know our own minds only in the way that we know what is going on in the minds of others. It grants that we have access to our own experience — our Jamesian flow — in a way that no one else does. But both first-person and third-person mental state ascription depend on something like the relation between a claim and the evidence for it. Even in the case of self-ascription, we proceed on the basis of evidence, not *direct* knowledge. If the mediated-access view is right, then the incorrigibility thesis is in trouble, since it depends on the immediacy of self-knowledge.

I raise these issues to point out that some of the discussions in the theory of mind literature seem to embrace a version of the mediated-access view. In fact, the ones I will consider here seem to embrace a very radical version of that view. Hobson (1991) cites the following remarks from Premack and Woodruff (1978):

In saying that an individual has a theory of mind, we mean that the individual imputes mental states to himself and to others (either to conspecifics or to other species as well). A system of inferences of this kind is properly viewed as a theory, first because such states are not directly observable, and secondly because the system can be used to make predictions, specifically about the behavior of other organisms.

This position has been challenged by Hobson because of its commitment to a *theory* of mind. It has been challenged by others who accept this commitment (the theoreticity claim) but take issue with Premack and Woodruff's minimalism in interpreting what makes something a theory. What has been overlooked, I think, is the fact that Premack and Woodruff — at least in this passage — make no distinction between ascription of mental states to oneself and ascription of such states to others. They talk of *both* as 'a system of inferences',* and characterize these mental states as 'not directly observable'. In a similar vein, Olson *et al.* (1988) say that if a theory of mind is the best explainer and predictor of action, then the entities posited by such a theory may be accepted as real. The implication is that the reality of beliefs, desires,

---

* It is not clear precisely which inferences they have in mind. Is it the set of inferences we make to figure out a theory of mind ('There must be these things, let's call them 'moods', and they affect . . .'), or is it the set of inferences made in the application of an already formed theory of mind to a particular case ('She must be upset because I forgot her birthday')?

and so on cannot be determined by our first-person awareness of them, but must be established in the way we establish the existence of something like genes—i.e., by showing that the larger theory works. So self-knowledge doesn't have a special status after all. Descartes thought that since he could be certain of 'I think', he could conclude 'I am'. The position we're investigating says that 'I think' is only as certain as the theoretical framework that supports it.

To be fair, perhaps Premack and Woodruff ought not to be subjected to this sort of textual exegesis. It could be that when they are talking about theory and inferences they really only have in mind the ascription of mental states to others. But the view that emerges from this passage is interesting in its own right. It may very well be correct. I said earlier that the Premack and Woodruff position is more radical than the philosophical view I sketched. That is so, I believe, because it not only says that self-knowledge is *mediated* (which I happen to think is right); it adds the claim that it is mediated by a *theory*. There is a crucial distinction here between the view that our self-knowledge is 'concept-mediated' and the claim that it is 'theory-mediated', but spelling it out clearly depends on our prior understanding of the distinction between concepts and theories in general, and this is a contentious and difficult issue. Here I will be satisfied with an example, and I'll return to the concept-theory distinction later.

Intuitively, we can develop or apply a set of concepts to make sense of the world in two different ways. Some conceptual frameworks are best viewed as taxonomies—as a grid of categories that we lay over experience to focus on the similarities we note and to abstract from the differences. This happens at every level of conceptualization. Take political concepts as an example. We talk of imperialism, socialism, fascism, totalitarianism, etc. We don't 'directly observe' fascism, and to the extent that the concept depends on what we call political theory, we can say that being fascistic is a *theoretical* property. But this is a weak sense of theoretical; it simply means that the higher-order concepts used to group the phenomena are remote from perception, perhaps that they are not easily operationalizable, that they are relatively abstract.

To say that a phenomenon is theoretical in this sense is not necessarily to say that it is theoretical in a stronger sense often discussed in the philosophy of science. Compare the political theory case with the philosophy of science favourite—viz, the hypothesis made in the early part of this century that genes account for the transmission of hereditary traits. Genes were then *unobserved entities* that were thought to play a causal role in the reproductive process. Is there a distinction between the theoreticity of genes and of fascism? I think that there is. Being fascistic is not some *further* state or property that we don't have the apparatus to detect, which works behind the scenes to *cause* all the observed properties of governments like that of Nazi

Germany.* It seems much more sensible to think of it as a theoretical category for a certain type of government. Genes, in contrast, *are* unobserved players in the causal process. If this intuitive contrast stands up, then there is a real difference between saying, on the one hand, that our mental concepts provide a framework that enables us to interpret our introspective experience, and the more radical claim that holds that our understanding of our own minds requires a theoretical leap on the order of the gene example, that our own mental states are unobserved prior causes; that they are *postulated to exist*. The first view is akin to the by-now tame insight that our concepts of physical objects like tables and chairs are constructs that enable us to interpret the flow of sensory–perceptual experience. The second would group our mental concepts not with tables and chairs, but with quarks and leptons. In the end, the issue is not over the theory–concept distinction, but of the abstractness of our mental notions; perhaps in how 'remote' the concept is from our 'raw' introspective experience.†

Let me summarize these last points. There is a labyrinth of issues that crops up when you think about the relation between the way we come to understand ourselves and the way we come to understand others. Common sense says these are very different. The history of philosophy seems to agree with common sense. But more recently, philosophers have raised questions about the most-favoured-judgement status of self-knowledge, and they have suggested that although the basis for self-knowledge is indeed different from the basis of our knowledge of others, both are mediated by a contingent conceptual framework, and neither is indubitable. We have also seen that some theory-of-minders (and there are philosophers who would join them) carry the argument a step further and claim that, even in our own case, we know ourselves via a theory. If we take a strong reading of 'theory' here, it suggests that we don't observe our mental states at all; that our concepts of belief, desire, and so on, are not just to some extent abstract, like our concepts of fascism and imperialism, but that they are like the unseen genes that give rise to our traits.

Returning to the theory of mind programme and its promise *vis-à-vis* autism, it is worth keeping both these competing views alive. Might it be that normal children have an intuitive and introspective grasp of the mental, but that autistic children need to actually develop an understanding of it more indirectly, as a theoretical postulate? This type of hypothesis, one that introduces different paths toward the same end-point, is well-suited to explain the *relative* deficits one finds in autism, and the *delays* in the

---

* This is not to deny that there are unobserved mental states that cause fascistic behaviour. It is only to deny that the political category 'fascist regime' denotes the mental causes.

† There is no a *priori* reason to think that our mental concepts are all equally abstract, in this sense. The relation between the flow of inner experience and the concept of pain, on the one hand, is different from its relation to the concept of frustration, on the other.

development of some of the relevant metacognitive skills. The general idea would be that even if one does not have the normal complement of mechanisms and knowledge to arrive at a 'natural' theory of mind (or to arrive at the theory 'naturally'), there might be general learning mechanisms that enable one to build some sort of rudimentary system that does some of the work of the natural system (see Leslie, this volume, Chapter 5).

This suggests an analogy to language learning after the critical period. It seems that if you begin to learn your first language at a late age, you can still reach a certain level of competence, but not the typical adult level. Assume for the moment that, just as in the language case, there is an innate element in our understanding of the mind. In the same way that in exceptional circumstances the language-learning faculty might not be triggered, perhaps the 'mind-understanding' faculty is not triggered in children with autism.* In such a case, the child would be without the significant clues and concepts that the rest of us have, but must still develop some sort of understanding of him/herself and others. Where normal children might at a certain age naturally ascribe mental states to others by analogy with their own mental states, the autistic child might approach the problem of making sense of others as a black-box problem: figure out the inner workings of an impenetrable machine using only observed output as your basis. Presumably, what can be acquired in this problem setting will be to some degree inadequate. Like the speech of the twelve-year-old, it will be 'off' in some way.

THE METAPHYSICS OF MIND AND ELIMINATIVE MATERIALISM

One difference between the 'theory-mediated' approach to theory of mind and the 'concept-mediated' approach is that the former—but not the latter—makes it a real question as to whether our theory of mind is all wrong and our whole framework for self-understanding is a mistake. Such a result would complete the destabilization of the mental that Freud initiated. Freud provided compelling evidence that our judgements about the actual contents of our minds—our beliefs and desires—were not the ultimate arbiter on what we really thought and really wanted. But if our whole taxonomy of the mental is mistaken, we might not even be right in thinking that we have beliefs and desires in the first place. More to the point, we might not even be right in thinking that we have minds.

Something like this line of reasoning has fuelled the recent revival of 'eliminative materialism' in the philosophy of mind—the thesis that all

---

* This hypothesis raises important questions about the modularity of the theory of mind that I will not discuss here.

physicalist attempts to reduce our talk of the mind to either behavioural, neurological, or functional notions is doomed to failure, *because there are no minds*. For eliminative materialists, the reductionist program is as ill-conceived as trying to reduce alchemical theories of the relation of the elements to gold to contemporary metallurgy. There may well be reasons why we will or should continue to use the common-sense theory in our day-to-day lives, but even if this were so, it would not have to constrain our scientific quest for a deeper understanding. Leibniz already noted that Copernicans might continue to talk of the sun's coming up and going down long after their astronomy abandoned the up–down perspective on which the locution is based.

The eliminative materialist urges us to start afresh — to forge a new scientific theory of how people work using the new theoretical concepts available in the sciences that study the brain (neurology, information theory, cognitive psychology, etc.), and to stop trying to make this new theory run in parallel with the supposed 'truths' about the mind that are embodied in the common-sense theory of mind. Skinner sometimes talked this way about behaviourism, and Quine followed his lead to some extent. More recently Paul Churchland (1981) has defended a similar view — the alchemy example is his — and has relied more explicitly on the theoreticity of the mental. In this vein, he urges us to consider our common-sense theory of mind as of the same ilk as other (past and present) 'common-sense theories' — for example, of motion, of illness, of witches and goblins, of the heavens. The lesson seems to be that, as J.L. Austin put it in another context, common sense may be the first word, but it is not to be taken as the last word.

What is the support for eliminative materialism? For some, the mere fact that the theory of mind is a common-sense theory is itself a strong reason for thinking that it is wrong. The argument would be that when we try to grind out the details, *no* common-sense theory has proved right. But typically, eliminative materialists like Churchland support their view with more substantive arguments aimed at the particular failings of our common-sense mentalism. At the heart of the (highly controversial) eliminativist approach is the (less controversial) view that our common-sense mentalism is part of our evolutionary legacy, and evolution is a satisficer, not an optimizer. The mechanisms and systems it selects don't need to give us the truth; they only need to work tolerably well. Common-sense mentalism does work tolerably well; but the lesson of the scientific revolution is that if you want deeper understanding you need to go further than common sense.[†]

[†] There is currently philosophical controversy as to whether this lesson is fully general; that is, are there domains in which deeper understanding is *in*compatible with scientific investigation. For an extended defense of the view that there are such domains, and that aspects of the human mind constitute one of them, see Nagel (1986).

Whether we need to abandon common sense and make a fresh start (as the eliminativist claims), or whether we simply need to refine and smooth the edges of our common-sense view is at this point a much-debated question.

In one sense, the philosophical controversy over eliminative materialism is beside the point for psychologists interested in the development of a theory of mind. After all, this research aims to discover how we conceptualize mind, and how that conceptualization develops. Whether that conceptualization is right or wrong is beside the point, and it is therefore beside the point whether some scientific conceptualization will replace it. But there is another sense in which it is very much to the point. Psychologists want to understand human beings, and if human beings don't have minds, they certainly want to hear about it quickly (not find it out from the *National Inquirer*).

But I've outlined the controversy here for another reason. This is an area where the philosophical issues might be settled to some measure by the results of the empirical research. This is so, I think, because the philosophical arguments ultimately turn on the precise character of our common-sense folk psychology. For example: if it turns out that our actual folk psychology is closer to a taxonomy with rough correlations between states, processes, etc. than to a causal theory postulating unobserved underlying structures, then the argument from the general failure of folk psychology is thereby weakened. We are ready to dismiss folk constructions like devil, witch, imp, pixie, etc as having no claim to reality at all. But folk *taxonomies* are more ontologically resilient and error-tolerant. The sun and the moon, trees and mountains, flesh and blood are all accepted as real despite the fact that they have been misunderstood and misconceived throughout most of our history. I would urge that the general reason for this is that these taxonomic notions are *not* validated only by their explanatory or predictive adequacy. Instead, they have their continuing place in our ontology in virtue of their role as referring expressions in every discourse. If, when the empirical dust settles, theory of mind shows more of the markings of a taxonomic theory than a postulational one, that will have a bearing on the philosophical controversy. This, then, would be a case of the interdisciplinary influence moving in the other direction.

IS A THEORY OF MIND INNATE?

The innateness question is psychologically and philosophically interesting, but as far as I know there is no compelling philosophical argument one way or the other. It is interesting to note the following historical precedent. Leibniz, the seventeenth-century defender of the innateness hypothesis against Locke's empiricist attacks, claimed that we are, as he put it, 'innate to ourselves'; that elements of our self-knowledge do not in any way depend upon experience of the world. One might interpret him as holding that the categories we use in self-conceptualization are innately given.

I have argued (Samet, in progress) that Leibniz tried to go farther and show that *all* our knowledge is actually based in some important way on our innate knowledge of these categories of selfhood. For instance, Leibniz held that (what we would refer to as) our conception of objects was grounded in our innate concept of Substance, and that the concept of Substance was available to us innately since we are 'innate to ourselves' as Substances. It is of further interest to note the family resemblance of Leibniz's idea to the contemporary thesis that our theory of mind serves as a model for understanding aspects of the non-mental world. I have in mind here Susan Carey's views about the role of intentional psychology in the child's developing understanding of the world (Carey 1985). Leibniz's view strikes us today as wildly speculative, but perhaps the empirical evidence will bear out his conception to some extent.

My own bias is that theory of mind is to a large extent innate. I am impressed by the evolutionary consideration mentioned by Fodor (1987). That is, given how important it is for the survival of the individual to understand herself and her social surroundings, it would not be surprising in the least if nature were to favour the development of such a cognitive system. But this sort of argument only provides a *prima facie* case, at best.

Wellman (1990) puts forth a hypothesis on the innateness question that is interesting and needs to be addressed. His view is that the child constructs a theory of mind—a belief–desire psychology—out of an earlier 'desire psychology'. Wellman draws on interesting differences between two- and three-year-olds that point us in this direction. I find this hypothesis interesting, but problematic, and I briefly want to say why.

The main problem, as I see it, is a problem that has been raised against the general Piagetian picture of development. That is, we don't really have an articulated model of how one might *construct* a concept of belief out of the concept of desire, and I do not find one in Wellman's discussion. Wellman is, I think, aware of this lacuna. He approvingly cites Carey's methodological distinction between a thesis about the *mechanisms* of development of a child's theory as opposed to one about the *origins* of the theory, and classifies his hypothesis as belonging to the second category. But it seems to me extremely difficult, if not impossible, to assess hypotheses about origins without some idea of the mechanisms that might mediate the transitions. Carey herself points out that hypotheses about theory-change need to specify *both* origins *and* mechanisms. Again I draw on the linguistic parallel. Nelson Goodman at one time countered Chomsky's innateness hypothesis with the suggestion that learning our first language draws on our prior mastery of gestural systems. But Chomsky's correct response was that unless we have some explanation of *how* universal grammar can be constructed out of knowledge about gestures, we do not yet have a serious alternative to consider.*

I would venture to say that the same is true here. Until we have at least a sketch of the mechanisms that manage this new construction, we are not yet at a point where we can judge between competing hypotheses. My criticism does not imply that there is no such mechanism to be found. Indeed, I find Wellman's speculations about the non-representational character of the primitive desire psychology he attributes to two-year-olds intriguing.† But I do not see how the the representational concept of mind that Wellman argues is at the centre of belief-desire psychology could be constructed by the child out of non-representational materials. The representational theory of mind might need to mature, but we still have no theory about how it could be learned from experience. Since Wellman rejects maturational and nativist accounts, he needs to provide more in the way of mechanisms.

I want to mention briefly two further points about the connection between nativism and the theory of mind research programme.

The first is the link to Jerry Fodor's radical concept nativism. Fodor has argued that (to a first approximation, at least) *all* our concepts are innate. As it is relevant here, this means that we are born with the concept of belief, desire, fear, person, mind, concept, love, hate, and all the concepts that are part of our theory of mind. Does this claim by itself imply that a theory of mind is innate? That depends on what one thinks a theory of mind is, and on how one thinks these concepts should be understood. One *might* hold that we have innate concepts of belief, desire, etc., but need to learn the relations that hold among these concepts. An alternative view would be that really to grasp these concepts is *already* to grasp the relations that hold between them, because these concepts are what they are because of the network of concepts they are embedded in. That is: if you don't know what a belief is, then you can't understand what a desire is, and so too for the rest of the network. Yet another view is that the concepts *are* specifiable independently of the details of the theory, but the concepts *and* the theory of mind are all innate. Again, I won't try to settle the matter here except to point out that this way of thinking about a theory of mind raises the

---

* See the exchange of views in Hook (1969).

† Intriguing but problematic. This is not the proper place to address those problems, but I will mention one puzzlement I have with the idea that desire can be primitively construed as a relation — a 'pro-attitude' — between a person and an object (a picture defended by Hobbes, by the way). The problem is that it seems to ignore the different modalities of desire: you want *to eat* an apple, *swim* in a pool, *scratch* an itch, *kill* a cockroach, etc. Analysis quickly reveals that the 'object' of desire is never an object. Are we dealing with the same attitude towards apples, pools, itches(?), and cockroaches? One might say that the two- and three-year-olds are not philosophers, and are allowed to have a very confused epistemology. But what would it mean to have *one* attitude that could be satisfied in such different ways?

possibility that one might have some or even *all* of the relevant mental concepts before acquiring the theory.

The second point takes us again to the patterns of nativist theorizing that become familiar from linguistics and psycholinguistics. In thinking about the acquisition of a theory of mind, we would do well to attend to the models that have been developed — some even discarded — in trying to plot out the relation between universal grammar and the grammar of a language. This comparison taps a rich lode of questions, not the least of which is whether there is one theory of mind (like a universal grammar) or many (like the grammars of individual languages). Some further questions to consider: If we in fact found that a theory of mind were universal, would that argue for an innate basis? What if we found that there was a universal core theory of mind; would that provide encouragement for the nativist? Might the innate element of a theory of mind have the structure of parameter theory, all or some of whose elements need to be set by relevant experience?

ASPECTS OF THEORY OF MIND

In this section and the next I want to say something more about the two key terms in the theory of mind literature — to wit, the notion of mind and the notion of theory. As we shall see, there are many things that ordinary adults know that could be classed as (part of) their theory of *mind*. For many purposes, lumping them all together is a good methodological strategy; but it does invite what I think is unnecessary controversy and misunderstanding, which could be avoided by making some distinctions explicit. This will be my aim in this section. In the next section I do something similar for *theory*. Some of the controversy over the theoreticity claim implicit in the theory of mind programme can be dissipated if we spell out carefully what we mean by theory and what we don't mean. Wellman (1990) tries to do just this for these key terms, and I will be extending, and sometimes criticizing, his insights on these matters.

There are many clusters of things that adults in our culture, and possibly in every culture, take for granted about people and minds. Here are a few. First, there are *existential* assumptions. Adults accept that there are minds, that all people have them, and that ordinary physical objects do not. Second, there is *taxonomic* knowledge. We all more-or-less share a general taxonomy of states (for example, beliefs), processes (for example, inferring), faculties (for example, memory), traits (for example, moodiness, curiosity), etc. Third, we know that the elements of mind are systematic; that they fit together into a system. We have a general understanding of some of the important relations between these states, processes, and faculties (for example, pain tends to make you unhappy; concentration helps you learn certain sorts of things), and we also know a good deal about the interactions of the mind-system with the environment. Fourth, we have a *practical* knowledge: we can often apply what we know; we can integrate our general knowledge about the mind with what we know about specific individuals or groups of individuals to form a conception of other people that goes beyond observed behaviour. This conception can lead to successful predictions and explanation. Fifth, we may have (what philosophers would call) implicit meta-theoretical views about the *metaphysical* interpretation of our theory of mind and about the nature of the states individuated. For example, some have argued that folk psychology is dualistic, and accepts the mind as non-physical; others have claimed that it is non-committal. In a related vein, a Cartesian might suggest that our theory of mind sees the self and the mind as identical (that I *am* my mind).

All of the above and more might reasonably be included under the rubric 'theory of mind', and in studying cognitive development it seems reasonable to look for answers to all the questions this partitioning raises: What exactly is the taxonomy and how are the elements organized into groups? How and when do children master it? Is there a point at which they have not yet ascribed a mind at all to others? How does the ability to apply the knowledge develop? What sorts of algorithms or inferential processes are employed in its application? Are some developmental psycho-pathologies distinctive in the failure to develop this or that aspect of the adult understanding? And so on.

Failing to keep the elements of this theory of mind grab-bag distinct can lead to futile arguments and talking at cross-purposes. For example: some, like Hobson,* have objected to the whole theory of mind research pro-gramme because they believe that, when it comes to mind, children do not need to develop a theory at all. Their view is that children know that people have minds without any theory-construction. My sense is that Hobson is right about this last point: normal children do not theorize *that there are minds*. But I would argue that this is a neutral point, and does not undermine the theory of mind research programme. All can agree, even if some have not, that the acceptance of the *existence* of the mind is not in any important sense theoretical. The claim is only that children need to develop a theory *about* the mind and the way it works—what I referred to above as tax-onomic, systematic, and practical knowledge. The belief in the *existence* of minds need not be supported by theoretical inference; the theory comes in when the child begins to understand the *nature* of the mind.

This point becomes clearer when we think about a parallel case. Consider the development of the child's theory of the motion of physical objects (see Spelke 1990 for a discussion of the differential development of pieces of

*See for example Hobson (1991).

the theory). The claim that there is such a theory does *not* imply that the concept of a physical object or the concept of motion is something that must be theoretically posited.* The belief in the existence of physical objects and the awareness of motion both seem to be independent of the articulated body of beliefs about the specific characteristics of these objects and the particular constraints on their motion. We should note, however, that there are deep philosophical problems that trouble this distinction. Specifically, if the notion of a physical object is partially constituted by this body of beliefs, how do we justify the claim that the child has a pre-theoretical commitment to *physical objects* — that is, that even before she discovers almost everything about physical objects, she believes that there are such things? What makes the existential commitment a belief involving the concept of a physical object? This problem is all the more difficult to resolve in light of the fact that the child at this stage does not have a linguistic term to pick out such things. The same problem, *mutatis mutandis*, looms over my earlier claim that children believe that there are minds before they develop a theory of what they are and how they work.

Hobson is right that much of the theory of mind literature invites us to think in terms of the black-box metaphor: of the child as starting with a body of behavioural data and arriving at the explanatory hypothesis that there must be these unobserved things — call them 'minds' — whose dynamic processes and states systematically cause the observed behaviour. This leaves the impression that the whole idea that there are minds is itself a 'posit', arrived at by a process of inference to the best explanation. But, as was mentioned above, we need not go this far; there are many less radical options. One might hold that the youngest children already operate on the supposition that behaviour is internally controlled, and the intellectual task is only to figure out how the control system operates. Or, to repeat a point made earlier, maybe the youngest children start out with an internal-control hypothesis and a rich innate taxonomy of mental states, and only need to discover a way to put the whole system to practical use. Or even the application procedure might be innately specified to some extent.

ASPECTS OF THE 'THEORY THEORY'

We now return to the issue of theoreticity. There are many discussions in the theory of mind literature about the appropriateness and implications of

* Yet another prominent example is the case of Chomsky's child-as-scientist model applied to the 'theory of language'. The phrase 'theory of language' picks out the domain the theory is supposed to cover. Given the thorough nativism of Chomsky's theory, it almost goes without saying that the child does not need to infer that there is such a thing as a language, that it is a rule-governed system, etc.

taking the child's developing conception of the mind to be a theory (Wellman 1990; Perner 1991; Morton 1980). Similar issues have been raised in the general developmental literature (Carey 1985). There is also a large and influential literature in the philosophy of science that addresses a set of related questions about the proper understanding of scientific theories. One might expect the latter to shed considerable light on the former, but, as in the case of the philosophy of mind, I think this is hoping for too much. For one thing, one does not find broad agreement among philosophers of science on the answers to these questions. But to see the deeper reasons we will need a one-minute history lesson in the philosophy of science.

Modern philosophy of science really begins with the logical positivists, who were very ambitious in trying to specify what theories were in very precise terms. To this end, they introduced hypotheses about the divisions in the canonical vocabulary of theories (observational vs theoretical), distinctions between different sort of generalizations within theories (theoretical laws, bridge laws, empirical generalizations), and formal relations that needed to hold between different elements of theories (deducibility, confirmation). The point of this reconstruction of the scientist's working notion of theory was to make precise certain key epistemic notions like evidence, reduction, confirmation, law, etc. One point, though, is that the effort was directed at understanding science, not theories. None of this work has much to tell us about Kripke's theory of proper names, or about theories of the development of Flemish art. But even its application to the scientific enterprise has been challenged, and contemporary philosophy of science is for the most part the playing out of this challenge. The backlash against positivism, represented by philosophers such as Kuhn, has more-or-less abandoned the idea that theories are the central creations of the scientific enterprise, and has emphasized broader notions like paradigms, frameworks, and research programmes within which successive and competing theories can be formulated.*

Although it has not been explicitly put in these terms, philosophers of science have generally come to recognize that 'theory' does not pick out a natural kind. 'Theory' is a term of art; it does not refer via some essential property. This is not to imply that talk of theories is vapid or fruitless or bankrupt. It only reminds us that in talking of theory or theoreticity we are characterizing in terms of some varied dimensions of similarity to certain

---

* I suspect that there is much to be learned by pursuing the possibility that the child's understanding of the mind is structurally more akin to Kuhn's paradigms, for instance; but I will not be able to follow this up here. Ironically, Kuhn acknowledges the influence of Piaget's theories of cognitive development on his view of scientific development. So we are talking here about an ironic kind of 'ping-pong' effect of mutual influence.

central cases. In this spirit, I want to provide a traditional time-honoured philosophical service by drawing some distinctions among a number of different dimensions of similarity that seem relevant; the different connotations of the term 'theoretical'. The point of the exercise is, as before, to help set aside disputes and controversies that turn on equivocations. Some *x* might be theoretical in one respect and not in another. After all is said and done, it is what the specific use of the term connotes that will be issue, not the term itself; so we do well to have these underlying connotations brought out into the open. Here then is a partial list of what we might be after in talking of something as 'theoretical'.

First, there is the *epistemogical* connotation. Here the point is that we *know x* or are warranted in positing the existence of some *x* because of the role such a posit plays in an explanatory theory. Quarks may be theoretical in this sense. If we didn't have a successful theory about subatomic interactions that said that there are such things, we would have no reason to believe that quarks existed. Here there is a common contrast between what must be inferred by appealing to a higher-order theory, and what we may know perceptually. This is the sense intended in the reference to Olson *et al.* (1988), which suggested that our accepting beliefs and desires as real rests in some way on the practical success of folk psychology. I have used the term 'postulational' for things that are theoretical in this sense.

Second, there is what we may call a *psychogenetic* connotation. Sometimes we call S's belief 'theoretical' to indicate that S arrived at the belief by a process of theorizing; that something like theory-construction — whatever we ultimately take that to be — is involved in the actual formation of S's beliefs. So consider Sherlock Holmes' theory about the theft of the crown jewels. In this sense of the term, what is theoretical for Holmes might be perceptual for Moriarty. The issue is not with the objective epistemological status of the belief-type — with what it is that ultimately makes it rational to accept it — but with the causal history of the particular *token* of the belief in a particular person's mind.

Third, we may distinguish a *pragmatic* connotation of the term. Here I have in mind the identification of the 'theoretical' with the provisional or hypothetical. This is unfortunately a very common use; consider the widespread acceptance of 'It's only a theory' as an excuse for irrational resistance to well-established beliefs or principles. This pragmatic connotation is no more than a connotation — there is no general argument from theoreticity to provisionality. Some theories are so well entrenched that it would be irrational to withold acceptance. On the other hand, many non-theoretical beliefs only have and only deserve provisional status.

Fourth, there is what we can call a *functional* sense of the term. This is a looser idea that combines elements of the others, and might be the guiding idea behind Premack and Woodruff's original application of the term 'theory of mind' to the chimp. The rough idea here is that something is a theory if it is used to explain observations and/or predict them. I say that this is 'roughly' the notion, because as it stands it includes much too much: not every explanatory basis is a theory. To cite an example due to Michael Scriven, a child might explain a stain on a carpet by citing the fact that ink spilled on it, and also predict a stain from such a spill. Nevertheless, she might have no theory about how ink and stains are related. She might simply appreciate the correlation between the two events. Translating this into our context, we might ask whether the four-year-old equipped with a theory of mind has anything more than a set of such correlations (albeit correlations of mental states and events). To take an obvious example: does the child, or for that matter anyone else, have a causal theory of *how* beliefs and desires can cause intentions and ultimately actions?

Fifth, there is a *structural* sense of the term, where to call something a theory is to say that it has the structure or organization of a theory. As I indicated earlier, there is considerable controversy about what exactly that structure *is*, or even if there is a common structure. But the positivist analysis discussed at the beginning of this section, and the distinctions it draws, are all structural in the sense I mean here. This structural notion of theory fixes on an internal property of a knowledge system, and I suspect that this notion is the most widespread among theory of mind researchers.

This last point about the primacy of the structural is borne out if we consider Wellman's recent proposal about the theory aspect of the theory of mind (Wellman 1990). According to Wellman, the child's understanding of mind counts as a theory because it is (i) coherent: the theoretical terms involved are interdefinable, but not definable in terms of other non-mental terms; (ii) ontologically specified: it provides 'an accepted conception of its objects'; and (iii) a causal–explanatory framework for the prediction and explanation of behaviour. The first two criteria are *structural*, internal, features of the knowledge system; the third is a combination of a structural feature – viz, that the laws are causal, and a *functional* aspect – viz, the idea that the whole structure can be put to use in explanation and prediction. Wellman's proposal must of course be judged on the basis of the substantive aspects of structure and function that he singles out as critical; but I will not consider those questions here.

We may now return to the generic claim that children develop a theory of mind. In what sense is there a theory here?

Let us start with the first epistemic sense. It seems to me that the theory of mind approach can remain neutral about whether the child's belief in the mental is validated in anything like the way in which the particle physicist's commitment to quarks is substantiated. The theory of mind approach is compatible with the view that some of my mental states are directly observable by me; that they are not theoretical posits in any full-blooded sense.

We have already discussed the second, psychogenetic, sense in the last section, and rejected the idea that the theory of mind approach is committed to there being an inference on the child's part that there are such things as minds. One can hold that the belief in the existence of minds is innate and triggered, for instance, without undermining the theory approach.

Both psychogenetic and pragmatic connotations are the focus of the attack on the theory approach by Hobson (1991). Hobson argues that the approach is misleading because it incorrectly suggests that the child's understanding of mind is theory-like in terms of (i) the depth of commitment that the child has to her conception of mind, and (ii) the sorts of mental processes involved in arriving at that conception. Hobson's counterclaims, in short, are that theories are provisional and the child's conception of mind is not, and that theories are arrived at by theorizing, and the child's conception is more directly experienced. These differences lead him to conclude that although there are other aspects of the child's understanding that do make it 'theory-like', it is not 'essentially theoretical' in nature. But I have argued that there is no notion of essential theoreticity to fall back on. One can grant the dissimilarities that Hobson points to with regard to psychogenesis and pragmatics without affecting the basic orientation of the theory-of-mind research enterprise.

There is a further objection of Hobson's that is related to the psychogenetic sense of theory. Hobson, if I understand him, makes the following argument. Theory of mind theorists think that children have a theory of mind. Therefore, the particular ascriptions of mental states and processes that the child makes must count as 'theoretical'. But if they are theoretical, then they are not perceptual; they must be based on some sort of indirect high-level inference. But we find that mentalistic categories play a role in perception itself: for instance, we *see* that someone is getting angry. So, the theory of mind approach must be mistaken.

I find this an interesting objection. It draws on a philosophical approach to mental-state ascriptions (originating with Wittgenstein and developed by Strawson) that Hobson has ably championed, and that I believe deserves more careful attention. Nevertheless, I think that the objection is misplaced here. The idea that we can often 'read' the mental states of others off their behaviour seems right. But this by itself does not tell us much about the ultimate role of theories in our understanding of mind. First of all, the assumption that perception does not involve any sort of higher-level inference is dubitable. But even if that were not so, even if we could distinguish the sort of inference that goes on in perceptual systems from more centralized sorts of inferential thinking (perhaps by recourse to something like Fodor's modularity hypothesis) the theory of mind theorist could cheerfully grant that the categories of a theory of mind are in fact embodied in a quasi-perceptual person-perception system. When experienced chess-players see a chessboard or remember a chess position, they are presumably applying 'theoretical' categories perceptually. I see no reason to exclude mental categories from functioning in the same way.

SIMULATION AND SELFHOOD

In the last two sections I have tried to provide a helpful philosophical service, and to let philosophy play a traditional role as handmaiden to the sciences. Now I want to be a little bit more aggressive, and conclude with two points that go to the heart of what we're after in talking of a theory of mind; one methodological, the other more substantial.

First, the methodological point. As I see it, the theory of mind research programme suffers greatly in not having an articulated common-sense theory of mind to work with. It is like trying to do psycholinguistics without having a candidate grammar to test and to suggest answers to questions. We take for granted that this common-sense theory of mind will involve concepts like belief and desire, emotions, moods, memory, dreams, and so forth. But, as we know, a taxonomy in itself does not make a theory. We actually need something that will work, something that will take inputs and give us reliable outputs – and we don't have such a thing. For this sort of articulation we used to look to the sort of self-discipline that would drive an old-fashioned AI simulation. But AI has lost much of its ambition in this area. Still, I believe that we need to approach the problem at this level of grain. Someone needs to design a programme that will use a theory of mind to explain why your mother will be angry at you if you forget your sister's birthday. I know of no such programmes, though there are valuable ideas in Churchland (1970) and Morton (1980). So I think that a high-priority item on the research agenda is to encourage someone to take on a task like this.

The substantive point I want to make has to do with the role of the concept of the *self* in theory of mind. Given that we have no explicitly formulated theory of mind in hand, researchers are forced to proceed on the basis of their best hunch about what notions and connections are central in this theory. Almost everyone who has worked on these issues – in philosophy and in psychology – has assumed that there are two central notions: belief and desire. Abstractly, these roughly correspond to the representation of the way the world is, and the way we want the world to be. In fact, philosophers often use the phrase 'belief–desire psychology' as another way to refer to what Premack and Woodruff christened 'theory of mind'.

The focus on belief and desire is correct as far as it goes, but I think it does not go far enough. There is more to our notion of mind than the conception of a belief–desire system. Common sense also adds the idea that every mind is a *person's* mind: that beliefs and desires are states of an enduring independent self that cannot be identified with the set of beliefs and desires. Consider this parallel: our folk ontology does not construe objects merely as abstract points at which sets of properties converge. We commonly think of objects as *things that have* the properties. This has caused no end of philosophical trouble—the career of the concept of substance in the history of Western thought is a record of that trouble. But be that as it may, our common-sense folk theory of objects is what it is, and the empirical project of characterizing it accurately is a different enterprise from the philosophical search for a satisfactory concept of objects. A psychological account of our theory of objects that does not acknowledge that our notion of an object is a notion of something over and above the properties of the object—even if that something is what Locke called an 'I-know-not-what'—is an incomplete account.

Much the same holds in spades—for our folk concept of mind. The philosophical 'bundle theory' of the self that we find defended in Hume and others, is—as even its defenders usually acknowledge—profoundly antagonistic to common sense. At the centre of our folk ontology—and at the centre of our moral, social, and cultural lives—is the concept of the self as the 'haver' of beliefs and desires, and our naïve conception of this underlying unity is of something much thicker than the unity that we think underlies ordinary physical objects. Notwithstanding the fact that this naïve conception has been challenged as philosophically incoherent, the empirical characterization that theory of mind research aims to provide cannot ignore it indefinitely. So far, it seems to me that it has been ignored.

Now one might respond that the self is not being ignored, but simply put off. Once we get the folk-psychology conception of beliefs and desires straight, we can turn to the common-sense conception of the self as the *subject* of these states. This might be sound policy in general; but I suspect that in the specific quest to understand people with autism, we need to explore that part of theory of mind that represents the self or personhood. We need to understand the sense of personal identity that enables normal individuals to tie their present self to the past and to the future, to care about themselves in a long-term way, to think about their existence as a life, and ultimately to identify other beings as the same in this regard. In ways we still do not understand very well, all this is implicated in the ability to build relationships with other selves, to feel guilt, pride, devotion, to have personal values, and to generally take oneself to be part of a social and moral community.

I grant that these last remarks are speculative and are not backed up by specific evidence. But there are elements in the lives of people with autism that seem to point us in this direction: I have in mind here especially the behaviour discussed in the joint-attention literature (see Chapters 4, 9, and 18, this volume), but also the strange remoteness of much of their social existence—a strangeness which still continues to evade adequate scientific conceptualization. Care, empathy, mutual recognition, attachment, as well as a number of other concepts that might be critical in our understanding of personhood involve both cognitive and affective elements. As such, these concepts are not easily integrated into models of mind designed primarily to account for our cognitive life. Beliefs and desires can be put into convenient 'boxes' in a simulation; it is not clear how attachment fits in. There are sound methodological constraints that cognitivist theories need to adopt to get anywhere, and these make it very hard to tolerate the level of vagueness one still must allow in thinking and talking about such aspects of the self.

The logical positivist Carl Hempel is said to have remarked that if you want to know why the chicken crossed the road, you don't really advance the matter by saying 'Let the chicken be "c" and let the road be "R".' Similarly, construing the need for attachment as just another desire represented by a two-place predicate that takes the self and the other as arguments does not advance the issue either. It seems to me that whether we like it or not, whether it lends itself to our favourite methodological paradigms or not, understanding this other dimension of theory of mind and what it presupposes about selfhood is a line of inquiry that must be pursued in one way or other. By my lights, these speculations do not undermine the cognitivist theory of mind approach to understanding people with autism that focuses on the belief–desire–action nexus. It remains true that a significant majority of these children fail the false-belief task, for instance, and it is likely that they do not have an adequate understanding of this aspect of mind. My sense is only that there is more to the disease than this.

Acknowledgement

I would like to thank Helen Tager-Flusberg, Josef Perner, Henry Wellman, and Deborah Zaitchik, as well as the participants in the 'Understanding Other Minds: Perspectives from Autism' workshop, Seattle, April 1991, for comments and discussion on an earlier draft of this paper.

REFERENCES

Carey, S. (1985). *Conceptual change in childhood*. MIT Press, Cambridge, Mass.
Churchland, P. (1970). The logical character of action explanations. *Philosophical Review*, **79** (no. 2).
Churchland, P. (1981). Eliminative materialism and the propositional attitudes. *Journal of Philosophy*, **77** (no. 2).
Fodor, J. (1987). *Psychosemantics*. MIT Press, Cambridge, Mass.
Hobson, P (1991). Against the theory of 'Theory of Mind'. *British Journal of Developmental Psychology*, **9**, 33–51.

Hook, S. (ed.) (1969). *Language and philosophy; a symposium*. NYU Press, New York.

Kaplan, M. (1991). Epistemology on holiday. *Journal of Philosophy,* **88**, (no. 3).

Kornblith, H. (ed). (1985). *Naturalizing epistemology*. MIT Press, Cambridge, Mass.

Morton, A. (1980). *Frames of mind*. Oxford University Press, New York.

Nagel, T. (1986). *The view from nowhere*. Oxford University Press, New York.

Olson, D. R., Astington, J. W., and Harris, P. L. (1988). Introduction. In *Developing theories of mind* (ed. J. W. Astington, P. L. Harris, and D. R. Olson). Cambridge University Press, New York.

Perner, J. (1991). *Understanding the representational mind*. MIT Press, Cambridge, Mass.

Premack, D., and Woodruff, G. (1978). Does the chimpanzee have a theory of mind? *Behavioral and Brain Sciences*, **1**, 515–26.

Quine, W. V. O. (1969). Epistemology naturalized. In *Naturalizing epistemology* (ed. H. Kornblith). MIT Press, Cambridge, Mass.

Samet, J. *Nativism*. MIT Press, Cambridge, Mass. (In progress.)

Spelke, E. (1990). On the infant's theory of physical motion. Paper presented at the annual meeting of the Society for Philosophy and Psychology, Johns Hopkins University, June 1990.

Wellman, H. (1990). *The child's theory of mind*. MIT Press, Cambridge, Mass.

20

Desire and fantasy: a psychoanalytic perspective on theory of mind and autism

LINDA MAYES, DONALD COHEN, AND AMI KLIN

The set of propositions defined by 'theory of mind' addresses a question which is at the core of very young children's developing ability to be engaged in the social world: how, and when, does a child understand, and act on, the knowledge that its parents have beliefs, emotions, and desires (i.e. mental states) which guide their actions towards their children and others? From the standpoint of child psychoanalytic theory, a young child's ability to be differentially related to others begins to develop in the first months of life, and reflects the workings of two interrelated processes, the construction of self (for example self–other differentiation), and two interrelated mental states, knowledge about and desire for the other. The theory of mind literature addresses one of these processes, the construction of the mind of another. The contribution of child psychoanalysis to this debate is to suggest that the *desire to be with others* is a necessary precursor to knowing about the mind of the other.

A central difference between the psychoanalytic and cognitive theories of mind concerns the roots and functions of *imagination* as a capacity and a process. Stated quite simply, psychoanalytic theory directs our attention to children's inner world of desires and wishes that underlies their understanding of the behaviours of others in the external world. Through the workings of fantasy and imagination, children create an inner world filled with mental representations of others. In this inner world, children play with different views of how others' beliefs and feelings influence their actions towards them (for example how a mother's caring actions reflect her thoughts and love). The child's desire for others motivates the workings of imagination and the creation of an inner world. This in turn influences how a child (or adult) views and responds to others' behaviours in the external world. Desire for others gives life to fantasies about others, which in turn bring a depth and cohesiveness to the cognitive capacity to attribute meaning to human interactions. The child's knowledge of and desire for another are mutually dependent and complementary. Impairments in either could lead to distortions in social relatedness.

In the present chapter, we address the interface between psychoanalytic views of desire and fantasy and the 'theory of mind' hypothesis. We shall consider several important issues: how the capacity to experience desire *vis-à-vis* another arises in the first years of life; why it varies so from individual to individual and within one individual across the life-span; and how basic impairments in the capacity to attribute meaning to others may reflect earlier impairments in the desire to be engaged and in the capacity to create an inner world. We have examined these questions through our studies both of the nature of the autistic individual's interpersonal world and of self–other differentiation in the first eighteen months.

THE INTERPERSONAL WORLD OF AUTISTIC INDIVIDUALS: A LONGITUDINAL VIEW

For many years, we, like others represented in this volume, have asked how and why autistic individuals often fail in the social world. We have investigated several different areas of social functioning in autism, including affective regulation, communicative functioning, social adaptability, and the capacities to perceive and understand others (Cohen 1980; Volkmar and Cohen 1988). Our observations and those of others reveal interrelationships in autism among selective attention, emotional responsivity, routines of shared activity in the first months of life, and social cognition and language (for example Caparulo and Cohen 1983; Paul *et al*. 1983; Tager-Flusberg 1989). There is marked individual variation in the degree of impairments autistic children show in these different functional areas (Volkmar 1987). Such a finding speaks not only to the multivariate nature of the autistic syndrome, but also to the apparent lack of a single, underlying dysfunction. That is, autism may represent the final behavioural outcome for disturbances of functioning in a number of capacities that are involved in socialization and human relationships. No single disturbance of function in any one area accounts for all autistic individuals.

On the other hand, autistic children do show impairments in two inter-related, complementary *processes* — that of attributing and understanding meaning in the behaviours and feelings of others, as proposed by the theory of mind hypothesis, and that of creating their own inner lives, as proposed by psychoanalytic views of internalization and fantasy construction (Cohen 1991). These two processes form the biological and psychological underpinnings of what it means to be a social human being. By studying autism from this viewpoint, we may come to understand individual variations in the development and differentiation of self and other and in the process of fully engaging with the social world (Cohen *et al.* 1976). Furthermore, these two processes are constantly interacting in so far as children's efforts to understand others bring a coherence to their own experiences of desire and love for their parents. In turn, having access to a coherent, organized inner world of representations and fantasies about others gives life to our actions to engage others and attribute meaning to their actions and affects.

Impairments either in the creation of an inner life or in the capacity to attribute meaning to others will lead to disturbances in social relatedness. Some individuals with these disturbances may meet the diagnostic criteria for autism. For all individuals, the maturation of these two processes reflects complicated interactions between constitutionally determined capacities and environmental experiences. These interactions may lead to the range of apparent behavioural, functional impairments in autism, that is, to the range of variation in the 'autistic phenotype'.

Understanding how these two processes fail to develop fully or even partially in autistic individuals is aided by the long-term study of a cohort of autistic children who were diagnosed in the 1950s and have now reached middle age. Some of these individuals have reached a sufficiently high level of function to be able to describe their own experiences *vis-à-vis* the social world (Volkmar and Cohen 1985), while others have remained isolated by their seeming lack of awareness of others and their lack of verbal language. Through these long-term studies we have begun to understand the vicissitudes of the autistic individual's relationships and thoughts about others, and how desire, an inner psychic state, and the capacity to attribute meaning, interact. Two case histories help us to clarify these ideas.

CASE HISTORIES

Herbert's parents became worried about his lack of social responsiveness when he was just six months old (Cohen 1991; Volkmar 1985). Herbert did not make eye-contact, and did not smile responsively. A diagnosis of 'possible autism' was made at a relatively early age, thirteen months, because his parents persisted in seeking help with their concerns. By this time, Herbert babbled little, showed almost no response to speech, and did not point with his hand or finger to draw his parents' attention or indicate his needs. He looked at his mother only in rapid sideways glances, did not imitate her gestures, and was unable to hug or kiss.

For the next several years, Herbert and his family were involved in a number of early childhood intervention and treatment programmes. As he grew older, Herbert became impossible to manage. If left unattended even for brief moments, he would climb on furniture, break windows, and unscrew any detachable object. By the age of three he had no expressive language, and looked through or past people. He responded to his mother, who was with him all day, with hardly more recognition than that usually shown a stranger. Only when he was upset would he seem to run to her intentionally. Efforts to make psychotherapeutic contact with Herbert by verbally labelling his actions or putting words to his presumed anxiety expressed as panic and wild behaviour were apparently unsuccessful. Herbert seemed generally bewildered by the reactions of others, and was unable to use other's actions to direct or modify his own behaviour. When he was placed in a nursery school for normal children, Herbert was unable to understand that the other children were frightened by his uncontrollable running and screeching. In turn, he was driven to states of confused excitement when his peers were engaged in active games.

By the age of seven Herbert was clearly a very retarded child, who drifted from place to place and activity to activity with little planning. He used others, including his siblings and parents, only to satisfy his basic needs, and would involve them by taking their hands or shoving them to what he wanted. When alone, he flicked his fingers or shook his hands before his eyes, rocked, and made guttural sounds or grimaced. When he was engaged in simple, repetitive tasks, his stereotypy decreased, and he seemed calmer. But by the time he entered puberty his face had become set into a mask-like dullness from years without normal affective responsivity.

By late adolescence Herbert could no longer be cared for at home, and his parents prepared for the time when he would move into a residential placement. They could not imagine what he would do without them. For eighteen years they had been constantly present for him—to protect him from running into the street, to try to comfort him when he was frightened by loud noises, to be always aware of where he was and to interpret for him what he needed. On the day they took him to the residential home, they were not prepared for his simply walking away, not looking back, not noticing their absence, not reflecting the meaning they presumed and hoped they had in his inner world. Herbert did not seem anxious without his parents, nor did he ever seem to miss them.

At the age of thirty, Herbert's life is one of quiet and orderly routines. He has no spontaneous play and no friendships, and lives a life of apparently calm structure without important others. Herbert's life history underscores not only his long-standing inability to respond to other persons, but just as poignantly his apparent lack of desire for interaction with others and his lack of longing in their absence. Herbert's frantic activity, his ease of excitability in the presence of others, and his bewildered responses to the affective behaviours of others suggest how confusing and disorganizing the social world was and is for him.

In contrast, Tony, who was able to tell his own story, gives us a closer view of the terrors of a social world he too could not interpret, but at the same time longed for (Volkmar and Cohen 1985). At twenty-two years of age, Tony returned to the Yale Child Study Center to find the records of an evaluation done when he was twenty-six months old. As he reconstructed his own story, he wrote down his account of the experience of autism.

When he was referred for an evaluation at twenty-six months, Tony's parents were most concerned about his lack of speech and his poor social relatedness. From the first weeks of life, Tony had avoided human contact, was difficult to hold, had never smiled responsively, and was preoccupied with spinning objects and with looking at his own hands. With intensive education and psychotherapy, Tony made significant gains developmentally and socially, and by three years of age he was beginning to communicate meaningfully. As an adolescent and young adult Tony was very aware of being different from other people, and felt socially isolated and unable to understand or empathize with others. In his own (unedited) account, he states:

'I was living in a world of daydreaming and Fear revolving aboud my self I had no care about Human feelings or other people. I was afraid of everything!' (p. 49). And later, after entering school, 'And [I] was and still [am] very insecure! I was very cold Harted too. I[t] was impossible for me to Give or Receive love from anybody. I often repulse it by turning people off. Thats is still a problem today and related to other people. I liked things over people and dint care about People at all . . . And was very Nervious about everything. And Feared People and Social Activity Greatly (p. 50).

That Tony was aware of 'this hellish disease' (p. 51) and how his difficulties set him apart and kept him isolated raises a number of questions about how autistic individuals create an inner world for themselves out of the bewildering, often terrifying, complexity they encounter in the object world. Though Tony wrote clearly of his preference for 'things' over people, he was nevertheless aware of his difficulties both in understanding and in desiring others. Through a variety of adaptive compromises, he found his way to a kind of responsiveness to others inasmuch as he was able to maintain a job and occasionally attempted again to be with people. He continued to find the social world bewildering, even hellish; but experienced a painful disappointment that his difficulties kept him isolated from it. We might reasonably ask was Tony in his own way, apparently unlike Herbert, able to experience longing for a social world; but, like Herbert, unable to be in it because he too could not make sense of it? The interplay between longing, desire, and the capacity to understand others is, in the most general sense, the contribution of psychoanalysis to the study of autism.

THE EMERGENCE OF DESIRE: FALLING IN LOVE

Initiating any social relationship requires the capacities to perceive, interpret, and act on the beliefs, feelings, and intentions of another. Yet enduring relationships also require the love of one for another and the understanding that such love is reciprocated. By love, we mean both the *capacity* and the *desire* to be deeply interested in the actions and feelings of another – for infants, in the actions and feelings of their parents. At some point in the first eighteen months of life, interest becomes too objective a word to describe infants' attraction to every detail of their parents' being and life. How parents look, smell, and sound, when they come and when they leave, how they can be brought near, and how sad it feels to have them far away consume infants' attention and motivate most, if not all, of their activities. Even those infant activities we label 'purely' cognitive or information-processing fall under the sway of infants' relationships with their parents, for without the motivating influence of an effectively nurturing relationship, infants do not develop and learn in the expected ways. Importantly, the act of loving and the process of forming attachments to others are not synonymous. Attachment indicates those behaviours which are manifestly observable that mark differentiated and special relationships between individuals. The act of loving involves the inner affective state underlying those attachments.

That love occurs is clear; but how remains a mystery. The central tasks of adaptation for the species are social attachment, engagement, and communication. These tasks set the stage for love. There are also species-specific and individually variable perceptual and neuroregulatory capacities which single out social stimuli as the most salient and attractive to the human infant. These set the lights on for love. But the actual drama that unfolds is not described entirely by phylogenetic heritage or constitutional endowments. Evolution and biology become individualized and personified through the effects of early caring relationships on the psychic life of the infant. Love matures out of these early relationships. In other words, species-specific adaptive needs, neuroregulatory and perceptual capacities, and desire are the necessary ingredients of the infant's love relationships and eventual full membership in the social world. Impairments at any one of these levels of function lead to disorders of socialization and communication, some of which will meet the diagnostic criteria for autism.

Desire is a complex phenomenon, which may seem more secure in literary discourse than in research. From the psychoanalytic point of view, desire involves more than intentionality or the attribution of motivation or intentionality to another. To say that someone seeks to obtain a toy because he or she wants it is a statement attributing a motivational state to an observable action, that is, that a person acted intentionally because of a motivating wish (or desire) for that toy. The psychoanalytic view of desire suggests first that the manifest action does not necessarily reflect an internal state of desire for something or someone. What we do in observable action is the end-stage of a number of interdependent intrapsychic processes or conditions, only one of which may be a wish to possess something or someone. A manifest action may be enacted as much in the service of denying an intrapsychic wish as it is to fulfil it. The context of the action does not unerringly show us the direction of the wish, and, more commonly than not, our actions represent complex compromises between one or multiple wishes and our conflicts over those wishes.

Secondly, not only is an inner state of desire not necessarily revealed or reflected in an individual's actions, but even more importantly, one need not engage in observable action to be acting on one's desire. Such are the nature and benefits of a fantasy life. We give our desires expression, if only to ourselves, and find ways to fulfil such desires through the inner psychic world of fantasy. Desire for others gains depth and complexity through the workings of fantasy, and the capacity to experience and express desire reflects a cohesive and active inner fantasy life. In our inner psychic lives we are able to create more variations on a theme, more scenarios to a story than would ever be possible if we depended on direct experiences with others and the memories of those experiences. Any active expression we give to our desires has been through many accounts and revisions in our inner psychic lives. In fantasy we take on the role of the other, and we experiment with the feelings of gratification, of frustration, of revenge, of immediate or delayed action. We write, produce, direct, and act in the screenplay of our choosing. And we learn about a person's desire most fully by learning about his or her inner fantasy life.

Thirdly, the psychoanalytic notion of desire for another is based on the cumulative effects of the individual's earliest experiences with others, that is, with parents. Desire is not an innate, constitutionally determined motivational state. It arises from the interaction between the infant's neurophysiological capacities for engaging others and the experiences that occur when the infant and others interact. The 'hardwiring' underlying desire involves the various constitutionally given perceptual and neuroregulatory functions which promote the salience and inherent attraction of social stimuli or even the perceptual attributes of social stimuli (Mayes and Cohen, in press).

For example, the earliest evidence of the capacities for social relationships is found in the newborn's preference for the human voice, and for a higher-pitched voice, over other sounds (DeCasper and Fifer 1980). Similarly, very young infants show a preference for curved lines over straight ones, for irregular patterns, for patterns with a high contour-density, and for symmetry (Olson and Sherman 1983) — each features of faces and of the various facial expressions. Normal newborns are also endowed with the capacity for eliciting parental attention — for example, to respond with change in states to optimal care-giving — and for perceptually and motorically orienting themselves toward their parents — to remain physically close to, reach toward, and grasp on to their mothers. Without such capacities or with specific impairments, infants would have greater difficulty engaging their parents — for example, congenitally blind infants demonstrate autistic-like impairments in their social relatedness (Rapin 1979). These capacities permit infants to be deeply engaged in trying to understand the world of their mothers' behaviour — where is she, how can I get her to me, why does she go, is this she or someone else, when does she come?

Without the desire to be engaged, these questions remain only cognitive puzzles. They become affective, deeply experienced dilemmas when the infant begins to experience desire for another. Then 'where is she' is more than an idle wondering; it is a question of marked affective intensity. In other words, these constitutional capacities alone are not sufficient for conditions of loving and intense involvement with others. How does desire emerge out of these early engagements between infant and parent, how do the emergence of desire and the differentiation of self relate to each other, and how do we recognize the workings of a state of desire, or, more to the issue, how do we know when a social dysfunction reflects a lack of desire?

THE DEVELOPMENT OF DESIRE FOR THE OTHER

To consider how desire emerges intrapsychically out of these early engagements between parent and infant, we need to consider the nature of the earliest interactions between parent and infant. What follows is a psychoanalytic conceptualization or metaphor of the mind of the infant that emphasizes the continuity of experience in a mind that is rapidly maturing and differentiating. In many ways, this metaphor is similar to the notions of interpersonal intersubjectivity discussed by Hobson and others in their considerations of the affective structure of the child's developing 'theory of mind' (Hobson 1989, this volume, chapter 10).

At the beginning of psychological life, there is not an infant, but a mother-infant matrix (Winnicott 1945). In the earliest period of the infant–mother matrix, psychic activity occurs as a fused event of the matrix, as one 'fleeting and very perishable mental entity' (Loewald 1977, p. 215), which is neither ego nor object. From the infant's point of view, the outside world consists mainly of mother, or perhaps of mother not as person but as the source of food and warmth. At this stage, 'mother' represents the world that acts contingently on the infant's needs; but she exists intrapsychically only inasmuch as the infant needs. She is neither a whole nor a differentiated object. The object, or interpersonal, world still lacks clear boundaries in the infant's psychic world. In the first weeks of life, needs are satisfied and social stimuli arouse pleasurable experiences without being attributed to the whole object 'mother'.

Such a stage cannot exist for long, since the infant soon experiences the frustration engendered by the mother's inevitable absences and delays. Desire for another is differentiated not through the pleasure of ever-present gratification but out of the frustration of unmet needs. Frustration and the ensuing discomfort represent a first break in the sensation of immediate gratification, and the first experience of the infant of the separateness between states of physical need and the satisfaction of such needs. Through such inevitably repeated experiences and the beginning feelings of separateness, desire for another takes shape — for it is that other who can alleviate discomfort. Because of frustration, the infant seeks out the mother and brings her closer by using those innate capacities for engagement. The contingency of a mother's nurturing acts on the infant's discomforts, and the association between her absence and the infant's sense of frustration establishes referencing links between the mother's behaviour and the affective experience for the infant. The infant's experience of mother oscillates between pleasure and discomfort, satisfaction and frustration, i.e. between a state of being united and a state of loss. It is out of this 'very polarity between separateness and union' (Fromm 1955, p. 32) that desire is born and reborn.

It is centrally important to understand that paradoxically desire for another is not based solely on gratifying experiences. Instead, the onset of desire reflects the experiences of disappointment, and of painful discomfort. Anger and frustration are as necessary a precondition for desire as blissful satisfaction. Desire and separation, whether physically or through delayed responsiveness, go hand in hand. The emergence of separation-distress and of stranger-anxiety represent the ways infants learn about the importance of others to their own security and comfort. Infants' dysphoric changes in state (crying or angry protests) on anticipated separation and excited pleasure on reunion are evidence that, at least in a rudimentary way, infants recognize their own feelings in relation to their parents' actions.

Such recognition is evident behaviourally not only in stranger- or separation-anxiety, but very early on in infants' reactions to their mothers' assuming a still or neutral face. In such situations, infants as young as three months protest or increase their positive engaging behaviours in reaction to their mothers' unanticipated change in responsivity (Mayes and Carter 1990; Tronick *et al.* 1978). Behaviours like these suggest that infants have a nascent 'theory' of mind about their own states and feelings *vis-à-vis* the behaviours of another. In particular, infants' increased use of positive engaging behaviours, even more than distressed responses, underscores their expectation that social engagement with mother is a condition to be desired. Longing for another is mapped on to these very early 'theories' of how others influence one's own affective states.

The recognition of the importance of the other to one's own comfort is also a powerful factor in the emergence of self (see Samet, Chapter 19, this volume, for a philosophical discussion of the place of the self). From the psychoanalytic point of view, self begins to be defined through the effects of cumulative experiences with mother in the first days, weeks, and months. While even very young infants have the perceptual capacities to discriminate the external boundaries of another and to discriminate, for example, their own movements or those of another, the internal sense of self is bound up in the memorial layering of schema and the creation of representations from multiple experiences with another. Through the repeated experiences in the presence of another, memories laden with affective traces are created, and the other of the inner world gradually takes shape. In a sense, the child's earliest experiences with another person are organized around affects or, as Hobson has proposed, an 'affective relatedness' (Hobson 1989, 1990). Thus, the differentiation of desire and of self are parallel, interdependent processes. The desire that evolves into love relationships begins to emerge when self begins to be distinct from other (Loewald 1977). When mother comes to represent the source of gratification which is not symbiotically available, she is becoming a separate and whole object, the object of love, a *desired* object. Similarly, when the infant experiences him or herself as an active agent in engaging the mother's interest and care, the process of individuation has begun (Mahler and Furer 1968; Mahler *et al.* 1975).

Individuation, and hence the onset of desire for a differentiated object, also marks the workings of *internalization* and the beginnings of an inner life. Through repeated interactions with mother around basic biological needs, endogenous events such as hunger and exogenous experiences such as mother's warmth are incorporated into the infant's psychic life as memorial traces (Loewald 1977). Gradually, good and bad memories and their associated affective currents are joined in the infant's inner world to define mother as a whole object toward whom one experiences a host of both angry and loving feelings (Klein 1957) Thus, when infants experience their own dysphoria and elation in response to another, not only is self distinct from other beginning to emerge, but also developing is the capacity to call on memories of past interactions to help with current affective states.

Others increasingly populate the infant's internal world. He or she not only imagines different scenes, but uses these fantasies to organize his own sense of self and responses to others. The infant is able to represent a remembered story such as 'I feel sad and angry to see her go but she will return and when she does, we will sit together, she will understand my anger and I will feel comforted as before.' In the remembering and rewriting of the story, the infant is capable not only of a higher order of mental activity, but also of using such activity for his or her own affective regulation. The anger and distress are not necessarily disorganizing. Fantasy has begun to assume an active role in caring for self and in deepening love for others.

Through these love relationships as they are represented in the psychic world and through the parallel, ongoing processes of desire and self-differentiation, the child (and adult) becomes more self-aware. We learn about ourselves in large part *vis-à-vis* our love relationships to others, and the vicissitudes of those relationships. Processes of internalization continue throughout the life-span. With every new love relationship, the process of making the other a part of one's inner world and the parallel process of increased self-awareness that comes with love occurs. It is not just the *early* separations that set the process of desire and self–other differentiation going. The sadness and the aggressive, hateful, destructive feelings mobilized by an actual or threatened loss of a loved one continue to be mutative processes for infants, children, and adults. Within limits, the experiences of threatened or actual loss deepen feelings of love for another and strengthen the sense that one's self persists even in the physical absence of others, that is, the others continue to exist in one's internal psychic world. Being able to grieve and mourn are the counterparts of being able to love. One does not occur without the other. Let us now turn to the relevance of this for autism.

THE ABSENCE OF DESIRE FOR THE OTHER IN AUTISM

It is our contention that autistic individuals neither fall in love readily, nor grieve or mourn the absence of presumably important others. When separated from parents or long-term caregivers, they may seem briefly unhappy, and conversely pleased if the person returns; but, generally, their responses to losses are as muted and brief as their displays of affection. It is as if their capacity to experience desire for another, to experience the full depth of love in their inner psychic lives is either absent or distorted. As one young autistic man said, responding to a picture of young woman resting her head on a man's shoulder, 'This is a husband and a wife embracing each other — typical of one's so-called human need for companionship' (Bemporad 1979, p. 189). Or as another young autistic man put it 'I never could have a friend. I really don't know what to do with other people, really.' (Cohen 1980, p. 388).

To this point, however, it is important to emphasize as before the diversity of responses within the diagnostic category, and the fact that the social dysfunction in autism is not an all-or-nothing impairment. With maturation, many autistic individuals are not entirely oblivious to others — they often make many attempts to engage in the social world, and show both pleasure and longing to be with others. They may be acutely aware of a sense of loneliness and estrangement, and feel sad about how such impairments keep them apart from others. They are able to understand that relationships are important to other people, that there are such things as feelings, and that there are certain customary ways to interact with others.

For those autistic individuals who do become more socially engaged, it is as if they find compensatory mechanisms or structures for achieving some semblance of relatedness *vis-à-vis* others, and are able to experience in some way longing for others. They may, for example, learn social conventions, for example, that 'How are you?' is generally followed by 'I am fine, and how are you?' They may attend social functions, and appear interested in various activities such as dating or games (Bemporad 1979). Yet, more often than not, even the brighter, more self-aware autistic individuals are puzzled by the depth of human relationships; they seem unable to get it, and their social relations rarely have the multi-dimensional quality of love, even early childhood love. Their relations with others lack the spontaneity, the affective variation, the empathic quality that characterizes relatedness learned early. Whatever the nature of the inborn disorder that interferes with the autistic child's development of the earliest patterns for affective relatedness, the early difficulties persist in the autistic adult's personality structure, and set limits on the building of an inner fantasy life. The latter in turn limits the depth of involvement in any loving relationships.

These limits are most apparent when autistic individuals have sufficient communicative capacities to allow a glimpse into their inner lives. More often than not, whether or not the individual experiences some longing for the social world, the inner psychic view of others is monochromatic, puzzling, even at times frightening to them, and they find comfort in the more predictable, stable, and concrete non-object world. The powerful range of feeling, including anger and hate as well as caring, which are set into play by loving relationships may prove overwhelming for the autistic individual.

For example, as a young adult, Peter was preoccupied by an image of his own crying face. From the first time the image had come to him, it remained unchanged and unremovable. 'It stung me like a bee, and I have never been able to be free of it.' From this image, he could describe how he imagined people to be angry and destructive, to look past him, spit on him, or never to like him (Cohen 1980, p. 388). Or even as they try to use their inner lives to imagine and learn about themselves, a certain lack of cohesiveness and easy fragmentation is apparent. As Leonard, trying to imagine what would happen if he were to take a new job and learn to drive, said, 'I can imagine myself doing it, my mind wanders in the air . . . but nothing like that is possible for me . . . Even the littlest things make me too anxious, too panicked. Unmistakably, unquestionably, if I tried to drive, I would just certainly fall apart.' (Cohen 1980, p. 389). Desire for others and the enduring capacity to love — founded in the matrix of the earliest love relationships — fosters and supports using the inner world for self-reflection and awareness.

THE INTERFACE BETWEEN PSYCHOANALYSIS AND THEORY OF MIND

The psychoanalytic view of mind draws our attention to the inner world of psychic representations and psychic action that is constantly behind the world of observable behaviours and capacities. Because what is observable in the external world may have multiple meanings in the psychic world, the psychoanalytic theory of mind complements the more cognitively oriented theory of mind conceptualized by several other contributors to this volume.

For example, Baron–Cohen's descriptions (this volume, chapter 4) abstract the cognitive prerequisites underlying social understanding, communicative skills, and pretend-play. They present a *competence model* of the ability to conceive of mental states in others and in oneself. The psychoanalytic view complements this model by adding the infant's *intrapsychic experiences* or inner world, and the affective and motivational *context* in which the infant's socio-cognitive development takes place. In other words, from the psychoanalytic frame of reference, the most central mental state to impute to others is the understanding that others reciprocate one's desire for them. Young children understand that their mothers appreciate and return their desire and love for them.

The psychoanalytic perspective also complements the developmental account of the origins of a cognitive theory of mind in several other respects. First, by encompassing the various intersubjective experiences evidenced in the infant–mother dyad before the advent of mental representations, it provides the continuity between innate social adaptive mechanisms and the momentous cognitive accomplishments described in the cognitive model. Second, by taking on board the multiple and individualized representations of the self and the other, it provides for the richness of individual inner lives and the variety of outcomes. Third, by contextualizing the competence model, it personalizes mental states, giving them the individual significance which is at the heart of human social experiences, and allows for the individual variation among autistic individuals' capacities to take the perspective of another.

There is, however, a broader issue highlighted by the differences between these two 'theories of mind'. In part, the categorical definition of autism defines what a theory of mind needs to explain. Metaphorically, a theory needing to account for a monochromatic visual world needs to explain far less complex phenomena than a theory accounting for the full range of the colour spectrum. Similarly, if autism is defined by all-or-nothing categories of social relatedness, the theory to explain such categories does not need to account for the diversity within the diagnostic grouping and the changes over maturation. But the individual variation in the social development of autistic individuals is strikingly apparent in the many longitudinal accounts now available. Even on the measures of the autistic child's capacity to understand the mental states of believing, knowing, or desiring, there is a range of impairment. Some autistic individuals are quite able to appreciate the beliefs of others, and to take their perspective, correctly (Baron–Cohen, this volume, Chapter 4).

What these observations of individual variation suggest is the need to understand the multiple ways infants and young children develop a theory of others and of mind. How do some children reach their third and fourth years of life apparently unable to understand the mental states of others? What are the multiple sources of variation in how children develop an inner sense of self and of the other and in how they come to desire and love others? Minimally, as we have outlined, the developmental emergence of an understanding of others involves the interaction in the first eighteen months between constitutionally given neuroperceptual and organizational capacities and the 'metabolism' and integration of social information received from the environment. Impairments are possible at several different levels, not all of which results in the social dysfunction seen in autism. For instance, children who are blind may show a number of social impairments early in life; but they most often develop over time alternative pathways to socialization that allow them affectively rich and integrated lives (Rapin 1979). How do we understand the 'physiology' of apparently normal relatedness, and what are the sources of these alternate pathways?

By its focus on the development of inner psychic lives, the psychoanalytic view of social development offers one window on how a theory of mind about others comes into being. Considering the earliest roots of social experiences in the first eighteen months focuses investigative attention on the very early and generalized social deficits exhibited by autistic children (Klin, Volkmar and Sparrow, 1992), which otherwise would lie outside the scope of the cognitive model. Understanding the various biological mechanisms of adaptations and the complexities of the infant–mother matrix permits us to investigate the heterogeneity of the autistic phenotype. The longitudinal view gained from seeing children in the first years and following their development also informs our appreciation of the social adaptations many autistic individuals are able to obtain despite their impairments. And, conversely, through appreciating the inner lives of many higher-functioning autistic individuals, we may understand how, despite a cognitive ability to conceive of other people's minds, they remain estranged from the social world because they cannot share their feelings (Bemporad 1979; Cohen 1980; Volkmar and Cohen 1985).

Traditionally, psychoanalysis has viewed mental functioning and adaptation along a developmental continuum, and has considered mental impairments along a spectrum instead of as a categorical grouping. Adopting such a view changes our frame of reference about the social development of autistic individuals, and emphasizes that autism represents a range of social impairments and adaptations. By taking the intrapsychic perspective of the growing infant, the psychoanalytic approach attempts to bridge the gap between modelled capacities and inner realities, between cognitive computations and affect-laden fantasies, and between the depersonified social environment and the infant–mother emerging love affair. The psychoanalytic view provides a new set of questions and empirical dilemmas about how all individuals learn about themselves and others and come to experience deep and enduring feelings of desire for another.

Acknowledgements

We gratefully acknowledge our collaboration with our colleague Dr Fred Volkmar, whose work with autistic individuals has greatly influenced the thoughts presented in this chapter.

REFERENCES

Bemporad, J.R. (1979). Adult recollections of a formerly autistic child. *Journal of Autism and Developmental Disorders*, **9**, 179–97.

Caparulo, B.K., and Cohen, D.J. (1983). Developmental language studies in the neuropsychiatric disorders of childhood. In *Children's language* (ed. K.E. Nelson). Gardner Press, New York.

Cohen, D.J. (1980). The pathology of self in primary childhood autism and Gilles de la Tourette syndrome. *Psychiatric Clinics of North America*, **3**, 383–402.

Cohen, D.J. (1991). Finding meaning in one's self and others: clinical studies of children with autism and Tourette's syndrome. In *Contemporary constructions for the child* (ed. F. Kessell and A. Sameroff), Erlbaum, Hillsdale, NJ.

Cohen, D.J., Caparulo, B.K., and Shaywitz, B.A. (1976). Primary childhood aphasia and childhood autism: clinical, biological, and conceptual observations. *Journal of the American Academy of Child Psychiatry*, **15**, 606–45.

DeCasper, A.J., and Fifer, W.P. (1980). Of human bonding: newborns prefer their mothers' voices. *Science*, **171**, 1174–6.

Fromm, E. (1955). *The sane society*. Holt, Rinehart, and Winston, New York.

Hobson, R.P. (1989). On sharing experiences. *Development and Psychopathology*, **1**, 197–203.

Hobson, R.P. (1990). On the origins of the self and the case of autism. *Development and Psychopathology*, **2**, 163–81.

Klein, M. (1957). *Envy and gratitude*. Hogarth Press, London.

Klin, A., Volkmar, F. and Sparrow, S. (1992). Autistic social dysfunction: some limitations of the theory of mind hypothesis. *Journal of Child Psychology and Psychiatry*, **33**, 861–76.

Loewald, H. (1977). Instinct theory, object relations, and psychic structure formation. In idem, *Papers on Psychoanalysis*. Yale University Press, New Haven.

Mahler, M., and Furer, M. (1968). *On human symbiosis and the vicissitudes of individuation.* International Universities Press, New York.

Mahler, M., Pine, F., and Bergman, A. (1975). *The psychological birth of the human infant.* Basic Books, New York.

Mayes, I. C., and Carter, A. S. (190). Emerging social-regulatory capacities as seen in the still face situation. *Child Development,* **61**, 754–63.

Mayes, L. C., and Cohen, D. J. The role of constitution in psychoanalysis. In *Concepts in psychoanalysis* (ed. B. Moore). (In press).

Olson, G. M., and Sherman, T. (1983). Attention, learning, and memory in infants. In *Infancy and developmental psychobiology,* (ed. M. Haith and J. J. Campos), Vol. 2. Wiley, New York.

Paul, R., Cohen, D. J., and Caparulo, B. K. (1983). A longitudinal study of patients with severe developmental disorders of language learning. *Journal of the American Academy of Child Psychiatry,* **22**, 525–534.

Rapin, I. (1979). Effects of early blindness and deafness on cognition. In *Congenital and acquired cognitive disorders* (ed. R. Katzman), Raven Press, New York.

Ritvo, S., and Provence, S. (1953). Form perception and imitation in some autistic children: diagnostic findings and their contextual interpretation. *Psychoanalytic Study of the Child,* **8**, 155–61.

Tager-Flusberg, H. (1989). A psycholinguistic perspective on language development in the autistic child. In *Autism: new directions on diagnosis, nature, and treatment* (ed. G. Dawson). Guildford Press, New York.

Tronick, E., Als, H., Adamson, L., Wise, S., and Brazelton, T. B. (1978). The infant's response to entrapment between contradictory messages in face-to-face interaction. *Journal of the American Academy of Child Psychiatry,* **17**, 1–13.

Volkmar, F. R. (1987). Social development. In *Handbook of autism and pervasive developmental disorders* (ed. D. J. Cohen and A. M. Donnellan). Wiley, New York.

Volkmar, F. R., and Cohen, D. J. (1985). The experience of infantile autism: a first-person account by Tony W. *Journal of Autism and Developmental Disorders,* **15**, 47–54.

Volkmar, F. R., and Cohen, D. J. (1988). Classification and diagnosis of childhood autism. In *Diagnosis and assessment in autism* (ed. E. Schopler and G. Mesibov). Plenum, New York.

Volkmar, F. R., Stier, D., and Cohen, D. J. (1985). Age of recognition of pervasive developmental disorder. *American Journal of Psychiatry,* **142**, 1450–2.

Winnicott, D. W. (1945). Primitive emotional development. In idem, *Through paediatrics to psychoanalysis.* Basic Books, New York.

21

The theory of mind deficit in autism: some questions for teaching and diagnosis

SIMON BARON-COHEN AND PATRICIA HOWLIN

In his scholarly review of the psychological literature on autism, Rutter (1983) included an account of his own clinical experiences with adults with autism. He wrote:

Several [adults with autism] have commented that they are distressed by their inability to understand what other people are thinking or feeling. One young man who has attended the clinic for a quarter of a century since he was first referred as a non-responsive non-speaking child put it most vividly when he came back a few years ago asking for help with his difficulties. He complained that he 'couldn't mind-read'. He went on to explain that other people seemed to have a special sense by which they could read other people's thoughts and could anticipate their responses and feelings; he knew this because they managed to avoid upsetting people whereas he was always putting his foot in it, not realizing that he was doing or saying the wrong thing until after the other person became angry or upset. (p. 526).

This account, by coincidence, picks up on the key impairment upon which this volume focuses: the theory of mind deficit, as it has since come to be known (see Baron–Cohen, this volume, Chapter 4 for a summary of the experimental findings relevant to this). Rutter's use of clinical description is, we think, a useful starting-point for setting this deficit into its everyday context. In this chapter, we begin by describing a set of examples drawn from our own clinical experience, so as to elaborate on Rutter's single case. We do this is in order to delineate the wide-ranging expression of the theory of mind deficit. This also stands as a backdrop against which to ask clinically relevant questions: First, what are the implications of the research on autistic children's theory of mind for teaching? In particular, can mental-state concepts be taught? If so, how would this affect the broad range of deficits listed in the clinical examples below? Secondly, could this research have any application for the diagnosis of autism, both in infancy and later?

THE THEORY OF MIND DEFICIT IN AUTISM: EXAMPLES FROM CLINICAL EXPERIENCE

In the following examples, our clinical anecdotes are presented under various categories of 'theory of mind' error. These have been taken from cases referred to the second author at the Maudsley Hospital.

1. *Insensitivity to other people's* **feelings**

Frederick is a twelve-year-old boy with autism. His parents were desperately anxious that he should be assimilated into his local secondary school, and were horrified to hear that in the first week he had approached the head teacher in Assembly and commented on how many spots he had on his face.

2. *Inability to take into account what other people* **know**

Jeffrey, an extremely able young man with autism who holds a responsible position in a computing company, is unable to appreciate that if he has witnessed an event, this knowledge may not be shared by others. He seems unable to comprehend that his experience is different to theirs, often referring to events without providing the essential background information necessary for his colleagues to understand what he is talking about.

3. *Inability to read* **intentions**

Samantha, a ten-year-old girl with autism attending a mainstream school, was deliberately teased by the children there, and frequently they would tell her to perform some unacceptable act, such as taking her clothes off in the playground. She was quite bewildered by the laughter that ensued (and the scolding by the teacher), believing that her compliance would result in them becoming 'her friend'.

4. *Inability to read the listener's level of* **interest** *in one's speech*

Robert, a twelve-year-old boy, also attending mainstream school, constantly irritated peers and teachers alike by his 'boring' monologues on the cubic capacity of Renault cars, structural details of the Severn Bridge, or albinoism. He would discuss just these three topics at length with anyone, and was quite unable to recognize that his enthusiasm for these arcane topics was in no way shared.

5. *Inability to anticipate what others might* **think** *of one's actions*

Joseph, although having done very well in many areas of his development, obtaining a university degree and various diplomas in computing, continued to have problems understanding what others might think of his actions. In particular, he had no sense of personal space, and would also tend to ask very intimate questions. Difficulties arose shortly after he started a job with a computer firm. He still showed no sense of personal space, and would hover over the desks of female employees or lean up against them in lifts or queues, etc. After some weeks of this the secretarial staff demanded his dismissal on the grounds of 'sexual harassment'.

6. *Inability to understand* **misunderstandings**

Michael, a young man with autism, was dismissed from his job after an incident in which he had attacked the cloakroom attendant. He showed absolutely no remorse for this, having hit her with his umbrella 'because she gave me the wrong ticket'. Being in the habit of doing everything meticulously himself, he simply could not understand that others might make mistakes. Long afterwards he still expressed bewilderment that he had lost his job whereas, by rights, he was convinced the cloakroom attendant should have lost hers.

7. *Inability to deceive or understand* **deception**

John, a twenty-five-year-old man with autism, had a job working in a jeweller's. Because his boss knew that he was absolutely honest and that he could be safely trusted with large quantities of money or valuables, he had access to the keys of the safe. However, his failure to understand deception left him open to exploitation by others, and a new night-watchman took advantage of the situation. Being asked casually for the keys one night John readily handed these over, and when the night-watchman, the keys, and the contents of the safe had disappeared, he was charged with being an accessory to the robbery. Although these charges were dropped, he could clearly no longer be employed in such a position of trust again.

8. *Inability to understand the* **motives** *behind people's actions*

David, a twenty-year-old man with autism and of normal intelligence, but with considerable social difficulties, was offered employment by his uncle. Taking into account David's particular pattern of social behaviour the uncle had, sensibly, found a niche for David in a quiet corner of the accounts office. Rather than being grateful for his uncle's efforts, David was outraged to learn that he had not instantly been made a managing director of the company. He walked out of the job after only a few days, and thereafter harboured intense resentment against the one person who had tried so hard to help him.

These instances of different theory of mind errors by no means exhaust the kinds of problems caused by a dysfunction in the development of a theory of mind, but they are sufficient to convey how people with autism often just 'miss the point' of another person's action or speech. What are the clinical implications of this deficit? In the next section, we consider this question with regard to the teaching of social understanding.

TEACHING SOCIAL UNDERSTANDING

We begin this section by reviewing existing methods for teaching social and communication skills in autism, and consider what they achieve. We then outline a new study that is under way, which adopts an approach to social-skills teaching aimed at facilitating the acquisition of mental-state concepts.

Social and communication skills teaching

A variety of approaches to social and communication skills teaching have been applied to autism. Most of these go under the heading of 'training'. We prefer the term 'teaching', as it avoids the implication of simply building in 'circus tricks'. In addition, teaching carries the implication of educating, that is, changing the child's understanding and way of thinking, and not just changing behaviour. In reviewing existing teaching methods, therefore, we also consider how far each of these does indeed change understanding, and not just behaviour.

Existing teaching methods include traditional behavioural techniques, advising caregivers, problem-solving, and role-playing techniques, group teaching, and the involvement of normal peers and siblings (see Schopler and Mesibov 1986; Groden and Cautela 1988; and Gaylord-Ross 1989). These approaches are summarized here:

i. Behavioural approaches

These employ techniques such as prompting, modelling, or shaping, together with differential reinforcement, to improve social and communication skills. They may concentrate on the teaching and development of socially appropriate behaviours, such as initiating or maintaining conversations, or increasing eye-contact, gesture, and facial expressiveness (Brady *et al*. 1987; Matson *et al*. 1988; Fantusso *et al*. 1989). Other programmes have focused on the removal of socially unacceptable behaviours by teaching basic rules (for example, not taking off clothing in public, not talking to strangers, not using inappropriate speech, etc). Teaching relaxation and self-control techniques, such as anger-management, has also been used to reduce difficulties resulting from confusion or anxiety in social situations (Favell 1983; Howlin and Rutter 1987).

ii. Advising caregivers

An alternative way of reducing the effects of the social deficit has been to educate caregivers about the specific ways in which autism affects social behaviour and development, and to advise them on methods of minimizing the problems that inevitably arise. The formulation of simple but explicit contracts with the person with autism, together with detailed timetables or work schedules, is used to ensure that basic rules are implemented and complied with and that tasks are completed within a set time or to a specified standard. Support of this kind seems particularly valuable when the goal is to maintain a person with autism in their school or job, when their social behaviour might otherwise have given rise to dismissal. It also provides important support for other members of the family (Howlin 1989).

iii. *Role-play and problem-solving techniques*

Role-play and drama techniques are sometimes used to teach new social skills (Dewey *et al.* 1988) or as a means of modelling and rehearsing strategies for dealing with difficulties (for example, teaching individuals with autism how to initiate and maintain conversations, or how to cope with teasing or anxiety, etc.). Video replay has also been used to provide feedback of, and attempt to reduce, abnormal behaviours such as inappropriate eye-gaze, facial grimacing, etc., which may cause other family members embarrassment, and affect social acceptance (Howlin and Rutter 1987). Because generalization to non-rehearsed situations is often limited, the teaching of more general problem-solving strategies is also used (Fagan *et al.* 1985; Plienis *et al.* 1987; Park and Gaylord-Ross 1989).

iv. *Group treatments*

The majority of studies of social-skills training of children with autism have been single-case investigations. Others, although involving larger groups, have provided only minimal data on the efficacy of treatment. Recently Williams (1989) used the social-skills training package developed by Spence (1980) with a group of ten children with autism. Three types of strategies were used (recreational games, role-play exercises, and modelling), with the emphasis being, on the development of effective social tactics rather than the learning of specific rules. This kind of programme represents an alternative to more rigid behavioural techniques.

v. *The involvement of normal peers*

A number of studies have explored the use of normal peers as 'social therapists' for children with autism (Strain *et al.* 1979; Brady *et al.* 1987). Most have focused on specific behaviours, such as the frequency of initiations, rather than on wider aspects of social interaction. Such approaches struggle with the difficulty of maintaining the enthusiasm of the normal children (Lord 1984), and with the difficulty of generalizing any changes to untrained peers (Breen *et al.* 1985). Schuler (1989) suggests that greater attention to the types of play activity involved may overcome some of these problems, by using more naturalistic interactions, and Lord's work confirms this.

What do such teaching techniques achieve?

Small sample-sizes and inadequate outcome measures have made it difficult to reach firm conclusions about the relative merits of these different procedures. The most common assessments used are *frequency* measures, focusing on decreases in perseverative or other inappropriate behaviours (Taras *et al.* 1988), increases in numbers of social interactions (Brady *et al.* 1987), or counts of specific behaviours such as smiles, eye-contacts, or utterances

(Fantusso *et al.* 1989; Matson *et al.* 1988). What has not been assessed is the pragmatic use of language and gesture (Howlin 1986).

Although there are a few programmes offering wider-ranging suggestions for developing social awareness and interactional skills (Mesibov 1984; Frankel *et al* 1987), these tend to lack objective outcome assessments. Williams' (1989) group study, mentioned earlier, is an exception to this, but relies on 'non-blind' teacher-evaluations that may affect the reliability of results. The potential of group-training studies remains to be fully explored, but doubts have been raised, for example, about whether a group exclusively comprising individuals with autism can be effective in increasing social skills when the group as a whole is so handicapped. The involvement of normal peers in structured but naturalistic settings may offer greater promise (Lord 1984; Schuler 1989).

It is clear that in order to evaluate the effectiveness of current treatment procedures, more socially valid, objective measures of outcome are needed. Nevertheless, despite these provisos, many studies do indicate that it is possible to increase specific behaviours, such as eye-contact or frequency of social interactions, using behavioural procedures. These are not insignificant achievements, as they help individuals to *appear* more 'normal', and this may affect how other people react to them. However, no studies have investigated if, when these specific behaviours 'improve', there is also an associated improvement in social *understanding*.

The outcome from verbal communication programmes suggests a similar picture to that emerging from studies of social interaction (Howlin 1989). That is, there has been moderate success in reducing inappropriate speech, increasing spontaneous utterances, building up vocabulary, and improving syntax, but no demonstrable improvements in the individual's understanding of what lies *behind* the other person's speech: the intended *meaning*. Given that problems in social understanding seem so central to autism, it is surprising that this has rarely been a major focus of either intervention or assessment in social and communication-skills training for this population.

Since cognition guides behaviour, it is of interest to ask whether teaching key aspects of social cognition is possible, and if so, whether such teaching affects social behaviour. In the next section, we describe an ongoing study which attempts to teach mental-state concepts to children with autism (Hadwin, Baron–Cohen, Howlin, and Hill, 1994). Whilst results are not yet available, we outline the framework we are using, in order to open discussion into these educational questions.

Can a theory of mind be taught?

Normal children do not seem to require explicit teaching in order to acquire a theory of mind. Indeed, they seem to develop this understanding irrespective of the particular form of parenting they receive (Avis and Harris 1991).

Whiten's (this volume, Chapter 17) thought-experiment into whether 'wild' children would develop a theory of mind pursues the same idea. However, it may be that a theory of mind can be explicitly taught to children who have failed to acquire it naturally. Such teaching might provide an alternative route into mentalistic understanding.

Consider the analogy with blind children learning to read: Braille gives an alternative way into the problem of learning to read written words. We are interested in whether there might be an alternative way into the problem of learning to 'read' minds. Clearly, this analogy is imperfect in at least one key respect: blind children have a sensory impairment, and also show some abnormalities in their language development, but there is no central cognitive deficit in their 'word recognition system'. Braille circumvents the sensory deficit, and since the necessary cognitive mechanisms for reading are not dependent on vision *per se*, reading can be achieved. In contrast, children with autism are postulated to have no sensory impairment, but to have a central cognitive deficit in their theory of mind. The task, then, in trying to teach them to employ a theory of mind, may be considerably harder than teaching a blind child to read, since changing understanding is involved.

The central questions our study is attempting to address are:

1. Can mental-state concepts be taught and, if so, which techniques facilitate this, how much teaching is necessary, and how long will such learning persist?

2. Are some mental-state concepts (for example, pretence, or desire) easier for children with autism to learn than others (for example, knowledge, and belief)? If so, are mental-state concepts only acquired in a strict sequence? That is, does acquisition of one concept (for example, pretence) *always* precede acquisition of another (for example, belief)? If so, is this because one is necessary for the other?

3. If mental-state concepts are acquired during understanding of particular examples of behaviour, do these generalize to allow the child to understand novel examples of behaviour? If so, are mental-state concepts that are acquired through explicit teaching used in the same way as those acquired more naturally?

4. Does acquisition of mental-state concepts lead to change in the child's own social and communicative behaviour, and, if so, which aspects of behaviour change?

5. Which factors might account for some children with autism acquiring mental-state concepts, and some not?

6. Finally, does teaching mental-state concepts have any incidental effects on the acquisition of concepts or reasoning?

One approach: teaching underlying principles governing mental states

Our study attempts to analyse mental-state understanding into simple *principles*, and then considers if these principles can be taught through intensive training with many examples, using a variety of media. This approach makes the assumption that mental-state understanding can indeed be reduced to simple principles. For a normal child, these principles do not appear to be explicitly taught, and they may not even be explicitly represented; but the good performance on tests of mental state comprehension provides evidence that they understand such principles (Wellman 1990). For children with autism, since they do not seem to acquire them naturally, such principles may need to be made explicit.

Examples of such principles for some fundamental mental states (know, desire, and pretend) are given here:

1. *Perception causes knowledge.* **A person will know x only if s/he saw or heard about x.** (Example: Snow White doesn't know the apple is poisoned because she didn't see the woman the poison into it.)

2. *Desires are satisfied by actions or objects.* **If a person wants x, s/he will look for or obtain x. Conversely, if a person doesn't want x, s/he will refuse or avoid x.** (Example: Hansel and Gretel want their father, so they look for him. They don't want the witch to catch them, so they run away when she comes.)

3. *Pretence involves object-substitution or outcome-suspension.* **When a person pretends x, s/he does x without the usual objects or consequences, just for fun.** (Example: Alan holds a banana to his ear. He is pretending to talk on the telephone.)

These are just some of the principles that govern different mental states. Others can be easily articulated by examining our own 'common-sense' folk psychology. Wellman (1990) gives a good survey of these. Such principles can be developed into more complex forms (for example, when you deceive someone *you make them think something false*; or, when you want something, you don't *always* try to obtain it *directly*.) We assume that a first attempt at teaching a theory of mind to children with autism should begin by teaching the principles in their *simplest* form, in as concrete a manner as possible, using a large number of examples, with the aim that each principle is learnt and generalizes to new instances. The examples sketched above give an idea of the form such simple principles might take.

In our study, we focus on teaching a range of mental states in this way, including belief, desire, knowledge, pretence, deception, and emotion. The principles governing each mental state are taught using a range of techniques, including doll-play, drama, language, pictures, and even computer-graphics, following Swettenham (1992), in order to maximize the possibility of one of these media being motivating for any given child. Only children with autism whose verbal mental age is above three- and-a-half years of age are being given this intensive tuition, as most of the techniques are derived from the developmental literature from normal three-year to four-year-olds. We are therefore not attempting to answer the question of whether mental-state concepts can be taught to children with a verbal mental age lower than this.

Aside from the questions listed earlier, we are also particularly interested to see if acquisition of a principle acts as a cornerstone in the construction of a *theory* of mind — that is, do mental-state concepts, if they are acquired, take on theory-like properties for the child, as Wellman (1990) has argued occurs with normal children? Some initial studies (Armstrong and Whiten 1991; Starr 1992; Stromm 1991; Swettenham 1992) suggest that, at least when the mental-state concept of belief is taught to children with autism, some progress is seen, although these studies are insufficient to answer many of the interesting questions outlined earlier. Similar studies with normal children suggest too that such teaching can lead to acquisition of the concepts of belief and knowledge at earlier ages than is usually seen (Taylor 1988; Swettenham 1992). Our own study involves the teaching of a range of different mental-state concepts. Those interested in further details of this study, and its results, should contact the authors directly.

THE THEORY OF MIND DEFICIT: IMPLICATIONS FOR DIAGNOSIS

The second area of clinical relevance we consider is to what extent the experimental work in autism and theory of mind may aid diagnosis. In addition, we consider how it may inform theories about precursors to theory of mind development.

Current diagnostic techniques are based exclusively on the presence or absence of *behavioural* criteria. DSM-IIIR (1987) criteria, for example, specify precisely the numbers of items that must be present before the diagnosis of autism can be made. Thus, a total of at least 8 out of 16 symptoms must be identified, 2 or more of which must relate to social impairments, 1 or more to communication deficits, and 1 or more to the presence of obsessional or ritualistic behaviours. While this reliance on behavioural criteria has produced clearer operational rules for diagnosis (Rutter and Schopler 1987), it nevertheless raises a number of problems. For example, although it is relatively easy to identify delays or 'absences', it is much more difficult to judge when a behaviour is *qualitatively* abnormal. And since behaviours can resemble each other while having entirely different cognitive bases, it also means that diagnostic systems that are exclusively behavioural in nature risk confusing apparently similar conditions.

It may be that these and other related problems could in part be overcome by using cognitive tests in conjunction with behavioural tests in diagnosis. Consider, for example, a child who is socially unresponsive. If we could determine the reason for the social unresponsiveness (is it due to anxiety, or prosopagnosia, or an impaired theory of mind, etc.?), this might add to diagnostic precision. Performance on a series of false-belief and other neuropsychological tests could be useful in this way (Prior and Hammond 1990; Ozonoff *et al.* 1991). It should be emphasized, however, that this is at present only a suggestion; at the time of writing no studies have attempted to include such specific cognitive (or neuropsychological) tests in psychiatric diagnosis. Nor have there been any studies on the sensitivity and specificity of theory of mind tests with different clinical populations, and these are much needed.

Of course, false-belief tests can only be used meaningfully with children whose mental age is above four years of age (Wimmer and Perner 1983; Baron–Cohen 1990). This does not rule out their diagnostic potential, but it reminds us that their role will be confined to relatively 'late' diagnosis. Since one aim of diagnosis should also be to improve early detection of disorders, it is of interest to consider if *precursors* of theory of mind deficits in infancy can also be used in early detection. In the next section we describe a recent attempt at this.

The use of 'theory of mind precursors' in detecting autism in infancy

Recent work has suggested that two possible precursors to the theory of mind deficit in autism are pretend-play impairments, and joint-attention deficits (see Leslie 1987; Baron–Cohen 1991; and see Chapters 4, 5, 9, and 18, this volume). In the case of pretend-play, this is identified as distinct from *functional* play, which is not specifically impaired in autism (Ungerer and Sigman 1981; Baron–Cohen 1987). In the case of joint-attention behaviours, these include gaze-monitoring, 'showing' gestures, and pointing (Sigman *et al.* 1986; Mundy *et al.*, this volume, Chapter 9.) One specific type of pointing, *protodeclarative* pointing, seems particularly impaired in autism, while *protoimperative* pointing is not (Baron–Cohen 1989).

Two important questions are: (a) Are pretend-play and protodeclarative pointing really precursors to a theory of mind? How can such claims about precursor status be tested? and (b) If such precursors do predict development of a theory of mind, do they also predict cases of autism? Baron–Cohen *et al.* (1992) investigated the second of these questions by employing a new instrument, the *Checklist for Autism in Toddlers (CHAT)*, shown in Fig. 21.1. This was administered by General Practitioners or Health Visitors

THE CHAT   (Medical Research Council Project)

To be used by GPs or Health Visitors during the 18-month developmental check-up.

Child's name: .... Date of birth: .... Age: ....

Child's address: ....

SECTION A: ASK PARENT:

1. Does your child enjoy being swung, bounced on your knee, etc.?   YES   NO

2. Does your child take an interest in other children?   YES   NO

3. Does your child like climbing on things, such as up stairs?   YES   NO

4. Does your child enjoy playing peek-a-boo/hide and seek?   YES   NO

5. Does your child ever PRETEND, for example, to make a cup of tea
    using a toy cup and teapot, or pretend other things?   YES   NO

6. Does your child ever use his/her index finger to point,
    to ASK for something?   YES   NO

7. Does your child ever use his/her index finger to point,
    to indicate INTEREST in something?   YES   NO

8. Can your child play properly with small toys (e.g.: cars or bricks)
    without just mouthing, fiddling, or dropping them?   YES   NO

9. Does your child ever bring objects to you (parent),
    to SHOW you something?   YES   NO

SECTION B: GP or HV OBSERVATION:

i. During the appointment, has the child made eye-contact with you?   YES   NO

ii. Get child's attention, then point across the room at an interesting object and say
    'Oh look! There's a (name a toy)!' Watch child's face.
    Does the child look across to see what you are pointing at?   YES[1]   NO

iii. Get the child's attention, then give child a miniature toy cup and teapot and say
    'Can you make a cup of tea?'
    Does the child pretend to pour out tea, drink it, etc.?   YES[2]   NO

iv. Say to the child 'Where's the light?' or 'Show me the light.'
    Does the child POINT with his/her index finger at the light?   YES[3]   NO

v. Can the child build a tower of bricks? (If so, how many?)   YES   NO
    (Number of bricks: ....)

[1](To record YES on this item, ensure the child has not simply looked at your hand, but has actually looked at the object you are pointing at.)

[2](If you can elicit an example of pretending in some other game, score a YES on this item.)

[3](Repeat this with 'Where's the teddy?' or some other unreachable object, if child does not understand the word 'light'. To record YES on this item, the child must have looked up at your face around the time of pointing.)

Figure 21.1. The Checklist for Autism in Toddlers (CHAT). Reproduced from Baron–Cohen *et al.* (1992) with permission.

during the routine 18-month-old developmental check-up (taking about 20 minutes to complete).

As can be seen, this schedule checks for the presence of pretend-play and joint-attention behaviours, among other things. This study found that, while some of a group of randomly selected toddlers at 18 months ($n = 50$, age 17–21 months) still lacked protodeclarative pointing, and some lacked pretend-play, none lacked both. In this study, a group of siblings of already diagnosed children with autism ($n = 41$, age 18–21 months) were also screened with the CHAT, on the assumption that 2–3 per cent of them would, for genetic reasons, themselves develop autism (Folstein and Rutter 1988). The key point of interest is that four children in this *high-risk* group lacked both pretend-play and joint-attention at eighteen months, and these went on to receive a diagnosis of autism at the age of thirty months. Overall, of the 91 toddlers screened, the other 87 were free of any psychiatric problems at thirty months, and none of these 87 cases had failed on both pretend-play and joint-attention at eighteen months.

These findings offer support for the claims that pretence and joint-attention may be useful in the early detection of autism. A larger, epidemiological study is now under way (screening 16 000 eighteen-month-olds in the south-east of England) to evaluate the diagnostic and predictive power of these behaviours (Baron–Cohen, Cox, Baird, Swettenham, Nightingale, Morgan, Drew, and Charman, 1994.) Of critical importance is the fact that this larger study is also *longitudinal*—those infants at eighteen months who fail the CHAT will be followed up at the age of four to five, to determine if these behaviours do stand as precursors in the development of a theory of mind. Such a prospective, longitudinal design is essential in testing precursor relationships (Bradley and Bryant 1983).

CONCLUSIONS

The theory of mind hypothesis has been used as an explanatory tool for understanding fundamental cognitive deficits in autism. The present chapter explores some issues of clinical relevance from this work. In particular, it considers questions about whether mental-state concepts can be taught, the effects of such teaching, the role of such cognitive tests in diagnosis, and the investigation of early precursors of theory of mind deficit. Such clinical research is really still in its infancy; but we hope that this chapter may stimulate further research into these questions.

Acknowledgements

The authors were supported during the writing of this work by grants from the Bethlem–Maudsley Research Fund and the Mental Health Foundation.

In addition, the first author was supported by a grant from the Medical Research Council. We are grateful to Cathy Lord, Donald Cohen and Julie Hadwin for their comments on an earlier version of this chapter.

REFERENCES

Armstrong, K., and Whiten, A. (1991). Training false-belief understanding in adults with autism. Unpublished MS, Department of Psychology, University of St Andrews.

Avis, J. and Harris, P. (1991). Belief–desire reasoning among Baka children: evidence for a universal conception of mind. *Child Development*, **62**, 46–7.

Baron–Cohen, S. (1987). Autism and symbolic play. *British Journal of Developmental Psychology*, **5**, 139–48.

Baron–Cohen, S. (1989). Perceptual role-taking and protodeclarative pointing in autism. *British Journal of Developmental Psychology*, **7**, 113–27.

Baron–Cohen, S. (1990). Autism: a specific cognitive disorder of 'mind-blindness'. *International Review of Psychiatry*, **2**, 79–88.

Baron–Cohen, S. (1991). Precursors to a theory of mind: understanding, attention in others. In *Natural theories of mind* (ed. A. Whiten). Basil Blackwell, Oxford.

Baron–Cohen, S., Allen, J., Gillberg, C. (1992). Can autism be detected at 18 months? The needle, the haystack, and the CHAT. *British Journal of Psychiatry*, **161**, 839–43.

Baron–Cohen, S., Cox, A., Baird, G., Swettenham, J., Nightingale, N., Morgan, K., Drew, A., and Charman, T. (1994). Screening for autism at 18 months of age. Unpublished MS, Institute of Psychiatry, London.

Bradley, L., and Bryant, P. (1983). Categorizing sounds and learning to read: a causal connection. *Nature*, **301**, 419–521.

Brady, M. P., Shores, R. E., McEvoy, M. A., Ellis, D., and Fox, J. J. (1987). Increasing social interactions of severely handicapped autistic children. *Journal of Autism and Developmental Disorders*, **17**, 375–90.

Breen, C., Haring, T., Pitts-Conway, V., and Gaylord-Ross, R. (1985). The training and generalisation of social interaction during breaktime at two job sites in the natural environment. *Journal of the Association of Persons with Severe Handicaps*, **10**, 41–50.

Dewey, D., Lord, C., and Magill, J. (1988). Qualitative assessment of the effect of play materials in dyadic peer interactions of children with autism. *Canadian Journal of Psychology*, **42**, 242–260.

DSM-IIIR (1987). *Diagnostic and Statistical Manual of Mental Disorders*, revised 3rd edn. American Psychiatric Association, Washington, DC.

Fagan, S., Gray, K., Factor, D. C. (1985). Foxx's 'Stack the Deck': teaching social skills to mentally handicapped adolescents. *Journal of Child Care*, **3**, 29–34.

Fantusso, J. W., Helgeson, D. C., Smith, C., and Barr, D. (1989). Eye-contact skill training for adolescents with developmental disabilities and severe behaviour problems. *Education and Training of the Mentally Retarded*, **24**, 56–62.

Favell, J. (1983). The management of aggressive behavior. In *Autism in adolescents and adults* (ed. Schopler E. and G. Mesibov). Plenum, New York.

Folstein, S. and Rutter, M. (1988). Autism: familial aggregation and genetic implications. *Journal of Autism and Developmental Disorders*, **18**, 3–30.

Frankel, R.M., Leary, M., and Kilman, B. (1987). Building social skills through pragmatic analysis: assessment and treatment implications for children with autism. In *Handbook of Autism and Developmental Disorders* (ed. D.J. Cohen, and A.M. Donnellan). Wiley, New York.

Gaylord-Ross, R. (ed.) (1989). *Integration strategies for students with handicaps.* P.II. Brookes, Baltimore.

Groden, J., and Cautela, J. (1988). Procedures to increase social interactions among adolescents with autism: a multiple-baseline analysis. *Journal of Behavior Therapy and Experimental Psychiatry*, **19**, 87–93.

Hadwin, J., Baron–Cohen, S., Howlin, P., and Hill, K. (1994). Concepts of emotion, belief, and pretence. To what extent can they be taught to children with autism? Unpublished MS, Institute of Psychiatry, London.

Howlin, P. (1986). An overview of social behavior in autism. In *Social behaviour in autism* (ed. E. Schopler and G. Mesibov). Plenum, New York.

Howlin, P. (1989). Changing approaches to communication training with autistic children. *British Journal of Disorders of Communication*, **24**, 151–68.

Howlin, P., and Rutter, M. (1987). *Treatment of autistic children*. Wiley, Chichester.

Leslie, A. (1987). Pretence and representation: the origins of 'theory of mind'. *Psychological Review*, **94**, 412–26.

Lord, C. (1984). Development of peer relations in children with autism. In *Applied Developmental Psychology* (ed. F. and D. Morrison Keating). Academic Press, New York.

Matson, J., Manikam, R., Coe, D., Raymond, K., and Taras, M. (1988). Training social skills to severely mentally retarded multiply handicapped adolescents. *Research in Developmental Disabilities*, **9**, 195–208.

Mesibov, G. (1984). Social skills training with verbal autistic adolescents and adults: a program model. *Journal of Autism and Developmental Disorders*, **14**, 395–404.

Ozonoff, S., Pennington, B., and Rogers, S. (1991). Executive function deficits in high-functioning autistic individuals: relationship to theory of mind. *Journal of Child Psychology and Psychiatry*, **32**, 1081–1106.

Park, H.S., and Gaylord-Ross, R. (1989). A problem-solving approach to social skills training in employment settings with mentally retarded youth. *Journal of Applied Behavior Analysis*, **22**, 373–80.

Plienis, A., Hansen, D., Ford, F., Smith, S., Stark, L., and Kelly, J. (1987). Behavioral small group training to improve the social skills of emotionally disordered adolescents. *Behavior Therapy*, **18**, 17–32.

Prior, M., and Hammond, W. (1990). Neuropsychological testing of autistic children through exploration with frontal lobe tests. *Journal of Autism and Developmental Disorders*, **20**, 581–90.

Rutter, M. (1983). Cognitive deficits in the pathogenesis of autism. *Journal of Child Psychology and Psychiatry*, **24**, 513–31.

Rutter, M., and Schopler, E. (1987, eds). *Autism: a reappraisal of concepts and treatment*. Plenum, New York.

Schopler, E., and Mesibov, G. (eds) (1986). *Social behaviour in autism*. Plenum, New York.

Schuler, A. (1989). The socialization of autistic children. Paper presented at the International Conference on Educational Issues in Autism, Mons, Belgium.

Sigman, M., Mundy, P., Ungerer, J., and Sherman, T. (1986). Social interactions of autistic, mentally retarded, and normal children and their caregivers. *Journal of Child Psychology and Psychiatry*, **27**, 647–56.

Spence, S. (1980). *Social skills training with children and Adolescents*. NFER–Nelson, Windsor.

Strain, P., Kerr, M., and Rauland, E. (1979). Effects of peer-mediated social initiations and prompting/reinforcement procedures on the social behavior of autistic children. *Journal of Autism and Developmental Disorders*, **9**, 41–54.

Stromm, E. (1991). *Training false-belief understanding with autistic children*. Unpublished MA thesis, Institute of Education, London.

Starr, E. (1992). *Theory of mind and autism: teaching the appearance–reality distinction*. Unpublished Ph.D. Thesis, University of Alberta, Canada.

Swettenham, J. (1992). *The autistic child's theory of mind: a computer-based investigation*. Unpublished Ph.D. Thesis, University of York.

Taras, M., Matson, J., and Leary, C. (1988). Training social interpersonal skills in two autistic children. *Journal of Behavior Therapy and Experimental Psychiatry*, **19**, 275–80.

Taylor, M. (1988). The development of children's ability to distinguish what they know from what they see. *Child Development*, **59**, 703–18.

Ungerer, J., and Sigman, M. (1981). Symbolic play and language comprehension in autistic children. *Journal of the American Academy of Child Psychiatry*, **20**, 318–37.

Wellman, H. (1990). *The child's theory of mind*. Bradford Books/MIT Press, Cambridge, Mass.

Williams, C. (1989). A social skills group for autistic children. *Journal of Autism and Developmental Disorders*, **19**, 143–55.

Wimmer, H., and Pemer, J. (1983). Beliefs about beliefs: representation and constraining function of wrong, beliefs in young children's understanding of deception. *Cognition*, **13**, 103–28.

22

Thinking and relationships: mind and brain (some reflections on theory of mind and autism)

MICHAEL RUTTER AND ANTHONY BAILEY

In 1985, Baron-Cohen, Leslie, and Frith reported evidence that autistic children lacked the ability to appreciate that other people's beliefs might differ from their own, and went on to argue that a theory of mind deficit might constitute the core of autism. Since their seminal paper, there has been a veritable explosion of research testing this hypothesis, the key findings of which are summarized in this volume. In many respects, one of the most notable features of the way in which the idea caught the imagination of a huge number of researchers is the extraordinary breadth of the field in which it excited immense interest. As this volume illustrates, the subject has attracted the attention of developmentalists quite as much as psychopathologists, and of social as well as cognitive investigators. It is no exaggeration to claim that theory of mind research has become almost an industry in its own right, with a resulting flow of books on the topic, as well as a flood of scientific papers. Why did the theory of mind hypothesis create such a stir, and what is its significance for the understanding of the nature of autism and of normal social development?

The reasons are not difficult to find. To begin with, autism is a psychiatric disorder with a wide range of symptomatology, and here was a suggestion that all of this might be explicable in terms of a single narrowly defined psychological process, albeit one with widely pervasive effects. For obvious reasons, the possibility of a simple explanation for all the complexity of autism had to be exciting. A particular attraction of the hypothesis was that it carried the promise that it might make sense of the *combination* of social and cognitive deficits in autism. Autism had been defined in terms of its distinctive and peculiar pattern of social impairment (Kanner 1943), and yet both psychological and clinical studies had emphasized the importance of cognitive deficits and abnormalities (Hermelin and O'Connor 1970; Rutter 1979). The theory of mind hypothesis was important because it claimed to account for the social abnormality in terms of a cognitive deficit.

These first two reasons perhaps explain why clinical investigators were so taken with the idea. However, the attraction for developmentalists had other roots. One of the problems in studying the mechanisms underlying the interconnections between different psychological processes in normal development is that they intercorrelate so highly. In order to examine mechanisms, it is necessary to pull the skills apart by searching for circumstances in which one cognitive skill is markedly advanced or retarded compared with others. Only then is it possible to test the effects of that skill on other functions. That is where the study of abnormal groups is so useful. This bringing together of clinical and developmental perspectives, as exemplified by the field of developmental psychopathology, is informative for the understanding of *both* normal development *and* psychiatric disorders (Sroufe and Rutter 1984; Rutter 1986, 1989, in press).

Yet, clearly there was nothing new about the suggestion that a basic cognitive deficit underlay the social abnormalities that characterize autism. Hermelin and O'Connor (1970) had introduced the idea some twenty years earlier; and, indeed, it is recognized that theory of mind researchers owe a great debt to these two pioneers, who were the first to show that supposedly 'untestable' autistic children could be tested if the tasks were presented in the right way; that an experimental approach to their psychological deficits was exceedingly fruitful; and that the findings indicated the need to reconceptualize the phenomena of autism in terms of the sequelae of abnormalities in developmental processes, rather than in terms of symptoms of an acquired mental illness or psychosis. Their findings were important in laying to rest the misguided view of autism as an unusually early form of schizophrenia (Rutter 1972). The origins of the theory of mind concept and of the style of investigation (using experimental methods derived from the study of normal development), therefore, were evident in much earlier research. Why, then, didn't that earlier research make the same impact?

Three main reasons may be put forward. First, the *Zeitgeist* was quite different. The fields of developmental psychology and child psychiatry were poles apart in the 1960s and 1970s (with a few notable exceptions), and investigators in the one tended not to read publications in the other. Also, in that era, psychology tended to remain remarkably enclosed within national boundaries. It is striking how little, at that time, North American psychologists referred to European research. Fortunately, that is no longer the case. There is a general recognition that science is, and must be, an international endeavour.

However, there is one crucial respect in which the theory of mind hypothesis constituted a substantial advance over earlier work. The Hermelin and O'Connor findings had highlighted the role of deficits in sequencing, abstraction, and the use of meaning. Clearly, such deficits were likely to have major implications for social functioning; but it was not at all apparent *how* they might account for the social impairment. Indeed, in the 1960s and

1970s, there had been very little investigation of the social features of autism, and it was not at all evident which particular social abnormalities had to be explained. By the 1980s that had changed radically. Stimulated by the work of Hobson (see Chapter 10, this volume) investigators had come to appreciate that the key challenge lay in the need to understand the nature of the social deficits, and that the implication was that attention needed to be paid to social cognitive processes. By a curious coincidence, Rutter's (1983) review paper putting that argument, published in the same year as the Wimmer and Perner (1983) study that set in motion the theory of mind work, used a clinical example of an autistic adult who complained that, unlike normal persons, he could not 'read minds'.

The theory of mind hypothesis constituted a major step forward just because it moved from general connections between cognition and social development to specific mechanisms. By the mid-1980s it had become apparent that a lack of social reciprocity was particularly characteristic of autism. It seemed to follow that, for such reciprocity to develop, children must be able to detect and understand other people's cues. If they could not do that, how could to-and-fro interactions in which each person's response built upon and developed the other person's communication, come about? Four main alternatives seemed to present themselves; an inability to understand the nuances of the meaning of language (Rutter 1968; Rutter *et al.* 1971); an inability to decipher socio-emotional cues (Hobson, this volume, Chapter 10); a failure to appreciate other people's thoughts—the theory of mind hypothesis (Baron-Cohen *et al.* 1985); and an impairment in early-appearing cognitive skills such as imitation (Rogers and Pennington 1991; Meltzoff and Gopnik, this volume, Chapter 16) or joint attention (Mundy *et al.*, this volume, Chapter 9). The relative merits and demerits of these alternatives are considered later in the chapter; but, for the moment, the key point is that the theory of mind notion seemed to go to the very heart of social reciprocity.

Of course, theory of mind was not new in its postulate that cognitive skills might underlie normal social development. For example, Kagan's (1982) research had shown how, towards the end of the second year of life, children first become aware of other people's standards and expectations. He argued, persuasively, that this cognitive transition was necessary for moral development. Naturally, the specific content of children's morality would be influenced by the culture in which they grew up; but the fact that moral development occurred at all was dependent on the availability of specific cognitive skills. One of the key requirements for future research will be to integrate theory of mind with the rest of the broader field of social cognition. As relative outsiders to this body of work, it seems to us unfortunate that so often advances are compartmentalized and separate from the rest of the field. Progress in science is characterised as much by the bringing together of apparently diverse approaches as by an 'in-depth' investigation of a narrow topic.

Finally, it is also worth noting that the theory of mind approach is part of a tradition in its bridging of normality and psychopathology. Hermelin and O'Connor (1970) had shown the way in relation to autism, and they had also done so in their studies of mental retardation (O'Connor and Hermelin 1963) and of blind and deaf children (O'Connor and Hermelin 1978). In addition, of course, to give but one example, memory researchers have done the same in their differentiations between episodic and semantic memory (Tulving 1983, 1984) and between short-term and long-term memory (Baddeley 1986, 1988). In both cases, the study of brain-injured adults was crucial in showing the functional independence of the different facets of memory. There remains a need to integrate these different uses of the evidence from pathologic groups in order to study normal psychological processes.

WHAT HAS TO BE EXPLAINED IN AUTISM

Perner (chapter 6, this volume) argues that 'the objective is not to explain the autistic syndrome as a whole but to explain what is specific to autism'. That narrowing of aims may be accepted provided that there is a careful analysis of the features that are indeed specific, and provided that the concept of specificity incorporates all those elements that seem to be intrinsic to the syndrome, and not just those that are found only in autism. These include social features, language abnormalities, stereotyped/repetitive behaviours, and mental handicap.

Social features

In that connection, it is necessary to appreciate both the characteristics of the social abnormalities in autism and also the other diagnostically-specific features. With respect to the former (see Lord, this volume, Chapter 14), three main aspects stand out. As has already been noted, the lack of social reciprocity seems particularly characteristic. Some autistic individuals are withdrawn, and some are intrusive; but all seem to lack social reciprocity, and all seem severely impaired in their ability to develop loving relationships on the basis of interpersonal interactions (Rutter *et al.* 1992). The quality of their attachment relationships is not normal, despite the fact that they do show attachment (Mundy and Sigman 1989). Second, autistic individuals are impaired in both their production of emotions and their understanding of other people's emotions. It is not difficult to see how a theory of mind deficit might account for most of these features, but it is not quite so apparent why it should interfere with the *expression* of emotions (Rutter 1991). Of course, much depends on the specifics of the deficit in emotional expression, and those have been little investigated so far. From the studies undertaken to date, it seems that autistic children *do* show a range of positive and negative emotions in their facial expressions and possibly in their tone of voice; but there is a deficit in those emotions that are more reliant on cognitive processes (such as embarrassment or shame); there is a poor integration of different facets of emotional expression; and the expression of all emotions tends to display both unusual qualities and a lack of fine discrimination (see Lord, this volume, Chapter 14, and Mundy *et al.*, this volume, Chapter 9).

The third key feature of the social abnormalities in autism concerns their timing and persistence. For obvious reasons, it is difficult to obtain good prospective data on the age at which the social deficits in autism are first apparent. Nevertheless, it is evident that most autistic children's social functioning is clearly abnormal by the age of two years, and in some cases it is so before the end of the first year. As Lord, and also Klin and Volkmar (this volume, Chapters 14 and 15) point out, that poses a problem for the theory of mind hypothesis, because it seems that is it is not until the age of four that children develop a clearly articulated appreciation that people's actions are necessarily framed by their beliefs, although major elements of this realization are evident by the age three (see Wellman, this volume Chapter 2). Of course, in their everyday social behaviour, even two-year-olds seem to show 'mentalizing' skills that are not evident on formal experimental testing (Dunn 1990). Also, it is evident that, even in the first year, children attend to people and objects, simultaneously gazing back and forth, and this appears to constitute a first step towards understanding intentionality. The emergence of pointing and showing towards the end of the first year and the developing ability to read others' desires and intentions speaks to the growth of young children's mentalizing skills.

However, all of this falls far short of what is meant by theory of mind and, in particular, it lacks the 'theory' component. It is these considerations that lead Mundy *et al.* (Chapter 9, this volume) to argue that the deficit in autism is more likely to lie in impaired joint attention, and to involve an affective, as well as a cognitive, component. The exponents of the theory of mind hypothesis get around this difficulty by arguing that a toddler's understanding of goals and attention constitutes a precursor of theory of mind (Baron-Cohen, this volume, Chapter 4). Such a precursor hypothesis requires further testing.

As Gómez, Sarria, and Tamarit (Chapter 19, this volume) point out (see also Hay and Angold, in press), the concept of precursor has two very different meanings. On the one hand, there is the notion of an event that simply precedes another in a meaningful sense. On the other hand, there is the very different sense of something that constitutes an earlier stage of what is essentially the same process (Gómez *et al*. discuss this in chemical terms; but, medically speaking, perhaps the different stages of carcinogenesis constitute a closer analogy). In order to test this latter type of precursor postulate, it would be necessary to show that the one is functionally dependent on the other, and that both show a similar pattern of correlates with external variables. As the memory research shows (Baddeley 1986, 1988; Tulving 1983, 1984) the hypothesis may also be put to the test by the use of functional brain imaging (to determine whether the skills are dependent on the same part of the brain). This research is possible, but has yet to be undertaken.

The second aspect of timing is provided by the evidence that autistic individuals of normal non-verbal intelligence continue to show serious and obvious social impairments in adult life (Ruter *et al*. 1992). The relevance of this finding is that at least some more intellectually able autistic individuals *do* pass four-year-old theory of mind tasks, yet fail second-order theory of mind tasks (see Baron-Cohen, this volume, Chapter 4). This indicates a continuing impairment in theory of mind skills, but leaves unexplained why autistic adults may lack social skills clearly evident in normal four-year-olds (Bailey and Rutter, in press). In seeking to understand the role of a theory of mind deficit in explaining the social abnormalities of autism, it should however be borne in mind that in some respects autistic children's theory of mind skills are deviant as well as impaired (see Baron-Cohen, and also Leslie and Roth, this volume, Chapters 4 and 5). That is, they show an order of difficulty with theory of mind tasks that differs from that seen in normal individuals.

Language abnormalities

In addition to the social impairments that characterize autism, it is obvious that autism involves serious abnormalities in language development. This was evident in Kanner's first papers (1943, 1946), and it has been confirmed repeatedly since (Rutter *et al*. 1975; Rutter 1978; Rutter *et al*. 1992; Tager-Flusberg, this volume, Chapter 7). Most autistic children show a marked delay in language development; they manifest an unusually wide impairment in language skills; and they exhibit a deviant pattern of language. Clinical reports have tended to emphasize qualitatively abnormal language features, such as pronominal reversal, neologisms, delayed echolalia, and idiosyncratic unusual usages of languages (what Kanner somewhat misleadingly term 'metaphorical' language).

Interestingly, Clark (1983) has described how young children, in the course of developing language, both overextend and underextend the meanings of words, and even develop their own made-up words. It seems that this constitutes part of a process of developing and trying out language concepts. Those that are confirmed are retained, and those that are disconfirmed are rapidly lost. In this way, so-called 'abnormal' language forms develop, but are quickly phased out as a result of feedback from other people. Rutter (1987) has suggested that it may be just because of autistic children's impaired ability to make use of feedback that those abnormal language forms persist. So far, the suggestion has not been put to the test in rigorous fashion; but certainly it is plausible that a failure of feedback processes might play at least some role.

However, it is also clear that the language deficit in autism does not mainly lie in the syntactic or semantic aspects of language, but rather in pragmatic and prosodic features (see Tager-Flusberg, and also Loveland and Tunali, this volume, Chapters 7 and 12). That is, autistic individuals find it very difficult to maintain an ongoing topic of conversation; they make little use of 'wh' questions; they seem to lack an understanding that conversation ought to entail an exchange of information; they are impaired in adapting their communication to different social contexts; and they are limited in their use of prosody to communicate social and affective information. They have been found to differ from Down's syndrome individuals in their calls for joint attention and their references to cognitive mental states (but, interestingly, not in references to the mental states of desire or emotion).

These findings are important in several different connections. First, the distinctive pattern argues against a straightforward delay in language development, and instead suggests the need to seek the explanation in an abnormality or impairment in non-linguistic cognitive processes. Second, the findings point to social and communicative aspects of language, and hence to the need to identify cognitive deficits that span language and socialization. A theory of mind deficit would seem to be in a strong position to make that link. However, once again, there is a need to invoke precursors of a theory of mind in order to account for language impairment in the second and third years of life.

Of course, before placing emphasis on the need for a theory of mind hypothesis to account for the language abnormalities seen in autism, it is necessary to ask whether there is evidence that the abnormalities are both specific to autism and basic to the disorder. The answer in both cases is a clear 'yes'. The pattern of language differs strikingly from that seen in either mental retardation or developmental disorders of receptive language (Tager-Flusberg, this volume, Chapter 7; Rutter 1978; Eales, in press). Moreover, language impairment is relatively resistant to therapeutic interventions (Howlin and Rutter 1987); the degree of language impairment is a strong predictor of outcome (Rutter 1979); and language deficits persist into adult life (Rutter *et al.* 1992) and also show a strong association with social functioning (Rutter *et al.* 1992). It is also pertinent that findings from both twin and family studies show that autism is genetically associated with both language and social abnormalities, and especially with their combination

(Bailey *et al.* in press; Rutter 1991; Rutter *et al.*, in press). The fundamental nature of that association is indicated by the finding that the degree of the verbal deficit is significantly associated with the strength of the familial loading.

It may be concluded with confidence that language impairment and abnormalities constitute an intrinsic specific feature of autism, and hence require to be accounted for by the theory of mind hypothesis. So far, this issue has not been tackled extensively in the empirical studies, though the initial studies (Perner *et al.* 1989; Baron-Cohen 1988; Tager-Flusberg, this volume, Chapter 7) suggest that the hypothesis may stand up to this challenge.

Stereotyped, repetitive patterns of behaviour

The third area of abnormality in autism concerns the pervasive tendency for affected individuals to exhibit stereotyped, repetitive patterns of behaviour. In his first description of the syndrome, Kanner (1943) referred to it as an obsessive desire for the preservation of sameness. This constitutes one aspect of the stereotyped behaviour shown by some autistic individuals; but subsequent research has revealed a wider range of stereotyped behaviours — including quasi-obsessive routines and rituals, abnormal attachments to objects, abnormal preoccupations, circumscribed interest patterns, particular motor stereotypies, and unusual idiosyncratic responses to sensory stimuli (Rutter and Lockyer 1967; Bartak *et al.* 1975; Cantwell *et al.* 1989; Le Couteur *et al.* 1989; Frith 1989).

Not only are these behaviours very common in autism; they also serve to differentiate autism from mental retardation, developmental disorders of receptive language, and other psychiatric disorders. In addition, an early study by Frith (1972) showed experimentally that autistic children tended to impose stereotyped patterns in their play. Boucher (1977), similarly, showed that autistic children had a tendency to follow repetitive rules in a maze task. Follow-up studies into adult life have shown that these perseverative features have a strong tendency to persist, although the particular manner of their manifestation may change (Rutter 1970; Goode, Rutter, and Howlin, unpublished data).

Once again, it must be concluded that these features constitute a specific and basic feature of autism that has to be explained by the theory of mind hypothesis. As Harris (Chapter 11, this volume) points out, not only have these features been subjected to remarkably little research, it is very difficult to see how they could be explained by a theory of mind deficit. As he notes, Baron-Cohen's (1989) suggestion that they serve to reduce the anxiety generated by a failure to understand social situations has so far not been tested empirically. Clark and Rutter (1979) did find that, when autistic children could not cope with cognitive tasks, they tended to adopt patterns of stereotyped responses. Perhaps this could be interpreted as indicating that anxiety might serve to intensify stereotyped behaviours. However, even if anxiety were found to increase stereotyped patterns (and it has not been shown that it does), it would still have to be shown that that anxiety was generated by the children's failure to understand the social world.

Frith (1989) suggested that autism may be associated with a weak central *cohesive* force. Ordinarily, people tend to view things as a whole, in an integrated fashion. Thus, faces are recognized in a holistic fashion rather than as a meaningless perceptual pattern, possibly explaining why it is surprisingly difficult to recognize faces when they are presented upside-down. Similarly, jigsaw puzzles are much easier to complete when you can see the parts of the picture on the puzzle pieces. However, compared with MA-matched controls, autistic children show superior skills on these tasks. Similarly, autistic children are better than controls in recognizing figures that are embedded in a pattern (Shah and Frith 1983). Also, whereas prior segmentation improves block design performance in non-autistic individuals, it does not do so in autistic people (Frith 1989). The degree to which the hypothesized weak central cohesive force is specifically characteristic of autism is not known, and even less is known about its possible explanatory role in stereotyped, repetitive behaviour.

Another feature of autism consists of the marked lack of spontaneous pretend-play (see Harris, this volume, Chapter 11). Leslie (1987) argued that this was intimately bound up with a theory of mind deficit in terms of an impaired decoupling mechanism. Harris (Chapter 11, this volume) argues that this is incorrect, and that, instead, both the stereotyped patterns and the lack of spontaneous pretend-play derive from a deficiency in internal controls that allows a detachment from environmental cues; in other words, a deficit in what has come to be termed 'executive planning'. It is by no means clear that executive planning constitutes a single psychological process. Indeed, it has been defined in terms of apparently diverse functions — including an appropriate problem-solving set for the attainment of a future goal, step-by-step planning, impulse control, inhibition of prepotent but incorrect responses, set maintenance, organized search, and flexibility of thought and action (Ozonoff, in press). Nevertheless, there is a growing body of evidence that executive-function deficits may be the most widespread and universal of cognitive impairments in autism, even more so than theory of mind deficits (Rumsey and Hamburger 1988, 1990; Prior and Hoffman 1990; Ozonoff *et al.* 1991*a, b*; Ozonoff, in press).

As Ozonoff (in press) has pointed out, there are problems still to be resolved in any executive-function deficit hypothesis for the basic of autism (see below). Nevertheless, it is clear that the findings from this approach constitute a major challenge for the theory of mind theory. Moreover, it is also apparent that the explanation of stereotyped, repetitive behaviours constitutes a key task for *all* theories of autism.

Cognitive patterns

Up to now, all psychological theories of autism have virtually ignored the mental retardation that is present in some three-quarters of cases of autism (Rutter 1979). The implicit assumption is that, because mental retardation is not specific to autism (it occurs in a wide range of conditions), it does not require explanation. The general notion seems to be that there is some form of general organic brain dysfunction that, depending on its degree and/or type, may cause mental retardation, but that it is only the deficit in the theory of mind that leads to autism. That view is open to major objections of several different kinds. To begin with, it leaves completely unexplained why the two should be associated; and, equally, it fails to explain the conditions in which they tend *not* to be associated (such as cerebral palsy or Down's syndrome, both of which are commonly associated with mental retardation, but only rarely with autism—see Wing and Gould 1979).

Clearly, there is no invariant association between IQ and autism. Equally clearly, there is no particular genetic association between autism and mental retardation. The family studies of Piven *et al*. (1990) and Gillberg *et al*. (1992) have been consistent in showing that there is no increase in mental retardation (except in the presence of autism) in the first-degree relatives of autistic probands. Not only do the twin-study findings show a very strong genetic component to autism (an estimated 97 percent heritability (Rutter 1991; Rutter *et al*. in press), but they also indicate that the genetic predisposition is specific to autism-type abnormalities and not to those that are variably part of mental retardation.

All these data would seem to justify the dismissal of low IQ as a feature that is not specific to autism, and the view that therefore it does not require explanation. However, that is an unwarranted assumption, based on a misunderstanding of the evidence. Most crucially, it leaves unexplained why some three-quarters of autistic individuals are also mentally retarded. The explanation would seem to require some sort of *two-hit mechanism*, one of them leading to autism and one to additional mental handicap. But what would the second of these be? Neither the twin nor the family findings provide any evidence of such a process (Rutter *et al*. in press). Indeed, the one possible contender, environmental damage brought about by obstetric complications, now seems to be excluded as a general explanation. Rather, the pattern of findings suggests that the obstetric complications result from a genetically abnormal fetus, and do not derive from an environmental hazard.

The second reason for regarding low IQ as an inherent (albeit not universal) feature of autism is that verbal IQ shows a substantial association with the severity of autism symptomatology and that both show a significant association with familial loading (Rutter *et al.*, in press). It is interesting that this was not the case with nonverbal IQ suggesting that verbal and nonverbal intelligence may play rather different roles in autism.

It is not only cognitive *deficits* that have to be accounted for; it is also the cognitive *skills* that are associated with autism. Reference has already been made to the finding that many autistic individuals have unusual talents in recognizing faces upside-down, in completing puzzles when the picture is hidden because the pieces have been placed the wrong way up, and in identifying embedded figures (Frith 1989). In addition, it is clear that many *idiots savants* are autistic, and that a substantial minority of autistic individuals have unusual circumscribed talents that are both out of keeping with their own general cognitive level and also superior to the average level of skills in the general population (Treffert 1989). While such extraordinary talents are not confined to autism, and are far from universally present, they are nevertheless relatively specific to autism, and therefore require explanation.

The studies of O'Connor and Hermelin (1988) have been usefully informative in their demonstration that the skills are 'real', and not just tricks. Nevertheless, it should not be assumed that this necessarily means that we must seek for some process that enhances cognitive performance. There is experimental evidence from studies in animals that *loss* of one skill may enhance another. For example, Otto *et al.* (1991) showed that entorhinal cortex lesions in rats facilitated their solution of odour-discrimination problems, presumably because they reduced cognitive interference under conditions that hindered direct comparisons among cues. Human evidence provides somewhat similar examples of circumstances in which it may be helpful *not* to be able to use particular cognitive strategies (O'Connor and Hermelin 1978). It is possible that such a mechanism, perhaps combined with the unusually circumscribed focus that characterizes much autistic behaviour, may account for the occurrence of unusual specific cognitive talents. However, what is all too apparent is that the topic has been scarcely touched by research so far.

Age of first manifestation

There is some uncertainty over the age when autism first becomes manifest. Clearly, in most cases abnormalities in social relationships, play and language are present by the age of 2 years (sometimes before age 1), although occasionally they seem not to become evident until between 2 and 3 years (see chapters 3, 11, 14, & 15, this volume). Also, there are cases in which parents report that the autistic child *lost* social or language skills that had been present at an earlier age (Kurita, 1985). In most cases, the skills that appeared to regress had been only weakly established—such as a few words, not sentences. Investigators have sometimes sought to infer when abnormalities must first have been evident by the nature of the deficits seen later (see Klin, Volkmar & Sparrow, 1992; also chapter 9, this volume). Thus, because joint attention is a feature evident by 12 months and because joint attention deficits are seen in older autistic children, it is inferred that if the children had been observed at 12 months they would have shown these abnormalities. It is important to appreciate that this is an uncertain inference, just as it does not follow that because skills have been gained and then lost then some event must have occurred at that age to damage them. The point is that there are several examples of psychological functions that change over the course of development in what 'drives' them. Thus, in early infancy, children's babble is not dependent upon sensory input whereas later it is (Aslin *et al.* 1983). Accordingly, deaf children babble normally at first but then the quality of their babble deteriorates around the age of 6 months as their lack of hearing prevents them receiving the necessary sensory input. The part of the brain subserving a particular psychological function may also change over time— as seems to be the case, for example, with Piaget's 'A not B' function (Goldman-Rakic *et al.* 1983).

THE NATURE OF THE COGNITIVE DEFICIT IN AUTISM

Various chapters in this volume consider the evidence in support of different psychological theories of autism. It is clear that it would be premature to draw conclusions on which one (if any) is 'correct'. Each has a body of supporting evidence, but each is open to certain rather fundamental objections. Nevertheless, important progress has been made, and some preliminary inferences may be drawn from the available findings.

Hobson (Chapter 10, this volume) deserves high credit for having been the first to provide a systematic research focus on the socio-emotional deficits associated with autism, and for drawing attention to the need to focus on social-cognitive processes. His research clearly demonstrated that autistic individuals are poor at the identification of socio-emotional cues, and it is reasonable to assume that this constitutes an important aspect of the autistic syundrome. His work created the focus on social cognition and the idea that cognitive deficits might underlie the problems in social relationships, which set the scene for the theory of mind hypothesis. Nevertheless, in our view, Hobson's hypothesis that autism is due to a lack of an innate direct appreciation of interpersonal emotions or intersubjectivity is not borne out by the evidence. He is, of course, right in rejecting the notion that cognition arises first, with a necessary primacy over emotions; but that is not an assumption of any major cognitive hypothesis.

The key reasons for doubting the Hobson emotional-primacy Hypothesis are:(1) between-group differences in emotional recognition deficits largely disappear once controls for language and cognitive skills have been introduced (see Baron-Cohen, this volume, Chapter 4); (2) such deficits have been found in other disorders, so that doubt is cast on diagnostic specificity (see Baron-Cohen, *ibid*.); (3) the non-autistic co-twins of monozygotic autistic probands show cognitive deficits early, but their socio-emotional deficits often only become obvious later, although doubtless they were present in more subtle form before that (Rutter 1991; Rutter *et al.,* in press); (4) autistic individuals do *not* differ from Down's syndrome individuals in their use of emotional terms, but *do* differ in their reference to cognitive mental states (Tager-Flusberg, this volume, Chapter 7); (5) verbal deficits predict social outcome (Rutter *et al*. 1992); (6) both twin and family twin data suggest that the autism phenotype includes cognitive deficits that are not accompanied by serious social impairments in infancy (Rutter *et al*. in press); and (7) there is no direct evidence that an affect deficit causes the degree and type of language delay that characterizes autism (Baron-Cohen 1988).

Of course, none of these pieces of evidence is either decisive or unequivocal. Thus, the emotional recognition deficits seen in other syndromes probably do not follow quite the same pattern as those found in autism (Rutter 1991). Also, controlling for language skills may inadvertently control for emotional skills (Hobson and Lee 1989). For these and other reasons, the emotional hypothesis cannot be firmly rejected, although empirical findings have certainly weakened it, and it no longer seems the main contender.

The theory of mind hypothesis has stood several critical tests. For example, the deficits do seem to be (at least relatively) specifc in the dual sense that they stand out from other aspects of cognitive functioning in autism, and that so far they appear diagnosis-specific. Moreover, they have the huge strength of providing a readily understandable mechanism by which a cognitive deficit *might* underlie the social impairments. There should not be too much concern over the apparent objection that some autistic or Asperger's syndrome individuals can pass theory of mind tests (Bowler 1992). This is because autistic individuals who pass such tests none the less often exhibit a lack of theory of mind *functioning* in their day-to-day lives. It is a commonplace observation in the social-skills training of autistic adults that often they are quite adept at saying what they *should* do in particular social circumstances, but are quite hopeless in doing what is needed when they actually encounter such circumstances. The observation is reminiscent of the finding that autistic children are markedly lacking in spontaneous pretend-play, in spite of the fact that they can show pretend-play in appropriately structured task situations (Lewis and Boucher 1988).

This disparity in performance has sometimes been interpreted as implying some kind of motivational lack. However, it is at least as likely that it represents two other features: (1) a tendency for the imposition of structure to alter task demands; and (2) a tendency for people to be less prone to use spontaneously skills that are weak. Our findings on socio-emotional tasks, in common with clinical observations, also suggest that autistic individuals may pass tasks using means that are different from those supposedly tapped by the task (Macdonald *et al*. 1989). Much the same has been found with empathy tasks (Yirmiya *et al*. 1992). In addition, the measurement error inherent in all testing must be taken into consideration; the follow-up study by Humphries and Baron-Cohen (in press) showed that while most autistic individuals showed unchanging performance over time on theory of mind tests, about one in nine who passed such tests first time round failed them seven years later.

However, there are several other findings that do appear more threatening to the theory of mind hypothesis. First, it is clear that a *theory* of mind in the sense tapped by tests ordinarily passed at a four-year level cannot account for the social deficits usually apparent in autistic children by the age of 2; the skills are simply too late-developing for this to be possible. Also, as Samet argues (chapter 20, this volume), it is quite likely that normal children do not begin by theorizing that other people have minds. Rather it is possible that initially they have an intuitive grasp of mentalizing, with theory coming only later when children begin to understand the nature of minds. By contrast, it is probable that autistic children need to develop their understanding of mentalizing more indirectly, if they do this at all. It may be suggested that it is not that the notion of a *theory* of mind is wrong but rather that mentalizing skills do not first begin that way. Be that as it may (and the matter is undecided), it is clear that if the social abnormalities derive from a cognitive deficit, it must be present before age 4, or even age 3 as proposed by Leslie and Roth (chapter 5, this volume). In any case, home movies strongly suggest that there *are* cognitive deficits that are detectable long before 4 and it is known that language delay is usually evident by age 2 (Massie & Rosenthal, 1984; Adrien *et al.* 1991, Eriksson & Chateau, 1992).

Of course, it is likely that 'theory of mind' skills do not develop *de novo* without any connection with earlier cognitive skills. It is entirely reasonable to argue, as does Baron-Cohen (chapter 4, this volume), that 'theory of mind' skills have their origin in the first year of life in toddler's understanding of goals and attention. Experimental studies indeed demonstrate that 5 to 6 year old autistic children differ markedly from normal toddlers and mentally retarded children in failing to use eye-to-eye gaze to monitor someone else's intentions when the cues are ambiguous (Phillips, Baron-Cohen and Rutter, 1992). As mentioned earlier they also show abnormalities in joint attention (Mundy *et al.*, chapter 9, this volume). However, it is one thing to suggest that these early skills are precursors of theory of mind (in the sense that they reflect the same basic cognitive function with the later skill representing a further phase of the earlier one), and it is quite another to demonstrate that functional connection. Such a demonstration is still lacking and a 'non-proven' verdict is necessary.

Nevertheless, clearly there is a great need to examine the interconnections between early-appearing and later-appearing skills, as well as between the different facets of cognition. As Mundy *et al.* (chapter 9, this volume) emphasize, it may well prove to be the case that different cognitive deficits are important at different phases of development. That could be because they all constitute different facets of the same modular function (as Leslie and Roth, and also Baron-Cohen, this volume, suggest), or because autism arises on the basis of several related, but different, cognitive deficits (as Harris, this volume, and Frith, 1989, suggest). The chapters of this volume bring out very clearly that autistic individuals show a range of varied cognitive problems as a key challenge concerns the question of how they interrelate.

As was mentioned earlier, a challenge to all theories is provided by the need to account for the language abnormalities, the stereotyped behaviours, the mental retardation, and the isolated talents. Theory of mind skills might account for the first of these (see Tager-Flusberg, Chapter 7, this volume), but it is not at all obvious how they could explain the other features. The inference would seem to be that it is very probable that a theory of mind deficit constitutes a key, specific, core aspect of autism; but it is much less certain that it constitutes a sufficient explanation of autism on its own.

At one time, the parallels between some of the behaviours shown by autistic children and those exhibited by children with severe developmental disorders of receptive language, together with the centrality of language delay and abnormalities in autism, led to the suggestion that language deficits might constitute part of the cognitive basis of autism (Rutter 1968). Research over the last quarter of a century has led to several reconceptualizations that make that hypothesis no longer tenable in the form in which it was originally put forward. First, although there is reason to suppose that normal language development is relatively separate from general cognitive skills (Bates *et al.* 1988), serious developmental language disorders involve such a range of cognitive deficits (Bishop 1992) that they cannot usefully be viewed as straightforward examples of 'pure' language delay. Second, it has become apparent that some developmental language disorders include particularly marked deficits in the semantic and pragmatic spects of language (Bishop and Rosenbloom 1987), and it is this group that seems to be nearest to autism (Bishop 1989; Lister-Brook and Bowler 1992). Third, it is evident that the language deficit in autism differs markedly from that in even severe developmental disorders of receptive language (Rutter 1978; Cantwell *et al.* 1989; Rutter *et al.* 1992). For all these reasons, it is no longer sensible to consider that autism might have arisen on the basis of a severe language deficit. Nevertheless, as has already been indicated, language deficits constitute a very fundamental feature of autism. They present early in life, they are very persistent, they have substantial predictive power, and they are intimately bound up with overall patterns of both cognitive and social functioning. Accordingly, any adequate psychological hypothesis about autism must account for the language problems.

The hypothesis that executive planning deficits underlie autism has a set of strengths and weaknesses that differs from that for theory of mind. Of course, it has been subjected to much less research, so that any inferences are necessarily more tentative. Nevertheless, those studies that have been undertaken have been consistent in showing strikingly marked deficits in a range of executive functions (see Ozonoff, in press). On the positive side, therefore, as evidenced in the studies of Ozonoff *et al.* (1991 *a*, *b*), executive-planning tests proved to be a better discriminator of autism than a theory of mind test. This does not necessarily mean that executive deficits are more likely than a theory of mind deficit to constitute the basis of autism. The former are more broad-ranging in the skills that they encompass, and the latter is more subject to ceiling effects. This breadth provides strength in discrimination, but, on the negative side, it may be a weakness in the search for an explanatory cognitive deficit (Ozonoff, in press).

Thus, autistic children are found not to be impaired on 'false photograph' tasks, in which an object is photographed in a given situation, and the object is then moved to a different situation, the picture on the photograph is covered up, and the child is asked where the object is in the picture (Leekam and Perner 1991). Perhaps surprisingly, autistic children are able to disengage from the situation that they can see as it is *now*, and to report the different situation as it was *previously* when photographed. Although this disengagement is part of what is usually regarded as the set of executive functions, it seems to be unimpaired in autism. This finding also perhaps casts doubt on the metarepresentation theory (Leekam and Perner 1991), although it does not challenge the central theory of mind notion.

A further problem for the executive-function hypothesis is that these skills have been found to be impaired in a large number of disorders other than autism, so that (at least as broadly conceived and measured) it lacks the diagnostic specificity of the hypothesis of theory of mind deficits. On the other hand, the breadth of functions overall makes it a better candidate than the theory of mind hypothesis for a possible explanatory cognitive basis for the stereotyped behaviours. The study of executive skills in autism constitutes a highly promising avenue for further research, but there will need to be a sharper focus before it can be viewed as a possible contender for a specific cognitive basis for autism.

Before leaving consideration of the nature of cognitive deficits in autism, it is necessary to return to the assumptions in the question. Is there reason to suppose that there is any cognitive deficit in autism? Research findings are quite unambiguous in showing that autism *does* involve cognitive deficits (Rutter 1979). However, what is less certain is whether they underlie the social and behavioural abnormalities. It seems reasonable to search for a deficit or deficits that could provide that bridge; and therein lies the attraction of both the theory of mind and executive-function hypotheses.

Nevertheless, it cannot be assumed that just *one* deficit will constitute the basis for the whole syndrome. It is parsimonious to start with the aim of seeing if such a core deficit can be found; but it could well turn out that a cluster of deficits are needed to provde the necessary explanatory power. As Goodman (1990) has pointed out, there are many examples in medicine of unitary disorders that nevertheless have diverse physiological consequences. It would be unwise to assume that there has to be just *one* key neuro-psychological deficit in autism; there could be several. Nevertheless, it is important to seek to identify the crucial deficit or deficits that serve(s) to define the conditions; that has been the way of progress in medicine as a whole, and there is every reason to suppose that it will also be so in the case of autism (Rutter and Schopler 1988). Until it proves possible to redefine autism on the basis of some underlying neurobiological or neuro-psychological deficit, controversies over the diagnostic boundaries of the condition are likely to continue and be difficult to resolve.

NEUROBIOLOGICAL–NEUROPSYCHOLOGICAL CONNECTIONS

Almost from the start, one of the attractions of the theory of mind hypo-thesis was the possibility that it might provide a lead to the neurobiological basis of autism. That is because the evidence suggested a substantial degree of modularity; in other words, the skill seems to be relatively independent of other aspects of cognition (see Baron-Cohen, this volume, Chapter 4 and Leslie and Roth, this volume, Chapter 5). Without doubt, there is a key need to span the domains of neuropsychological, neurobiological, genetic, and clinical investigations. Until that has been accomplished, no explanation of autism will be complete. In that connection, several points need to be made.

To begin with, it must be appreciated that, in the case of lesions incurred very early in life, it is rarely warranted to use the pattern of psychological deficits to infer the part of the brain that must be damaged. That is because it has been found that early lesions do *not* show the differentiated psycho-logical patterns evident with later-acquired lesions (Woods and Carey 1979; Vargha-Khadem *et al.* 1992). Accordingly, the evidence that autistic children usually have verbal IQ scores well below their performance IQ scores (Prior 1979) does not necessarily mean that a left-hemisphere lesion is likely; nor does the presence of socio-emotional deficits necessarily imply a right-hemisphere óne (Fein *et al*. 1987). Similarly, although executive-planning tests generally reflect frontal lobe function in adults (Ozonoff, in press), there may not be such an unambiguous connection in child-hood. Neurobiological–neuropsychological connections must be demon-strated directly in autism, and not just assumed on the basis of findings in adults.

It might be thought that the availability of structural brain imaging over the last decade, together with a few neuropathological studies, should have settled the issue; but it is obvious that that is far from the case (Bailey, in press). Most brain-imaging studies have shown structural abnormalities of one sort or another; but the findings are highly inconsistent both within and across investigations. This inconsistency stands in sharp contrast to the rather high level of consistency in the psychological findings, with such few differences as there have been mainly being in terms of either the inferences drawn or the adjusted findings after controls for possible confounding variables.

The reasons for the inconsistency in brain-imaging findings remain a mat-ter for speculation, but several possibilites are likely to have played a part. These include: heterogeneity in the clinical groups studied; lack of adequate attention to methodological issues in the imaging and in quantification of the findings; and lack of systematic coverage of the whole of the brain. In addi-tion, three other considerations must be taken into account. First, the struc-tural brain abnormalities found are as likely to index the phase of fetal life when development went awry as the part of the brain damaged in autism. If there is some variability in timing, it is likely to lead to some variation in the structural abnormalities found.

Second, the evidence from studies of focal brain lesions in early develop-ment indicates that it is not very likely that a unilateral localized lesion could give rise to a disorder like autism, with such pervasive and persistent han-dicaps (Rutter 1983). Rather, it is more likely that autism derives from some system abnormality. If so, any abnormalities found may represent, not the basic underlying affected part of the brain, but rather some abnormality that is, as it were, some way downstream from the brain region that is mainly responsible for the functional deficit seen in autism.

Third, it is known already that there is a degree of aetiological hetero-geneity in autism. This is evident from the association with the fragile X anomaly in some 2 to 3 per cent of cases (Bailey *et al*. in press), and from the less frequent association with other medical conditions (Folstein and Rutter 1988; Smalley *et al*. 1988; Reiss *et al*. 1986). It may well be that these asso-ciated medical conditions will give rise to their own structural brain abnor-malities for reasons unconnected with autism. Also, it cannot be assumed that the neurobiological basis of these 'secondary' cases of autism will be the same as that in idiopathic varieties.

Three main ways forward may be suggested, all of which rely on the crossing of domains of investigation. First, there is a need for more systematic neuropathological and structural brain-imaging studies, with a broader coverage than the investigations so far, *and* for systematic attempts to link the findings with the particular pattern of clinical findings (as well as neuropsychological findings in so far as they are available). The point is that the key associations may apply to particular subgroup patterns, rather than to the syndrome as a whole. Second, neuropsychological studies need to be extended to the broader phenotype, as demonstrated in the genetic studies. If there is a distinctive pathognomonic cognitive deficit in autism, it may be expected that it should apply to the relatives in family studies who appear clinically affected, even though they do not exhibit the syndrome of autism as traditionally diagnosed. However, for this strategy to be effective it will be essential to develop cognitive tests (both theory of mind and executive-function tests) that are appropriate for use with intelligent adults, and which are discriminatory in this apparently normal group.

Third, functional brain imaging needs to be used in conjunction with 'challenge' cognitive tasks that tap skills shown to be impaired in autism. The potential value of this strategy has been evident for at least a decade (Rutter 1982); but it has to be said that the benefits have yet to be seen in child psychiatry. Part of the difficulty lies in the development of highly focused tasks that are of sufficiently short duration to be used in this way, and part in the need to assume that the autistic subjects actually engage in the tasks using the skills that the tasks are meant to tap. In addition, until recently, the functional imaging techniques have been quite demanding of the cooperation of subjects over relatively prolonged periods of time. Furthermore, both positron emission tomography (PET) and single-photon emission tomography (SPET) involve radiation hazards (albeit small ones). Magnetic resonance imaging (MRI) does not have this disadvantage and, in the long term, probably has the greatest potential. However, at present it has a poor power of resolution, and it remains to be seen how well its potential will be realized (Guze 1991).

CONCLUSIONS

It is too early to say whether the theory of mind deficit will provide an answer to the riddle of autism. As this volume brings out very well, there are numerous matters of detail over which real and important controversies remain to be resolved. Nevertheless, what is clear is that the hypothesis, whether or not it proves correct, has taken autism research a major step forward in its extremely valuable attempt to bridge the social and cognitive deficits that characterize this fascinating and puzzling condition. At least as important is that it also carries the potential to bridge the gap between neurobiological and neuropsychological studies. That constitutes the essential next step, and the research findings to which that research will give rise should get us nearer still to an understanding of the nature of autism.

REFERENCES

Adrien, J.L., Faure, M. Perrot, A. Hameury, L. Garreau, B. Barthelemy, C. and Sauvage, D. (1991). Autism and family home movies: preliminary findings. *Journal of Autism and Developmental Disorders*, 21, 43–49.

Aslin, R.N., Pisoni, D.B. and Jusczyk, P.W. (1983). Auditory development and speech perception in infancy. In *Infancy and Developmental psychobiology, Vol. 2, Mussen's Handbook of Child Psychology (4th edn)* (ed. M.M. Haith and J.J. Campos). Wiley, New York.

Baddeley, A.D. (1986). *Working memory*. Oxford University Press.

Baddeley, A.D. (1988). Cognitive Psychology and human memory. *Trends in Neurosciences (Special Issue)*, 11, 177–81.

Bailey, A. Biology of autism. *Psychological Medicine*. (In press.)

Bailey, A. and Rutter, M. Autism. *Science Progress*. (In press.)

Bailey, A., Bolton, P., Butler, L., Le Couteur, A., Murphy, M., Scott, S., *et al.* (in press). Prevalence of the Fragile X anomaly amongst autistic twins and singletons. *Journal of Child Psychology and Psychiatry*.

Baron-Cohen, S. (1988). Social and pragmatic deficits in autism: cognitive or affective? *Journal of Autism and Developmental Disorders*, 18, 379–402.

Baron-Cohen, S. (1989). Do autistic children have obsessions and compulsions? *British Journal of Clinical Psychology*, 28, 193–200.

Baron-Cohen, S., Leslie, A.M., and Frith, U. (1985). Does the autistic child have a 'theory of mind'? *Cognition*, 21, 37–46.

Bartak, L., Rutter, M., and Cox, A. (1975). A comparative study of infantile autism and specific developmental receptive language disorder. I. The children. *British Journal of Psychiatry*, 126, 127–45.

Bates, E., Bretherton, I., and Snyder, L. (1988). *From first words to grammar: individual differences and dissociable mechanisms*. Cambridge University Press.

Bishop, D.V.M. (1989). Autism, Asperger's syndrome and semantic–pragmatic disorder: where are the boundaries? *British Journal of Disorders of Communication*, 24, 107–21.

Bishop, D.V.M. (1992). The underlying nature of specific language impairment. *Annual Research Review of the Journal of Child Psychology and Psychiatry*, 33, 3–66.

Bishop, D.V.M. and Rosenbloom, L. (1987). Childhood language disorders: classification and overview. In *Language development and disorders* (ed. W. Yule and M. Rutter). MacKeith Press, London.

Boucher, J. (1977). Automation and sequencing behaviour, and response to novelty in autistic children. *Journal of Child Psychology and Psychiatry*, 18, 67–72.

Bowler, D. M. (1992). 'Theory of mind' in Asperger's syndrome. *Journal of Child Psychology and Psychiatry*, **33**, 877–93.

Cantwell, D. P., Bakker, L., Rutter, M., and Mawhood, L. (1989). Infantile autism and developmental receptive dysphasia: a comparative follow-up into middle childhood. *Journal of Autism and Developmental Disorders*, **19**, 19–31.

Clark, E. V. (1983). Meanings and concepts. In *Mussen's handbook of child psychology (4th edn), Vol. III: Cognitive development* (ed. J. H. Flavell and E. M. Markman). Wiley, New York.

Clark, E. V. and Rutter, M. (1979). Task difficulty and task performance in autistic children. *Journal of Child Psychology and Psychiatry*, **20**, 271–85.

Dunn, J. (1990). Understanding others: evidence from naturalistic studies of children. In *The emergence of mindreading* (ed. A. Whiten). Blackwell, Oxford.

Eales, M. Pragmatic impairments in adults with childhood diagnosis of autism or developmental receptive language disorder. *Journal of Autism and Developmental Disorders*. (In press.)

Eriksson, A. S. and de Chateau, P. (1992). Case report: A girl aged two years and seven months with autistic disorder videotaped from birth. *Journal of Autism and Developmental Disorders*, **22**, 127–31.

Fein, D. Pennington, B., and Waterhouse, L. (1987). Implications of social deficits in autism for neurological dysfunction. In *Neurobiological issues in autism* (ed. E. Schopler and G. B. Mesibov).

Folstein, S. and Rutter, M. (1988). Autism: familial aggregation and genetic implications. *Journal of Autism and Developmental Disorders*, **18**, 3–30.

Frith, U. (1972). Cognitive mechanisms in autism: experiments with colour and tone sequence production. *Journal of Autism and Childhood Schizophrenia*, **2**, 160–73.

Frith, U. (1989). *Autism: explaining the enigma*. Basil Blackwell, Oxford.

Gillberg, C. Gillberg, I. C., and Steffenburg, S. (1992). Siblings and parents of children with autism: a controlled population-based study. *Developmental Medicine and Child Neurology*, **34**, 389–98.

Goldman-Rakic, P. S., Isseroff, A., Schwartz, M. L. and Bugbee, N. M. (1983). The neurobiology of cognitive development. In *Infancy and developmental psychobiology, Vol. 2, Mussen's Handbook of Child Psychology (4th edn)* (eds. M. M. Haith and J. J. Campos). Wiley, New York.

Goodman, R. (1989). Infantile autism: A syndrome of multiple primary deficits. *Journal of Autism and Developmental Disorders*, **19**, 409–24.

Guze, B. H. (1991). Magnetic resonance spectroscopy: a technique for functional brain imaging. *Archives of General Psychiatry*, **48**, 572–4.

Hay, D. and Angold, A. *Precursors and causes of development and psychopathology*. Wiley, Chichester. (In press.)

Hermelin, B. and O'Connor, N. (1970). *Psychological experiments with autistic children*. Pergamon, Oxford.

Hobson, R. P. and Lee, A. (1989). Emotion–related and abstract concepts in autistic people: evidence from the British Picture Vocabulary Scale. *Journal of Autism and Developmental Disorders*, **19**, 601–23.

Howlin, P. and Rutter, M. (1987). *Treatment of autistic children*. Wiley, Chichester.

Humphries, S. and Baron-Cohen, S. How far can people with autism go in developing a theory of mind? *Journal of Autism and Developmental Disorders*. (In press.)

Kagan, J. (1982). The emergence of self. *Journal of Child Psychology and Psychiatry*, **23**, 363–81.

Kanner, L. (1943). Autistic disturbance of affective contact. *Nervous Child*, **2**, 217–50.

Kanner, L. (1946). Irrelevant and metaphorical language in early infantile autism. *American Journal of Psychiatry*, **103**, 242–6.

Klin, A., Volkmar, F.R. and Sparrow, S.S. (1992). Autistic social dysfunction: Some limitations of the 'theory of mind' hypothesis. *Journal of Child Psychology and Psychiatry*, **33**, 861–76.

Kurita, H. (1985). Infantile autism with speech loss before the age of 30 months. *Journal of the American Academy of Child Psychiatry*, **24**, 191–96.

Le Couteur, A., Rutter, M., Lord, C., Rios, P., Robertson, S., Holdgrafer, M., et al. (1989). Autism Diagnostic Interview: a standardized investigator-based instrument. *Journal of Autism and Developmental Disorders*, **19**, 363–87.

Leekam, S.R. and Perner, J. (1991). Does the autistic child have a metarepresentational deficit? *Cognition*, **40**, 203–18.

Leslie, A.M. (1987). Pretence and representation: the origins of 'theory of mind'. *Psychological Review*, **94**, 412–26.

Lewis, V. and Boucher, J. (1988). Spontaneous, instructed and elicited play in relatively able autistic children. *British Journal of Developmental Psychology*, **6**, 315–24.

Lister-Brook, S. and Bowler, D.M. (1992). Autism by another name? Semantic and pragmatic impairments in children. *Journal of Autism and Developmental Disorders*, **22**, 61–82.

Macdonald, H., Rutter, M., Howlin, P., Rios, P., Le Couteur, A., Evered, C., et al. (1989). Recognition and expression of emotional cues by autistic and normal adults. *Journal of Child Psychology and Psychiatry*, **30**, 865–77.

Mundy, P. and Sigman, M. (1989). Specifying the nature of social impairment in autism. In *Autism: new perspectives on diagnosis, nature, and treatment* (ed. Dawson). Guildford Press, New York.

O'Connor, N. and Hermelin, B. (1963). *Speech and thought in severe subnormality. An experimental study*. Pergamon, Oxford.

O'Connor, N. and Hermelin, B. (1978). *Seeing and hearing and space and time*. Academic Press, London.

O'Connor, N. and Hermelin, B. (1988). Low intelligence and special abilities. *Journal of Child Psychology and Psychiatry*, **29**, 391–6.

Otto, T., Eichenbaum, H., Schottler, F., Staubli, U., and Lynch, G. (1991). Hippocampus and olfactory discrimination learning: effects of entorhinal cortex lesions on olfactory learning and memory in a successive-cue, go–no–go task. *Behavioral Neuroscience*, **105**, 111–19.

Ozonoff, S. Executive functions in autism. In *Learning and cognition in autism* (ed. E. Schopler and G.B. Mesibov). Plenum, New York. (In press.)

Ozonoff, S., Rogers, S.J., and Pennington, B.F. (1991a). Asperger's syndrome: evidence of an empirical distinction from high-functioning autism. *Journal of Child Psychology and Psychiatry*, **32**, 1107–22.

Ozonoff, S., Pennington, B.F., and Rogers, S.J. (1991b). Executive function deficits in high-functioning autistic individuals: relationship to theory of mind. *Journal of Child Psychology and Psychiatry*, **32**, 1081–1105.

Perner, J., Frith, U., Leslie, A.M., and Leekam, S.R. (1989). Exploration of the autistic child's theory of mind: knowledge, belief and communication. *Child Development*, **60**, 688–700.

Phillips, W. Baron-Cohen, S. and Rutter, M. (1992). The role of eye-contact in goal-detection: evidence from normal toddlers and children with autism or mental handicap. *Development and Psychopathology*, **4**, 375–84.

Piven, J., Gayle, J., Chase, G., Fink, B., Lande, R., Wzorek, M., *et al*. (1990). A family history study of neuropsychiatric disorders in the adult siblings of autistic individuals. *Journal of the American Academy of Child and Adolescent Psychiatry*, **29**, 177–83.

Prior, M.R. (1979). Cognitve abilities and disabilities in infantile autism: a review. *Journal of Abnormal Child Psychology*, **7**, 357–80.

Prior, M. and Hoffman, W. (1990). Neuropsychological testing of autistic children through an exploration with frontal lobe tests. *Journal of Autism and Developmental Disorders*, **20**, 581–90.

Reiss, A.L., Feinstein, C., and Rosenbaum, I.C. (1986). Autism and genetic disorders. *Schizophrenia Bulletin*, **12**, 724–38.

Rogers, S.J. and Pennington, B.F. (1991). A theoretical approach to the deficits in infantile autism. *Development and Psychopathology*, **3**, 137–62.

Rumsey, J.M. and Hamburger, S.D. (1988). Neuropsychological findings in high-functioning men with infantile autism, residual state. *Journal of Clinical and Experimental Neuropsychology*, **10**, 201–21.

Rumsey, J.M. and Hamburger, S.D. (1990). Neuropsychological divergence of high-level autism and severe dyslexia. *Journal of Autism and Developmental Disorders*, **20**, 155–68.

Rutter, M. (1968). Concepts of autism: a review of research. *Journal of Child Psychology and Psychiatry*, **9**, 1–25.

Rutter, M. (1970). Autistic children: infancy to adulthood. *Seminars in Psychiatry*, **2**, 435–50.

Rutter, M. (1972). Childhood schizophrenia reconsidered. *Journal of Autism and Childhood Schizophrenia*, **2**, 315–37.

Rutter, M. (1978). Language disorder and infantile autism. In *Autism: a reappraisal of concepts and treatment* (ed. M. Rutter and E. Schopler), Plenum, New York.

Rutter, M. (1979). Language, cognition and autism. In *Congenital and acquired cognitive disorders* (ed. R. Katzman). Raven Press, New York.

Rutter, M. (1982). Developmental neuropsychiatry: concepts, issues and problems. *Journal of Clinical Neuropsychology*, **4**, 91–115.

Rutter, M. (1983). Cognitive deficits in the pathogenesis of autism. *Journal of Child Psychology and Psychiatry*, **24**, 513–31.

Rutter, M. (1986). Child psychiatry: the interface between clinical and developmental research. *Psychological Medicine*, **66**, 151–69.

Rutter, M. (1987). The 'what' and 'how' of language development: a note on some outstanding issues and questions. In *Language development and disorders*, Clinics in Developmental Medicine (ed. W. Yule and M. Rutter). MacKeith Press–Blackwell Scientific, London–Oxford.

Rutter, M. (1989). Epidemiological approaches to developmental psychopathology. *Archives of General Psychiatry*, **45**, 486–500.

Rutter, M. (1991a). Autism as a genetic disorder. In *The new genetics of mental illness* (ed. P. McGuffin and R. Murray). Heinemann Medical, Oxford.

Rutter, M. (1991b). Autism—pathways from syndrome definition to pathogenesis. *Comprehensive Mental Health Care*, **1**, 5–26.

Rutter, M. Developmental psychopathology as a research perspective. In *Longitudinal research on individual development: present status and future perspectives* (ed. D. Magnusson and P. Casaer). Cambridge University Press. (In press.)

Rutter, M. and Lockyer, L. (1967). A five to fifteen year follow-up study of infantile psychosis: I. Description of sample. *British Journal of Psychiatry*, **113**, 1169–82.

Rutter, M. and Schopler, E. (1988). Autism and pervasive developmental disorders. In *Assessment and diagnosis in child psychopathology* (ed. M. Rutter, A. H. Tuam, and I. S. Lann). Plenum, New York.

Rutter, M., Bartak, L., and Newman, S. (1971). Autism—a central disorder of cognition and language. In *Infantile autism: concepts, characteristics and treatment* (ed. M. Rutter). Churchill Livingstone, Edinburgh.

Rutter, M., Bartak, L., and Cox, A. (1975). A comparative study of infantile autism and specific developmental receptive language disorder. I. The children. *British Journal of Psychiatry*, **126**, 127–45.

Rutter, M., Mawhood, L., and Howlin, P. (1992). Language delay and social development. In *Specific speech and language disorders in children* (ed. P. Fletcher and D. Hale). Whurr Publishers, London.

Rutter, M., Bailey, A., Bolton, P., and Le Couteur, A. Autism: syndrome definition and possible genetic mechanisms. In *Nature, Nature and Psychology* (ed. R. Plomin and G. E. McLearn). American Psychological Association Press, Washington, DC. (In press.)

Shah, A. and Frith, U. (1983). An islet of ability in autistic children: a research note. *Journal of Child Psychology and Psychiatry*, **24**, 613–20.

Smalley, S. L., Asarnow, R. F., and Spence, M. A. (1988). Autism and genetics. A decade of research. *Archives of General Psychiatry*, **45**, 953–61.

Sroufe, L. A. and Rutter, M. (1984). The domain of developmental psychopathology. *Child Development*, **55**, 17–29.

Treffert, D. A. (1989). *Extraordinary people*. Bantam Press, London.

Tulving, E. (1983). *Elements of episodic memory*. Oxford University Press.

Tulving, E. (1984). Précis of 'Elements of Episodic Memory'. *The Behavioral and Brain Sciences*, **7**, 223–68.

Vargha-Khadem, F., Isaacs, E., van der Werf, S., Robb, S., and Wilson, J. (1992). Development of intelligence and memory in children with hemiplegic cerebral palsy: the deleterious consequences of early seizures. *Brain*, **115**, 315–29.

Wimmer, H. and Perner, J. (1983). Beliefs about beliefs: representation and constraining function of wrong beliefs in young children's understanding of deception. *Cognition*, **13**, 103–28.

Wing, L. and Gould, J. (1979). Severe impairments and associated abnormalities in children: epidemiology and classification. *Journal of Autism and Developmental Disorders*, **9**, 11–29.

Woods, B. T. and Carey, S. (1979). Language deficits after apparent clinical recovery from childhood aphasia. *Annals of Neurology*, **6**, 405–9.

Yirmiya, N., Sigman, M. D., Kasari, C., and Mundy, P. (1992). Empathy and cognition in high-functioning children with autism. *Child Development*, **63**, 150–60.

Index

action(s)
  agents 188
    imitation of, *see* imitation
    motives behind 468
  other people's, inability to anticipate
    467
active but odd individuals 47, 302
adaptive skills 303–4
adolescents 174
adulthood, social outcome in 46–7
advising caregivers 469
affect 186–8, 194, 212–22, 273
  cognition and, *see* cognition
  communication of, *see* communication
  expression of, *see* expression
  joint-attention and 186–8, 194, 195, 196,
    197
  meaning of 212
  perception of 198
  prosocial behaviour and 298
  role of 212–16
  see also emotions
affective attitudes 215
affective contact, lack of 43, 204
affective relatedness 214
  impairment in 216–21
affordances 417
  social 262
age
  two years of, *see* two-year-olds
  three years of, *see* three-year-olds
  four years of, theory of mind at 13–18
  of onset of autism 491–2
agent(s)
  attention to 105, 107
  attention with, joint 105, 106
  attitudes and 89–90, 93
  of contemplation 188
  person as 407–8
alliance formation in non-human primates
  371
aloneness, autistic 45, 204
aloof individuals 47
amygdala, abnormalities in 327
anecdotes 249–50, 254
animate vs inanimate distinction 20, 68
antisocial behaviour 298–9

anxiety
  separation 458
  from unpredictability of other people's
    action 233
apes
  linguistic 402
  requesting behaviour 415
  wild, mentalism vs behaviourism
    in 367–70
  see also chimpanzees; gorillas
appearance and reality distinction 71
arousal
  general 341
  self-regulation 194
Asperger's syndrome 241, 242, 243, 288
  antisocial behaviour in 299
  interpersonal functioning 288
attachment 299–300, 484
  theories of 19
attention 105–6, 181–203, 410, 418–20
  to/with agents, *see* agents
  early understanding of 22, 23, 23–4,
    72–4, 77
  to emotional expression 23–4, 217–18
  establishment of contact 410
  joint 105, 106, 144–5, 181–203, 400–1
    affect and 186–8, 194, 195, 196, 197
    communication and 144–5, 148, 149,
      150, 400–1
    debate about 181–3
    diagnosis of autism 475–7
    non-verbal 185–9
    play development and 189–93
    symbolic skills and 188–93 *passim*,
      193–7
  monitoring of 410
  in non-human primates 377, 418–20
    expression 418–20
    vs infants 410
  rapid shifts in 131
  infant's goal of 411–12, 413
  in chimpanzees 377
attention–goal psychology 77
attitudes
  affective 215
  agents and 89–90, 93
  propositional 121–4, 214